International Conference on Port, Coastal and Offshore Engineering 2024 (ICPCO 2024)

AA007160

IOP Conference Series: Earth and Environmental Science
Volume 1464

Bandung, Indonesia
28 November 2024

ISBN: 979-8-3313-1585-6
ISSN: 1755-1307

Printed from e-media with permission by:

Curran Associates, Inc.
57 Morehouse Lane
Red Hook, NY 12571

Some format issues inherent in the e-media version may also appear in this print version.

Institute of Physics
Dirac House, Temple Back
Bristol BS1 6BE UK

Phone: 44 1 17 929 7481
Fax: 44 1 17 920 0979

techtracking@iop.org

Additional copies of this publication are available from:

Curran Associates, Inc.
57 Morehouse Lane
Red Hook, NY 12571 USA
Phone: 845-758-0400
Fax: 845-758-2633
Email: curran@proceedings.com
Web: www.proceedings.com

TABLE OF CONTENTS

IOP Conf. Series: Earth and Environmental Science **1464** (2025) 011001 doi:10.1088/1755-1315/1464/1/011001

International Conference on Port, Coastal, and Offshore Engineering 2024 (ICPCO 2024) - Energizing Ocean Development and Sustainability, 28th November 2024

Pullman Bandung International Convention Center, Bandung, West Java, Indonesia

1. Event Overview

The International Conference on Port, Coastal, and Offshore Engineering 2024 (ICPCO 2024), organized by Ocean Engineering Program of Institut Teknologi Bandung (ITB), signifies a pivotal moment for collaboration and knowledge dissemination in ocean engineering. The conference held November 28th, 2024, in Pullman Bandung International Convention Center, Indonesia, aim to enhance cross-disciplinary and cross-sector collaboration in the development of port, coastal, offshore technologies, infrastructures, and environmental sustainability.

Remarks from Chairman ICPCO 2024

A comprehensive presentation from keynote speakers, including expert and leaders from Indonesian state-owned enterprises such as Pertamina International Shipping (PIS), Pelabuhan Indonesia (PELINDO), and Perusahaan Listrik Negara (PLN), as well as prominent professors from University of Massachusetts Amherst (USA) and Delft University of Technology (The Netherland).

Several topics were discussed such as efficient port and waterway operations, sustainable infrastructure for ports and coastal regions, marine-based energy development, and energy & resource management in the marine sector.

IOP Conf. Series: Earth and Environmental Science **1464** (2025) 011001 doi:10.1088/1755-1315/1464/1/011001

Following the keynote speakers, parallel conference sessions are held in separate rooms. Each of the parallel sessions will have a specific topic of discussion, which encompasses the following subtopics: Coastal Engineering and Management, Offshore Engineering and Technology, Marine Energy and Sustainability, Marine Technology and Sensing, Port Engineering and Logistics, Ocean Environmental and Coastal Process Modeling.

Enthusiastic in Keynote Session ICPCO 2024

An Instance of Parallel Room Presentation in ICPCO 2024

IOP Conf. Series: Earth and Environmental Science **1464** (2025) 011001 doi:10.1088/1755-1315/1464/1/011001

Gala Dinner Ambience in ICPCO 2024.

Impressive 46 papers are presented from 37 institutions around the world, with a total of 176 contributors from 7 countries. Invited expert speakers from Yokohama University, Japan, and PELINDO are presented their works on Coastal Engineering Management and Port Engineering Logistics. Enthusiastic participants foster discussion and collaboration in each specific area.

Several exhibitors also showcased cutting-edge technologies and solutions from industry sectors, offering participants hands-on demonstrations and networking opportunities. Several partnerships emerged from ICPCO 2024, including collaborations between universities and industry stakeholders to address real-world challenges.

As a highlight of the ICPCO 2024 experience, the Gala Dinner provided an exceptional opportunity for participants to unwind and strengthen professional connections in a relaxed, elegant setting. This featured a captivating cultural performance "Rampak Kendang" that celebrated the rich heritage of West Java. Attendees were treated to traditional music and dance, offering a unique glimpse of the local culture and traditions.

Closing from chairman of ICPCO 2024, remark the importance of fostering global partnerships in the maritime sector. Additionally, awards were presented to best student paper. The Gala Dinner concluded with a group photo and a vote of thanks from the organizing committee, leaving participants with cherished memories and strengthened professional relationships.

The organizing committee extends its gratitude to all speakers, participants, our main sponsor PELINDO, Pertamina International Shipping (PIS), all sponsors, and all the contribution towards a resounding success of ICPCO 2024. The organizing committee look forward to hosting the next ICPCO, continuing advancing port, coastal, and offshore engineering. Promising greater platform for innovation and collaboration. For updates future events, visit https://icpco.itb.ac.id/ and http://ocean.itb.ac.id.

Photo session after Gala Dinner

2. Paper Submission

The total count of paper submission was 52 for IOP EES. A peer review for selection of papers was followed and 46 (fourty-six) papers were accepted for paper in IOP EES.

3. Committees

Scientific Committee:
Editor and Chairman : Prof. Ir. Andojo Wurjanto, Ph.D (Institut Teknologi Bandung, Indonesia)
Prof. Nobuhisa Kobayashi, Ph.D. (University of Delaware, USA)
Prof. Dr.Eng. Takayuki Suzuki (Yokohama National University, Japan)
Dr. Ravindra Jayaratne (University of East London, United Kingdom)
Dr. Kadir Orhan, M.Sc. (Christian-Albrechts-Universität zu Kiel, Germany)
Dr. Mohamad Hasif Bin Osman (Universiti Teknologi Malaka, Malaysia)
Prof. Yinghui Tian (University of Melbourne, Australia)
Prof. Ir. Ricky Lukman Tawekal, M.S.E, Ph.D. (Institut Teknologi Bandung, Indonesia)
Prof. Ir. Harman Ajiwibowo, M.Sc., Ph.D. (Institut Teknologi Bandung, Indonesia)
Prof. Ir. Muslim Muin, M.SOE, Ph.D (Institut Teknologi Bandung, Indonesia)
Ir. Entin Agustini Karjadi, MCE, Ph.D. (Institut Teknologi Bandung, Indonesia)
Ir. Irsan Soemantri Brodjonegoro, Ph.D. (Institut Teknologi Bandung, Indonesia)
Ir. Sri Murti Adiyastuti, M.Sc., Ph.D. (Institut Teknologi Bandung, Indonesia)
Ir. Rildova, Ph.D. (Institut Teknologi Bandung, Indonesia)
Dr. Ir. Hendriyawan M.T. (Institut Teknologi Bandung, Indonesia)
Dr.Eng. Ir. Hendra Achiari, S.T., M.T. (Institut Teknologi Bandung, Indonesia)
Dr.Eng. Ir. Nita Yuanita, S.T., M.T. (Institut Teknologi Bandung, Indonesia)
Dr.Eng. Ir. Paramashanti (Institut Teknologi Bandung, Indonesia)
Ir. Ahmad Mukhlis Firdaus, DPhil. (Institut Teknologi Bandung, Indonesia)
Ir. Alamsyah Kurniawan, S.T., Ph.D. (Institut Teknologi Bandung, Indonesia)
Ir. Heri Setiawan, Ph.D. (Institut Teknologi Bandung, Indonesia)
Ir. Eko Charnius Ilman, Ph.D. (Institut Teknologi Bandung, Indonesia)
Farid Putra Bakti, Ph.D. (Institut Teknologi Bandung, Indonesia)
Dr. Ir. Jessica Rikanti Tawekal (Institut Teknologi Bandung, Indonesia)
Dr. Eng. Ir. Febri Zukhruf S.T M.T. (Institut Teknologi Bandung, Indonesia)
Sugeng Krisnanto S.T M.T Ph.D. (Institut Teknologi Bandung, Indonesia)

Mohammad Bagus Adityawan S.T M.T Ph.D. (Institut Teknologi Bandung, Indonesia)
Prof. Dr. Indra Jaya, M.Sc. (IPB University, Indonesia)
Dr. Ir. Erwandi, M.Eng. (National Research and Innovation Agency, Indonesia)
Faizal Immaddudin Wira Rohmat S.T., M.T., Ph.D (Institut Teknologi Bandung, Indonesia)
Mohammad Farid S.T., M.T., Ph.D (Institut Teknologi Bandung, Indonesia)

Organizing Committee from Institut Teknologi Bandung, Indonesia:
Chairman : Ir. Ahmad Mukhlis Firdaus, S.T., M.T., D.Phil.
Co-Chairman : Ir. Alamsyah Kurniawan, S.T., Ph.D.
Secretary : Dr. Eng. Ir. Hendra Achiari, S.T., M.T.
Treasurer : Dr. Ir. Jessica Rikanti Tawekal, S.T., M.Sc.
Fundraising : Heri Setiawan, S.T., M.T., Ph.D., Ir. Eddy Rachman Gandanegara, S.T., M.T.
Publication and Proceeding : Ir. Eko Charnius Ilman, Ph.D.
Event : Farid Putra Bakti, S.T., M.Sc., Ph.D.
Logistics : Lurohman Mamin Masturi, ST., M.Sc.
Promotion : Ir. Entin Agustini Karjadi, MCE, Ph.D.
Secretariat: Anita Fitriani, S.T. M.T., Juneard Stevelana, A.Md., SAB.
Assistants: Dega Damara Aditramulyadi, S.T., M.T., Rajin Sihombing, S.T., M.T., Devi Ulumit, S.T., M.T., Maulidiyah Pravi, S.T., M.T., Eka Suendra Putra, S.T., AMIMartEST, M.T., Meiya Narulita Suyasman, S.T., Tegar Putra Perdana, ST.

4. Sponsors

PT Pelabuhan Indonesia (Diamond)
PT Pertamina International Shipping (Diamond)
British Petroleum (Platinum)
Boskalis (Gold)
PT PLN (Gold)
PT Promisco Sinergi Indonesia (Silver)
OE ITB Alumnae - ALKA ITB (Silver)
PT LAPI Ganesha Utama (Bronze)
Mores Strategics (Bronze)
Royal Haskoning DHV (Bronze)
Pertamina Marine Engineering (Bronze)
OCEnviro (Bronze)
Waskita Toll Road (Special Spot)

5. Participants affiliations

ICPCO 2024 was attended physically by participants from various universities and organizations:
Dong-A University, Republic of Korea
Geo Oceanics Pte Ltd,, Singapore
Institut Teknologi Bandung, Indonesia
Institut Teknologi Kalimantan, Indonesia
Institut Teknologi Sepuluh Nopember, Indonesia
Institut Teknologi Sumatera, Indonesia
Institut Transportasi dan Logistik Trisakti, Indonesia
IPB University, Indonesia
Java Offshore, Malaysia
Jember University, Indonesia
MuTeknologi Software, Indonesia
Pelindo, Indonesia
Republic of Indonesia Defense University, Indonesia
Royal Haskoning DHV, Indonesia
The University of Edinburgh, UK
Universitas Pertamina, Indonesia
Universitas Tanjungpura, Indonesia
University of Exeter, UK
University of Kiel, Germany
Yokohama National University, Japan
ALKA, Indonesia

IOP Conf. Series: Earth and Environmental Science **1464** (2025) 011001 doi:10.1088/1755-1315/1464/1/011001

Alumni Teknik Kelautan (ALKA ITB), Indonesia
Asosiasi Energi Laut Indonesia, Indonesia
Boskalis, Indonesia
Designbagus, Indonesia
Directorate of Port Affairs, Indonesia
Direktorat Logistik Kementerian Kelautan dan Perikanan, Indonesia
Direktorat Sungai dan Pantai Kemen PU, Indonesia
Dit Kepelabuhanan Ditjen Hubla Kemenhub, Indonesia
Dosen Tamu Teknik Kelautan, Indonesia
FLOT Indonesia, Indonesia
HAPI, Indonesia
Ikatan Ahli Fasilitas Minyak dan Gas Bumi Indonesia, Indonesia
Institut Teknologi Bandung, Indonesia
Institut Teknologi Sumatera, Indonesia
Institut Transportasi dan Logistik Trisakti, Indonesia
ITENAS, Indonesia
Kemenkumham, Indonesia
Kementerian PPN/Bappenas, Indonesia
Kementerian PU, Indonesia
Marine Advanced Design Solution, Indonesia
NZMATES, Indonesia
Package 8 Patimban Port Development Project, Indonesia
Persatuan Insinyur Indonesia, Indonesia
Pertamina Marine Engineering, Indonesia
PHE ONWJ, Indonesia
PIANC, Indonesia
Pixartistic, Indonesia
Politeknik Kelautan dan Perikanan Karawang, Indonesia
Politeknik Kelautan Dan Perikanan Pangandaran, Indonesia
Politeknik Kelautan dan Perikanan Pangandaran, Indonesia
Private Legal Counselor & Contract Advisor, Indonesia
PRTT, Indonesia
PT Bhumi Warih Geohydromatics, Indonesia
PT Catur Manunggal Elektromekanik, Indonesia
PT Meindo Elang Indah, Indonesia
PT Pelindo Solusi Logistik, Indonesia
PT Pertamina Port & Logistic (Subholding PT PIS), Indonesia
PT Perusahaan Gas Negara, Indonesia
PT PGN, Indonesia
PT Waskita Toll Road, Indonesia
PT Weichai Indonesia Utama, Indonesia
PT Wimarine Sindo Teraprima, Indonesia
PT Worley SEA Indonesia, Indonesia
PT. Adhi Karya (Persero) Tbk., Indonesia
PT. Atrya Swascipta Rekayasa, Indonesia
PT. Dredgng International Indonesia, Indonesia
PT. Karya Cipta Madani, Indonesia
PT. Krakatau Pipe Industries, Indonesia
PT. Mineering Energi Internasional, Indonesia
PT. Prima Springs Mfg, Indonesia
PT. Rayasurverindo Tirtasarana, Indonesia
Pusat Pendidikan Hidro-Oseanografi Angkatan Laut, Indonesia
Royal HaskoningDHV, Indonesia
Sekolah Tinggi Teknologi Angkatan Laut, Indonesia
Trisakti Institute of Transportation And Logistics, Indonesia
Universitas Jendral Ahmad Yani, Indonesia
Universitas Muhammadiyah Yogyakarta, Indonesia

IOP Conf. Series: Earth and Environmental Science **1464** (2025) 011002 doi:10.1088/1755-1315/1464/1/011002

Peer Review Statement

All papers published in this volume have been reviewed through processes administered by the Editors. Reviews were conducted by expert referees to the professional and scientific standards expected of a proceedings journal published by IOP Publishing.

- **Type of peer review:** Double Anonymous
- **Conference submission management system:** Morressier
- **Number of submissions received:** 52
- **Number of submissions sent for review:** 52
- **Number of submissions accepted:** 46
- **Acceptance Rate (Submissions Accepted / Submissions Received × 100):** 88.5
- **Average number of reviews per paper:** 2.91
- **Total number of reviewers involved:** 20
- **Contact person for queries:**
  Name: Dr. Eko Charnius Ilman
  Email: eko.ilman@itb.ac.id
  Affiliation: Institut Teknologi Bandung

IOP Conf. Series: Earth and Environmental Science **1464** (2025) 012001 doi:10.1088/1755-1315/1464/1/012001

Sea tolls as Indonesia's blue economy strategy

Evi Apriani[1*], Syachrul Arief[2] and Asep Adang Supriyadi[3]

[1] Maritime Security Study Program, Republic Indonesia Defense University, Jakarta, Indonesia
[2] Training and Education, Geospatial Information Agency, Bogor, Indonesia
[3] Sensing Technology Study Program, Republic Indonesia Defense University, Jakarta, Indonesia

*E-mail: evi.apriani@kn.idu.ac.id

Abstract. As a maritime country, Indonesia has great potential to develop its sea area. The sea is one of the sources of Indonesia's economy. To improve the national economy, the Indonesian government initiated the sea toll program as part of the blue economy. The term "blue economy" describes the sustainable use of the sea to develop the economic sector by taking into account the sustainability of the ecosystem. The sea toll is a sea transportation network that connects regions in Indonesia through various ports. The sea toll program has been running for almost ten years. This program has been implemented since the first period of President Joko Widodo's administration. The purpose of this research is to examine the alignment of sea tolls in the blue economy strategy towards Indonesia's economic expansion for the long-term sustainability of the maritime industry. This article uses a qualitative approach to understand the implementation of sea tolls on various strategic routes in Indonesia. The data collection process was carried out by searching for data from books, scientific articles, news, journals, agency reports, policy documents, and other sources of information. The results showed that the sea toll has great potential in the economic sector as a manifestation of Indonesia's blue economy strategy. The implications of the sea toll program can improve inter-island connectivity, reduce price disparities among regions, reduce logistics costs, preserve marine ecosystems, and increase national and regional economic growth. One of the positive impacts on the economy is to improve the welfare of local communities and entrepreneurs. Owing to this program, ships going back to Java have backloads of commodities produced by the community, such as copra. As an effort to continue to support the Sea Toll program, it requires collaboration between relevant stakeholders, including the government, private sector, and community. The Sea Toll is one of the important keys to realizing Indonesia's vision as a world maritime axis by always improving regulations, developing infrastructure, empowering communities, and engaging in maritime diplomacy with various countries in the world.

1. Introduction

This document outlines the "Sea Toll" program as one of Indonesia's blue economy strategies. As a maritime country, Indonesia has a lot of potential in the marine sector that has not been optimally utilized, such as the fisheries industry [1], [2], biodiversity [3], [4], renewable and non-

IOP Conf. Series: Earth and Environmental Science **1464** (2025) 012001 doi:10.1088/1755-1315/1464/1/012001

renewable energy resources [3], [5], [6], [7], [8], [9], marine tourism [3], [4], [7], [8], [9], marine industry [7], [8], marine services [7] and marine transportation [3], [7], as well as many other potentials. However, due to inadequate infrastructure and the growth gap that exists between remote and urban areas, this potential has not been fully utilized optimally [10]. Indonesia has a tremendous opportunity to build a maritime industry as an important pillar in sustaining the country's economy, which is supported by abundant marine resources [11], [12], [13]. If managed properly, the maritime sector has the potential to support the country's economy, to encourage national economic growth [4], [8], [11], [13], [14], [15], [16]. Regarding Indonesia's maritime potential, as mentioned earlier, various jobs can be created in the maritime sector to increase people's income [13], [17]. In addition, improving the maritime sector can strengthen Indonesia's position as the world's maritime axis [18], [19], [20], which is one of the main visions of the current government.

An environmentally friendly approach to monitoring and utilizing Indonesia's marine resources is realized through a blue economy. The concept of "blue economy" highlights the importance of maintaining a balance between economic activities and the protection of marine habitats [18], [21], [22]. By using this strategy, Indonesia can sustainably maximize its marine potential, thereby improving the welfare of society in the long run [4]. Thus, Indonesia can utilize the ocean to support various economic sectors while ensuring that the marine ecology is maintained and protected for future generations.

The "Sea Toll" program aims to improve inter-island connectivity, reduce inter-regional price disparities, and lower logistics costs. Tol Laut is a key initiative of President Joko Widodo, which was launched on November 4, 2015 [23] for sea logistics transportation, which aims to connect major ports in Indonesia. This port-to-port connection can facilitate the easy distribution of products to remote locations [24]. It is expected that the Sea Toll will lead to equitable economic development [16], especially in eastern Indonesia [10], which has historically had limited access. Maritime infrastructure development is essential to drive equitable and sustainable economic growth across the archipelago [10], [16], as stated in Indonesia's blue economy policy, which includes this initiative.

Several previous studies discussing the Sea Toll have been conducted. The results of the research [25] stated that the Sea Toll program has reduced national logistics costs from 24.5% of gross domestic product to 19% with baseline data and 11% to 20% for basic commodities. In addition, the Sea Toll also enables trade between countries in the South Pacific and East Asia [26] stated that the resilience and competitiveness of ports are affected by the "Sea Toll" in terms of frequency of ship calls and dependability, to remain competitive In addition, the Sea Toll also affects strategic partnerships. In his research, [27] obtained new findings that can minimize 50% of the total cost of the Sea Toll network and save 60% of the current subsidy with the Butterfly Hub option. Based on several previous studies, no one has discussed the Sea Toll program in relation to the blue economy. Thus, this paper aims to examine the linkages between the Sea Toll program and Indonesia's blue economy strategy for the long-term sustainability of the maritime industry.

2. Research Method

This research uses a qualitative approach by examining secondary data from various sources, such as books, scientific articles, reports, and government policies. In qualitative research, an issue or social phenomenon is investigated by collecting narrative or descriptive data, which is then scrutinized for trends, themes and relationships. The literature review serves as a foundation for the discussion of this article. The literature helps to strengthen the research problem in qualitative research. In a literature review, studies on a topic are found and summarized. Several steps were taken in the literature review, including identifying keywords, then searching and finding articles relevant to the topic. The next step is to gather information and start creating a literature map by organizing important concepts [28]. Thus, data relevant to the needs of the discussion related to the Sea Toll as Indonesia's blue economy strategy was obtained. Figure 1 illustrates the brainstorming of this research.

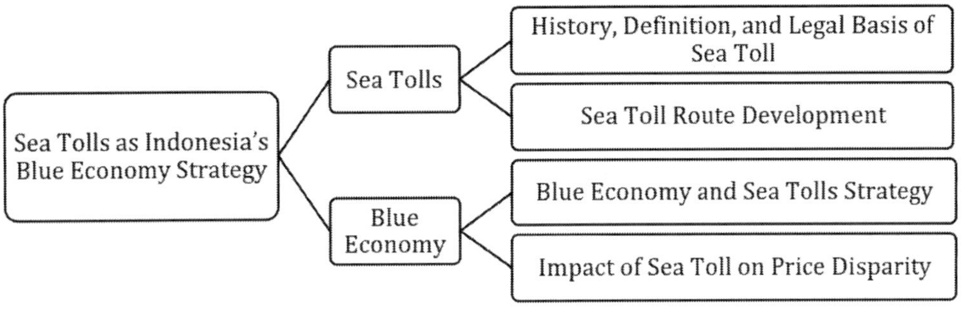

Figure 1. Brainstorming.

3. Result and Discussion

The Sea Toll is a concrete effort to realize a blue economy in Indonesia. Through this program, commodity distribution across Indonesia will be strengthened, inter-island connectivity will be improved, and logistics costs will be reduced [10], [29]. The Sea Toll provides many changes in the field of Indonesia's maritime economy.

3.1 History, definition, and legal basis of sea tolls

As a country consisting of various islands, Indonesia must be able to unite its entire territory through an integrated shipping system. The shipping sector as maritime connectivity is a crucial system to support the needs of regions and areas in Indonesia, especially the 3TP (Frontier, Outermost, Remote, and Border) areas. Considering that Indonesia is internationally recognized as an archipelago, the government plays a very big role in a big leap for the prosperity of its people. This is the origin of the "Sea Toll" program. The Sea Toll program was first revealed in a speech by Indonesian President Joko Widodo [30], [31] on August 16, 2016, stating that strategic sectors of the country's economy must be mobilized to achieve economic growth [30]. Thus, this strategy is used to develop the development of Eastern Indonesia in order to reduce price disparities and social welfare gaps. Sea Toll is an integrated system that includes safety and security related to shipping, ports, and waters [30].

With ships sailing from the west to the east of Indonesia in a regular and planned manner, the Sea Toll provides efficient maritime connectivity. The abbreviations "total logistics" and "sea" that are part of the sea transportation mode are the source of the word "Sea Toll". Sea Toll, in a broad sense, connects ports to facilitate the flow of people and products, while in a more specific sense, the program provides subsidies for freight transport to reduce price differentials for basic goods necessary for daily life across Indonesia [30].

The Sea Toll Program was first initiated by President Joko Widodo in 2015 as an effort to unify eastern Indonesia through maritime connectivity. The regulations underlying this program are Presidential Regulation No. 106/2015 and Presidential Regulation No. 70/2017 [30].

3.2 Sea toll route development

Implementation of the Sea Toll program began in 2015 and continues to be developed in stages. The first fiscal year for Sea Toll implementation was 2016 for freight traffic and 2015 for livestock transportation.

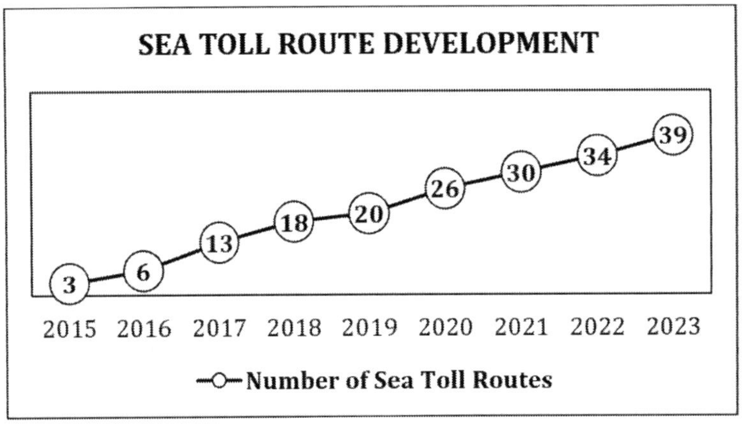

Figure 2. Sea toll route development [32].

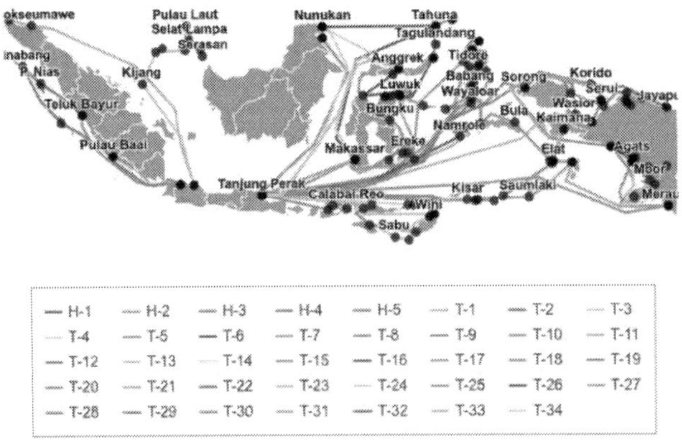

Figure 3. Sea toll routes in 2023 [32].

Figure 2 shows that the Sea Toll routes have grown each year. There were only three Sea Toll services operating in the first year of 2015. The number of Sea Toll routes doubled to six in the following year. There were thirteen Sea Toll routes in 2017, and this number increased to eighteen routes in 2018. There was only an increase of 2 routes to 20 routes in 2019. The number of sea toll routes increased in 2020, 2021, and 2022, to 26, 30, and 34 routes respectively. As seen in Figure 3, the number of sea toll routes increased dramatically to 39 routes in 2023 [32]. Decree of the Director General of Sea Transportation No. AL.108/6/2/DJPL-15, which outlines the Network of Routes for the Implementation of Public Service Obligations for Freight Transport on Tracks in the Framework of Implementing Sea Tolls in Fiscal Year 2015, states that six (6) routes have been established in 2015. PT Pelayaran Indonesia (Persero) was assigned by the Ministry of Transportation to run this Sea Toll program. However, at that time only 3 (three) ships were operated for 3 (three) routes with a subsidy value of 30 billion due to the limited time of 2 (two) months and the availability of the PT Pelni fleet. Volume I of the Sea Toll initiative, which ran from 2015 to 2019, successfully connected public services related to Sea Toll logistics. Based on information from [33] the three routes are listed in Table 1 below:

Table 1. Sea toll routes in 2015 [33].

No.	Route Code	Route Networks
1.	T-1	Tanjung Perak–Tual–Fak fak–Kaimana–Timika–Kaimana–Fak fak–Tual–Tanjung Perak
2.	T-4	Tanjung Priok–Biak–Serui–Nabire–Wasior–Manokwari–Wasior–Nabire–Serui–Biak–Tanjung Priok
3.	T-6	Tanjung Priok–Kijang–Natuna–Kijang–Tanjung Priok

3.3 Blue economy and sea tolls strategy

The idea of a "blue economy" aims to preserve the marine environment and coastal areas while promoting social inclusion, economic growth, and the maintenance or improvement of livelihoods [21], [34], [35]. With more than 17.000 islands and the sixth-largest exclusive economic zone (EEZ) in the world, Indonesia is an archipelago with a thriving maritime economy that has the potential to continue growing in the future [4], [11], [12]. Indonesia's seas present business prospects in various fields, such as industry, tourism, transportation, logistics, and marine biological and non-living resources [36], [37], [38], [39]. To be considered sustainable, economic development must be inclusive, environmentally sound and conducted in a way that avoids depleting the natural resources on which civilization ultimately depends [40].

A fundamental element of the blue economy is the requirement to achieve a balance between the social, economic, and environmental aspects of sustainable development concerning the ocean [21], [41], [42]. It is expected that Indonesia's shift towards a blue economy will reduce reliance on exploitation-based industries by acting as a prototype for the development of sustainable ocean-based industries. A dedicated document outlining the vision and strategy for Indonesia's ocean-based industry development is available. For the social, environmental and economic sustainability of Indonesia's ocean-based industries, this document acts as a policy foundation. Indonesia's blue economy development framework will support global efforts to move towards a more sustainable blue economy by conserving and responsibly using ocean and coastal resources for the benefit of current and future generations. To steer the medium and long-term roadmap for Indonesia in achieving high and sustainable productivity, six key strategies have been developed (National Long-Term Development Plan (RPJPN) and National Medium-Term Development Plan (RPJMN)). The development of a new capital city, strengthening domestic economic integration, strengthening the implementation of the green economy, increasing economic productivity, digital transformation, and improving competitive human resources are the six main strategies [36].

IOP Conf. Series: Earth and Environmental Science **1464** (2025) 012001 doi:10.1088/1755-1315/1464/1/012001

In particular, the goal to preserve oceans, seas and coastal areas for sustainable development is highlighted in Indonesia's economic transformation strategy, which includes strengthening the implementation of the green economy. The blue economy aims to accelerate ocean-based economic activities, and improve productivity and national welfare [16], [22], [30]. Unlike the green economy, the blue economy emphasizes the management of non-terrestrial resources. With great potential, Indonesia sees the blue economy as an important part of its economic transformation and diversification [30]. In this case, the Sea Toll is one of the blue economy strategies implemented in Indonesia to realize equity between the eastern and western parts of Indonesia. Ships, ports, logistics systems, and institutional interactions are the four supporting components of the Sea Toll implementation. The Sea Toll, has so indicated an improvement in the Indonesian economy, as seen by the 50% decrease in transportation costs [30]. The following types of goods can be loaded onto ships as part of the sea toll program [43], among others:

1. Basic necessities include agricultural products, industrial products, and livestock and fishery products.
2. Essential goods include seeds, fertilizer, 3-kilogram LPG gas, cement, plywood, mild steel and construction steel.
3. Other goods according to transportation mode (according to the needs of the 3TP area).

As part of Indonesia's blue economy strategy, the Sea Toll program is critical to strengthening the sea transportation industry that is the backbone of the Indonesian economy. In 2015, Indonesia generated the most value-added from sea freight at USD2.6 billion and sea passenger transport at USD2.2 billion among The Association of Southeast Asian Nations (ASEAN) countries [44]. Moreover, according to the Organisation for Economic Co-operation and Development (OECD) estimates, ocean-based sectors contributed around USD31.7 billion to Indonesia's global value added in the same year. This data confirms that the sea transportation sector has a significant contribution to the Indonesian economy, and the Sea Toll serves as a catalyst to improve inter-island connectivity, lower logistics costs, and encourage efficiency in the distribution of marine and fisheries products. Thus, the Sea Highway not only supports maritime economic growth but also strengthens Indonesia's role in a sustainable blue economy.

3.4 Impact of sea toll on price disparity

The Sea Toll has had a significant impact on the Indonesian economy, contributing to a 6.9% reduction in national food price disparities, although the impact is still relatively small. Problems such as inaccurate ship schedules, long travel times, and limited port infrastructure are contributing factors [23]. In addition, distance also affects price disparity, which means that prices in each place are different as the distance increases. Rice prices differ between the cities of Surabaya and Kalabahi, which are 733 Nm apart. In Surabaya, the price of rice is Rp. 9.442/kg, but due to distance and profit margins, the price in Kalabahi is Rp. 10.662/kg, Rote Rp. 10.679/kg, and Moa Rp. 10.689/kg [45]. Nevertheless, the Sea Toll program has become a new breakthrough for the Indonesian people in reducing prices in eastern Indonesia which are quite high. Comparison of prices before and after the Sea Toll in several regions of Indonesia on several commodities can be seen in Table 2 [23].

In Table 2, the price comparison refers to the 2012-2015 price period which shows prices before the Sea Toll and the 2016-2019 price period shows prices after the Sea Toll. Six regions including North Sumatra, West Sumatra, Bengkulu, Jakarta, Bangka Belitung, and Riau Islands are included in Hub 1. While Hub 2 includes East Java, East Nusa Tenggara, Gorontalo, East Kalimantan, North Kalimantan, North Sulawesi, Central Sulawesi, South Sulawesi, Southeast

Sulawesi, Maluku, Papua, North Maluku, and West Papua. In general, the price disparity between the eastern and western regions of Indonesia improved in the 2016-2019 period [23].

Table 2. Price comparison before and after the sea toll [23].

Group	Price Disparities (%): Target < 13%							
	Rice		Sugar		Frying Oil		Chicken	
	Before	After	Before	After	Before	After	Before	After
Hub 1	9,68	5,71	10,99	4,90	7,56	6,13	10,45	11,46
Hub 2	17,18	13,51	8,80	6,48	10,97	8,63	20,27	16,21
National	12,31	11,41	8,69	5,99	11,57	8,39	16,14	13,63

Prices of various commodities in the western region or hub 1 show a significant decline after the implementation of the Sea Toll. For example, for rice, prices fell from 9,68% to 5,71%. Similar declines were seen in sugar, cooking oil, and chicken meat. Likewise, prices of the same commodities in the eastern region or hub 2 have decreased after the Sea Toll. Although not as big as the decline in the western region. For example, rice prices fell from 15,33% to 11,31%. Smaller decreases occurred in the commodities of cooking oil and chicken meat. Before the Sea Toll, commodity prices in Hub 2 tended to be higher than commodity prices in Hub 1. After the Sea Toll, the price disparity between the two regions was reduced.

Table 3. Prices of commodities transported by sea toll in 2015 and 2019 [23].

Commodities	Average Price Before Sea Toll (2015)				Average Price After Sea Toll (2019)			
	Natuna	Tahuna	Timika	Saumlaki	Natuna	Tahuna	Timika	Saumlaki
Rice	Rp.15.000	Rp.13.000	Rp.18.000	Rp.13.000	Rp.13.000	Rp.12.000	Rp.12.600	Rp.11.750
Cooking Oil	Rp.16.000	Rp.13.500	Rp.17.000	Rp.17.500	Rp.13.000	Rp.13.000	Rp.13.250	Rp.16.000
Sugar	Rp.14.000	Rp.14.900	Rp.16.000	Rp.15.000	Rp.13.000	Rp13.000	Rp.13.750	Rp.11.400
Breed Chicken	Rp,36.000	Rp.40.000	Rp.38.000		Rp.13000	Rp.35.000	Rp.36.000	

Report results from [23], Table 3 illustrates that the prices of basic goods transported by the Sea Tolls were found to have decreased in the survey area, which is the port city of the Sea Toll destination. The average price of several commodities (rice, cooking oil, sugar, and broiler chicken meat) before the Sea Toll (2015) and after the Sea Toll (2019) has decreased in the price of each type. Table 3 shows a comparison of several basic commodities in the Natuna, Tahuna, Timika, and Saumlaki regions before and after the Sea Toll program. The price of rice dropped significantly in the four areas traversed by the sea highway. This was particularly noticeable in Timika, where prices fell from IDR 18.000 to IDR 12.600/kg. Following the Sea Toll, cooking oil prices tended to be stable or slightly lower in some regions, with Tahuna experiencing the largest drop, from IDR 13.500 to IDR 13.000. All regions experienced a decrease in sugar prices, with Timika experiencing the largest decrease, from IDR 16.000 to IDR 13.750. Chicken meat prices fell in Natuna and Tahuna, but remained the same in Timika and Saumlaki. A decrease in the price of basic necessities in the Sea Toll destination area, as the results of research showing a decrease of 11% to 20%. [24] reflects the positive impact of this initiative. Overall, the Sea Toll has succeeded in reducing price disparities in various regions, especially in areas that previously had very high commodity prices. While this price reduction may not be huge, it still shows that the Sea Toll program, as part of Indonesia's blue economy strategy, has a positive impact on the national economy.

As an effort to improve the economy through ports, the Ministry of Transportation not only builds sea transportation infrastructure in the form of ports and other sea transportation facilities but also prepares a fleet of ships to support the strengthening of the economy. Matui Jailolo Port, West Halmahera, Maluku, is one of the ports that has the largest and always crowded Sea Toll return cargo, with return cargo in the form of copra. This route has been included in the official Toll Route since early 2020, which has a route of Surabaya-Makassar-Jailolo-Morotai. From Morotai, it goes directly to Surabaya as a return route [46]. Copra farmers in Maluku, one of which is on Kei Besar Island, feel the increasing price of copra, which continues to increase thanks to this program. Previously, the price of copra was only around Rp. 3.000/kg, now it has reached Rp. 8.000/kg [47], [48].

In addition, the backload from Matui Jailolo Port to Java on KM Log Nusantara 3 had a total of 35 containers, of which 29 were copra containers, 1 squid container, 1 fish container, 1 squid and fish container, 1 coconut container, 1 clove and nutmeg container, and 1 people container. The Sea Toll program is very significant in decreasing price disparities, where industrial raw materials are very clearly visible price changes [46]. Not only does the Sea Toll initiative impact the agriculture and fisheries sectors, it also affects national tourism. One of the goals of this initiative is the revitalization of people's shipping. This has been done through a concept called "Kampung Pinisi-Nusantara" that was organized in Bulukumba, South Sulawesi. This concept is a blending of Indonesia's historical and cultural dimensions related to the maritime context [49]. Obviously, the people who live in the area gain benefits because they are also involved as part of the management of the Pinisi ship local wisdom development program. Thus, the Sea Toll program is very helpful to improve the economy and welfare of local communities and entrepreneurs in various regions, especially Eastern Indonesia.

4. Conclusion

The Sea Toll plays an important role as a tangible manifestation of efforts to realize a blue economy in Indonesia. The initiative has successfully improved inter-island connectivity, lowered logistics costs and improved commodity distribution, all of which have benefited Indonesia's maritime economy. The Sea Toll has increased economic growth in previously marginalized areas and reduced the price differential between eastern and western Indonesia by unifying the shipping industry and ensuring access to 3TP areas. The significant reduction in commodity costs in various locations demonstrates the tangible benefits of the Sea Toll to the local and national economy. Furthermore, ships returning to Java from the eastern part of Indonesia no longer return with nothing but backloads of cargo produced by the local community. Hence, the prosperity of the local community has increased thanks to this initiative. The Sea Toll is therefore an important component in Indonesia's future economic transformation and diversification as it plays an important role in the blue economy strategy that emphasizes social, economic and environmental sustainability.

References

[1] H. W. Palm *et al.*, "Anisakis (Nematoda : Ascaridoidea) from Indonesia," *Dis. Aquat. Organ.*, vol. 123, pp. 141–157, 2017.

[2] D. A. A. Sari and E. Latifah, "Revitalization of traditional fisheries rights of indigenous people in sustainable fisheries management in Indonesia," *IOP Conf. Ser. Earth Environ. Sci.*, vol. 724, no. 1, pp. 0–7, 2021, doi: 10.1088/1755-1315/724/1/012117.

[3] S. M. Badriyah *et al.*, "The responsibility of private pawnshops in facilitating ecologically-friendly marine economies," *Environ. Ecol. Res.*, vol. 9, no. 6, pp. 362–369, 2021, doi: 10.13189/eer.2021.090604.

[4] Office of Assistant to Deputy Cabinet Secretary for State Documents & Translation, "Indonesia Has Enormous Potential to Become Global Maritime Axis, President Jokowi Says," Setkab.go.id. Accessed: Aug. 12, 2024. [Online]. Available: https://setkab.go.id/en/indonesia-has-enormous-potential-to-become-global-maritime-axis-president-jokowi-says/

[5] J. Langer, A. A. Cahyaningwidi, C. Chalkiadakis, J. Quist, O. Hoes, and K. Blok, "Plant siting and economic potential of ocean thermal energy conversion in Indonesia a novel GIS-based methodology," *Energy*, vol. 224, p. 120121, 2021, doi: 10.1016/j.energy.2021.120121.

[6] H. Ajiwibowo, K. S. Lodiwa, M. B. Pratama, and A. Wurjanto, "Field measurement and numerical modeling of tidal current in larantuka strait for renewable energy utilization," *Int. J. GEOMATE*, vol. 13, no. 39, pp. 124–131, 2017, doi: 10.21660/2017.39.98548.

[7] H. Hastuti, A. Muhidu, R. Rastin, and E. Agus Mokodompit, "Indonesia's Marine Economic Potential As A Maritime Country," *Int. J. Sci. Technol. Manag.*, vol. 4, no. 4, pp. 813–825, 2023, doi: 10.46729/ijstm.v4i4.897.

[8] D. A. Yeremia and N. Istifadah, "The Role of Maritime Sector in the Economic of the World's Largest Developed Archipelagic Country," *Int. J. Manag.*, vol. 11, no. 6, pp. 2083–2091, 2020, [Online]. Available: http://www.iaeme.com/IJM/index.asp2083http://www.iaeme.com/IJM/issues.asp?JType=IJM&VType=11&IType=6http://www.iaeme.com/IJM/issues.asp?JType=IJM&VType=11&IType=6

[9] P. Choudhary *et al.*, "Empowering blue economy: From underrated ecosystem to sustainable industry," *J. Environ. Manage.*, vol. 291, no. April, p. 112697, 2021, doi: 10.1016/j.jenvman.2021.112697.

[10] R. Ernawati, "Pembangunan Jaringan Rute Multiport Tol Laut Wilayah Papua Bagian Utara Guna Memperlancar Arus Barang Serta Menjaga Keamanan dan Keselamatan Pelayaran di Indonesia Bagian Timur," Universitas Pertahanan, 2024. (In Indonesia)

[11] Y. Rochwulaningsih, S. T. Sulistiyono, N. N. Masruroh, and N. N. Maulany, "Marine policy basis of Indonesia as a maritime state: The importance of integrated economy," *Mar. Policy*, vol. 108, no. July, p. 103602, 2019, doi: 10.1016/j.marpol.2019.103602.

[12] World Bank Group, "A Sustainable Ocean Economy is Key to Indonesia's Prosperity," World Bank Group. Accessed: Oct. 22, 2024. [Online]. Available: https://www.worldbank.org/en/news/press-release/2021/03/25/sustainable-ocean-economy-key-for-indonesia-prosperity

[13] World Bank Group, "Blue Economy," World Bank Group. Accessed: Oct. 22, 2024. [Online]. Available: https://www.worldbank.org/en/topic/oceans-fisheries-and-coastal-economies

[14] Prime News, "2025-2045 RPJPN Emphasizes Potential of Indonesia's Maritime Resources," Bappenas. Accessed: Oct. 22, 2024. [Online]. Available: https://www.bappenas.go.id/en/berita/optimalkan-potensi-kemaritiman-indonesia-dalam-visi-rpjpn-2025-2045-xkBEv

[15] P. Saha, "Indonesia's potential as a maritime power," *Marit. Aff.*, vol. 12, no. 2, pp. 28–41, 2016, doi: 10.1080/09733159.2016.1232951.

[16] Salma, "Indonesia's Maritime Highway: Experts Reflect on Progress, Barriers, and Solutions," Universitas Gadjah Mada. Accessed: Oct. 22, 2024. [Online]. Available: https://ugm.ac.id/en/news/21830-indonesia-s-maritime-highway-experts-reflect-on-progress-barriers-and-solutions/

[17] Editor, "Guide to Indonesia's maritime industry," Maritime Fairtrade. Accessed: Oct. 22, 2024. [Online]. Available: https://maritimefairtrade.org/guide-to-indonesias-maritime-industry/

[18] J. Duha, Sutrimo, and S. A. Purwantoro, *Forecasting Geopolitik Blue Economy*. Bogor: UNHAN RI Press, 2023.

[19] O. S. Suharyo, "Indonesia's maritime strategy for realizing the world's axis facing China's maritime silk road policy," *Glob. J. Eng. Technol. Adv.*, vol. 13, no. 3, pp. 045–052, 2022, doi: 10.30574/gjeta.2022.13.3.0206.

[20] S. Suwandi, S. Suryoko, and R. Marlina, "Optimization Defense Industry Master To Strengthen Defensemaritime in the Order To Realize Indonesia As World Maritime Axis," *J. Asro*, vol. 12, no. 02, pp. 31–42, 2021, [Online]. Available: http://asrojournal-sttal.ac.id/index.php/ASRO/article/view/400

[21] M. Vierros and C. De Fountaubert, "The Potential of the Blue Economy," 2017. doi: 10.1596/26843.

[22] A. A. Supriyadi *et al.*, "Kebijakan Ekonomi Ketahanan Pangan dengan Strategy Blue Economy Menghadapi Ancaman Perubahan Iklim," *J. Innov. Res. Knowl.*, vol. 2, no. 4, p. 12, 2022. (In Indonesia)

[23] R. E. Danny D., "Kajian Dampak Tol Laut Terhadap Disparitas Harga," 2019. (In Indonesia)

[24] PELNI, "Tol Laut," PT. PELNI. Accessed: Aug. 12, 2024. [Online]. Available: https://www.pelni.co.id/jasa-logistik-tol-laut. (In Indonesia)

[25] A. A. Juliadi, A. S. Oktirellia, G. C. N. Fadhilah, I. Shaffiyah, and V. Dwiyanti, "Indonesia Toll Roads: Its Effect on The Trade Routes in ASEAN," *ASEAN J. Sci. Eng.*, vol. 3, no. 1, pp. 57–62, 2023, doi: 10.17509/ajse.v3i1.33765.

[26] L. T. W. N. Kusuma and F. S. Tseng, "Analysis of the impact of the 'sea toll' program for seaports: Resilience and competitiveness," *Appl. Sci.*, vol. 9, no. 16, 2019, doi: 10.3390/app9163407.

[27] M. A. Zamal, "Liner Shipping Network Design in Indonesia 'Sea-Toll' Agenda: Tanjung Perak Corridor," in *International Conference on Operations and Supply Chain Management*, 2018, pp. 1–11.

[28] J. W. Creswell and J. D. Creswell, *Qualitative, quantitative and mixed methods research*. 2018. doi: 10.4324/9781315707181-60.

[29] I. T. Yunianto, H. I. Nur, E. W. Ardhi, and B. P. Adhitya, "Optimalisasi Model Jaringan Rute Multiport Tol Laut di Negara Kepulauan: Studi Kasus Evaluasi Rute di Maluku dan Papua Bagian Selatan," *J. Penelit. Transp. Laut*, vol.

21, pp. 83–95, 2019. (In Indonesia)
[30] Direktorat Jenderal Perhubungan Laut, *Tol Laut: Konektivitas Visi Poros Maritim Indonesia*, 1st ed. Jakarta: Penerbit Buku Kompas, 2020. (In Indonesia)
[31] Biro Komunikasi dan Informasi Publik, "Program Tol Laut Fokus Angkutan Balik," Kementerian Perhubungan Republik Indonesia. Accessed: Oct. 22, 2024. [Online]. Available: https://dephub.go.id/post/read/program-tol-laut-fokus-angkutan-balik. (In Indonesia)
[32] Kementerian Perdagangan, "Trayek," Sistem Informasi Gerai Maritim. Accessed: Aug. 17, 2024. [Online]. Available: https://geraimaritim.kemendag.go.id/trayek. (In Indonesia)
[33] Direktorat Jenderal Perhubungan Laut, "Program Tol Laut Resmi Diluncurkan," Dephub. Accessed: Aug. 17, 2024. [Online]. Available: https://dephub.go.id/post/read/program-tol-laut-resmi-diluncurkan. (In Indonesia)
[34] D. Hunter, "The blue economy: Growth, opportunity and a sustainable ocean economy," *Sea Technol.*, vol. 56, no. 1, pp. 41–42, 2015, doi: 10.2307/j.ctvgc60f0.9.
[35] A. Midlen, "What is the Blue Economy? A spatialised governmentality perspective," *Marit. Stud.*, vol. 20, no. 4, pp. 423–448, 2021, doi: 10.1007/s40152-021-00240-3.
[36] D. D. Pane *et al.*, *Blue Economy Development Framework for Indonesia's Economic Transformation*, vol. 1, no. 2. 2021.
[37] L. A. A. T. Sambodo *et al.*, *Indonesia Blue Economy Roadmap*. Jakarta: Ministry of National Development Planning/National Development Planning Agency (BAPPENAS), 2023.
[38] G. Auad and B. D. Fath, "Towards a flourishing blue economy: Identifying obstacles and pathways for its sustainable development," *Curr. Res. Environ. Sustain.*, vol. 4, no. November, p. 100193, 2022, doi: 10.1016/j.crsust.2022.100193.
[39] L. A. Pace, O. Saritas, and A. Deidun, "Exploring future research and innovation directions for a sustainable blue economy," *Mar. Policy*, vol. 148, no. November 2022, p. 105433, 2023, doi: 10.1016/j.marpol.2022.105433.
[40] M. A. Ali, M. Kamraju, and D. B. Sonaji, "Economic Policies for Sustainable Development: Balancing Growth, Social Equity, and Environmental Protection," *ASEAN J. Econ. Econ. Educ.*, vol. 3, no. 1, 2023, [Online]. Available: https://ejournal.bumipublikasinusantara.id/index.php/ajeee
[41] K. H. Lee, J. Noh, and J. S. Khim, "The Blue Economy and the United Nations' sustainable development goals: Challenges and opportunities," *Environ. Int.*, vol. 137, no. October 2019, p. 105528, 2020, doi: 10.1016/j.envint.2020.105528.
[42] H. J. Niner *et al.*, "Issues of context, capacity and scale: Essential conditions and missing links for a sustainable blue economy," *Environ. Sci. Policy*, vol. 130, no. May 2021, pp. 25–35, 2022, doi: 10.1016/j.envsci.2022.01.001.
[43] M. Fuady, "Evaluasi Kebijakan Tol Laut," *J. Law, Adm. Soc. Sci.*, vol. 3, no. 2a, pp. 284–293, 2023, doi: 10.54957/jolas.v3i2a.628. (In Indonesia)
[44] OECD, "Sustainable Ocean Economy Country Diagnostics of Indonesia," 2021.
[45] H. I. Nur, T. Achmadi, and A. Verdifauzi, "Optimalisasi Program Tol Laut Terhadap Penurunan Disparitas Harga: Suatu Tinjauan Analisis," *J. Penelit. Transp. Laut*, vol. 22, no. 1, pp. 1–12, 2020, doi: 10.25104/transla.v22i1.1315. (In Indonesia)
[46] Biro Komunikasi dan Informasi Publik, "Tol Laut dan Upaya Mendongkrak Perekonomian Indonesia Timur," Kementerian Perhubungan Republik Indonesia. [Online]. Available: https://dephub.go.id/post/read/tol-laut-dan-upaya-mendongkrak-perekonomian-indonesia-timur. (In Indonesia)
[47] Valid News, "Tol Laut Dongkrak Harga Kopra Di Maluku Tenggara," Valid News. [Online]. Available: https://validnews.id/ekonomi/tol-laut-dongkrak-harga-kopra-di-maluku-tenggara. (In Indonesia)
[48] S. Yanyaan, "Petani kopra di Maluku Tenggara rasakan manfaat dari Tol Laut, jayalah para petani," AntaraMaluku. [Online]. Available: https://ambon.antaranews.com/berita/115713/petani-kopra-di-maluku-tenggara-rasakan-manfaat-dari-tol-laut-jayalah-para-petani. (In Indonesia)
[49] H. Putra and T. Rusata, "Konsep 'kampung pinisi-nusantara' pengembangan ethno-maritime tourism Sulawesi Selatan-IKN yang inklusif dan terkoneksi," *Archeol. Nexus J. Conserv. Cult.*, vol. 1, no. 1, pp. 47–58, 2024. (In Indonesia)

Transportation challenges for onshore wind turbine power plant development in Indonesia

Maria Angelin Naiborhu[1*], Dwi Rahayu Purwanti[2], Silvia Dewi Kumalasari[1], Muhamad Kemal Idris[1], Cokorda Bagus Purnama Dwisa[1], Ali Sumara Elkhomei[1], Yani Nurita Purnawanti[3]

[1]Port Engineering Program, Faculty of Transportation and Logistik Engineering, Institut Transportasi dan Logistik (ITL)-Trisakti, Jakarta, Indonesia
[2]Construction Engineering and Management Program, Faculty of Civil and Environmental Engineering, Institut Teknologi Bandung (ITB), Bandung, Indonesia,
[3]Politeknik Kelautan dan Perikanan, Sorong, Indonesia

*E-mail: mariaangelin@lecturer.itltrisakti.ac.id

Abstract. Wind energy for power generation in Indonesia is dominated by onshore wind turbine farm power plants (WPP), where the potential onshore installation located in southern West Java, South Sulawesi, and Eastern Indonesia. Most of the prospective WPP are located in distant area with mountainous terrain. This circumstance raises transportation concerns as the main components of wind turbines are manufactured in large sizes. In the coming years, wind turbine technology will likely grow in size. The expected increase will ultimately lead to higher handling and transportation costs. By considering the size and weight for the wind turbine components, and required unloading equipment as well as sufficient space for manoeuvring during unloading process, then limited sea ports are available to handle such huge cargo size, specifically when the WPP will be developed in Eastern Indonesia. In addition, existing road condition also a major consideration in transporting the components. This paper discusses current WPP technology development, handling and transportation schemes and challenges for onshore installation, particularly in Indonesia. General description of transportation scheme from previous Tolo 1-WPP project in Jeneponto Regency-South Sulawesi Province is also included in this paper to provide insight into logistic complexity of the WPP project.

1. Introduction

The utilization of wind energy for electricity generating, which is known as a Wind Power Plant (WPP) or *Pembangkit Listrik Tenaga Bayu* (PLTB) has good potential in Indonesia for both onshore (also called as land-based) and offshore installation as shown in Figure 1. The construction of two large onshore WPPs in Sidrap and Jeneponto (South Sulawesi) with capacity 30 x 2.5 Mega Watt (MW) and 20 x 3.6 MW respectively, has set the stage for the participation of wind energy resources in Indonesia which is now expected to be followed in other regions with good wind energy potential.

According to the Electricity Supply Business Plan (RUPTL) 2021-2030 of the State Electric Company (PLN), Indonesia government has planned potential onshore WPPs development in various regions in Indonesia (several location of WPPs are plotted in Figure 1), where the largest potential development will be in Java with capacity 1,138 MW, followed by Sulawesi (458 MW) and Sumatera (180 MW) Regions.

Indonesia has unique challenges in developing wind turbine power plants as most of the potential development of WPPs are located remotely at high altitude with mountainous topography and some of them are located near the coastal area with harsh wave condition. Such conditions raise transportation issues as the main components of current wind turbine technology are produced in large sizes. Several technical requirements such as height clearance, road width, slope, bend and bearing capacity during land transportation shall be filled. Furthermore, there are not many commercial ports are available for loading-unloading of wind turbine components as specific unloading equipment, vehicles, maneuvering space and temporary storage area are required.

This paper aims to provide insight into the current development of wind energy-based power plant as part to support the Indonesia government program in renewable energy uses. With the highlight on the transportation process of wind turbine components, this paper is expected to assist the reader in understanding the challenges encountered during the development of WPP in Indonesia, taking into account the existing condition of public infrastructures.

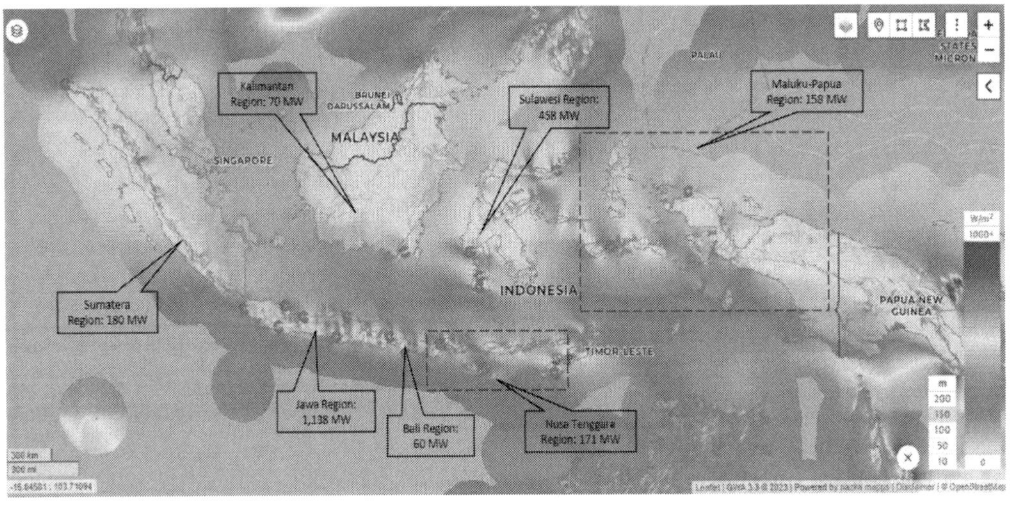

Figure 1. Mean power density in Indonesia at 150 m above ground [1] and several potential developments of WPP in Indonesia based on PLN's RUPTL 2021-2030 [2].

2. Wind Turbine Size Development

Wind turbines have grown in size and generated more energy [3]. In 2018, the average diameter of an onshore wind turbine rotor was approximately 110.40 meters. Between 2022 and 2025, reaching approximately 170 meters, with generation ratings increasing as well [4] as shown in Figure 2. Table 1 summarizes the size and weight ranges for wind turbine components with capacities of 4.0-6.0 MW from various sources. This figure is expected to increase as the recent technologies are intending to generate wind energy in areas with low wind potential. Such dimension and weight obviously raise transportation issues in many countries, including Indonesia.

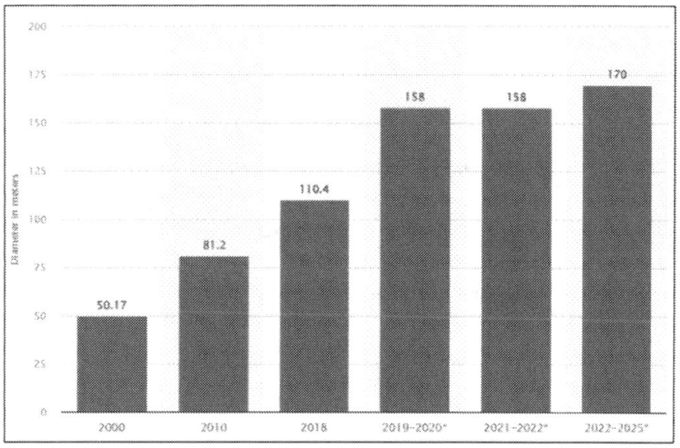

Figure 2. The average rotor diameter of onshore wind turbines worldwide from 2000 to 2025 [4].

Table 1. The size and weight range of wind turbine components with capacity 4.0-6.0 MW.

Component	Length (m)	Diameter or Height (m)	Weight (Ton)
Tower Section	10.0-35.0	3.5-6.0	67.0-117.0
Nacelle	12.0-13.0	3.0-4.0	90.0-130.0
Hub	4.0-5.5	3.5-4.5	34.0-67.0
Blade	64.0-85.0	3.0-4.0	17.0-35.0

3. Transportation Challenges

Most of the wind turbine components, specifically the blades, nacelle, tower sections and hub, are manufactured abroad. It must be delivered by using sea-going vessel from the origin country to the designated port in Indonesia, then unloaded and hauled to the cluster by using special vehicles. Wind turbine components are classified as oversized cargoes shipment. The blades present the largest challenge logistically [7] as they cannot be dismantled for shipping and hauling, unlike the other components, such as the tower. Typically, the blades are transported individually on truck, whereas the tower sections and nacelle may be shipped on several trucks.

To facilitate the transportation and safety of components of such dimensions and weight through inland road, the general requirements outlined in Table 2 shall be met, which frequently requires major development work on existing infrasctructures. Reffering to Table 1 and 2, there are five major factors must be assessed in order to achieve an appropriate transportation plan in Indonesia, as illustrates in Figure 4. Those factors are related to the existing infrastructure, availability of cargo handling equipment, permitting, environmental and social condition.

Evaluation to various aspects as mentioned above is necessary to estimate the cost for overall transportation process. According to the NREL report in 2023, wind turbine transportation costs 172 USD/kW in the United States, which accounts for 9.9% of the total CapEx (Capital Expenditure) for land-based WPP as illustrated in Figure 3 [5]. The component of transportation cost includes rare-special-purpose cranes, trailer, permitting, improvement for existing infrastructure and development new infrastructure [6]. Though the cost may vary widely between countries, it is worthwhile to have a certain benchmark value for early stage estimation.

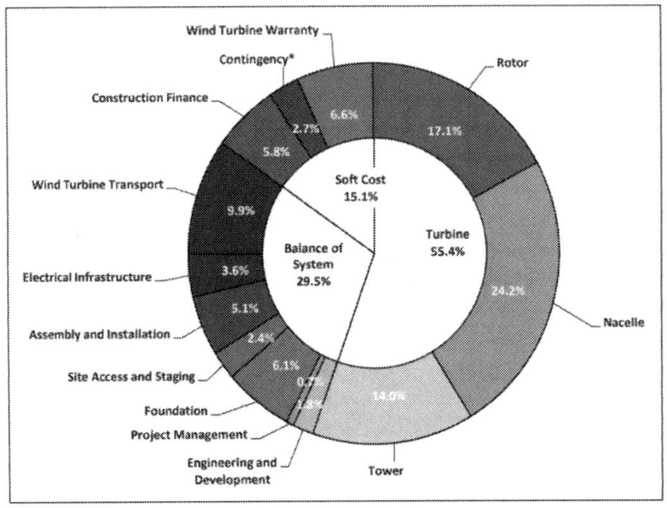

Figure 3. Land-based WPP project component cost breakdown [5].

Table 2. General inland transportation requirement for wind turbine component.

Constraining Factor	Min. Required Value
Height Clearance	7.5 m
Road Width	4.5 m
Road Longitudinal Slope	8.0-14.0%
Road Bend	70.0 m
Road Load Bearing Capacity	200 kN/m^2

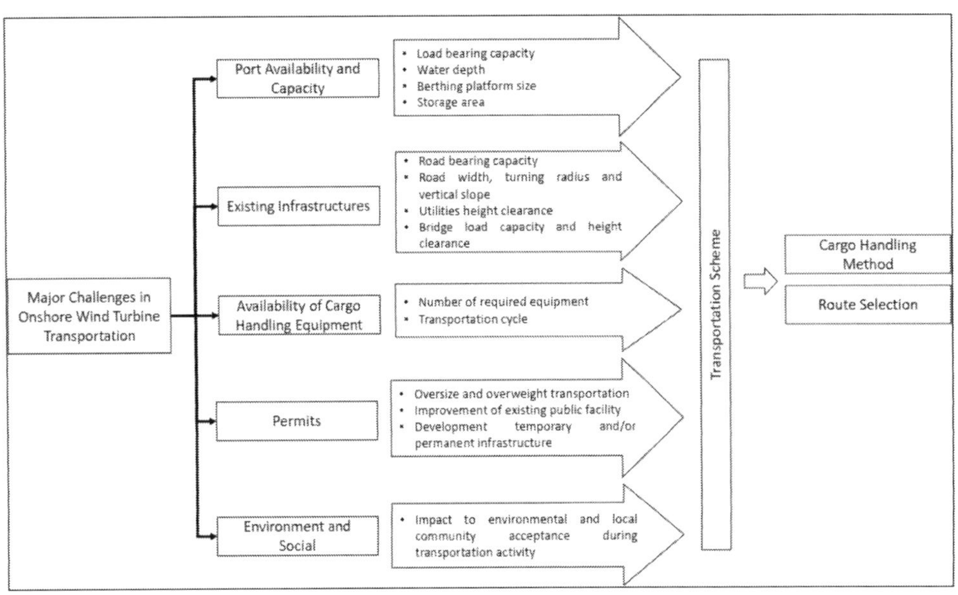

Figure 4. Evaluation factors for transportation scheme planning of wind turbine components.

3.1 Port availability

Given the size and weight of the wind turbine components, then large and various unloading equipment, as well as ample maneuvering space are required to transfer the components from the sea-going vessel to the multi-wheeled truck in the designated port. Commercial major ports are available in northern Sumatera, northern and western Java, western and southern Sulawesi, as well as southern Kalimantan as shown in Figure 5. However, such a heavy load during the unloading process and the size of the wind turbine components may exceed the design load capacity of existing structures and capacity of port unloading equipment. Also, the existing berthing platform size may not be sufficient for unloading process and equipment's maneuvering space.

Figure 6 shows general overview for the arrangement of the wind turbine components and required space at Makassar Port during unloading and transportation process of Tolo 1 WPP. Based on measurement in google earth, the temporary storage area is 2.5 Ha. Makassar Port is located approximately 90 km from the WPP cluster and it took 6 months to transport all the components to the cluster.

Finding the suitable port and as near as possible to the WPP cluster is the first challenge for the transportation process, specifically for the WPP which is located in southern Java, Maluku and Nusa Tenggara Regions, as well as other small and remote islands. Alternatively, a temporary unloading structure nearby the WPP cluster can be an option, though it requires high development costs and is not economically viable for small WPP capacity.

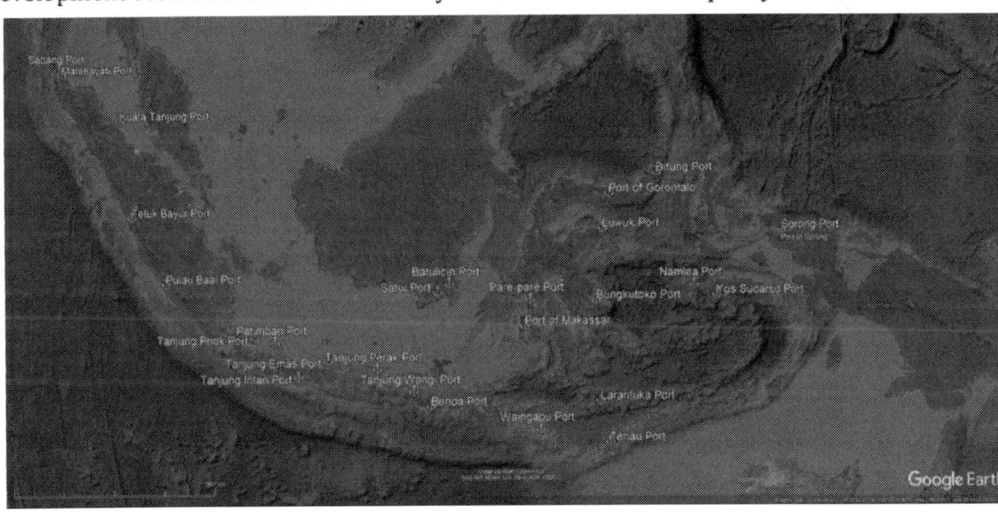

Figure 5. Several existing commercial ports surrounding the potential development area of WPP [8].

3.2 Road availability

The current size of wind turbine blades will not fit into typical trailers, hence specially designed trailers are required for road transport. The height of the blades and the diameter of the tower section may also raise issues with road transport, as some underpasses, bridge frames, tunnels, overhead distribution power lines and other overhead utilities crossing the road may not have sufficient height clearance as required in Table 2. Furthermore, the existing road bearing and bridge capacity, width, slope, turning radius and intersections along the route from the unloading port to the WPP cluster are the other major aspects to consider during transportation.

Figure 6. Wind turbine component temporary storage for Tolo 1 WPP at the Makassar port [8], [9].

Trailers with steer axles are frequently used to help navigate sharp turns and intersections during transportation. The trailers length and number of axle are dictated by the weight and length of the cargoes. For the example, the generators for the Tolo 1 WPP project weigh 74 tonnes each, while the tower sections weigh 86 tonnes. They were loaded onto 11 axle lines of conventional trailers and 6 axle lines of clamp tower trailers. For the 64-meter-long blades, 12 axle lines of specialty extendable blade trailers were used [10]. Longer blade requires longer trailer and wider turn path. Another example for the profile dimension of the trailer and requires turning paths for blade length 70.0 m is further described in [7]. Furthermore, it is important to ensure that the fatigue impact generated by the loading condition during transport is insignificant and negligible in the structural verification of the blade [11].

In Indonesia, maximum allowable axle load for road is defined by its class, which is either Class I, II, or III. Class I has maximum allowable axle load 10 ton, while Class II and III have a maximum allowable axle load of 8 tons. As a result, the dimensional requirement of the trailers and loads during transportation may requires alterations and improvement to the existing road.

New road infrastructure often necessary for the WPP which is located in remote area where the existing road network is not available. As a WPP cluster will be on a large area and consists of many turbines, then a long access road is required to connect each turbines within the WPP cluster. For the example presented in Figure 7, the road network length within Tolo 1 WPP is approximately 13 km. Roads within the WPP cluster are typically gravel-paved and built to enable access to each turbines and associated infrastructure, such as the plant substation. These roads must be designed to withstand the loads stipulated by turbine manufacturer. If there are existing access roads within the area of the wind farm, these roads are likely be used as much as possible, but improvement may be needed to accommodate delivery vehicle weights and sizes.

3.3 Permits and social aspects

Aside from planning the appropriate trucks and trailers for transportation, the other critical aspects of transportation are obtaining road transport permits and social condition along the transportation route. As oversize load transportation, a heavy transport permit shall be applied and obtained as well as other associated permits to improve the existing road and construct a new access road from the Ministry of Public Works and Public Housing and the related authorities.

For transportation scheme involving temporary jetty, then associated permits for port construction and operation shall be obtained from the Ministry of Transportation, as well as associated permits for handling import cargoes in the berthing facility from the Customs Clearance.

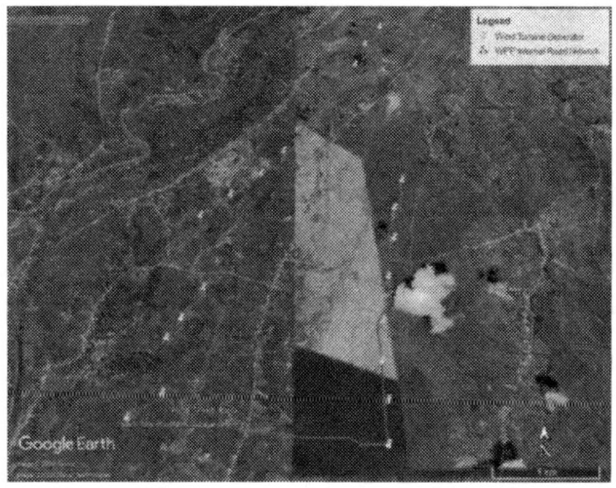

Figure 7. Road Network Within Tolo 1 WPP [8].

The social aspect shall also be considered for many reasons as each stage of the transportation process may have both positive and negative impact to local community along the selected route, such as: impact due to road improvement work and land acquisition in order to provide sufficient space for swept path, width and slope for transportation; and disturbance to local community as the impact of road closure during transportation process.

4. Transportation Schemes

Obviously, the assessment in determining the transportation scheme for wind turbine components is carried out based on the size and weight of the components and the location where the turbine will be installed. Given the aforementioned challenges, there are two transportation schemes to consider: i. barging-road transportation, ii. road transportation. Figure 8 introduces the workflow and evaluation factor for both schemes.

For the first scheme, the wind turbine components (also called as cargoes) will be transferred from the main vessel to a barge at the designated commercial port and will then be delivered to the selected temporary unloading area (also called as the Landing Point) near the WPP cluster, unloaded and then delivered to the cluster through the improved/modified road by using suitable transportation vehicles. Cargoes trans-shipment should take place inside the designated port area to ensure sufficient protection against the wave and current during trans-shipment. This scheme is beneficial when the commercial port is far and the access road from the port to the WPP cluster requires major development or improvement, as well as height clearance constraints. Though this scheme may shorten the road transportation route and lessen the constraints associated with road transportation, it might be costly as sufficient loading/unloading facilities must be provided at the selected Landing Point and double handling may risk the cargoes. Double handling can be avoided if the Landing Point area has sufficient depth for the main sea-going vessel which is normally difficult to find without dredging work. It is also important to ensure that the wave condition during unloading at the Landing Point area remains within the safe unloading procedure parameters.

Meanwhile, for the second scheme, cargoes will be unloaded at the selected commercial port and then directly delivered to the WPP cluster via road transportation, as was done for Tolo 1 WPP. This scheme is advantageous if the port has sufficient area for unloading and storage, as previously stated, and the transportation distance is relatively short.

As seen in Figure 8, road transportation is applicable for both schemes, thus it shall be optimized in order to obtain the most efficient routes. Evaluation and optimization of route selection involve analyzing of various factors such as vehicle size and capacity, number of vehicles, delivery schedule, existing traffic condition, required development works, road/bridges capacity and overhead clearances. The dashed line indicates the inputs for the specific stage.

Finally, both schemes should be evaluated through extensive transportation surveys and analysis to identify and mitigate the potentials constraints. Table 3 proposes four major parameters to evaluate which transportation scheme would be suitable based on three key parameters: cost, time, and risk. The impact factor score is defined from 1 to 5, where 1 being the lowest (has the highest impact, less benefit); 5 being the highest (has the lowest impact, high benefit).

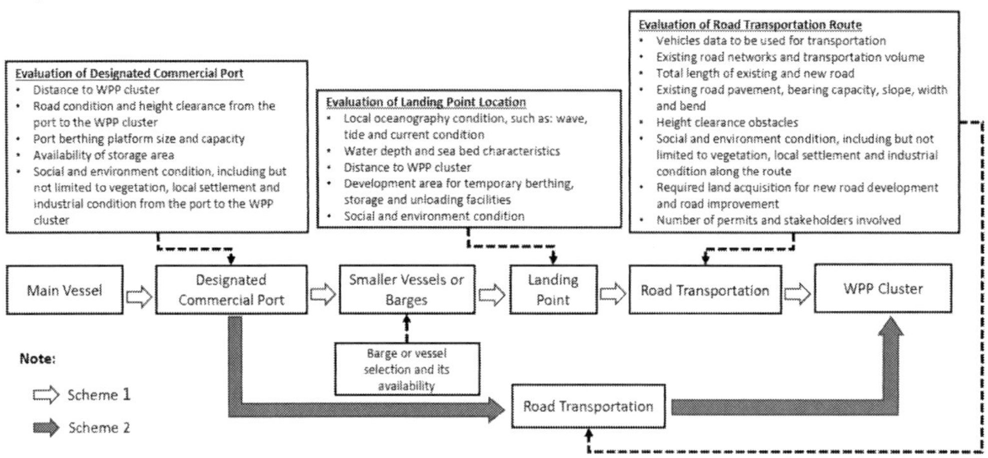

Figure 8. Workflow and evaluation factors for each transportation schemes.

Table 3. Major parameters and impact factor for the selection of onshore wind turbine components transportation schemes.

<table>
<tr><th rowspan="2">Parameter</th><th colspan="5">Impact Factor Score</th></tr>
<tr><th>1</th><th>2</th><th>3</th><th>4</th><th>5</th></tr>
<tr><td>Certainty around delivery schedule (by avoiding/minimizing congested areas, planned development works on the route and barging operation, delivery time due to limitation of vehicle)</td><td colspan="5">High certainty in delivery schedule gives higher score</td></tr>
<tr><td>Minimizing development works to be done and its complexity on the selected route from the designated commercial port or Landing Point to cluster (subject to verification of bearing capacity of bridges and road; overhead clearance; sharp bends and steep slope)</td><td colspan="5">Less development works and complexity give higher score</td></tr>
<tr><td>Minimizing delivery and development works permits, and negotiations with private land owners</td><td colspan="5">Less permits and stakeholders involvement give higher score</td></tr>
<tr><td>Minimizing risk of damaging cargo by avoiding additional handling of cargo</td><td colspan="5">Less cargo transfer process gives higher score</td></tr>
</table>

5. Risk Analysis

Transportation of wind turbine farm power plants (WPP) components certainly have its own risks, particularly when carried out in remote area. Therefore, the risks that occur for each transportation scheme shall be mitigated and minimized through risk management. The scheme which has lower risk and lower impact in term of cost and time should be chosen.

The Project Management Institute (PMI) on PMBOK Guide [12] defines a part of the risk management process as risk identification. Risk identification is performed after the risk management planning process. This identification can be managed through risk categorization by decomposing potential sources of risk into layers of increasing detail known as the Risk Breakdown Structure (RBS). The RBS can be divided into several levels. According to the information from previous projects, the proposed RBS for transportation of onshore WPP components is shown in Figure 9. This scheme shows 14 risk variables classified as RBS Level 3 to be identified in risk register. The risk variables include the risk related to time and cost, which have impact to the transportation process and conjunctively may affect the whole project.

There are several risk response strategies that are used to deal with this kind of transportation activity such as avoiding, mitigating, and accepting. Example of avoidance actions may include removing the cause of a threat or risk. In risk mitigation, action is taken to reduce the probability of occurrence and or impact of a threat or risk. If the risk is accepted, a contingency reserve should be established for the entire activities. Risk analysis in this paper only include risk identification, whereas the risk register and risk mitigation are excluded.

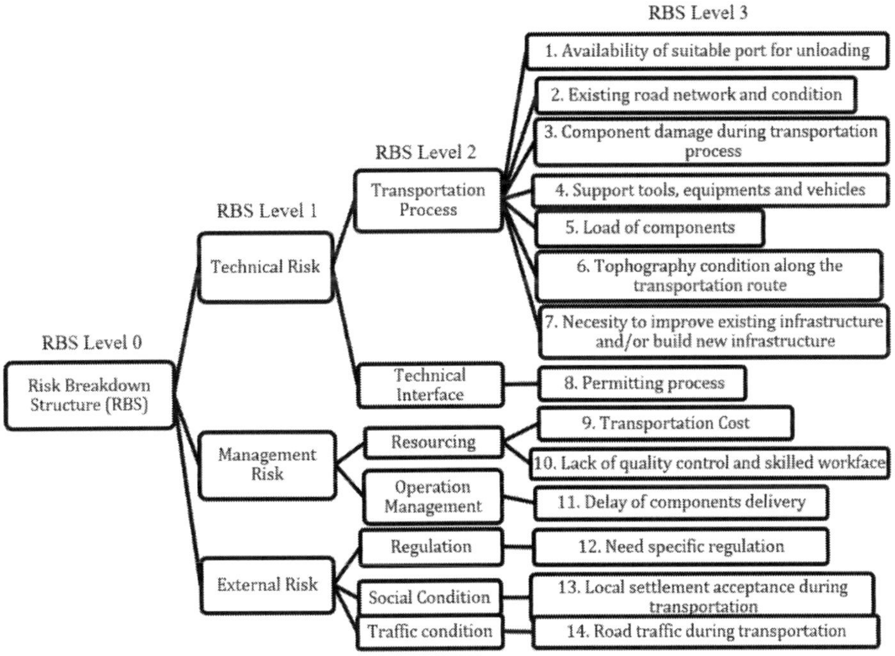

Figure 9. Risk Breakdown Structure (RBS) of transportation process for onshore wind turbine components.

6. Conclusions

Based on literature review, onshore wind turbine technology tends to have bigger size in coming years and its components transportation is not likely to subside in the near future. Several challenges are identified for onshore WPP development in Indonesia as follow:

- Onshore WPP development opportunities in Indonesia primarily located in remote and mountainous terrain. This condition increases the complexity and difficulty of handling and transporting the wind turbine components from the unloading port, storage area and to the WPP cluster.
- Handling and transporting wind turbine components typically require large and special equipment or vehicles. In Indonesia, only certain ports and highways are suitable for such activity. In addition, there are also several concerns for the transportation through highway and other public road, such as: height restrictions, existing above ground crossing road utilities, turning radius, road width, slope and bearing capacity.
- Taking into account the economics of WPP, the expected increase in size will ultimately lead to higher transportation costs in Indonesia as it requires improvement of existing infrastructure or even new infrastructure, specifically in remote area and/or islands. Hence, the risks also increase.
- A combination of transportation modes and route selection are necessary for handling and transporting the wind turbine components from the arrival port for unloading, storage area, and to the WP cluster in early development stage.

Finally, infrastructure development and readiness for WPP development should be organized together among WPP developers, government, financial institutions, and other stakeholders in creating the right conditions for investment in infrastructure to enable the next generation of onshore WPP projects.

References

[1] ESMAP (2021). Global Wind Atlas. https://globalwindatlas.info/en/area/Indonesia.
[2] PT PLN (Persero) (2021). Electricity Supply Business Plan 2021-2030
[3] Hartman, L. (2022). Wind Turbine: The Bigger, The Better. https://www.energy.gov/eere/articles/.
[4] Fernández, L. (2023). Average global onshore wind turbine rotor diameter 2000-2025. https://www.statista.com/.
[5] Stehly, T., Duffy, P., & Mulas Hernando, D. (2023). 2022 Cost of Wind Energy Review.
[6] Mooney, M., & Maclaurin, G. (2016). Transportation of Large Wind Components: A Review of Existing Geospatial Data. www.nrel.gov/publications.
[7] AECOM (2017). TC01-D Route Analysis Wind Turbine Blade Transport
[8] Google earth (February 18, 2018). Port of Makassar.
[9] PT Energi Bayu Jeneponto (2021). Environmental and Social Monitoring Report-Indonesia: Eastern Indonesia Renewable Energy Project (Phase 1).
[10] Redmelon. (n.d.). ALE reaches half-way of transport milestone for Indonesian wind project. Heavy Lift News. https://www.heavyliftnews.com/ale-reaches-half-way-of-transport-milestone-for-indonesian-wind-project/
[11] DNVGL (2017). Transport and installation of wind power plant, http://www.dnvgl.com.
[12] PMI. (2017). A Guide to the project management body of knowledge-PMBOK Guide Sixth Edition, Project Management Institute, Pennsylvania, 69-90.

IOP Conf. Series: Earth and Environmental Science **1464** (2025) 012003 doi:10.1088/1755-1315/1464/1/012003

Brackish peat water desalination treatment using a combination of continuous electrocoagulation and filtration technique

Mochammad Meddy Danial[1*], Fitri Imansyah[2], Stefanus Barlian Soeryamassoeka[3], Rizki Purnaini[4] Rozin Dhiyaul Haq[4], Muhammad Zidan Zain[4]

[1] Department of Ocean Engineering, Universitas Tanjungpura, Pontianak, Indonesia
[2] Department of Electrical Engineering, Universitas Tanjungpura, Pontianak, Indonesia
[3] Department of Civil Engineering, Universitas Tanjungpura, Pontianak, Indonesia
[4] Department of Environmental Engineering, Universitas Tanjungpura, Pontianak, Indonesia

*E-mail: meddydanial@civil.untan.ac.id

Abstract. Coastal regions possess a distinct water quality known as brackish peat water, resulting from a blend of peat and seawater. Unfortunately, this water type cannot serve as a freshwater source without treatment. Hence, it's crucial to explore effective and cost-efficient treatment methods. This study focuses on evaluating the efficacy of continuous electrocoagulation and filtration techniques in converting brackish peat water, characterized by peat soil, into freshwater. Desalination via electrocoagulation supports key elements of the blue economy, including water availability, economic growth, and community resilience. Samples were collected from the Kakap river mouth. Six parameters were examined before and after treatment: salinity, temperature, total dissolved solids (TDS), pH, iron (Fe), manganese (Mn). Salinity, pH, temperature, and TDS were measured using a calibrated multi-functional water quality tester. Iron (Fe) and manganese (Mn) levels were assessed according to SNI 6980.84:2019 standards, conducted at the laboratory of the Institute of Industrial Standardization and Services (Baristand) in Pontianak. The results show that the electrocoagulation and filtration can reduce salinity and TDS ranging from ~40 to ~83%, where the best outcome for desalinating brackish peat water through EC and filtration occurs when the EC process runs for 4 hours, with a filtration detention time of 30 minutes. Additionally, statistical analysis using ANOVA revealed the p-value is smaller than 0.05, indicating a significant effect of salinity reduction. Morover, the electrocoagulation and filtration substantially lower iron levels, while manganese concentration initially increases after 1 hour of EC. However, during 2 and 3 hours of EC, manganese levels decrease to meet acceptable standards. For effective salinity reduction in brackish peat water, it is essential to integrate electrocoagulation with filtration process. Power consumption for the electrocoagulation ranges from 0.0629 to 0.28 kWh/m^3.

1. Introduction

Residents of coastal regions and island communities inherently face a critical demand for freshwater resources, given the unique geographical and environmental challenges that characterize these locales. They typically gather rainwater or rely on river water to fulfil the need of water for cooking, drinking, bathing, laundry, and other sanitation purposes [1]. However, climate change has resulted in reduced rainfall and frequent drought conditions. Moreover, rainwater itself is often subject to various contaminants, making it unsuitable for direct use without proper treatment [2]. These factors cause coastal communities to be vulnerable to freshwater crises.

Waters in coastal rural areas are typically characterized by low salinity levels [3–5]. These regions often are found in swamp, peatland, and wetland, which are rich in natural organic matter, including humic acid, as well as minerals such as iron (Fe) and mangan (Mn) [6–8]. Therefore, waters in coastal rural areas often presents as brackish peat water [9]. This unique type of water, due to its composition and availibility, holds significant potential as a supplementary water source alongside seawater [10]. Understanding the composition and characteristics of brackish peat water is essential for effective management and utilization, particularly in regions where freshwater resources are scarce and alternative sources must be leveraged [11].

Indonesia is globally acknowledged for having the world's second-largest expanse of peatlands, exceeded only by Brazil. [12,13]. In addition to their environmental importance, coastal peatlands located in estuary have significant potential as a valuable natural resource, particularly in the coastal areas where freshwater is limited [14]. The brackish peat water stored in coastal peatlands can be connected as a source of clean water, providing a sustainable solution to the challenges of water scarcity in the coastal regions. By implementing advanced water treatment technologies, the water from peatlands can be purified and made suitable for consumption, addressing the needs of local communities and contributing to the overall improvement of public health.

Utilizing brackish peat water as a water source in coastal peatlands areas represents a promising opportunity for sustainable development. As peoples in these regions continue to grow, the demand for clean water increases, and peatlands offer a naturally occurring reservoir that can be tapped into with proper management and treatment. This approach not only provides a reliable source of water but also helps to reduce the pressure on other freshwater resources, which are often overexploited and vulnerable to contamination.

Moreover, the integration of coastal peatlands into water management strategies aligns with the environmental issues and conservation goals. By recognizing the value of coastal peatlands as both a carbon sink and a water resource, the country can develop policies and practices that promote their sustainable use while protecting these ecosystems from degradation. This includes efforts to prevent the draining and burning of peatlands, which can lead to significant environmental damage, including the release of stored carbon and the loss of biodiversity.

Various water treatment methods in coastal areas have been extensively researched, including conventional methods such as coagulation with chemical additives [15,16], aeration, and filtration [17], as well as non-conventional methods like continuous electrocoagulation [18] and desalination using reverse osmosis membranes [3,8]. However, both conventional and non-conventional methods mentioned above have drawbacks, such as lengthy processing times and high costs associated with reverse osmosis membrane treatment.

Mitigating the crisis of clean water disasters should be addressed with technologies that are easy to construct, inexpensive, and can be applied to produce water quality performance in accordance with the standards of Minister of Health Regulation Number 2 of 2023. Therefore, there is a need for a rapid, practical, cost-effective method of treating brackish and brackish water for coastal areas.

One of the most promising methods of reducing brackish peat water is using electrocoagulation (EC). EC is physical-chemical method that not need an additive of chemical in water treatment [19]. Electrocoagulation aims to treat various pollutants including suspended solids, heavy metals, ink, organic matter, oils and fats, ions, and radionuclides. Additionally, electrocoagulation is effective in treating pharmaceutical-containing water, such as medical drug residues, and is suitable for reducing pollutants or contamination in brackish water in peatland or lowland areas [6]. Moreover, electrocoagulation is more effective compared to chemical substances like PAC [20]. Electrocoagulation also offers several benefits, including no use of chemicals in the water treatment process, faster coagulation processes, and the ability to treat small-sized colloidal particles [21]. Another advantage is that electrocoagulation devices can reduce energy consumption [6].

Absorption has several advantages, such as the ability to remove organic compounds like carbonyl compounds that cannot be effectively eliminated by electrocoagulation [6]. Another advantage of using activated carbon in absorption is its ability to prevent membrane fouling in RO systems [22].

Employing electrocoagulation as a desalination technique aligns with the principles of blue economy, promoting the sustainable use of brackish peat water as part of the ocean resources, and simultaneously fostering economic development and improving environmental health.

2. Methodology

2.1 Sample collection

Brackish peat water is classified as brackish due to its salinity levels, which range between 1,000 and 15,000 mg/L. This type of water is characterized by the presence of humic acid—a complex amalgamation of amorphous organic matter that varies in color from yellowish-brown to black [14,23]. The presence of humic acid imparts unique chemical properties to the water, influencing its acidity, color, and interaction with metals and pollutants [11].

The brackish peat water was collected in the coastal rural areas of Sungai Itik, Sungai Kakap district, around 20 km away from Pontianak city. This location is connected to the river mouth facing the open sea, resulting in brackish peat water characteristics. The water sample was collected during the flood tide, which increased the salinity level as it mixed with the peat soil of the rural river. The salinity level was tested before pumping the water into a 2000-liter capacity tank and transporting it to the Ocean Technology Laboratory at the Faculty of Engineering, Tanjungpura University.

2.2 Design of electrocoagulation and filtration

The design concept for continuous electrocoagulation builds on prior research that utilized a zigzag plate arrangement within a treatment chamber [24,25], with dimension of 0.4 m x 0.47 m x 0.67 m. This configuration optimizes the flow dynamics and enhances the interaction between the brackish water and the electrodes. The electrodes in this design are spaced 5 to 10 centimeters apart and are operated within a voltage range of 1 to 5 volts. The system is designed to run for varying durations—specifically 1, 2, 3, and 4 hours [26,27]. One significant advantage of this continuous electrocoagulation approach is its efficiency in handling large volumes of water, making it particularly suitable for applications requiring extensive water treatment. The zigzag arrangement of the plates promotes effective mixing and increases the surface area contact between the water and the electrodes, thereby improving the overall performance of the electrocoagulation process [25]. This design not only enhances treatment efficacy but also provides scalability for larger-scale operations.

The addition of filtration components following the electrocoagulation device involves using chamber of 0.5 m x 0.5 m x 0.6 m, containing three layers arranged from the top to the bottom of greensand, zeolite, and activated carbon with the layer depth of 0.2 m, 0.3 m, and 0.05 m, respectively. This setup aims to reduce salinity levels in brackish peat water environments [6,26]. The design of continuous electrocoagulation and filtration device can be seen in Figure 1.

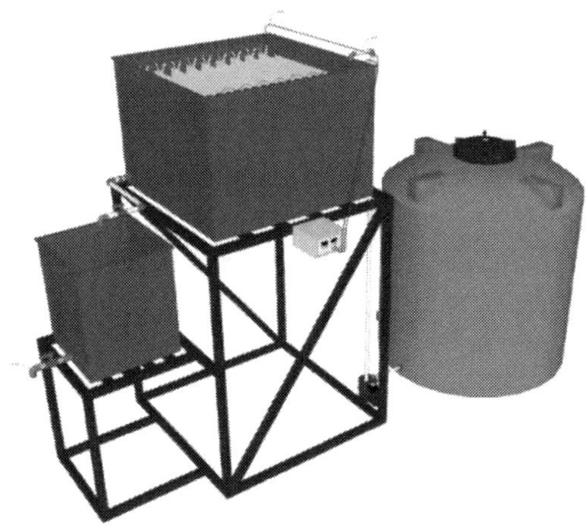

Figure 1. Design continuous electrocoagulation and filtration.

2.3 Experiment method

Performance testing of electrocoagulation and filtration involves measuring the water quality before and after treatment. The parameters tested on the water samples include salinity, TDS (Total Dissolved Solids), pH, temperature. A multi-functional water quality tester was used to measure salinity, TDS, pH, and temperature. Additionally, the iron (Fe), manganase (Mn), were tested based on SNI 6980.84:2019, which were conducted at Laboratory of The Institute of Industrial Standarization and Services (Baristand) in Pontianak.

Raw water samples from the water tank are initially directed through a pipe using a low-pressure pump into the inlet of the electrocoagulation chamber. The water flow rate and level within the chamber are precisely regulated by an adjustable valve. Upon reaching the desired water level, the electrocoagulation process is initiated by activating the DC power supply, configured to 36 V and 5 A. This power supply is connected to a set of electrodes, with the positive and negative currents linked to the anode and cathode, respectively. This setup ensures that the electrocoagulation process commences under controlled conditions, facilitating effective treatment of the water sample.

In this study, the primary objective is to analyze the reduction in salinity and TDS levels resulting from the combined processes of electrocoagulation and filtration. To evaluate the effectiveness of salinity and TDS reduction achieved through these treatments, the following equation (Eq. 1) is employed as a measurement tool[14]. This equation allows for a precise

IOP Conf. Series: Earth and Environmental Science **1464** (2025) 012003 doi:10.1088/1755-1315/1464/1/012003

quantification of the decrease in salinity and TDS, providing a clear indicator of the treatment's efficiency.

$$\text{Percentage reduction (\%)} = \frac{X_t - X_0}{X} \times 100\% \quad (1)$$

where X_0 represents the initial level of the parameter, while X_t denotes the parameter level after treatment. The parameter X can be substituted with salinity, TDS, pH, or temperature, depending on which water quality metric is being assessed. By applying this equation, the study provides a precise measurement of the decrease in each parameter, offering a clear indicator of the treatment process's efficiency. In addition, to investigate the effect of electrocoagulation and filtration in lowering the salinity and TDS level, an statistical approach namely analysis of variance (ANOVA) was used to confirm the impact of pre-treatment [14].

3. Results

The experimental study aimed at reducing salinity in brackish peat water through electrocoagulation (EC) and filtration processes has been thoroughly examined. Fig. 2 presents the data patterns for salinity, total dissolved solids (TDS), pH, and temperature following various durations of EC treatment—specifically, 1 hour, 2 hours, 3 hours, and 4 hours—and subsequent filtration with detention times of 30 minutes, 60 minutes, and 90 minutes. Each section presents the data for salinity, TDS, pH, and temperature, allowing for a comparative analysis of how varying treatment times impact these parameters. This layout provides a comprehensive view of the treatment's effectiveness over different durations.

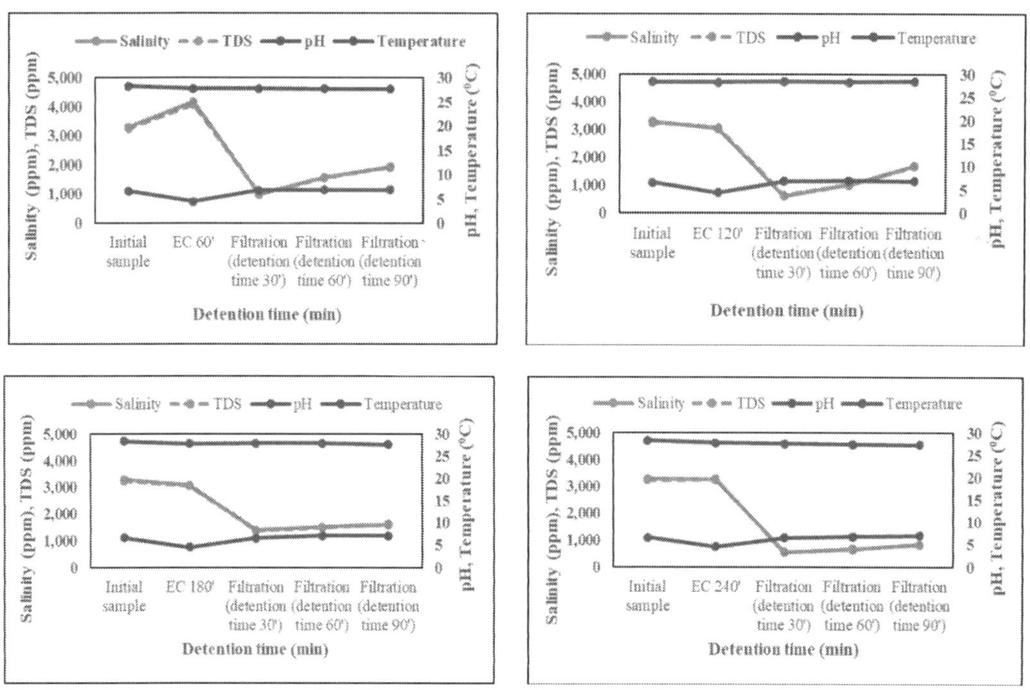

Figure 2. The pattern of salinity, TDS, pH, and temperature before and after brackish peat water treatment using electrocoagulation at different durations (top left) 1 hr, (top right) 2 hr, (bottom left) 3 hr, (bottom right) 4 hr, continued to filtration process with detention time of 30, 60, and 90 min.

The results indicate that while the salinity and TDS levels show a modest decrease following the EC treatment, there is a significant reduction in these parameters after the filtration process. Conversely, the pH parameter demonstrates a notable decline immediately after the EC treatment but exhibits an increase once the filtration process begins. This suggests that the filtration process has a restorative effect on the pH levels, potentially counteracting the acidic or alkaline shifts caused during the EC phase.

Additionally, the study reveals a significant drop in the voltage of the power supply, from an initial 5 V to just 2 - 2.38 V, maintained at a constant level throughout the treatment. This voltage reduction occurred despite the initial salinity level of the brackish peat water is at 329 ppm, indicating a possible decrease in electrical resistance or a change in the efficiency of the EC process over time.

Fig. 3 illustrates the reduction efficiency in several water quality parameters following the treatment of brackish peat water. Specifically, it shows the percentage reduction of salinity level, total dissolved solids (TDS), pH levels, and temperature that calculated by using Eq. 1. Those graphics allows for a comparative analysis of how effectively the treatment process mitigates these factors.

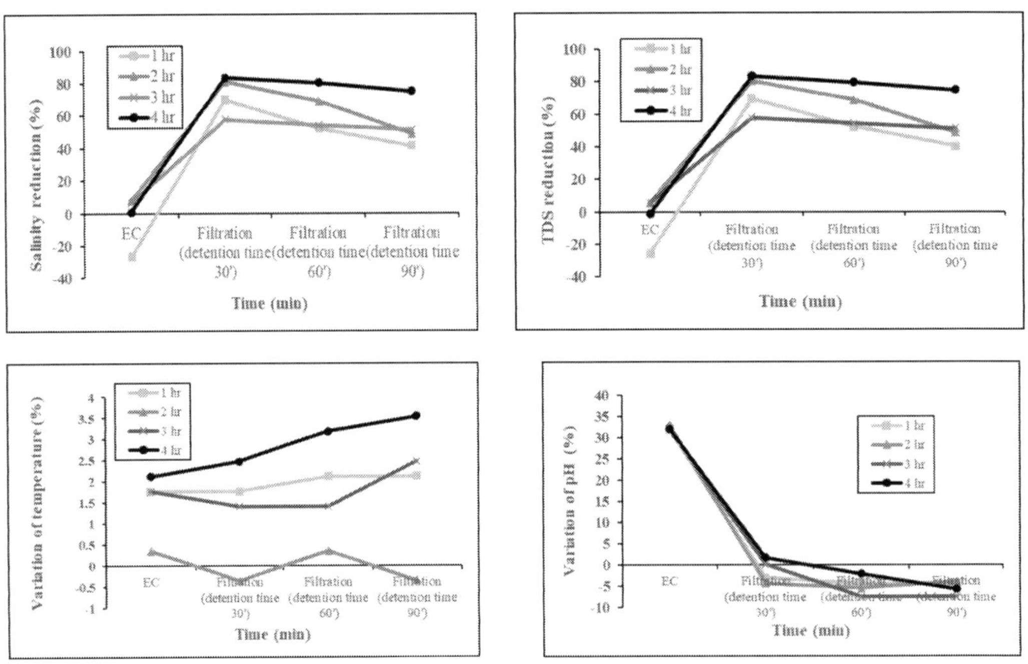

Figure 3. (top left) Percentage of salinity reduction after EC and filtration, (top right) Percentage of TDS reduction after EC and filtration, (bottom left) Variation of temperature after EC and filtration, (bottom right) Variation of pH after EC and filtration.

It can be seen from Fig. 3 (top left and top rigth) that both the electrocoagulation (EC) and filtration processes significantly reduce the total dissolved solids (TDS) and salinity levels at the percentage range from 40–83.5%. The most pronounced reductions of salinity and TDS, achieving

~83%, are observed when the electrocoagulation is applied for 4 hours, followed by a 30-minute detention period after the filtration process. This indicates that longer EC treatment times and appropriate post-filtration detention periods are effective in achieving optimal reductions in the parameters. Additionally, the temperature levels show a similar trend of reduction ranging from 1.4 to 3.5%, in response to these processes throughout the treatment process as shown in Fig 3(bottom left). The pH level shows a different trend during the treatment process compared to other parameters as depicted in Fig. 3 (bottom right). Initially, during the electrocoagulation process, the pH experienced a significant jump, increasing by 32% to a level of 4.4. However, as the treatment continued to the filtration stage, the pH level experienced further adjustment that increased at the range of 6.5 to 8.5, which denotes an acceptable level according to drinking water criteria.

Table 1 presents the ANOVA analysis assessing the effectiveness of electrocoagulation and filtration in desalinating brackish peat water. The *F* value for electrocoagulation exceeds the critical *F* value, and the *P*-value is less than 0.05, indicating that filtration treatment has a more significant impact than electrocoagulation. Furthermore, the Kruskal-Wallis test, with a K value of 1.42 (> 0.05), also supports the significance of the treatment in reducing salinity and TDS levels.

Table 1. ANOVA analysis for the effectivy of electrocoagulation and filtration treatment

Source of Variation	*SS*	*df*	*MS*	*F*	*P-value*	*F crit*
Electrocoagulation	1,302,124.55	3	434041.5	4.10	0.032151	3.49
Filtration	2,845,162.80	4	5711291	54.01	1.42E-07	3.26
Error/Residual	1,268,909.20	12	105742.4			
Total	25,416,196.55	19				

Contribution of filtration layer does not directly affect the salinity level reduction. In a separate experiment compraing processess with and without electrocoagulation, the results shows that electrocoagulation must be coupled with the filtration. As can be seen in Fig. 4, filtration without electrocoagulation does not affect the salinity level reduction. In contrast, when electrocoagulation is combined with filtration, salinity levels significantly decrease. This is due to the formation of aluminium hidroxide during the electrocoagulation process, which acts as a coagulant. This coagulant can bind or adsorb various contaminants, including both organic and inorganic pollutans, as well as salts, leading to the formation of flocs [3,28]. These flocs can either precipitate or float during the electrocoagulation process, making them easily removed using filtration process[3,28].

After the EC process, filtration process plays important role in further purifying the water. In this experiment, a multi-layer filter consisting of greensand, zeolite, and activated carbon is employed. Greensand is mainly utilized to remove iron and manganese, while both zeolite and activated carbons serve as effective adsorbents in wastewater treatment, removing organic compounds, residual disinfectants, turbidity and suspended solids. Moreover, activated carbon is frequently used as a pre-treatment steps to protect reverse osmosis membrane from biological foulling [5,25]. Due to its extensive surface area, acivated carbon provide an efficient means for contaminants to be attached and removed from the water [29]. As previously mentioned, the flocs generated during EC tends to be much larger and more stable, contains less bound water, acid-resistant, and hence they are easily separated faster during filtration [30,31].

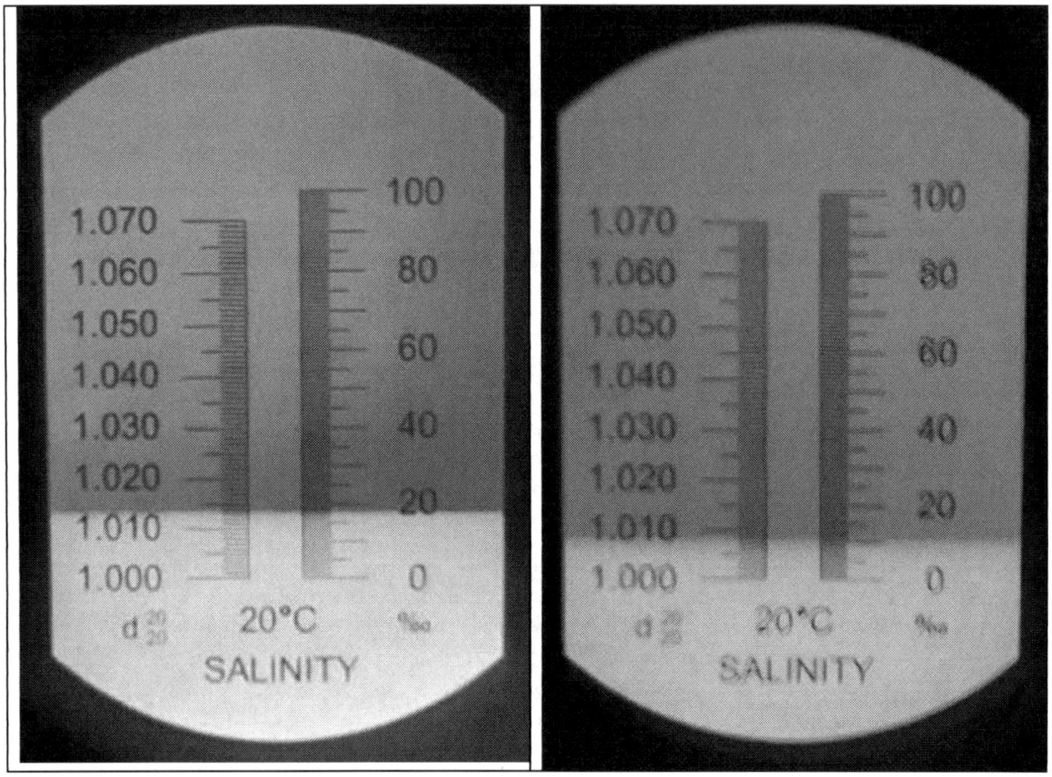

Figure 4. (left) Salinity reduction using filtration without EC (top right) Salinity reduction using EC and combined with filtration.

The energy consumption of the electrocoagulation process is primarily influenced by three key parameters, including electrical current, cell voltage, and treatment time [9,32]. When operating at a current of 5 A and a cell voltage ranging from 0.03 to 1.4 volts, the system exhibits a power consumption between 0.0629 and 0.28 kWh/m^3. Consequently, over one hour duration with an electricity tariff of Rp. 1,500 per kWh, the electrical current cost for the electrocoagulation process amounts to Rp Rp. 94.5 and Rp. 420, respectively.

The EC is deemed as an environmentally sustainable technology in water treatment because of chemical-free operation and reduced sludge volume [33]. The EC process can generates sludge disposal due to precipitation and flotation mechanisms [34], although the amount of sludge produced is less than that from traditional chemical coagulation. Proper treatment of this sludge is important before disposal. Several treatement can be proposed, such as chemical additives, sludge dewatering, sludge collection, microfiltration membrane, and landfill disposal [26,33,35,36].

Table 2 shows the results of laboratory tests for iron (Fe) and manganese (Mn) before and after treatment. Filtration process can reduce significantly the concentration of Iron (Fe), while manganese (Mn) concentration showed an increase after undergoing the EC process for 1 hr. Hoewever, during the EC process for 2 and 3 hrs, the manganese concentration showed a decrease that meet the acceptable standard. This can happen probably due to several factors, such as the instability of manganese concentration during the initial phase of the treatment, which might have been influenced by fluctuations in water chemistry, particularly pH levels. For manganese removal to be effective, it is crucial that the pH remains within the optimal range of 6 to 8. [37–39].

Manganese parameter may appear after electrocoagulation process because of several reasons. The EC process can create condition that promote the release of manganese from sediment or other sources within the system. Manganese probably exists at very low levels that are undetectable in initial test, but become more soluble due to changes in pH or oxidation-reduction process during EC [40].

Table 2. Laboratory test result of Iron and Manganese before and after treatment

Parameter	Initial value	EC 1 hr	EC 1 hr and filtration	EC 2 hrs	EC 2 hrs and filtration	EC 3 hrs	EC 3 hrs and filtration
Iron (Fe)	0.653	<0.037	<0.037	<0.037	<0.037	<0.037	<0.037
Manganese (Mn)	<0.016	0.376	0.376	<0.016	0.808	<0.016	1.02

4. Conclusions

An experimental study of desalinating brackish peat water has been carried out by utilizing electrocoagulation and filtration process. Both the electrocoagulation and filtration processes effectively reduce total dissolved solids (TDS) and salinity by up to 83%, from the initial level of 3,290 ppm to only 540 ppm. The most significant reductions observed after 4 hours of EC treatment, followed by a 30-minute post-filtration detention period. Statistical analysis by using ANOVA and Kruskal-Wallis test also support these findings. Electrocoagulation must be combined with filtration to obtain a significant result to reduce salinity level of brackish peat water.

This study has also noted that temperature also decreases slightly, while pH levels rise from 4.4 to an acceptable range of 6.5 to 8.5. Moreover, the electrocoagulation and filtration can significantly reduce the iron levels throughout the process, although the manganese concentration removal still shows some instability patterns. Overall, these findings indicate that electrocoagulation and filtration can offer a promising approach for the pretreatment of brackish peat water in the future.

Power consumption for the electrocoagulation process ranges from 0.0629 to 0.28 kWh/m^3. Over one hour, at an electricity tariff of Rp. 1,500 per kWh, the cost amounts to Rp. 94.5 and Rp. 420, respectively.

Acknowledgement

The authors wish to express their gratitude to the Faculty of Engineering at Tanjungpura University and LPPM UNTAN for their continuous support throughout this study. This research is part of a major project of fundamental research, funded by the Directorate of Research, Technology, and Community Service of the Directorate General of Higher Education, Research, and Technology (DRTPM), under contract number 096/E5/PG.A2.00.PL12024, June 11, 2024.

References

[1] J.N. Hakizimana, B. Gourich, M. Chafi, Y. Stiriba, C. Vial, P. Drogui, J. Naja, Electrocoagulation process in water treatment: A review of electrocoagulation modeling approaches, Desalination. 404 (2017) 1-21. https://doi.org/10.1016/j.desal.2016.10.011.

[2] H.C. Yang, M. Jusoh, Z.Y. Zakaria, A. Rosli, Rainwater to Potable Water: Mini Review, Chem. Eng. Trans. 106 (2023) 775–780. https://doi.org/10.3303/CET23106130.

[3] N.A. Rahman, C. Jose Jol, A. Albania Linus, F.L. Dampam, N.S. Abdul Jalal, N. Baharudin, W.W.S. Wan Borhan, Desalination of Borneo tropical brackish peat water with adsorption process in continuous electrocoagulation treatment, Desalination. 527 (2022) 115574. https://doi.org/https://doi.org/10.1016/j.desal.2022.115574.

[4] L. Gosch, M. Janssen, B. Lennartz, Impact of the water salinity on the hydraulic conductivity of fen peat, Hydrol. Process. 32 (2018) 1214–1222. https://doi.org/https://doi.org/10.1002/hyp.11478.

[5] L.F. Greenlee, D.F. Lawler, B.D. Freeman, B. Marrot, P. Moulin, Reverse osmosis desalination: Water sources, technology, and today's challenges, Water Res. 43 (2009) 2317–2348. https://doi.org/10.1016/j.watres.2009.03.010.

[6] M. Qadafi, D.R. Wulan, S. Notodarmojo, Y. Zevi, Characteristics and treatment methods for peat water as clean water sources: A mini review, Water Cycle. 4 (2023) 60–69. https://doi.org/10.1016/j.watcyc.2023.02.005.

[7] N. Li, T. Huang, X. Mao, H. Zhang, K. Li, G. Wen, X. Lv, L. Deng, Controlling reduced iron and manganese in a drinking water reservoir by hypolimnetic aeration and artificial destratification, Sci. Total Environ. 685 (2019) 497–507. https://doi.org/https://doi.org/10.1016/j.scitotenv.2019.05.445.

[8] Y. Mao, S. Cotterill, D. Morgan, S. Regan, Y. Zhao, Electrocoagulation of peatland runoff: Statistical optimization and economic analysis, J. Water Process Eng. 49 (2022) 103113. https://doi.org/https://doi.org/10.1016/j.jwpe.2022.103113.

[9] N. Abdul Rahman, C. Jose Jol, A. Albania Linus, W.W.S. Wan Borhan, N.S. Abdul Jalal, N. Baharudin, S.N.A. Samsul, N. Abdul Mutalip, A.A. Jitai, D.F.A. Abang Abdul Hamid, Continuous electrocoagulation treatment system for partial desalination of tropical brackish peat water in Sarawak coastal peatlands, Sci. Total Environ. 880 (2023). https://doi.org/10.1016/j.scitotenv.2023.163517.

[10] I.G. Wenten, K. Khoiruddin, A.K. Wardani, P.T.P. Aryanti, D.I. Astuti, A.A.I.A.S. Komaladewi, Preparation of antifouling polypropylene/ZnO composite hollow fiber membrane by dip-coating method for peat water treatment, J. Water Process Eng. 34 (2020) 101158. https://doi.org/https://doi.org/10.1016/j.jwpe.2020.101158.

[11] K.H. Tan, Humic Matter in Soil and the Environment: Principles and Controversies, 1st editio, CRS Press, New York, 2003. https://doi.org/10.2136/sssaj2015.0004br.

[12] F. Ali, D.L. Lestari, M.D. Putri, Peat Water Treatment as an Alternative for Raw water in Peatlands Area, IOP Conf. Ser. Mater. Sci. Eng. 1144 (2021) 012052. https://doi.org/10.1088/1757-899x/1144/1/012052.

[13] I. Budiman, R. Januar, D. Ayunda, S. Wicaksono, E.N.N. Sari, D. Affandi, Transformation of Post-disaster Governance of Indonesian Peatland Wildfires BT - Post-Disaster Governance in Southeast Asia: Response, Recovery, and Resilient Societies, in: A.N.R. Mardiah, R.B. Olshansky, M.B.F. Bisri (Eds.), Post-Disaster Gov. Southeast Asia, Springer Singapore, Singapore, 2022: pp. 87–119. https://doi.org/10.1007/978-981-16-7401-3_5.

[14] N. Abdul Rahman, C. Jose Jol, A. Albania Linus, W.W.S. Wan Borhan, N.S. Abdul Jalal, N. Baharuddin, S.N.A. Samsul, N. Abdul Mutalip, Statistical analysis of salinity reduction in Borneo tropical brackish peat water with continuous electrocoagulation treatment system, J. Hazard. Mater. Adv. 10 (2023) 100265. https://doi.org/10.1016/j.hazadv.2023.100265.

[15] A.Y. Bagastyo, E. Nurhayati, S.P.H. Manah, A.A.W.R. Iswari, A. Yulikasari, I.D.A.A. Warmadewanthi, T.F. Lin, The role of aeration and pre-chlorination prior to coagulation-flocculation process in water treatment: A laboratory and field research in Indonesia, Case Stud. Chem. Environ. Eng. 7 (2023). https://doi.org/10.1016/j.cscee.2023.100352.

[16] Q.H. Malik, Performance of alum and assorted coagulants in turbidity removal of muddy water, Appl. Water Sci. 8 (2018) 1–4. https://doi.org/10.1007/s13201-018-0662-5.

[17] N.N. Arifianingsih, Y. Zevi, Q. Helmy, S. Notodarmojo, H. Fujita, Y. Shimayama, M. Kirihara, Peat water treatment using oxidation and physical filtration system and its performance in reducing iron (Fe), turbidity, and color, E3S Web Conf. 148 (2020) 1–5. https://doi.org/10.1051/e3sconf/202014807011.

[18] K. Hendaoui, M. Trabelsi-Ayadi, F. Ayari, Optimization of continuous electrocoagulation-adsorption combined process for the treatment of a textile effluent, Chinese J. Chem. Eng. 44 (2022) 310–320. https://doi.org/https://doi.org/10.1016/j.cjche.2020.10.047.

[19] L.W.M. Zailani, N.S.M. Zin, Application of Electrocoagulation in Various Wastewater and Leachate Treatment-A Review, IOP Conf. Ser. Earth Environ. Sci. 140 (2018). https://doi.org/10.1088/1755-1315/140/1/012052.

[20] B.H. Prabowo, U. Jenderal, A. Yani, H. Hendriyana, U. Jenderal, A. Yani, L. Nurdini, U. Jenderal, A. Yani, METODE KOAGULASI DAN ELEKTROKOAGULASI DENGAN PENAMBAHAN HIDROGEN PEROKSIDA PADA PROSES PENGOLAHAN LIMBAH CAIR Metode Koagulasi dan Elektrokoagulasi dengan Penambahan Hidrogen

Peroksida pada Proses Pengolahan Limbah Cair Buangan Laundry Coagulation and Ele, (2019). https://doi.org/10.31315/e.v16i2.2812.
[21] Masthura, Penerapan Metode Elektrokoagulasi Sebagai Alternatif Pengolahan Air Bersih, 2019.
[22] M. Bidari, M.A. Putri, S. Nasir, Pengaruh karbon aktif terhadap fouling membran reverse osmosis pada pengolahan air terproduksi, J. Tek. Kim. 28 (2022) 100–106. https://doi.org/10.36706/jtk.v28i3.991.
[23] M. Grzegorzek, K. Majewska-Nowak, The influence of humic acids on desalination process with the use of electrodialysis, E3S Web Conf. 17 (2017). https://doi.org/10.1051/e3sconf/20171700027.
[24] M. López-Guzmán, M.A. Flores-Hidalgo, L. Reynoso-Cuevas, Electrocoagulation process: An approach to continuous processes, reactors design, pharmaceuticals removal, and hybrid systems—a review, Processes. 9 (2021). https://doi.org/10.3390/pr9101831.
[25] N.S. Graça, E. Rodrigues, clean technologies The Combined Implementation of Electrocoagulation and Adsorption Processes for the Treatment of Wastewaters, Clean Technol. 4 (2022) 1020–1053.
[26] F. Akhter, S.A. Soomro, Pollutant Removal Efficiency of Electrocoagulation Method from Industrial Wastewater : Comparison with Other Treatment Methods and Key Operational Parameters — a Comparative Study Review, Water Air Soil Pollut. 232 (2021) 1–13. https://doi.org/10.1007/s11270-021-05022-5.
[27] W. Oktiawan, B.P. Samadikun, Junaidi, I.G.N. Bramahesa, T.A. Taqiyya, M.R. Amrullah, C. Basyar, Effect of electrode configuration and voltage variations on electrocoagulation process in surfactant removal from laundry wastewater, IOP Conf. Ser. Earth Environ. Sci. 896 (2021). https://doi.org/10.1088/1755-1315/896/1/012049.
[28] A.A. Al-raad, M.M. Hanafiah, Removal of inorganic pollutants using electrocoagulation technology : A review of emerging applications and mechanisms, J. Environ. Manage. 300 (2021) 113696. https://doi.org/10.1016/j.jenvman.2021.113696.
[29] B. Edward E, Water Treatment Plant Design, 4th ed., McGraw-Hill, New York, 2005.
[30] D.T. Moussa, M.H. El-naas, M. Nasser, M.J. Al-marri, A comprehensive review of electrocoagulation for water treatment : Potentials and challenges, J. Environ. Manage. 186 (2017) 24–41. https://doi.org/10.1016/j.jenvman.2016.10.032.
[31] M.Y.A. Mollah, R. Schennach, J.R. Parga, D.L. Cocke, Electrocoagulation, J. Hazard. Mater. 84 (2001) 29–41. https://doi.org/https://doi.org/10.1016/S0304-3894(01)00176-5.
[32] B. Abdulhadi, P. Kot, K. Hashim, A. Shaw, M. Muradov, R. Al-Khaddar, Continuous-flow electrocoagulation (EC) process for iron removal from water: Experimental, statistical and economic study, Sci. Total Environ. 760 (2021) 143417. https://doi.org/10.1016/j.scitotenv.2020.143417.
[33] D.D. Salinas-Echeverría, L.C. Sánchez-De La Cruz, L.A. Zambrano-Intriago, J.M. Rodríguez-Díaz, K.A. Sanoja-Lopez, R. Luque, K.J. Fernández-Andrade, Y. Gómez-Salcedo, R.J. Baquerizo-Crespo, Evaluation of a continuous flow electrocoagulation reactor for turbidity removal from surface water, Chem. Eng. Res. Des. 198 (2023) 478–488. https://doi.org/10.1016/j.cherd.2023.08.049.
[34] N. Marsidi, H.A. Hasan, S. Rozaimah, S. Abdullah, Journal of Water Process Engineering A review of biological aerated fi lters for iron and manganese ions removal in water treatment, J. Water Process Eng. 23 (2018) 1–12. https://doi.org/10.1016/j.jwpe.2018.01.010.
[35] M.F. Eskibalci, M.F. Ozkan, Comparison of conventional coagulation and electrocoagulation methods for dewatering of coal preparation plant, Miner. Eng. 122 (2018) 106–112. https://doi.org/10.1016/j.mineng.2018.03.035.
[36] A. Tahreen, M.S. Jami, F. Ali, Role of electrocoagulation in wastewater treatment: A developmental review, J. Water Process Eng. 37 (2020) 101440. https://doi.org/10.1016/j.jwpe.2020.101440.
[37] K. Nkele, L. Mpenyana-Monyatsi, V. Masindi, Challenges, advances and sustainabilities on the removal and recovery of manganese from wastewater: A review, J. Clean. Prod. 377 (2022) 134152. https://doi.org/https://doi.org/10.1016/j.jclepro.2022.134152.
[38] Clack Corporatiom, Manganese Greensand, Clack Corp. (2008) 2–3. https://www.watertreatmentspecialists.com/wp-content/uploads/2015/03/manganese-greensand.pdf (accessed August 27, 2024).
[39] I. Srinati, T.B. Prijanto, S. Lingkungan, J.K. Lingkungan, P. Kemenkes Bandung, VARIASI WAKTU KONTAK PADA FILTER MANGANESE GREENSAND DAN PASIR SILIKA TERHADAP PENURUNAN MANGAN (Mn) DALAM AIR BERSIH Variation of Contact Time on Manganese Greensand and Silica Sand Filter on Reduction of Mangan (Mn) In Clean Water, J. Ris. Kesehat. 16 (2024) 344–353. https://doi.org/10.34011/juriskesbdg.v16i2.2529.
[40] T.W. Shaddox, J.B. Unruh, J.K. Kruse, N.G. Restuccia, Solubility of Iron, Manganese, and Magnesium Sulfates and Glucoheptonates in Two Alkaline Soils, (2016). https://doi.org/10.2136/sssaj2015.10.0382.

IOP Conf. Series: Earth and Environmental Science **1464** (2025) 012004 doi:10.1088/1755-1315/1464/1/012004

Seismic risk assessment on pile-supported wharf using reliability-based approach

Aris Aryanto[1*] and Ersadi Wiguna[1]

[1] Faculty of Civil and Environmental Engineering, Institut Teknologi Bandung, Bandung, Indonesia

*E-mail: aris_ary@itb.ac.id

Abstract. Past worldwide seismic events have shown that pile-supported wharves are vulnerable and experience severe damage to collapse under earthquake loads. Significant economic loss has also been recognized, especially when wharves are built at a strategic port for sea transportation. This study assessed existing pile-supported wharves using the reliability-based approach of wharf structures under seismic loads. The procedure for seismic risk assessment for pile-supported wharf structures is explained and presented. This study uses typical container wharf structures at Jakarta Port as a case study. A nonlinear wharf platform is modeled, including soil-spring models to represent the soil-structure interaction. Nonlinear time history analysis with incremental seismic was performed to obtain the fragility curves using selected seismic events that matched the site-specific spectrum of current standards to represent the demand. The probability of collapses and reliability indexes for any performance levels of the existing wharves were obtained. For the case study, the reliability index at the life safety performance level is below the target reliability index for the most common structures, indicating the existing wharf structure needs to be enhanced. As a result, it demonstrates that the reliability-based approach can reasonably be used in decision-making to repair or strengthen existing wharves depending on seismic demand and required performance levels.

1. Introduction

Port is an important transportation hub of sea or marine transportations. It facilitates logistics movement within a worldwide trade network. Due to geographical conditions, many strategic ports in Indonesia are located in seismically active regions. This raises concerns regarding the seismic safety and operation of port infrastructure. Because ports are crucial for a country's economy and lifeline logistics, it is logical to demand that port structures remain in operation immediately after the earthquake.

The past earthquakes of Kobe in 1995 [1], Haiti in 2010 [2], and Tohoku in 2011 [3] showed that port structures are vulnerable to earthquakes and have substantial economic loss. For example, two wharves of Port-au-Prince, which are pile-supported, are reported to have substantial damage during the 2010 Haiti earthquake [2].

Port facilities, such as pile-supported wharves, frequently fail during earthquakes due to several factors that induce structural deformation [1]: (1) resulting from the inertial force acting on the deck; (2) caused by a horizontal force exerted by the retaining backwall; or (3) caused by the lateral spreading of loose subsoil. The primary factor leading to the breakdown of pile-supported wharves is the formation of plastic deformation, also known as hinges, in the top piles or in-ground piles. This deformation is caused by the inertial force exerted on the deck during an earthquake. This study primarily focuses on analyzing this collapse mechanism.

Due to the unexpected nature of earthquake ground motion, a reliability-based approach has emerged as the ideal approach for conducting seismic design or assessment of wharf structures. Nowadays, this technique has attracted considerable attention from researchers. The study evaluates the likelihood of collapse for various pre-established damage conditions, which function as benchmarks for performance. A fragility curve was used as a visual representation of the probability of collapse taking place. A collapse happens when the seismic demand (D) exceeds the seismic capacity (C). To develop fragility curves, a nonlinear time-history analysis (NLTHA), known as the most accurate analysis, was performed with incremental seismic demand to assess the seismic risk condition of the existing wharf structure.

2. Reliability-Based Approach

In a reliability-based approach, it is crucial to have a correct method for defining seismic hazard and seismic risk. Seismic hazards primarily focus on natural phenomena or occurrences of earthquakes, whereas seismic risks are particularly related to the harm produced by these seismic hazards [4].

The risk integral technique [5] is presented and adopted in this study. This concept integrates two probability curves, specifically the fragility curve and the hazard curve, to determine the reliability and probability of structural collapse. The fragility curve represents the capacity distribution of a structure. Thus, it may be shown as:

$$f_{capacity}(c) = \frac{1}{\zeta c\sqrt{2\pi}} exp\left[-\frac{1}{2}\left(\frac{\ln c - \lambda}{\sigma}\right)^2\right] \tag{1}$$

The symbols λ and ζ is the average value and variability lognormal distributions. The given equation can be used to calculate the probability of collapse using the risk integral approach.

$$P[collapse] = \int_0^\infty P[PGA > c] f_{capacity}(c)\, dc \tag{2}$$

P[PGA>c] means seismic hazard, the concept of a certain magnitude of earthquake force (in Peak Ground Acceleration/PGA) that can cause damage to the building, while $f_{capacity}$ (c) means seismic risk, the concept of designing the capacity of buildings that refers to the probability of loss and damage caused by an earthquake. Either seismic hazards or seismic risks will develop a fragility curve. Luco et al. [6] have formulated the following equation to calculate the probability of collapse over a time frame (Y years).

$$P[collapse\ in\ Y\ years] = 1 - (1 - P[collapse])^Y \tag{3}$$

$$Reliability = 1 - P[collapse\ in\ Y\ years] \tag{4}$$

IOP Conf. Series: Earth and Environmental Science **1464** (2025) 012004 doi:10.1088/1755-1315/1464/1/012004

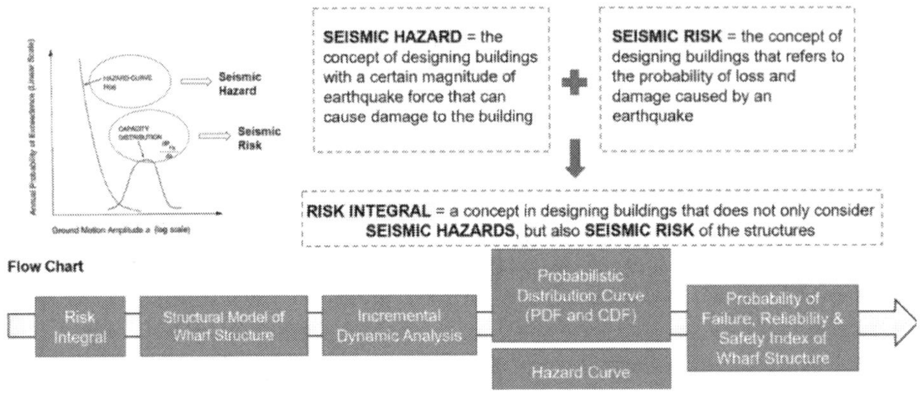

Figure 1. Seismic risk integral procedure.

3. Case Study of Seismic Risk Assessment

A case study was conducted on a typical pile-supported wharf located at the Tanjung Priok Port in Jakarta (Figure 2(a)). The wharf construction took place in 1995, thereby according to the outdated building regulations. This port was selected based on its status as a strategic port for international and domestic trading in Indonesia, which plays a crucial role in the country's economy. Furthermore, the port itself is located in a highly seismic region. Figure 2(b) displays the wharf's cross-section and dimensions. This container wharf has dimensions of 112.5 m in length and 40 m in width. There are 28 frames evenly spaced 4 m apart in the longitudinal direction. It has eight rows of prestressed piles in the transverse direction. The spacing between piles ranges from 4.6 to 5.4 m. The concrete deck supported by these piles has a 0.5 m thickness.

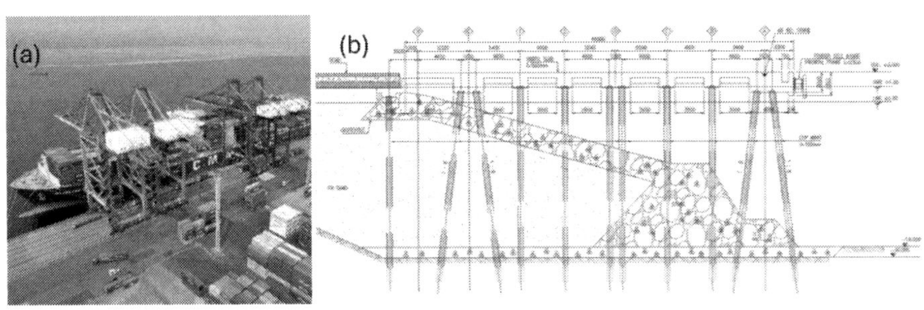

Figure 2. (a) Tanjung Priok Container port (b) Typical wharf structure cross-section for case study.

3.1 Wharf Modelling

A 2D model of the middle frame is created because it has a typical pattern along the lengthwise axis of the wharf. The wharf deck is modeled using linear elastic beam elements, whereas nonlinear elements are employed to represent the piles. The connection between the piles and the deck is a rigid linkage. Figure 3 depicts a model frame that was chosen according to the layout of the wharf construction. A pile-supported wharf is loaded with dead and live loads, including the weight exerted by cranes.

IOP Conf. Series: Earth and Environmental Science **1464** (2025) 012004 doi:10.1088/1755-1315/1464/1/012004

To conduct a nonlinear time history analysis, plastic hinges are assigned throughout the whole length of the pile to induce plastic deformation of the piles. The pile's nonlinearity behavior is determined by a moment-rotation curve called a backbone curve (Figure 4). The backbone curve is determined by taking into account the nonlinear material characteristics of both concrete and steel. The plastic rotation is calculated by multiplying the curvature with the hinged plastic length (Lp) of 0.208 m and 1.2 m for the top piles and in the ground, respectively, as specified in the POLB 2012 [7]. Figure 4 also displays several points representing the performance levels of piles, determined by the rotation limit as defined by FEMA 356 [8]. FEMA 356 [8] categorizes structural performance into three main points: immediate occupancy (IO), life safety (LS), and collapse prevention (CP). Performance levels like IO, LS, and CP also represent the damage level. Immediate occupancy (IO) is a condition where non-structural elements suffer minor damage, there is no structural damage, and the structure continues to function normally. Life safety (LS) is characterized by structural damage without collapse, which occurs when non-structural components are damaged. However, the structure remains usable following repairs. Collapse prevention (CP) is when all structural and non-structural elements are damaged, the structure is on the threshold of collapsing, and accidents involving death may occur. Based on ASCE 61-14 [9], life safety (LS) will be the designated level of performance.

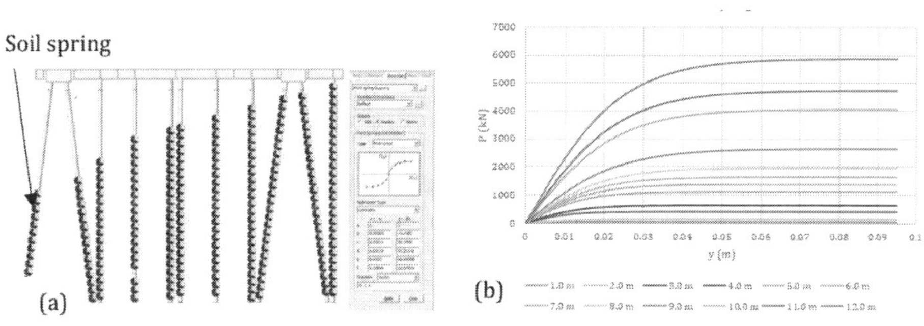

Figure 3. (a) 2D FE model (b) Lateral soil spring model (p-y curves).

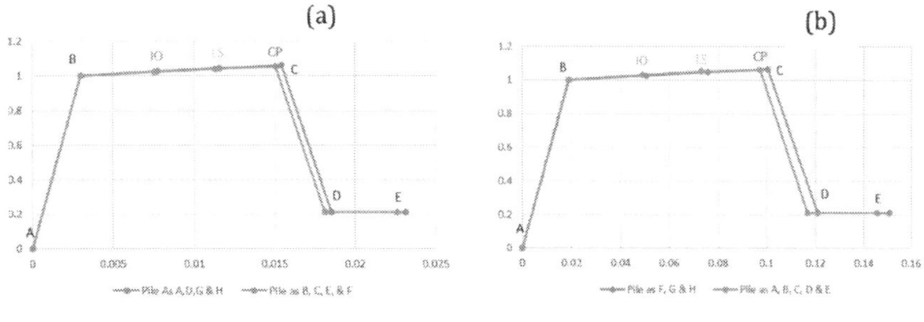

Figure 4. Normalized moment-rotation curves of piles (a) near pile head (b) in the ground.

The soil layers beneath the wharves display various properties, indicating stratification. A precise evaluation of the soil's reaction to lateral and vertical pressure is essential for determining soil-structure interaction. The interaction between soil and piles is demonstrated by horizontal soil springs, p-y curves, vertical spring, t-z, and the tip resistance, Q-z, developed according to API RP 2GEO [10].

3.2 Seismic ground motion input

The seismic load is calculated using the given ground acceleration for nonlinear time history analysis (NLTHA). NLTHA simulates how a structure behaves under a given earthquake ground motion, accounting for nonlinearities in the structure's geometry, deformations, materials, and systems. The selection of ground motion depends on their compatibility with the geological parameters of the site. This study utilizes data evaluated by Sengara [11] and is grounded on the synthesis technique of analyzing ground motion acceleration in Jakarta City. The study employed a total of seven different sets of ground motion data, which are displayed in Table 1.

Seven sets of ground motion are adjusted and matched with the spectral design spectrum of the site and are utilized in the nonlinear time history analysis. The ground acceleration was with the SNI 1726:2012 [12] design spectra because the original design spectra used when constructing the wharf were unavailable.

Table 1. Selected ground motion.

No	Catalogue Code	Earthquake Mechanism	Direction	Magnitude	Epicentral Distance (km)	Earthquake
1	MEL	Shallow Crustal	90-180	7.28	138.49	Landers Earthquake 28 June 1992
2	ABY	Shallow Background	0-90	7.28	75.2	Landers Earthquake 28 June 1992
3	TCU-120	Benioff	W-N	7.62	25.57	Chi - chi Earthquake 20 September 1999
4	TCU 136	Benioff	E-N	7.62	48.75	Chi - chi Earthquake 20 September 1999
5	MYG013	Megathrust	EW-NS	9	170	Tohoku Earthquake 11 March 2011
6	212V5	Megathrust	90-180	7.68	42.85	Sitka Earthquake 30 March 1972, ALASKA
7	TCU089	Megathrust	N-W	7.62	7.04	Chi - chi Earthquake 20 September 1999

4. Seismic Risk Assessment Results

4.1 Fragility curve results for all performance levels

The fragility curve illustrates the relationship between the peak ground acceleration (PGA) and the probability of attaining particular performance levels. This is done by doing non-linear time history analysis (NLTHA), and then we gradually scale up the PGA, known as incremental dynamic analysis (IDA). These analyses are adequate when each level of performance is gained. Table 2 presents the results of scaled Peak Ground Acceleration (PGA).

Table 2. Incremental dynamic analysis result.

No	Earthquake	PGA domain (g)	Immediate Occupancy (IO)		Life Safety (LS)		Collapse Prevention (CP)	
			Scale Factor	PGA Scaled (g)	Scale Factor	PGA Scaled (g)	Scale Factor	PGA Scaled (g)
1	LandersABY	0.15704	1.54	0.24176	2.14	0.33617	2.70	0.42370
2	LandersMEL	0.20523	1.35	0.27637	1.81	0.37229	2.29	0.47036
3	ChiChi120	0.18445	1.46	0.26843	2.03	0.37445	2.50	0.46066
4	ChiChi136	0.19547	1.55	0.30382	2.06	0.40272	2.50	0.48916
5	Tohoku	0.1177	2.28	0.26800	3.14	0.36906	3.76	0.44270
6	Sitka	0.19196	1.50	0.28886	2.09	0.40127	2.65	0.50830
7	ChiChi089	0.15779	2.00	0.31633	2.68	0.42340	3.25	0.51277

Subsequently, the scaled peak ground accelerations (PGA) are subjected to statistical evaluation to ascertain the suitable distribution. The Kolmogorov-Smirnov test was performed on a log-normal distribution, as indicated in Table 3.

Table 3. Kolmogorov-Smirnov test for data distribution.

Number of Data		7		
Error Value (Dn)		0.16211	0.18923	0.16018
Significance Level	Maximum Error Value	IO Level	LS Level	CP Level
α	(Dnα)	Check	Check	Check
0.2	0.38148	Ok	Ok	Ok
0.1	0.43607	Ok	Ok	Ok
0.05	0.48342	Ok	Ok	Ok
0.02	0.53844	Ok	Ok	Ok
0.01	0.57581	Ok	Ok	Ok

To generate the distribution parameters presented in Table 4, the values of the parameters depicted in the aforementioned data distribution are modified to incorporate a 10% uncertainty constraint and include seven ground motion data points.

Table 4. Distribution parameter

	Immediate Occupancy	Life Safety	Collapse Prevention
μ	0.2805	0.3828	0.4725
Ω'	0.1378	0.1281	0.1250
σ'	0.0387	0.0490	0.0591
λ'	-1.2806	-0.9685	-0.7574
ζ'	0.1372	0.1276	0.1245

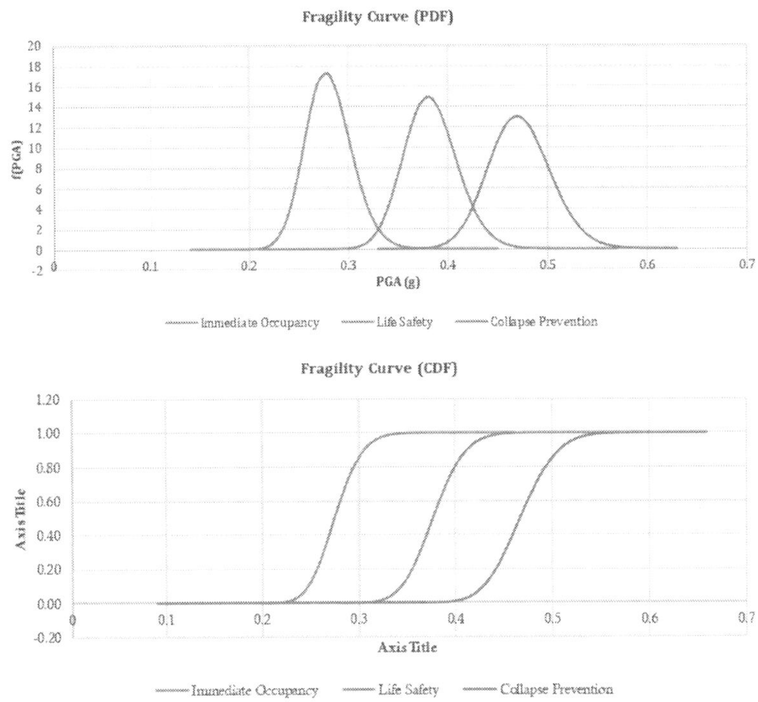

Figure 5. Fragility curves: PDF and CDF at immediate occupancy (IO), life safety (LS), collapse prevention (CP) levels.

The symbols μ and λ' reflect the average value of normal and lognormal distributions, correspondingly, while σ' and ζ' show the measure of variability of normal and lognormal distributions, correspondingly. The statistical parameters are used to construct the probability density function (PDF) and cumulative density function (CDF) distributions for each level of performance, as seen in Figure 5.

4.2 Probability of Collapse and Reliability for All Performance Levels

Sengara et al. [13] analyzed the hazard curve for all structures in Jakarta. The hazard curve in this research corresponds to earthquakes with a 500-year return period, as designated by the design earthquake. A polynomial equation (Equation 5) was derived to model the hazard curve accurately by utilizing reference points. Consequently, the equation representing the function of the damage curve is as follows:

$$P(PGA > x) = \exp\left(0.0001\,(\ln PGA)^6 + 0.0033\,(\ln PGA)^5 + 0.0262(\ln PGA)^4\right. \quad (5)$$

Calculating the integral of equations (5) and (1), the chance of collapse for several timeframes, such as annually, 50 years, and 25 years of remaining lifespan, as well as the dependability of each performance level, are computed and presented in Table 5.

Table 5. Probability of collapse and reliability of pile-supported wharf structure for all performance levels.

Performance Level	Immediate Occupancy	Life Safety	Collapse Prevention
P[Collapse] (Annual)	0.14%	0.04%	0.02%
P[Collapse] of 50 Years	6.89%	2.07%	0.80%
P[Collapse] of 25 Years	3.50%	1.04%	0.40%
Reliability of Remaining Lifetime (25 years)	96.50%	98.96%	99.60%

From Table 5, at the life safety level, the annual probability of collapse is 0.04%, and the probability of collapse for 50 years (structure lifetime) is 2.07%. For the remaining lifetime of 25 years (where this investigation was taken), the probability of collapse is 1.04%, and the reliability is 98.96%.

5. Conclusions

This article uses a reliability-based methodology to examine the seismic risk assessment of an existing typical pile-supported wharf. Some findings are summarized as follows:

1. The reliability-based approach can be used as an assessment tool to identify the seismic risk of the existing wharf at various levels of performance targets.
2. life safety (LS) performance level is targeted for this case study, as mentioned in ASCE 41-17 [14]. The probability of collapse for a structure lifetime (50 years) is 2.07%. This probability of collapse exceeds the acceptance criteria of SNI 1726: 2012 subchapter 6.10.2.1, said 1%.
3. This demonstrates an increase in seismic loads within the SNI 1726:2012 code, affecting older structures, such as the wharf in this case study constructed in 1995, which was insufficiently reliable and required further strengthening analysis.

References

[1] Permanent International Association for Navigation Congresses (PIANC) (2001) Seismic design guidelines for port structures. Grafisch Produktiebedrijf Gorter, Steenwijk, The Netherlands

[2] Werner S, Mccullough N, Bruin W, et al. (2011) Seismic performance of Port de Port-au-Prince during the Haiti earthquake and post-earthquake restoration of cargo throughput. Earthq. Spectra 27(S1):S387–S410.

[3] Sugano T, Nozu A, Kohama E, Shimosako K, Kikuchi Y. (2014). Damage to Coastal Structures, Soils and Foundations, 54(4):883–901

[4] Wang, Zhenming (2009). Seismic Hazard vs Seismic Risk. Seismological Research Letters 80(5):673-674

[5] Sidi I. D., (2017). Probabilistic Modeling of Seismic Risk Based Design for a Dual Systems Structure, J. Eng. Technol. Sci., 49(2): 179-192

[6] Luco et al. (2007). Risk-Targeted versus current seismic design maps for the conterminous United States, Structural Engineers Association of California 2007 Convention Proceedings, pp. 163-175.

[7] POLB (2012). Port of Long Beach Wharf Design Criteria. Port of Long Beach. California.

[8] FEMA 356 (1997). FEMA356/NEHRP 1997. Prestandard and Commentary on Seismic Rehabilitation of Buildings.

[9] ASCE 61-14 (2014). Seismic Design of Piers and Wharves. American Society of Civil Engineers.

[10] API R. 2geo (2011). Recommended practice for geotechnical foundation design consideration. American Petroleum Institute

[11] Sengara, I. W. (2012). Investigation on Risk-Targeted Seismic Design Criteria for a High-rise Building in Jakarta-Indonesia. 15th World Conference on Earthquake Engineering, Lisbon, Portugal.

[12] SNI 1726:2012. Standard Design of Seismic Resistance Structure for Buildings, National Standardization Agency. (in Indonesian)

[13] Sengara, I. W., Yulman, M. A., & Mulia, A. (2015). Seismic Time-History Ground-Motions for A Specific Site in Jakarta. Jurnal Teknologi, 77(11).

[14] ASCE 41-17 (2017) Seismic Evaluation and Retrofit of Existing Buildings. American Society of Civil Engineers.

The resistance of Traditional Phinisi Boats: an examination the effect of bilge keel length using computational fluid dynamics (CFD)

Eka Suendra Putra[1], Eko Charnius Ilman[1*]

[1] Ocean Engineering Program, Institut Teknologi Bandung, Bandung, Indonesia

*E-mail: eko.ilman@itb.ac.id

Abstract. In the contemporary marine tourism industry, the necessity to convey a larger number of passengers has required that conventional vessels be both user-friendly and secure. On the other hand, traditional shipbuilders have seldom contemplated the integration of bilge keels as a compliance measure for the ship's stability system. In this paper, the potential for an increase in resistance that may arise from the incorporation of a bilge keel into traditional Phinisi boats is discussed. A computational fluid dynamics (CFD) methodology is implemented to analyse the resistance values. This investigation endeavours to ascertain whether variations in bilge keel length in 5-degree bilge keel slope angles are influenced by varying forward velocities. By incorporating bilge keels into the Phinisi boat, the BK05-L5 model demonstrated a substantial improvement in rolling motion reduction, with a reduction of up to 76%. Moreover, the ship's resistance increased by only an insignificant 9.5% in comparison to the model without a bilge keel. This configuration appears in the nearly even keel condition produced by the 0.25 m reduction in draft and the reduced pitch value at high speed. This study encourages the implementation of bilge keel geometry and position that has no significant effect on ship resistance to improve the operational safety and performance efficiency of traditional Phinisi boats.

1. Introduction

The modernization of traditional Phinisi boats has enabled their widespread integration into marine tourism operations, extending beyond their original roles in cargo transport and fishing activities. However, it is commonly observed that many traditional boat builders in Indonesia still depend primarily on practical experience without a thorough technical understanding. This scenario emphasise the pressing need for a comprehensive hydrodynamic evaluation of the Phinisi boat's various characteristics, a necessity that is encouraged by earlier studies [1].

In terms of minimising rolling motion, the utilisation of bilge keels as a fundamental component offers substantial advantages, thereby enhancing the stability and safety of the vessel. Specifically, the implementation of bilge keels may result in an increase in hydrodynamic resistance under specific operating conditions. The computational fluid dynamics (CFD) has been implemented to precisely evaluate the influence of bilge keels on ship resistance. These cutting-edge methodologies guarantee the accuracy of resistance calculations, particularly at high Froude Numbers. These methods employed a Reynolds-averaged Navier-stokes (RANS) calculation, which effectively simulate the effects of free surface phenomena, for example wave formation that results from the interaction between the ship's hull and the surrounding water. This thorough examination is indispensable for comprehending the trade-offs associated with the utilisation of bilge keels and for optimising ship design to maximise performance while minimising resistance. This method significantly improves the precision of ship resistance measurements, as evidenced by prior research [2], [3].

Previous studies by [4], have utilized CFD to comprehensively analyse the total resistance of conventionally constructed vessels. Furthermore it found a minimal increase in resistance of approximately 1.17% [5]. Suggesting that such an increase negligible impact on high-speed vessels that employ bilge keels. This consideration is crucial for further assessments.

However, the resistance value of conventional Phinisi boats increased by an average of 11% in a previous study [6]. The study was conducted by supplying a fixed length of 5 meters and various bilge keel angles between 15 and 25 degrees. Conversely, the experiment involving the intended patrol ship hull equipped with a side keel [6], demonstrated a more substantial increase in ship resistance of 14%. Furthermore [7], has analysed resistance and propulsion characteristics of various merchant ships. They have employed CFD to identify the resistance and propulsion components of each ship, thereby establishing a reference for the design of future vessels. The optimal design can be used to determine the precise measurement of ship resistance by analysing the configuration of the bilge keel with fluid flow. Furthermore, frictional resistance is primarily the consequence of the combination of the surfaces between the hull ship and the bilge keel surface [5], [8], [9].

The resistance characteristics of a traditional Phinisi boats are predicted in this research paper through a comprehensive computational investigation. The objective of this investigation is to assess the resistance value (R_T) of a Phinisi botas in placid water. The simulation was conducted in its entirety analysed the variations in forward speed and length of a variety of hull keels. The objective of this study is to examine the relationship between vessel design and hydrodynamics through the application of CFD capabilities.

2. Methodology

2.1 CFD solver

The CFD solver is based upon the second-order finite volume method [3], [10]. The RANS equations were resolved through the application of turbulence modelling and acceptable boundary conditions. The free surface boundary is generated by employing the volume of fluid (VoF) procedure [11].

$$\frac{\partial}{\partial t}\int_V \rho \, \mathrm{d}V + \int_S \rho(\mathbf{U}-\mathbf{U}_d) \cdot \mathbf{n} \, \mathrm{d}S = 0 \quad (1)$$

$$\frac{\partial}{\partial t}\int_V \rho U_i \, \mathrm{d}V + \int_S \rho U_i(\mathbf{U}-\mathbf{U}_\mathrm{d}) \, . \mathbf{n} \, \mathrm{d}S = \int_S \left(\tau_{ij} I_j - p I_i\right) \cdot \mathbf{n} \, \mathrm{d}S + \int_V \rho g_i \, \mathrm{d}V \quad (2)$$

$$\frac{\partial}{\partial t}\int_V c_i \, \mathrm{d}V + \int_S c_i(\mathbf{U}-\mathbf{U}_\mathrm{d}) \, . \mathbf{n} \, \mathrm{d}S = 0 \quad (3)$$

In Equation (1), (2), and (3) given for integral forms of mass, momentum, and volume fraction, respectively. The gravity acceleration is g, p denoted for pressure, ρ is density of fluid, **U** is for relative velocity and for controlling velocity incorporate volume given as $\mathbf{U}_d$, normal condition of vector was denoted by S, **n** is the control of surfaces, and t is for time. Moreover, the equation of Reynolds stresses was denoted by τ_{ij}, that I_j is a tensor identity, and the final term, also the c_i, represents the volume fraction.

2.2 Resistance calculation

Residual resistance (R_R) and total friction resistance (R_F) are the components of the total resistance (R_T). The hull surface was subjected to a tangential force, which resulted in the accumulation of shear tension due to the fluid viscosity. The viscous pressure resistance (R_{VP}) and wave-making resistance (R_W) were generated by this force, and they were perpendicular to the force [12], are illustrated in Eq. (4) and (5).

$$R_T = R_F + R_R + R_A \tag{4}$$

$$R_T = R_F + R_{VP} + R_W + R_A. \tag{5}$$

3. CFD computation

A comprehensive summary of the analysis is provided in this section, which includes the domain measurment and boundary layers, meshing program, cell grids analysis, and principle dimension of the boat. The complete resistance quantity will be discernible as the ship's speed increases.

3.1 Principal dimension

The focus of this investigation is the Phinisi boat, which is equipped bilge keel of the ship's hull. Table 1 summarises the principal dimensions of the boat.

Table 1. Principals ship dimension.

Components	Symbols/units	Value
Displacement	Δ, tonnes	76.65
Length over All	LoA, *m*	30.00
Length of Water Line	LWL, *m*	20.93
Breadth	B, *m*	5.80
Depth	H, *m*	3.20
Draught	T, *m*	2.00
Wetted Surface Area	WSA, m^2	131.06

3.2 Parametric study

The CFD resolved the force comparison between model without bilge keel and other models with bilge keels of varying lengths at a 5-degree angle of inclination. These models are referred to as

BK05-L0, BK05-L1, BK05-L2, BK05-L3, BK05-L4, and BK05-L5, respectively, and they operate at a variety of forward velocities. Specific details regarding the scenario are given in Table 2.

Table 2. Parameter bilge keels model.

Model code	Length (m)	Width (m)	Velocities (knots)		
			7	10	12
Bare hull (without bilge keel)	-	-	√	√	√
BK05-L0	5.00		√	√	√
BK05-L1	6.00		√	√	√
BK05-L2	6.30	0.45	√	√	√
BK05-L3	6.50		√	√	√
BK05-L4	7.50		√	√	√
BK05-L5	10.0		√	√	√

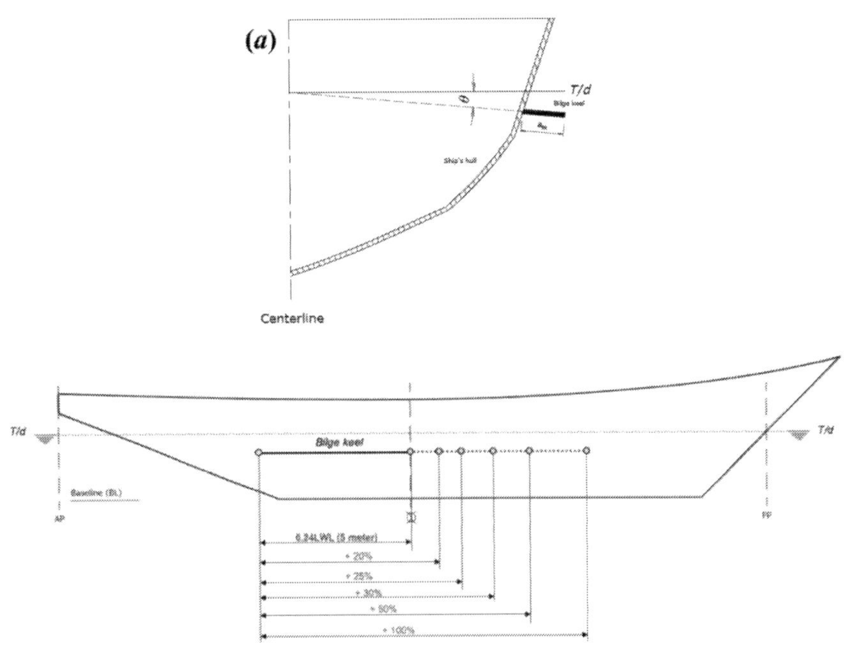

Figure 1. Definition of bilge keels position: *(a)* transverse section of hull; *(b)* longitudinal location of bilge keel.

3.3 Boundary condition

The configuration is predicated on the origin point, which is from the ship's length, as illustrated in Table 3. In Figure 2 demonstrates the boundary condition that is enforced within the domain [13].

Table 3. Computational domain.

Description	Plot origin point	Boundary	Domain
X-negative	3.0LoA	Far field	External
X-positive	LoA	Far field	External
Y-negative	0	Symmetry	Mirror
Y-positive	1.5LoA	Far field	External
Z-negative	1.5LoA	Prescribed pressure	External
Z-positive	1.5LoA	Prescribed pressure	External

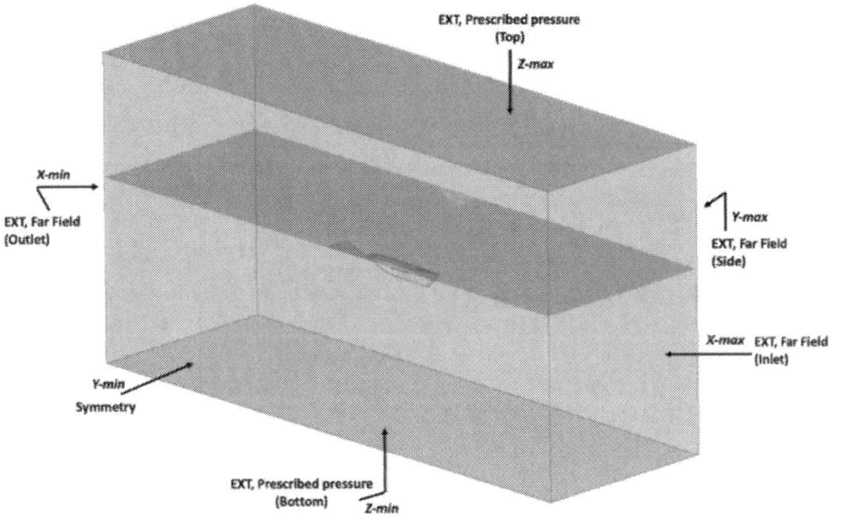

Figure 2. Domain setup and boundary conditions setup.

The external boundary type of the computation domain about velocity and pressure conditions. The boundary's upper and lower limits were established per the prescribed pressure. The hydrostatic pressure at the top and bottom of the boundary has been revised through the implementation of this computation. A mirror image of the boundary is used to represent only half of the hull to reduce the computational complexity and intensity.

3.4 Grid independence study

To achieve precise results within an adequate time for computation, a mesh study was implemented at a ship speed of 9 knots [2], [13]. As defined the percentage error is computed by increasing the number of mesh cells within each simulation. These simulations were generated by Putra et al. (2023) that were depict in Table 4 and Figure 3.

Table 4. Mesh independence value.

Number of computation (n)	Cells number (million)	Total resistance, R_T (N)
1	0.5	7146.311
2	1.0	6966.384
3	2.0	6887.445
4	3.0	6833.943
5	4.0	6824.332

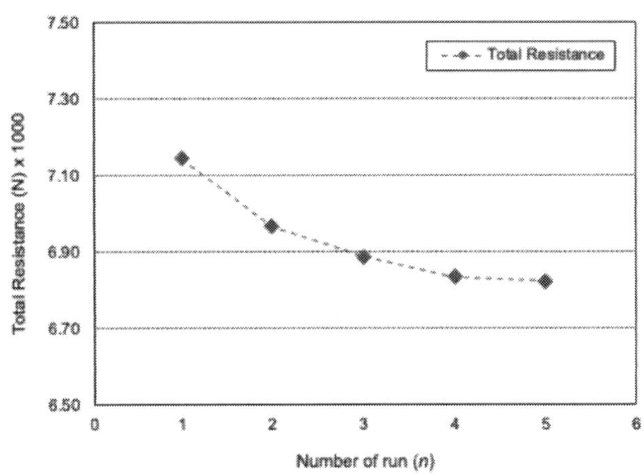

Figure 3. Mesh sensitivity study.

The workstation PC was equipped to perform four parallel computations with a time step value of 0.017s which 60 to 70 hours estimate the duration of each computation in this simulation. The cell number's sensitivity to probable values was also assessed through the grid independence study.

In Table 4 and Figure 3, the total resistance decreases progressively as the cell number in simulation was increased. Correlation for this computation was less than 2% due to the mesh condition, which employed 3.0 million approximate cells that not significantly impact the resistance further [2], [12].

4. Results and Discussion

This study employs the computation to assess resistance force in various bilge keel length configurations. The results demonstrate a direct correlation between ship speed and resistance, with resistance increasing as speed rises. The analysis focuses on the influence of bilge keels added to the ship's hull, showing that their impact on resistance is negligible at lower speeds [14], [6] but becomes pronounced at higher speeds. Consequently, it is important to optimising the design in order to reduce the resistance effects that are associated [15], [4].

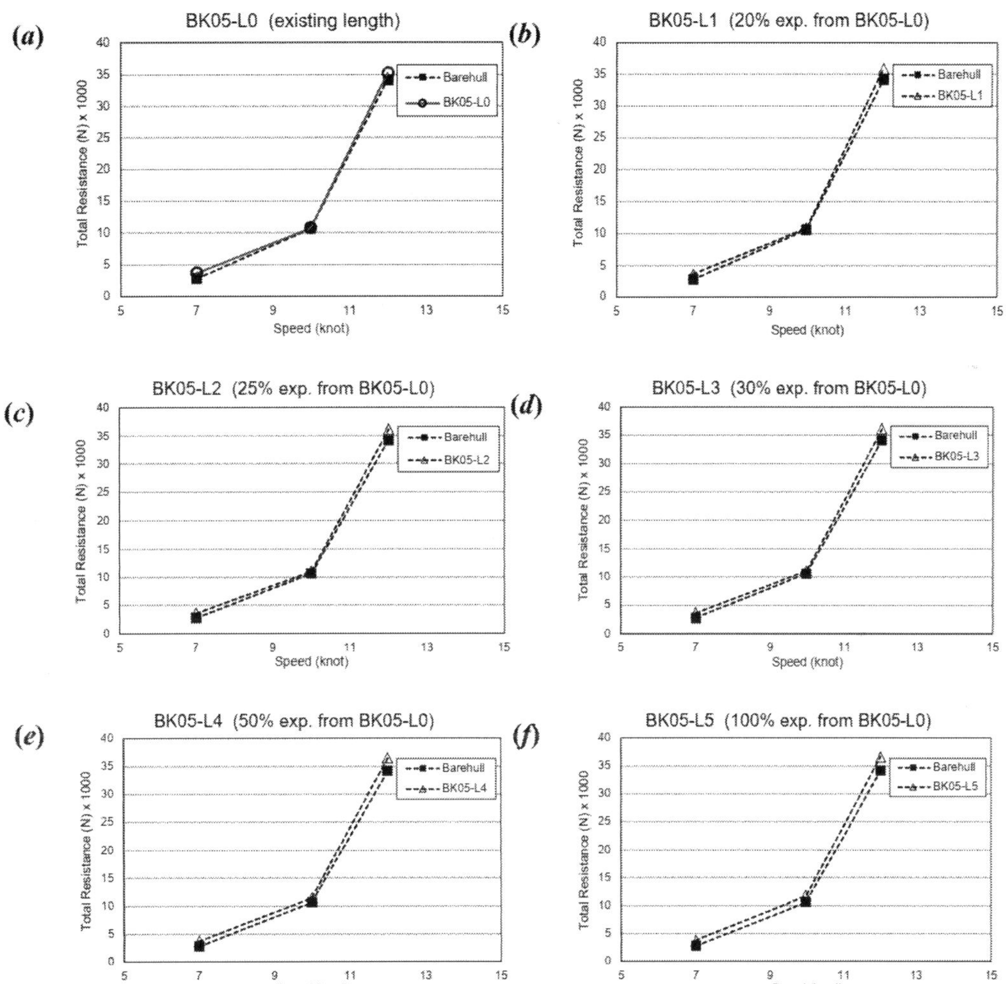

Figure 4. Total resistance compares with Bare hull model in various bilge keel length: *(a)* BK05-L0; *(b)* BK05-L1; *(c)* BK05-L2; *(d)* BK05-L3; *(e)* BK05-L4; *(f)* BK05-L5, respectively.

The total resistance increased insignificantly because of the bilge keel's length variations. The transition between velocities of 10 and 12 knots was the primary time for observing significant changes in total resistance. The Figure 4*(f)* exhibited the most significant increase in resistance, reaching 9.5%. However, this increase appeared insignificant in comparison to the bare hull model. The study examined the effects of bilge keel length modifications on heave and pitch motions. The resistance remained largely unchanged; however, the BK05-L5 model showed a heave deviation of over 0.25 meters at 12 knots, indicating greater vertical displacement at high speeds compared to other models. Additionally, the BK05-L5 model had a reduced pitch motion at high speeds, bringing the Y-axis rotation closer to zero, achieving an even keel or zero trim.

Minor resistance increases were observed by [16], particularly at higher velocities, as the bilge keel's length was increased. The increments were consistently less than 10% across a variety of hull types. Their investigation revealed that these minor resistance increases were frequently tolerable, as they were outweighed by improved stability, particularly for vessels that prioritised roll damping. Additional research has explored how the length of bilge keels affects resistance across different hull shape configurations. These studies found that longer bilge keels tend to result in moderate increases in resistance, with increases reaching up to 12%, especially at higher speeds. This impact was modest for streamlined vessels that were designed for high-speed performance, and it closely matched the resistance trends observed in the BK05-L5 model.

In contrast, the investigation examined a variety of bilge keel shapes and locations along the hull and discovered that the profile and location of the bilge keel influenced resistance changes. However, resistance increases were consistently only modest when length adjustments were made alone. The BK05-L5 result is in line with the assertion that the overall resistance is only marginally influenced by the lengthened bilge keels, as identified [17].

Moreover, this investigation suggested that resistance increases are moderate but manageable, typically occurring within the 10–12% range, and are more pronounced at higher velocities due to the increase in bilge keel-length.

5. Conclusions

Based on the analysis of ship resistance with variations in bilge keel by increasing length dimensions, favourable results were obtained, providing a sound basis for bilge keel design considerations. By applying bilge keels to the Phinisi boat, an effective reduction of rolling motion by up to 76% was achieved with the BK05-L5 model indicating a significant improvement. Additionally, the increase in ship resistance was not substantial, with only a 9.5% increase [18] compared to the model without a bilge keel.

Acknowledgement

The Research, Community Service, and Innovation Program, Faculty of Civil and Environmental Engineering, Institut Teknologi Bandung, Indonesia, were acknowledged for their support in this work

CRediT authorship contribution statement

Eka Suendra Putra: Conceptualization (equal), Formal Analysis (lead), Methodology (equal), Validation and Investigation (lead), Visualization (lead), Writing – Original Draft Preparation (lead). **Eko Charnius Ilman**: Conceptualization (lead), Supervision (lead), Methodology (lead), Funding Acquisition (lead), Validation and Investigation (equal), Resources (lead), Writing – Review & Editing (lead).

References

[1] M. Johny, M. Jinca, P. Herman, and S. Abrar, "Traditional Shipping Transport Safety Case Study: Phinisi Fleet," 2013.
[2] I. Suastika, A. Fauzi, A. Saputra, I. Utama, A. Firdhaus, and M. Hariadi, "Benchmark Tests Of FINE/Marine CFD Code for The Calculation of Ship Resistance at High Froude Numbers," presented at the Proceedings of International Conference Royal Institution of Naval Architects, Royal Institution of Naval Architects, 2021, pp. 106–112.
[3] P. Voxakis, "Ship hull resistance calculations using CFD methods," 2012.
[4] W. Liu *et al.*, "Bilge keel design for the traditional fisihing boats of Indonesia's East Java," *Int J Nav Arch. Ocean Eng*, pp. 1–16, 2018, doi: 10.1016/j.ijnaoe.2018.07.004.
[5] R. Bhattacharyya, *Dynamics of Marine Vehicles*. in A Wiley-Interscience publication. Wiley, 1978.
[6] A. Maimun, A. Priyanto, K. Wong, M. Pauzi, and M. Rafiqul, "Effects of side keels on patrol vessel safety in astern waves," *Ocean Eng.*, vol. 36, no. 3–4, pp. 277–284, 2009, doi: 10.1016/j.oceaneng.2008.12.003.
[7] J. Choi, K.-S. Min, J. Kim, S. Lee, and H. Seo, "Resistance and propulsion characteristics of various commercial ships based on CFD results," *Ocean Eng.*, vol. 37, no. 7, pp. 549–566, 2010, doi: 10.1016/j.oceaneng.2010.02.007.
[8] A. Lloyd, "Seakeeping: ship behaviour in rough weather," *Admir. Res. Establ. Haslar Gosport Publ. Ellis Horwood Ltd John Wiley Sons ISBN 0 7458 0230 3*, 1989.
[9] D. W. Taylor, *The speed and power of ships*. BoD–Books on Demand, 2013.
[10] A. Fitriadhy, N. Razali, and N. AqilahMansor, "Seakeeping performance of a rounded hull catamaran in waves using CFD approach," *J. Mech. Eng. Sci.*, vol. 11, no. 2, pp. 2601–2614, 2017, doi: 10.15282/jmes.11.2.2017.4.0238.
[11] C. W. Hirt and B. D. Nichols, "Volume of fluid (VOF) method for the dynamics of free boundaries," *J. Comput. Phys.*, vol. 39, no. 1, pp. 201–225, 1981, doi: 10.1016/0021-9991(81)90145-5.
[12] R. C. Ariesta, M. Aliffrananda, S. Riyadi, and I. Utama, "An Investigation into the Justification of the Service Speed of Ro-Ro Ferry with Block Coefficient 0.8 Based on the Resistance and Seakeeping Performance," presented at the Proceeding of International Conference Royal Institution of Naval Architects. no. November 2021, pp. 19–20.
[13] E. S. Putra, E. C. Ilman, and A. Fitriadhy, "Computational Investigation into Bilge Keel Effect on a Traditional Phinisi Boat's Resistance," *Semarak Ilmu*, 2023.
[14] A. Maimun, "Incorporating side keels to an existing planing hull vessel," presented at the Proceedings of the 4th Osaka Colloquium on Seakeeping Performance of Ships, paper: P2000-11 Proceedings., 2000.
[15] Z. Liu, W. Liu, Q. Chen, F. Luo, and S. Zhai, "Resistance reduction technology research of high-speed ships based on a new type of bow appendage," *Ocean Eng.*, vol. 206, p. 107246, Jun. 2020, doi: 10.1016/j.oceaneng.2020.107246.
[16] A. F. Molland, S. R. Turnock, and D. A. Hudson, *Ship resistance and propulsion*, 2nd ed. Cambridge university press, 2017.
[17] K. H. Son and K. Nomoto, "On the Coupled Motion of Steering and Rolling of a High-Speed Container Ship," *J. Soc. Nav. Archit. Jpn.*, vol. 1981, pp. 232–244, 1981.
[18] E.S. Putra, "Analysis of the Effect Of Bilge Keel on the Performance of 70 GT Pinisi Ship Motion Using Computational Fluid Dynamics (CFD) Approach", Master Thesis, Institut Teknologi Bandung, 2023.

IOP Conf. Series: Earth and Environmental Science **1464** (2025) 012006 doi:10.1088/1755-1315/1464/1/012006

Determination of optimal vessel for Tol Laut Container Program: T-2 Tol Laut route case study

Muhammad Weldy Hermawan[1*] and and Muslim Muin[1]

[1] Ocean Engineering Program, Institut Teknologi Bandung, Bandung, Indonesia

*E-mail: weldyhermawan@gmail.com

Abstract. Container transportation is one of the transportation activities served by Tol Laut program. An increase of container volume activities will impact on adjusting the size of vessels serving the container transportation activities. This adjustment of vessel size for container transportation needs to consider the existing conditions and port development needed, such as berthing facilities and dredging activities. In this research, the determination of container vessel size is reviewed through a case study of optimal vessel requirements to meet the container demand of 250,000 TEU on the T-2 Tol Laut transport network which serves the Sumatera area and stops at Port of Teluk Bayur, Port of Sikakap, Port of Pulau Baai, Port of Gunung Sitoli, and Port of Sinabang. Determination of the optimal vessel size for the network will be reviewed from the requirements for berthing facilities development, dredging activities, as well as cost requirements for the port development activities. The research results show that to meet the demand for 250,000 TEU containers on this route, a 5,000 DWT vessel with 370 TEU capacity is the optimal vessel size to serve the container transportation activities based on development needs that need to be carried out at the destination ports on this route.

1. Introduction

Tol Laut Program is an Indonesian government program that is organized providing shipping routes for goods transport by sea. This Sea Toll Program aims to accelerate national economic growth and ensure connectivity between islands in remote areas in Indonesia[1]. One of the Tol Laut program is using container. The increase in container activity can impact the container services in terms of adjusting the vessel size. Increasing the vessel size can also have an impact on increase in the vessel's specifications such as draft and length[2].

The adjustment of vessel size to accommodate container activities affect the need for port facilities in form of pier and water depth needed. Optimization of a port's pier can be accommodated by increasing the number of berthing number vessel[3]. And to accommodate the water depth, dredging work is required so the vessel operations is not disrupted[4].

Port investment is a key issue in modern port economics with respect to planning port development[5]. In this study, the optimal size of the vessel that needed to accommodate container services at a container transport network is reviewed through a case study of a Tol Laut T-2 route[6] that stops at Port of Teluk Bayur, Port of Sikakap, Port of Pulau Baai, Port of Gunung Sitoli, and Port of Sinabang. Assessment of optimal vessel conditions is reviewed through the port development needed in each port through the berthing structure development, dredging work activities, as well as cost requirements for the port development activities.

2. Description of Study Area

This study focuses on Tol Laut T-2 Route that stops at Port of of Teluk Bayur, Port of Sikakap, Port of Pulau Baai, Port of Gunung Sitoli, and Port of Sinabang. The location of the port reviewed in this study can be seen in the image below.

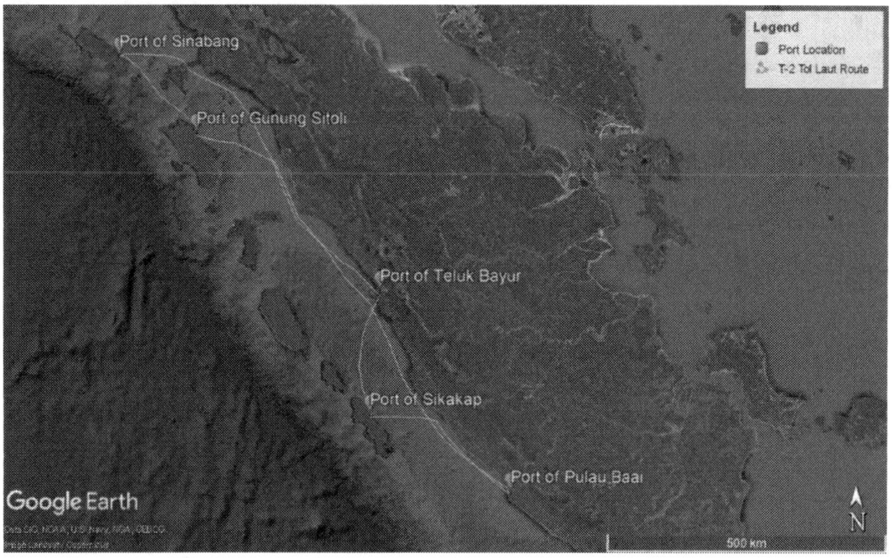

Figure 1. Location and satellite (Google Map) view general layout of the ports study location.

The general information and existing berthing facilities for container activities in each ports shown in Table 1 and Table 2 below.

Table 1. General information of ports study location.

No	Port Location	Geographical Location	Port Hierarchy[7]	Tol Laut Role[8]
1	Port of Teluk Bayur	0°59'58.24"S; 100°22'14.73"E	Main Port	Hub Port
2	Port of Sikakap	2°46'38.41"S; 100°12'54.69"E	Collector Port	Spoke Port
3	Port of Pulau Baai	3°54'26.97"S; 102°18'15.33"E	Collector Port	Spoke Port
4	Port of Gunung Sitoli	1°18'20.38"N; 97°36'35.66"E	Collector Port	Spoke Port
5	Port of Sinabang	2°28'17.40"N; 96°23'6.83"E	Regional Feeder Port	Spoke Port

Table 2. Existing of berthing facilities for container activities in ports study location.

No	Port Location	Existing Berthing Facility Dimension	Water Depth
1	Port of Teluk Bayur	348 x 20 m^2	-10 m
2	Port of Sikakap	127.5 x 8 m^2	-8 m
3	Port of Pulau Baai	126.5 x 20.5	-7 m
4	Port of Gunung Sitoli	153 x 20 m^2	-6 m
5	Port of Sinabang	348 x 20 m^2	-8 m

IOP Conf. Series: Earth and Environmental Science **1464** (2025) 012006 doi:10.1088/1755-1315/1464/1/012006

3. Research Method

The flow chart of this study can be seen from Figure 2.

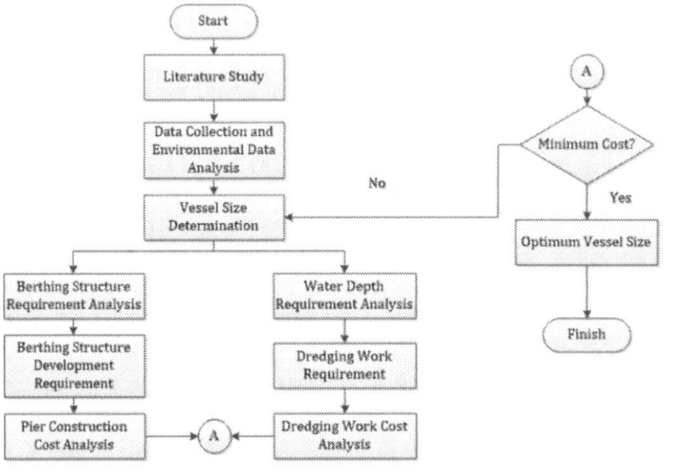

Figure 2. Flow chart of determination of optimum vessel size.

3.1. Berthing structure requirements analysis

The requirements of length of berthing structure/pier is influenced by the design number of berthing vessel to accommodate the demand in a port. The calculation of the requirement of pier length used the following formula.

$$L_p = nLOA \times 1.2 \tag{1}$$

Where L_p is the required length of pier, n is the number of berthing vessels, and LOA is vessel's length overall. The number of berthing vessels determined by the estimated number of vessels that berth in the port and Berth Occupancy Ratio (BOR) value. The number of berthing vessels calculated using the following formula.

$$n = \frac{T_{WTC} \times S_{CS}}{W_D W_H \times BOR} \tag{2}$$

Where T_{WTC} is the duration of loading and unloading, S_{CS} is the total number of berthing vessels in a week, W_D is working day in a week, and W_H is working hours in a day. The standard BOR value is 70%.

3.2. Dredging work requirements analysis

The required water depth to accommodate vessel in a port influenced by the draft of the vessel where the minimum depth required is about 1.1 of the vessel's draft. In case of the existing water depth does not meet the required water depth, dredging work need to be conducted. The calculation of the required dredging depth should also accommodates the sedimentation rate in each port, so the water depth can be used for 5 year design. This condition can be expressed through the following formula.

$$d_{dr} = d - d_{ex} + (Sed \cdot t) \tag{3}$$

Where d_{dr} is dredging depth, d is required water depth, d_{ex} is the existing water depth, Sed is the estimated sedimentation rate, and t is the design year. This minimum water depth shall be accommodated in the channel, basin, and berthing area.

IOP Conf. Series: Earth and Environmental Science **1464** (2025) 012006 doi:10.1088/1755-1315/1464/1/012006

The estimated sedimentation rate is calculated using longshore sediment calculation method that was given in the Shore Protection Manual[9] as can be seen below.

$$Q_l = K\left(\frac{\rho\sqrt{g}}{16\kappa^{\frac{1}{2}}(\rho_s - \rho)(1-n)}\right)H_b^{\frac{5}{2}}\sin(2\alpha_b) \tag{4}$$

Where Q_l is longshore sediment transport, K is given coefficient 0.39, ρ is the density of seawater, κ is the breaker index 0.78, ρ_s is the density of sediment, H_b is the breaking wave height, and α_b is the wave breaking angle. To get the estimated sedimentation rate due to longshore sediment transport, the volume of longshore sediment transport can be divided by the breaking zone area as the area in which the sediment deposited.

$$Sed = \frac{Q_l}{A_b} \tag{5}$$

3.3. Cost analysis

The cost for port development needed due to accommodate design vessel analysis is calculated for the Capital Expenditure, in which in this study is for berthing structure development and the dredging work required. The minimum total cost from this port development to accommodate the design vessel is indicating the optimum vessel used in this analysis. The cost analysis in this study for the berthing structure construction is using this following formula.

$$Cost_{br} = \left((N_B \times 1{,}2L_{OA}) - L_{ex}\right) \times B_q \times UP_{br} \tag{6}$$

Where UP_{br} is the unit price of berthing structure construction. The calculation for dredging work is calculated using this following formula.

$$Cost_{dr} = [\{(1{,}1 \times T) - d_{ex}\} + Sed \cdot t] \times \left[\{L_{ch} \times 4B\} + \{\pi L_{OA}{}^2\} + \{(1{,}25B) \times (N_B \cdot 1{,}5L_{OA})\}\right] \times UP_{dr} \tag{7}$$

Where T is the vessel's draft, L_{ch} is the length of navigational channel, T is the vessel's breadth, and UP_{dr} is the dredging work unit price.

4. Estimation of Longshore Sediment Transport

Environmental data calculated is the estimated longshore sediment in each port. This estimated longshore sediment is used to further analysis of dredging for port development. To calculate the estimated longshore sediment transport, wave height condition is needed. The wave height data was calculated using the hindcasting method given in the Shore Protection Manual[9] using hourly ERA5 wind data acquired from the *European Centre for Medium-Range Weather Forecast* (ECMWF) [10] ranging from 2016 to 2021. The point of wind data acquisition for each port is shown in the table below.

Table 3. Wind data acquisition point.

No	Port Location	Wind Data Acquisition Point
1	Port of Teluk Bayur	1°03'45" S; 100°18'45" E
2	Port of Sikakap	2°45'00" S; 100°15'00" E
3	Port of Pulau Baai	3°52'30" S; 102°15'00" E
4	Port of Gunung Sitoli	1°18'45" N; 97°41'15" E
5	Port of Sinabang	2°33'45" N; 96°26'15" E

The calculation of estimated longshore sediment is calculated for the estimated breaking zone area as the estimated area affected by sedimentation. The estimated breaking zone area for each port in this analysis shown in figure below.

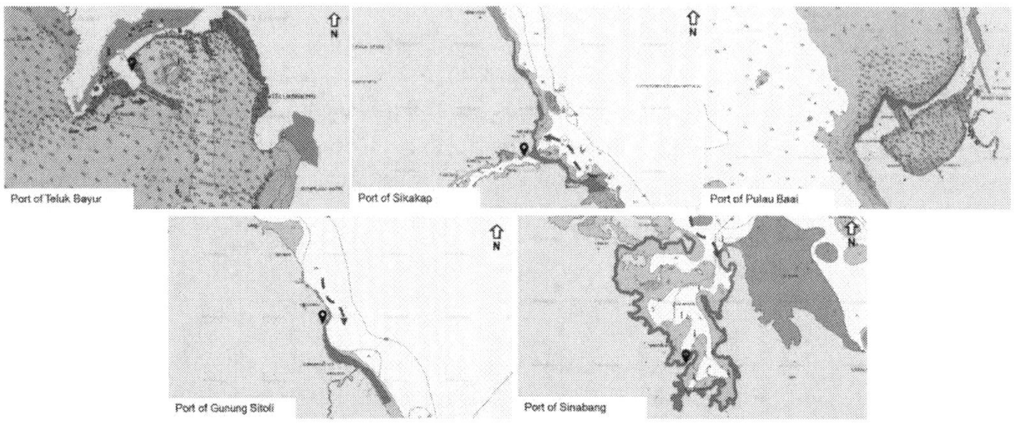

Figure 3. Observed breaking zone area for longshore sediment transport calculations.

The above breaking zone area indicate the area in which the sedimentation occurred in the port area. The results of the estimation of yearly sedimentation rate due to longshore sedimentation in each port can be seen on Table 4.

Table 4. Estimated yearly sedimentation rate due to longshore sediment transport.

No	Port	Estimated Yearly Sedimentation Rate (m/year)
1	Port of Teluk Bayur	0.065
2	Port of Sikakap	0.022
3	Port of Pulau Baai	0.156
4	Port of Gunung Sitoli	0.160
5	Port of Sinabang	0.057

5. Analysis of Port Development

5.1. Container demand

Container demand in port is one of the main factors for requirements for port development. The increase in container demand can affect the increase in ship calls and even the increase in vessel size. In this study, the analysed demand for containers on the T-2 route is using the assumption of 250,000 TEUs/year. The Port of Teluk Bayur as the hub port serves all the container demand, while the demand for the spoke ports is distributed based on the total population in each port's hinterlands. Table 5 shows the container demand for each port location used in this study.

5.2. Design vessel

Design vessels in this analysis is using the existing vessel used in Tol Laut Program, with several additional design vessels taken from OCDI[11]. Table below shows the specifications for container vessels used in this analysis. The specifications for design vessels used in this study shown in Table 6 below.

Table 5. Container demand for each port location.

No	Port	Container Demand (TEUs/year)
1	Port of Teluk Bayur	250,000
2	Port of Sikakap	7,116
3	Port of Pulau Baai	163,678
4	Port of Gunung Sitoli	71,657
5	Port of Sinabang	7,549

Table 6. Specifications of design vessels.

DWT	Capacity (TEUs)	Vessel Specifications (m)		
		Length Overall	Beam	Draft
1,500*	120	74.05	17.20	4.90
3,000*	220	80.00	15.00	5.00
5,000*	370	107.00	18.00	5.00
10,000	900	138.00	22.20	7.90
20,000	1,700	175.00	27.00	10.20
30,000	2,500	201.00	30.30	11.90
40,000	2,800	228.00	31.80	11.90
50,000	4,300	269.00	32.30	12.80

*Existing Tol Laut Program vessel

5.3. Required length of berthing structure

Table 7 shows the required pier length for each vessel designs to accommodate the container demand in each port while Table 8 shows the required pier development in each port. The total pier development required to accommodate each vessel if used for the T-2 Tol Laut route shown in Figure 4 .

Table 7. Required pier length for each vessel designs.

Vessel Capacity (TEUs)	Ship Call (Call/Week)	Required Pier Length (m)				
		Port of Teluk Bayur	Port of Sikakap	Port of Pulau Baai	Port of Gunung Sitoli	Port of Sinabang
120	33	450	90	450	450	90
220	18	480	100	480	480	100
370	11	390	130	390	390	130
900	5	500	170	500	500	170
1,700	3	630	210	630	630	210
2,500	2	490	250	490	490	250
2,800	2	550	280	550	550	280
4,300	1	330	330	330	330	330

Table 8. Required pier length development for each vessel designs.

Vessel Capacity (TEUs)	Ship Call (Call/Week)	Required Pier Length Development (m)				
		Port of Teluk Bayur	Port of Sikakap	Port of Pulau Baai	Port of Gunung Sitoli	Port of Sinabang
120	33	102.0	0.0	323.5	297.0	0.0
220	18	132.0	0.0	353.5	327.0	0.0
370	11	42.0	2.5	263.5	237.0	8.4
900	5	152.0	42.5	373.5	347.0	48.4
1,700	3	282.0	82.5	503.5	477.0	88.4
2,500	2	142.0	122.5	363.5	337.0	128.4
2,800	2	202.0	152.5	423.5	397.0	158.4
4,300	1	0.0	202.5	203.5	177.0	208.4

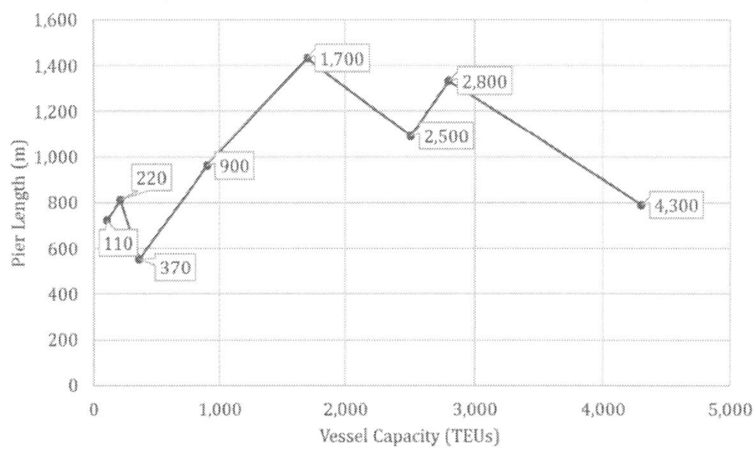

Figure 4. Required pier length development for each vessel designs.

From the calculations above, it can be seen that container vessel with the capacity of 370 TEUs is the vessel design that required the minimum volume of pier development.

5.4. Required dredging work

The volume of required dredging work to accommodate each vessel designs in each port can be seen in Table 9 below with the total volume of dredging work for each vessel shown by Figure 5.

Table 9. Required dredging volume for each vessel designs.

Vessel Capacity (TEUs)	Ship Call (Call/Week)	Required Dredging Volume (m^3)				
		Port of Teluk Bayur	Port of Sikakap	Port of Pulau Baai	Port of Gunung Sitoli	Port of Sinabang
120	33	0,00	0,00	0,00	0,00	0,00
220	18	0,00	0,00	0,00	0,00	0,00
370	11	0,00	0,00	0,00	0,00	0,00
900	5	0,00	17.232,75	671.293,68	51.698,25	5.744,25
1,700	3	245.578,80	245.578,80	3.129.915,00	159.468,75	1.728.071,33
2,500	2	1.354.851,64	599.048,44	5.283.547,57	182.709,00	3.854.333,30
2,800	2	1.555.225,41	762.006,21	5.778.387,67	217.512,00	4.205.329,76
4,300	1	2.426.637,29	1.283.265,74	7.216.186,39	146.621,81	5.677.011,90

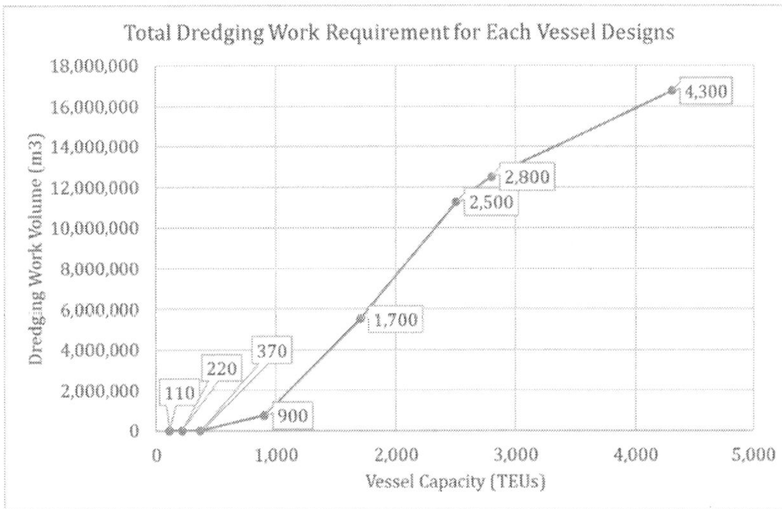

Figure 5. Total dredging work for each vessel designs.

From the calculations, it can be seen that container vessel with the capacity of 110 to 370 TEUs is the vessel design that required the minimum volume of dredging work.

6. Analysis of Capital Expenditure of Port Development

Port development is closely related to the Capital Expenditure (CAPEX) requirements. The CAPEX of port development in this study determined by the construction cost for the berthing facility requirement and the cost of dredging work that need to be conducted to accommodate the design vessel. The calculation of CAPEX in this study used the unit price at the Ministry of Transportation of Indonesia[12] with the escalation using inflation location factor. The total required CAPEX for port development in Tol Laut T-2 route for each vessel designs can be seen in Table 10 and Figure 6.

Table 10. Port development CAPEX for each vessel designs.

Vessel Capacity (TEUs)	Construction Week (Billion IDR)		Total CAPEX
	Construction of Pier	Dredging Work	
120	287.60	0.00	**287.60**
220	324.86	0.00	**324.86**
370	216.09	0.00	**216.09**
900	372.47	40.76	**413.23**
1,700	553.70	313.39	**867.09**
2,500	399.55	644.66	**1.044.21**
2,800	488.89	717.21	**1.206.10**
4,300	248.72	963.85	**1.212.57**

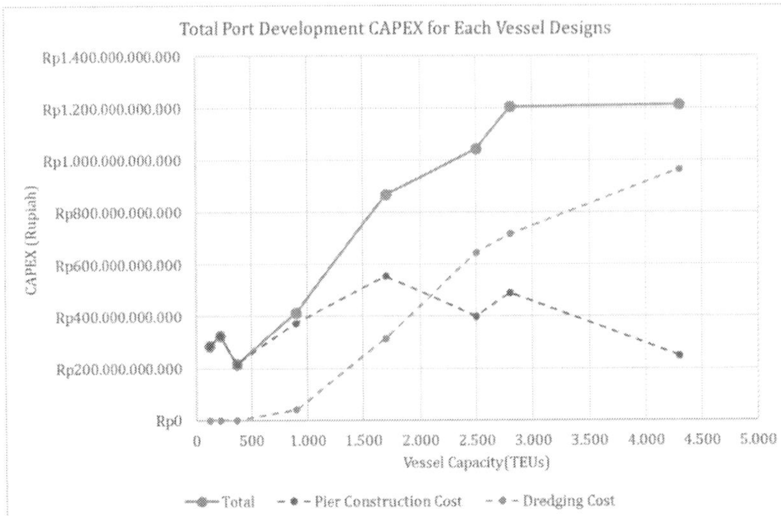

Figure 6. Total port development CAPEX for each vessel designs.

From the calculations above, it can be seen that container vessel with the capacity of 370 TEUs is the vessel design that required the minimum CAPEX of port development to accommodate the container demand in the route.

7. Result and Discussion

This study shows the relationship between the requirements of port development for each vessel designs. On the reviewed Tol Laut T-2 route, a designed demand of 250,000 TEUs can be accommodated with the optimal container vessel design of 370 TEUs capacity or 5,000 DWT. The design vessel is considered optimal based on the least CAPEX cost requirements compared to the other vessel design.

This relationship can further be used to determine the optimal vessel design in relation with port development to accommodate container transportation in a network or route for a designed or planned container demand. Further study can be developed to determine the most optimal vessel design by considering other port operational variables such as port hydro-oceanographic conditions, port equipment needs, land facility needs, port operational plans, and vessel operational plans.

References

[1] Hardjono S 2018 Identifikasi Rasio Dimensi Utama Kapal Kontainer Kelas Small Feeder Untuk Toll Laut Indonesia *Warta Penelitian Perhubungan* **28**

[2] Muin M, Sihombing R, Hermawan M W, Muslim A B and Alam H N 2022 Role of Non-Orthogonal Hydrodynamic and Sedimentation Model for Port Facilities Analysis on Port of Kuala Tanjung *IOP Conf Ser Earth Environ Sci* **1065** 012032

[3] Imanuel Pau D 2015 Optimasi Kebutuhan Dermaga Pada Pelabuhan Maumere **1** 60–4

[4] Erftemeijer P L A and Robin Lewis R R 2006 Environmental impacts of dredging on seagrasses: A review *Mar Pollut Bull* **52** 1553–72

[5] Musso E, Ferrari C and Benacchio M 2006 Port Investment: Profitability, Economic Impact and Financing *Research in Transportation Economics* **16** 171–218

IOP Conf. Series: Earth and Environmental Science **1464** (2025) 012006 doi:10.1088/1755-1315/1464/1/012006

[6] Kementerian Perhubungan Republik Indonesia 2021 *Keputusan Direktur Jenderal Perhubungan Laut KP.199/DJPL/2021 tentang Penetapan Jaringan Trayek Penyelenggaraan Kewajiban Pelayanan Publik Untuk Angkutan Barang di Laut Tahun Anggaran 2021* (Kementerian Perhubungan Indonesia)
[7] Kementerian Perhubungan Republik Indonesia 2017 *Lampiran Rencana Induk Pelabuhan Nasional KP 432 Tahun 2017*
[8] Yuwono N 2022 *Evaluasi dan Optimasi Penyelenggaraan Tol Laut Pada Masa yang Akan Datang* (Yogyakarta)
[9] Departement of The Army Coastal Engineering Research Centeer 1984 *Shore Protection Manual* vol I (Washington D.C.: U.S. Government Printing Office)
[10] Copernicus EU and European Centre for Medium-Range Weather Forecast 2022 ERA5 hourly data on single levels from 1940 to present
[11] Committee for Propagating Japanese Technical Standards Abroad 2020 *Technical Standards and Commentaries For Port And Harbour Facilities In Japan* (Tokyo: The Overseas Coastal Area Debelopment Institute of Japan)
[12] Kementerian Perhubungan Republik Indonesia 2013 *PM 75 Tahun 2013 tentang Standar Biaya Tahun 2014 di Lingkungan Kementerian Perhubungan* (Kementerian Perhubungan Indonesia)

IOP Conf. Series: Earth and Environmental Science **1464** (2025) 012007 doi:10.1088/1755-1315/1464/1/012007

Wave shoaling and its interaction with mangrove

N A Christien[1,2] **and I Magdalena**[1,3,4*]

[1]Center for Coastal and Marine Development, Institut Teknologi Bandung, Bandung, Indonesia

[2]Climate Resilience Team, PT Witteveen Bos Indonesia, Jakarta, Indonesia

[3]Faculty of Mathematics and Natural Sciences, Institut Teknologi Bandung, Bandung, Indonesia

[4]Center for Mathematical Modelling and Simulations, Institut Teknologi Bandung, Bandung, Indonesia

*E-mail: ikha.magdalena@itb.ac.id

Abstract. This study investigates wave shoaling phenomena and its interaction with mangroves. We propose a numerical model based on shallow water equations, consisting of two equations: mass conservation and momentum balance. We will modify the equations to include the interaction of waves with mangroves. A staggered finite volume approach is implemented to numerically determine the reduced wave shoaling coefficient. For validation, we compare the numerical results with analytical solutions. This study provides insights into optimizing strategies for sustainable coastal defense.

1 Introduction

Wave shoaling refers to the increase in wave height as waves move into shallower waters, a phenomenon governed by the shallow water equations (SWE) [1, 2]. These hyperbolic partial differential equations describe fluid motion under certain topographical conditions, where the vertical scale of the flow is relatively small compared to the horizontal scale [3]. For simplified analysis, the linearized shallow water equations (LSWE) are often used, offering an effective framework for modeling coastal processes [4]. LSWE are particularly relevant for studying the dynamic interactions between waves and coastal features, including the protective role of natural barriers like mangroves [5].

Numerical methods such as the finite difference method, Riemann solvers, and the finite volume method have been developed to simulate these equations [6, 7]. Among these, the finite volume method is well-regarded for its robustness, particularly in coastal modeling [8]. This method is closely associated with the leapfrog method, which employs a staggered grid—a technique crucial to accurately capture the complexities of wave behavior [9]. The staggered grid technique has proven to be effective in various fluid dynamics applications, including weather prediction models [10] and oceanographic simulations [11].

In coastal regions, mangroves play a significant role in mitigating wave impacts, including shoaling [12, 13]. Their dense root systems and above-ground structures interact with incoming waves, dissipating energy and reducing wave height [14]. This wave attenuation capacity of mangroves has been studied in various geographic contexts, demonstrating their importance in coastal protection strategies [15, 16]. In this paper, we study the interaction between wave shoaling and mangroves by applying an alternative staggered grid to solve the one-dimensional (1-D) LSWE. We compare analytical and numerical results from simulations that include both wave attenuation and shoaling phenomena, specifically assessing how mangroves influence these processes.

Section 1 provides the motivation and objectives of this study. Section 2 details the formulation of the 1-D LSWE and the simulation settings. In Section 3, we derive the analytical solution, focusing on the reduced shoaling coefficient, which quantifies the reduction in wave shoaling [17]. Section 4 introduces the proposed numerical method, utilizing the new staggered grid to solve the equations. We then simulate wave attenuation and reduced shoaling in the presence of mangroves in Section 5, comparing the results with analytical solutions. The paper concludes in Section 6 with recommendations for future research on the interaction between wave dynamics and coastal ecosystems.

2 Mathematical Model

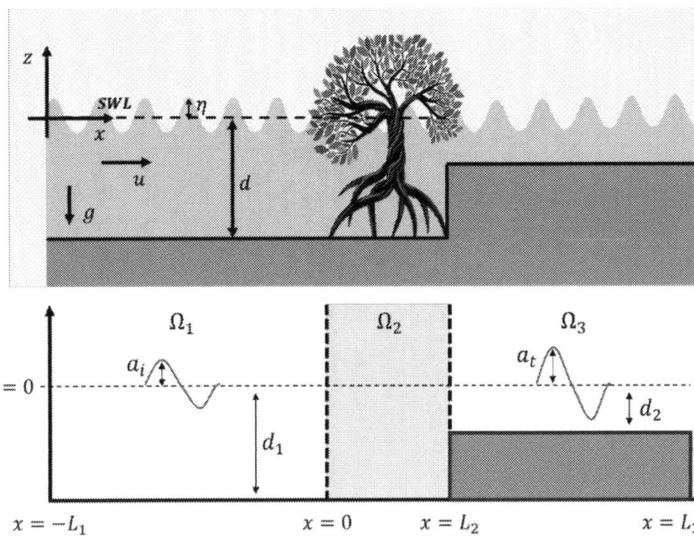

Figure 1: Sketch of the 1-D Observation Domain

Consider a basin where x and t respectively represent the spatial and temporal variables as shown in Figure 1. The basin has a depth $d(x)$, which remains constant over time and is defined by the following equation:

$$d(x) = \begin{cases} d_1 & ,x \in \Omega_1 \cup \Omega_2 \\ d_2 & ,x \in \Omega_3, \end{cases}$$

The 1-D LSWE with a smooth bottom can be described as follows

$$\eta_t + (hu)_x = 0, \tag{1}$$

$$u_t + g\eta_x = 0. \tag{2}$$

In this context, $\eta(x,t)$ represents the wave height, $h(x,t)$ denotes the water thickness, $u(x,t)$ signifies the horizontal water wave velocity, and g stands for the gravitational acceleration. The total water thickness, defined as $h(x,t) = \eta(x,t) + d(x)$, can be approximated as $h(x,t) \approx d(x)$, assuming the vertical flow is negligible compared to the horizontal domain of the motion.

Now, consider the presence of a porous structure that attenuates the water waves. The momentum equation, as given in Equation (2), is modified by introducing a friction term, $c_f\omega u$, where c_f and ω respectively denote the friction coefficient and the wave frequency. This friction term allows the evaluation of wave attenuation due to friction by vegetation, utilizing the periodic wave characteristics [18]. Consequently, the momentum equation for each domain is adjusted accordingly:

$$u_t + g\eta_t + F_f\omega u = 0 \tag{3}$$

where

$$F_f = \begin{cases} 0 & ,x \in \Omega_1 \cup \Omega_3 \\ c_f & ,x \in \Omega_2 \end{cases} \tag{4}$$

and

$$\Omega_1 = \{x \in \mathbb{R} \mid -L_1 \leq x < 0\}$$
$$\Omega_2 = \{x \in \mathbb{R} \mid 0 \leq x < L_2\}$$
$$\Omega_3 = \{x \in \mathbb{R} \mid L_2 \leq x \leq L_3\}.$$

3 Analytical Solution

In this section, Equations (1) and (3) are solved analytically to obtain the reduced shoaling coefficient K_{rs}. It is defined as

$$K_{rs} = \left|\frac{a_t}{a_i}\right| \tag{5}$$

with a_i denotes the initial amplitude of wave which enters region Ω_1 and a_t denotes the transmitted wave amplitude in region Ω_3.

Firstly, Equations (1) and (3) can be reduced to an ordinary differential equation. We let η and u follow the ansatz:

$$\eta(x,t) = F(x)e^{i\omega t} \tag{6}$$
$$u(x,t) = G(x)e^{i\omega t} \tag{7}$$

which represents an oscillating monochromatic wave with an angular frequency of ω. By substituting Equations (6)–(7) to Equations (1) and (3), we obtain

$$i\omega F(x) + d_x G(x) + dG_x(x) = 0 \tag{8}$$
$$G(x) = \frac{-g}{i\omega + c_f\omega}F_x(x). \tag{9}$$

By substituting Equation (9) to Equation (8) and noting that the depth values are constant in each region, we obtain the ordinary equation:

$$F_{xx}(x) + \frac{\omega^2(1-ic_f)}{dg} = 0 \tag{10}$$

which has the general solution

$$F(x) = C_1e^{-ikx} + C_2e^{ikx} \tag{11}$$

with $k^2 = \frac{\omega^2(1-ic_f)}{dg}$. Substituting Equation (11) to (9) gives the general solution for $G(x)$, that is

$$G(x) = \frac{\omega}{kd(x)}\left(C_1e^{-ikx} - C_2e^{ikx}\right).$$

By using the solution of $F(x)$ and $G(x)$,the solution for η and u over each region will be derived. For the first region, a particular solution is chosen by letting a_i and a_r as the initial and the reflected amplitude of the wave. Thus, the solutions for domain Ω_1 are

$$\eta_1(x,t) = a_ie^{-i(k_1x-\omega t)} + a_re^{i(k_1x+\omega t)} \tag{12}$$
$$u_1(x,t) = \frac{\omega}{k_1d_1}\left(a_ie^{-i(k_1x-\omega t)} - a_re^{i(k_1x+\omega t)}\right) \tag{13}$$

with $k_1 = \dfrac{\omega}{\sqrt{d_1g}}$.

Next, the solutions over domain Ω_2 are

$$\eta_2(x,t) = C_3e^{-i(k_2x-\omega t)} + C_4e^{i(k_2x+\omega t)} \tag{14}$$
$$u_2(x,t) = \frac{\omega}{k_2d_1}\left(C_3e^{-i(k_2x-\omega t)} - C_4e^{i(k_2x+\omega t)}\right) \tag{15}$$

with $k_2 = \omega\sqrt{\dfrac{1-ic_f}{d_1g}}$ and C_3, C_4 as undetermined coefficient.

IOP Conf. Series: Earth and Environmental Science **1464** (2025) 012007 doi:10.1088/1755-1315/1464/1/012007

For the final region, we assume that the domain is connected to a shore which can completely absorb the wave. We let a_t as the transmitted wave amplitude in the final region. Therefore, the solutions over domain Ω_3 are

$$\eta_3(x,t) = a_t e^{-i(k_2(x-L_2)-\omega t)} \tag{16}$$
$$u_3(x,t) = \frac{\omega}{k_3 d_2}\left(a_t e^{-i(k_2(x-L_2)-\omega t)}\right) \tag{17}$$

with $k_3 = \dfrac{\omega}{\sqrt{d_2 g}}$.

Before the solutions are used to obtain the reduced shoaling coefficient, the previous solutions need to follow flow continuity which governs

$$\eta_1(0^-,t) = \eta_2(0^+,t), \tag{18}$$
$$u_1(0^-,t) = u_2(0^+,t), \tag{19}$$
$$\eta_2(L_2^-,t) = \eta_3(L_2^+,t), \tag{20}$$
$$d_1 u_2(L_2^-,t) = d_2 u_3(L_2^+,t). \tag{21}$$

By substituting Equations (18)–(21), we obtain

$$a_i + a_r = C_3 + C_4, \tag{22}$$
$$\frac{k_2}{k_1}(a_i - a_r) = C_3 - C_4, \tag{23}$$
$$C_3 e^{-ik_2 L_2} + C_4 e^{ik_2 L_2} = a_t, \tag{24}$$
$$\frac{k_3}{k_2}(C_3 e^{-ik_2 L_2} - C_4, e^{ik_2 L_2}) = a_t. \tag{25}$$

Finally, by solving the system of Equations (22)–(25), the reduced shoaling coefficient is obtained from:

$$\frac{a_t}{a_i} = 0.5(1+\alpha+(1-\alpha)\gamma)e^{-ik_2L_2} + 0.5\left(1-\alpha+(1+\alpha)\gamma\right)e^{ik_2L_2} \tag{26}$$

with

$$\alpha = \sqrt{1-ic_f},$$
$$\beta = \sqrt{\frac{d_1}{d_2(1-ic_f)}},$$
$$\gamma = \frac{-(1+\alpha)(1-\beta)e^{-ik_2L_2} - (1-\alpha)(1+\beta)e^{ik_2L_2}}{(1-\alpha)(1-\beta)e^{-ik_2L_2} + (1+\alpha)(1+\beta)e^{ik_2L_2}}.$$

From Equation (26) we can see that the reduced shoaling coefficient is dependent only to c_f, L_2, and $\frac{d_1}{d_2}$. In other words, it is affected only by the properties of the mangroves, that is its friction coefficient and length, and also the ratio between different basin's depth.

4 Numerical Method

In this section, the finite method on a new staggered grid is proposed to solve the LSWE. We use the first order of upwind approximation and centered dicretization to compute the numerical fluxes.

Consider the observation domain to be $\Omega = [0, L]$ and the time interval $t = [0, T]$. The spatial domain will be partitioned into $N_x = \lfloor \frac{L}{\Delta x} \rfloor + 1$ grid points with interval Δx. Similarly, the time interval is partitioned into $N_t = \lfloor \frac{T}{\Delta t} \rfloor + 1$ time steps with the step size of Δt. The spatial domain is discretized to $0 = x_1, x_{3/2}, x_2, ..., x_{N_x}, x_{N_x+1/2} = L$ as seen in Figure 2.

The mass conservation governed by Equation (1) is evaluated in cells centered at x_j, while the momentum balance governed by Equation (3) is evaluated in cells centered at $x_{j+1/2}$. This means that the wave height η is evaluated at full grid x_j and the wave velocity u is evaluated at half grid $x_{j+1/2}$. By applying the forward time central space (FTCS) method, the numerical scheme is written as

$$\eta_j^{n+1} = \eta_j^n - \frac{\Delta t}{\Delta x}(q_{j+1/2}^n - q_{j-1/2}^n) \tag{27}$$

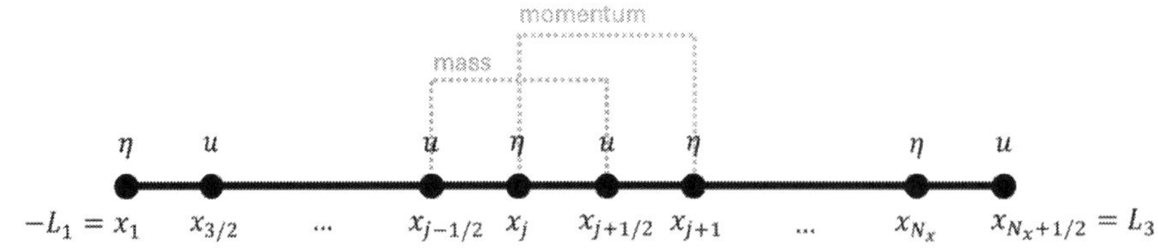

Figure 2: Staggered Grid Scheme

for $j = 2, 3, ..., N_x$ and

$$u_{j+1/2}^{n+1} = \frac{u_{j+1/2}^{n} - g\frac{\Delta t}{\Delta x}(\eta_{j+1}^{n+1} - \eta_{j}^{n+1})}{1 + \Delta t c_f \omega} \tag{28}$$

for $j = 1, 2, ..., N_x - 1$ with $n = 1, 2, ..., N_t$ and $q_{j+1/2}^n = {}^*h_{j+1/2}^n u_{j+1/2}^n$. The ${}^*h_{j+1/2}$ represents the approximation of the water thickness values–which is related closely to the wave height η–in the half grid. With the upwind method, it is approximated as

$${}^*h_{j+1/2}^n = \begin{cases} h_j^n & , u_{j+1/2}^n \geq 0 \\ h_{j+1}^n & , u_{j+1/2}^n < 0 \end{cases} \tag{29}$$

where the value follows the wave motion's direction.

5 Results and Discussion

5.1 *Simulating Wave Attenuation and Shoaling Phenomena*

In this study, we simulated wave attenuation and shoaling phenomena in a spatial domain defined by $-L_1 < x < L_3$ with parameters $L_1 = 5$, $L_2 = 5$, and $L_3 = 10$. The initial condition for the simulation was a still water state, given by

$$\eta(x, 0) = u(x, 0) = 0 \tag{30}$$

for all $x \in \Omega$, where $\eta(x, t)$ represents the water surface elevation and $u(x, t)$ is the velocity. To generate a periodic wave input with amplitude of 0.1 at the left boundary, we applied

$$\eta(-L_1, t) = 0.1 \sin(\omega t), \tag{31}$$

with $\omega = 2\pi$. For the right boundary, the absorbing condition was implemented as

$$u(L_3, t) = \sqrt{\frac{g}{d_2}}\eta(L_3, t), \tag{32}$$

to ensure that waves leaving the domain were effectively absorbed. To utilize the numerical schemes, we used $\Delta x = 0.01$ and $\Delta t = \dfrac{\Delta x}{g \max(d_1, d_2)}$. The simulation is run for $T = 2.5$.

In the first simulation, we investigated wave attenuation due to friction in the second region. The depth of the water was kept uniform across the entire domain, with $d_1 = d_2 = 5$. We introduced friction with a coefficient of $c_f = 0.25$ in the second region.

Figure 3 illustrates the effect of friction on wave attenuation. It can be seen that the wave amplitude arriving at the third region is reduced to 0.06. This result confirms that the simulation effectively captures wave attenuation due to friction.

For the second simulation, we focused on wave shoaling. The friction coefficient was set to $c_f = 0$, with depths $d_1 = 10$ and $d_2 = 8$. Shoaling describes the phenomenon where wave height increases as waves move from deeper to shallower water. By setting these parameters, we aimed to observe how waves steepen and grow in height as they encounter shallower regions.

The shoaling phenomena is illustrated in Figure 4. When there is no friction, the wave amplitude increases to 0.11–confirming that the shoaling phenomena has been captured. Additionally, when mangrove is introduced by setting $c_f = 0.03$, it can be seen that the wave amplitude stays as 0.10 when reaching the last region of simulation. This indicates that the simulation has captured the reduction of shoaling effect by mangrove.

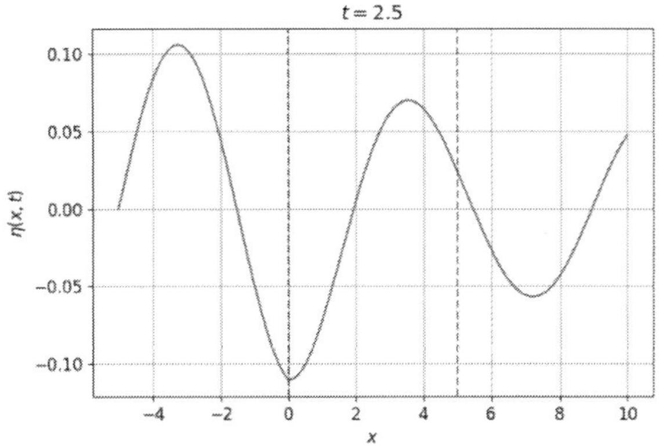

Figure 3: Wave Attenuation with $c_f = 0.25$

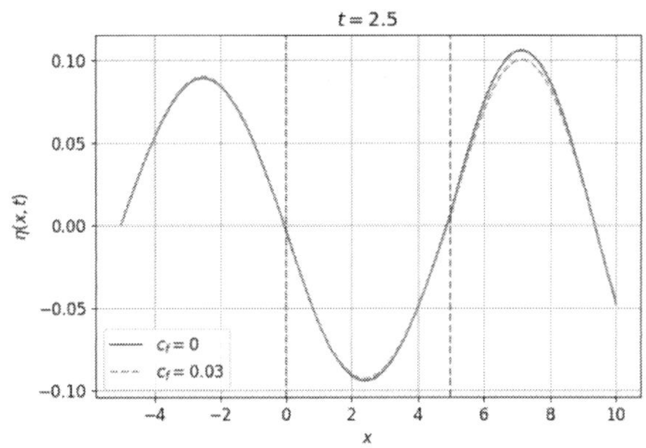

Figure 4: Shoaling Phenomena with $c_f = 0$ and Reduced Shoaling Phenomena with $c_f = 0.03$

5.2 Comparing Numerical and Analytical Results

In this section, we perform a detailed comparison between the numerical results obtained from our simulations and the analytical solution for the reduced shoaling coefficient, as derived in Section 3. The goal is to assess the accuracy and reliability of the numerical model in capturing key wave phenomena. We keep the spatial and temporal domain as defined in the Subsection 5.1 with small modification. We set $L_3 = 12$ and keep the basin's depth with $d_1 = 10$ and $d_2 = 8$.

Firstly, we investigate the impact of varying the friction coefficient, c_f, on wave attenuation. Figure 5 presents a comparison between the numerical results and the analytical solution across different values of c_f. The Normalized Root Mean Square Error (N-RMSE) calculated for these results is 3.2%, indicating a good agreement between the numerical and analytical approaches. The figure clearly shows that as the friction coefficient increases, the wave amplitude decreases. This trend is consistent with the physical expectation that higher friction leads to greater energy dissipation within the wave, thereby reducing its amplitude as it propagates through the medium.

Next, we explore the influence of varying the length of the porous media domain, denoted as L_2, on wave transmission. This investigation is crucial for understanding how the extent of the porous region affects the overall behavior of wave propagation. In Figure 5 (Right), we compare the numerical results with the analytical solution for different values of L_2. The N-RMSE for this comparison is calculated to be 1.4%, which further supports the validity of our numerical model. As the length L_2 increases, the interaction between the wave and the porous media is prolonged, leading to a more significant attenuation effect. This is reflected in the reduced wave amplitude observed in the numerical results, which aligns

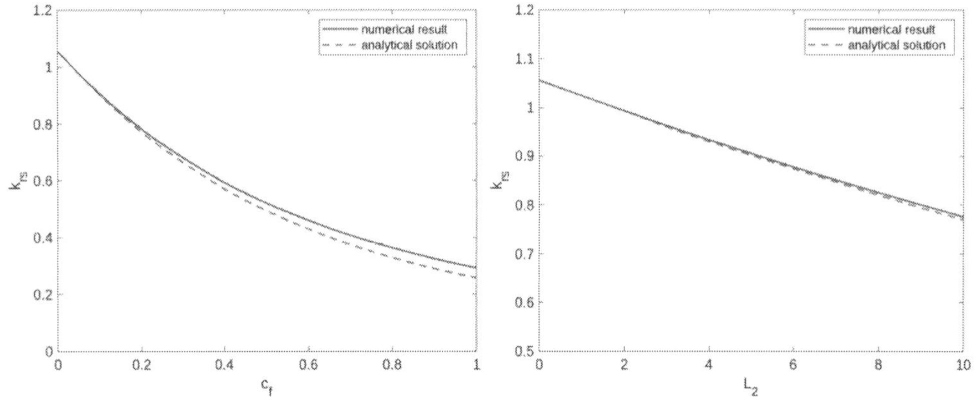

Figure 5: Comparison of Numerical and Analytical Results for Different Values of c_f (left) and L_2 (right)

closely with the predictions from the analytical model.

The consistency between the numerical and analytical results in these simulations demonstrates the effectiveness of our proposed numerical method in accurately modeling wave attenuation and transmission in porous media. These findings are essential for the design and optimization of coastal protection structures, where understanding the interaction between waves and porous materials is crucial.

6 Conclusion

This paper has explored the interaction between wave shoaling and mangroves using the linearized shallow water equations (LSWE) in a one-dimensional domain. By employing an alternative staggered grid method, we were able to accurately model and simulate wave attenuation and shoaling phenomena in coastal regions where mangroves are present. The results demonstrate that mangroves significantly influence wave dynamics, effectively reducing wave heights and mitigating the impacts of shoaling. The comparison between analytical and numerical solutions highlighted the effectiveness of the proposed grid method in capturing the complexities of wave-mangrove interactions. The reduced shoaling coefficient derived from the analytical model provided valuable insights into the extent of wave energy dissipation by mangrove ecosystems.

The findings underscore the importance of integrating natural barriers like mangroves into coastal protection strategies. As climate change continues to intensify coastal hazards, the role of mangroves in enhancing coastal resilience becomes increasingly critical. This study lays the groundwork for future research that can further refine numerical models and expand our understanding of how coastal ecosystems interact with wave dynamics. In conclusion, the study not only advances the application of staggered grids in coastal modeling but also contributes to the broader discourse on sustainable coastal management. Future work may involve extending this approach to two-dimensional domains, exploring more complex topographies, and incorporating additional ecological factors to develop comprehensive models that support the conservation and restoration of mangrove ecosystems.

Contribution Statement

N A Christien: Data Curation (equal), Formal Analysis (equal), Visualization (lead), Validation and Investigation (equal), Writing – Original Draft Preparation. **I Magdalena:** Conceptualization (lead), Supervision (lead), Writing – Review & Editing (equal), Data Curation (equal), Formal Analysis (equal), Visualization (lead), Funding acquisition (lead), Project administration (lead), Methodology (lead), Validation and Investigation (equal).

References

[1] Chiang C. Mei, Michael Stiassnie, and D. K. P. Yue. *Theory and applications of ocean surface waves.* World Scientific, 2005.

[2] Maarten W. Dingemans. *Water wave propagation over uneven bottoms.* World Scientific, 1997.

[3] Michael B. Abbott. *Computational hydraulics.* Pitman Publishing, 1979.

[4] Robert G. Dean and Robert A. Dalrymple. *Water wave mechanics for engineers and scientists.* World Scientific, 1991.

[5] Tien Quang Bao. Effect of mangrove forest structures on wave attenuation in coastal vietnam. *Oceanologia*, 53(3):807–818, 2011.

[6] Randall J. LeVeque. *Finite volume methods for hyperbolic problems.* Cambridge University Press, 2002.

[7] Eleuterio F. Toro. *Shock-capturing methods for free-surface shallow flows.* John Wiley & Sons, 2001.

[8] Ami Harten, Peter D. Lax, and Bram van Leer. On upstream differencing and godunov-type schemes for hyperbolic conservation laws. *SIAM Review*, 25(1):35–61, 1983.

[9] Akio Arakawa and V. R. Lamb. Computational design of the basic dynamical processes of the ucla general circulation model. In *Methods in Computational Physics*, volume 17, pages 173–265. Elsevier, 1977.

[10] Fedor Mesinger and Akio Arakawa. Numerical methods used in atmospheric models. Technical report, GARP Publication Series, 1974.

[11] Alan F. Blumberg and George L. Mellor. A description of a three-dimensional coastal ocean circulation model. In *Three-dimensional coastal ocean models*, volume 4, pages 1–16. American Geophysical Union, 1987.

[12] Sjaak Quartel, Aart Kroon, Paul G. E. F. Augustinus, Peter van Santen, and Nguyen Hieu Tri. Wave attenuation in coastal mangroves in the red river delta, vietnam. *Journal of Asian Earth Sciences*, 29(4):576–584, 2007.

[13] Yoshihiro Mazda, Masahiro Magi, Makoto Kogo, and Phan Nguyen Hong. Mangroves as a coastal protection from waves in the tong king delta, vietnam. *Mangroves and Salt Marshes*, 1(2):127–135, 2006.

[14] Stanislaw R. Massel, Kazuaki Furukawa, and Rob M. Brinkman. Surface wave propagation in mangrove forests. *Fluid Dynamics Research*, 24(4):219–249, 1999.

[15] Keryn B. Gedan, Matthew L. Kirwan, Eric Wolanski, Edward B. Barbier, and Brian R. Silliman. The present and future role of coastal wetland vegetation in protecting shorelines: answering recent challenges to the paradigm. *Climatic Change*, 106(1):7–29, 2011.

[16] Siddharth Narayan, Michael W. Beck, Patrick Wilson, Curtis J. Thomas, Anantha Guerrero, Christine C. Shepard, and Kelly A. Burks-Copes. The effectiveness, costs and coastal protection benefits of natural and nature-based defences. *PLoS One*, 11(5):1–17, 2016.

[17] Wiktor Skrzypiński and Stanislaw R. Massel. Modeling wave attenuation in mangrove forests as a new approach for effective management of coastal zones. *Oceanologia*, 59(4):422–430, 2017.

[18] Ikha Magdalena, Vivianne Kusnowo, Moh Ivan Azis, and Widowati. 1D–2D numerical model for wave attenuation by mangroves as a porous structure. *Computation*, 9(6):66, 2021.

IOP Conf. Series: Earth and Environmental Science **1464** (2025) 012008 doi:10.1088/1755-1315/1464/1/012008

Experimental and numerical evaluation of tsunami wave reduction through the implementation of sea dikes and trenches

Hany Qoshirotur Rif'atin[1,2,*], Ikha Magdalena[1,3,4], Widjo Kongko[5], Mohammad Farid[1,2]

[1]Center for Coastal and Marine Development, Bandung Institute of Technology, Bandung, Indonesia
[2]Faculty of Civil and Environmental Engineering, Bandung Institute of Technology, Bandung, Indonesia
[3]Faculty of Mathematical and Natural Sciences, Bandung Institute of Technology, Bandung, Indonesia
[4]Center for Mathematical Modelling and Simulations, Bandung Institute of Technology, Bandung, Indonesia
[5]National Research and Innovation Agency, Yogyakarta, Indonesia

*Corresponding author: Hany Qoshirotur Rif'atin (hany.qoshirotur@gmail.com)

Abstract. A tsunami is a sequence of ocean waves that can occasionally cause water to surge onto land at heights of more than 100 feet (30.5 meters). When these water walls break ashore, they have the potential to do extensive damage to the surrounding areas. Therefore, mitigation systems are required to at least minimize the damage due to tsunami waves in the surrounding coastal areas. One of the most commonly implemented tsunami mitigation systems is building a coastal protection structure (e.g., breakwater, seawalls, sea dike, trench, revetment, etc.). In this study, we investigate the implementation of a coastal protection system in the form of a sea dike, a trench, and their combination, in which all structures are located onshore. This study is conducted using a mathematical model based on the Nonlinear Shallow Water Equations. The model is then solved numerically using the Finite Volume Method on a staggered grid. To validate the proposed model and method, the simulation results are compared to experimental data in which the initial wave is reproduced based on the characteristics of the real 2004 Indian Ocean's tsunami wave. The main goal of this study is to determine the optimal combination of the two structures (sea dike and trench) such that the resulted tsunami wave reduction is maximum.

1 Introduction

Tsunamis are catastrophic natural events characterized by a series of ocean waves lasting minutes to hours and reaching heights of tens of metres [1]. In contrast to regular waves, which naturally are driven by wind or tides, tsunamis are triggered by an extensive and unexpected disruption of the sea surface, which is typically initiated by an submarine earthquake and landslide, or a volcanic eruption [2]. Once the tsunami strikes, it might come onto the shore like a rapidly rising flood and hit with deadly intensity. In human history, the 2004 tsunami wave centered at the Indian Ocean was among the greatest natural disasters, killing or missing at least 230,000 people throughout 14 surrounding countries [3]. At 2024, an earthquake beneath Japan's Noto Peninsula triggered the first large tsunami warning in Japan since the Tohoku earthquake in 2011, measuring a 6.58 m tsunami along the Sea of Japan coastline [4]. Indonesia's

Meteorology, Climatology, and Geophysics Agency recently issued a warning on probable megathrust earthquakes to occur in the Sunda Straits and Mentawai-Siberut Straits that could result in another tsunami occurrence [5]. This comes after the 7.1 magnitude earthquake that devastated Japan on August 8, 2024. Therefore, studies regarding tsunami wave and its effective mitigation strategies are crucial to reduce the impact of such waves on coastal communities.

In the effort to mitigate tsunami wave attacks, studies regarding the wave itself, including its characteristics and evolution, have been conducted [6, 7]. In later studies, these have became the basis to find and evaluate possible solutions to mitigate the event. One of the solutions studied or proposed the most is implementing coastal protection structures, including breakwaters [8, 9], seawalls or sea dikes [10, 11], trenches [12, 13], and revetments [14, 11], have been commonly investigated. These structures have been proven to provide protections to coastal areas to some degree when implemented individually. However, combinations of these structures as coastal protections were rarely studied, while it is possible that combining them may enhance their effectiveness. Therefore, in this study, we are interested in investigating whether or not a combination between two structures, particularly a sea dike and a trench (both positioned onshore), is able to reduce tsunami wave better than each structure independently.

In examining the effectivity of a combined protection system of sea dikes and trenches in mitigating tsunami's impact, we applied two approaches, which are experimental and numerical approaches. The experiments were carried out at the Laboratory of Coastal Dynamics, National Research and Innovation Agency, Yogyakarta, Indonesia. For the experiments, the scaled-down 2004 Indian Ocean tsunami wave was used. The computational model is developed by solving the Nonlinear Shallow Water Equations (NSWEs) utilizing the staggered Finite Volume Method (FVM). This numerical method has been used previously and considered a robust and relatively accurate method to simulate other wave occurrences [15, 16, 17, 18, 19, 20]. To validate the developed numerical model, its results are then compared to the data measured in the mentioned experiments. This validation ensures that the model accurately represents real-world scenarios and provides reliable predictions. One of the main goals of this research is to examine the wave reduction rate due to its interaction with a trench, a sea dike, and their combination. Further, this information is used to determine the optimal combination of the structures (only trench, only sea dike, or their combination) to maximize tsunami wave attenuation and minimize potential damage. By analyzing the synergistic effect of these structures, this research aims to advance coastal protection strategies and contribute to more resilient coastal infrastructure.

2 Mathematical Model

Here, we explain the mathematical model utilized to simulate tsunami wave amplitude reduction by sea dikes and trenches, which is the Nonlinear Shallow Water Equations (NSWEs). It is well-known that NSWEs are quite capable in simulating nonlinear phenomena effectively, as mentioned in [21, 22, 23]. Incorporating the mass conservation equation and momentum equation, the NSWEs are written respectively as:

$$\eta_t + (hu)_x = 0, \tag{1}$$

$$u_t + uu_x + g\eta_x = 0. \tag{2}$$

In this context, η represents wave elevation, u represents the average horizontal wave velocity, while g stands for the gravitational acceleration. The water thickness is given by $h = \eta + d$, in which d denotes the depth of the water relative to the still water level. The water depth, d, follows the experimental set-up that will be explained in details on the later section. The illustration of the water depth, including the profile of the dike and trench, is shown in Figure 1.

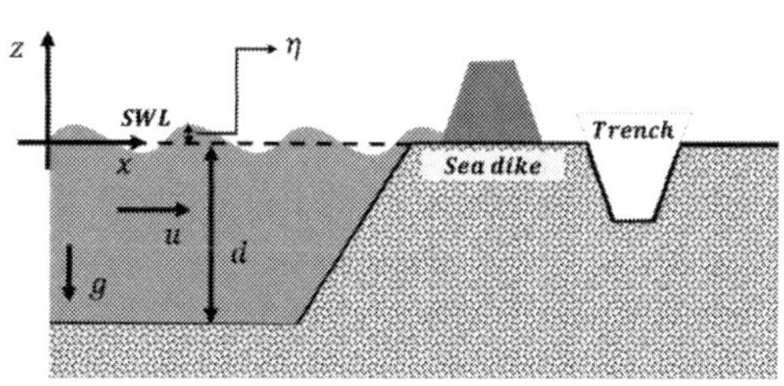

Figure 1: Illustration for the mathematical model set-up.

3 Numerical Method

In solving the model computationally, we utilize the staggered finite volume method, in which the computational domain $\Omega_S = [0, L]$ is partitioned into half-grids and full-grids with a step length of Δx. The observation time $\Omega_T = [0, T]$ is also divided with a time step of Δt. On the half-grids, denoted by $[x_{i-\frac{1}{2}}, x_{i+\frac{1}{2}}]$, the mass conservation equation is calculated. Meanwhile, the momentum equation is computed on the full-grids, symbolized by $[x_i, x_{i+1}]$. Figure 2 illustrates this scheme. The wave horizontal velocity $u(x, t)$ and elevation $\eta(x, t)$ are defined at the half-grid and full-grid points, respectively.

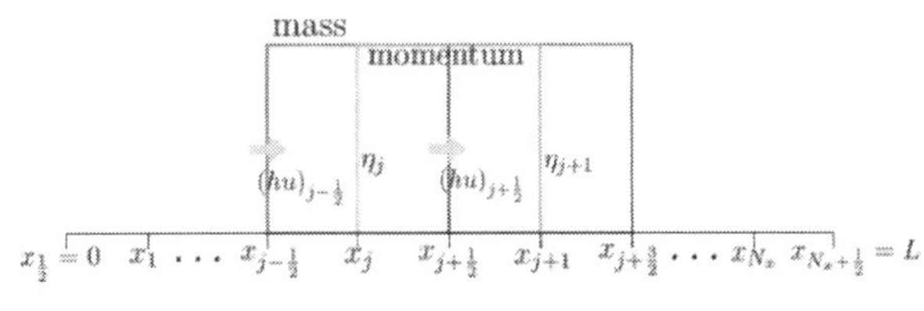

Figure 2: A Staggered Grid Illustration.

By implementing the staggered finite volume method with a forward-in-time and center-in-space scheme, the discrete formulation for mass conservation equation (1) and momentum equation (2), respectively, can be expressed as

$$\frac{\eta_i^{n+1} - \eta_i^n}{\Delta t} + \frac{h^{*n}_{i+\frac{1}{2}} u^n_{i+\frac{1}{2}} - h^{*n}_{i-\frac{1}{2}} u^n_{i-\frac{1}{2}}}{\Delta x} = 0. \tag{3}$$

$$\frac{u^{n+1}_{i+\frac{1}{2}} - u^n_{i+\frac{1}{2}}}{\Delta t} + g\frac{\eta^{n+1}_{i+1} - \eta^{n+1}_i}{\Delta x} + (uu_x)^n_{i+\frac{1}{2}} = 0. \tag{4}$$

To solve η at the mass conservation (3), which is calculated on the full grid, it is necessary to approximate h on the half grid. For this purpose, we employ the upwind method, relying on the flow's direction. The approximation can be expressed as

$$h^{*n}_{i+\frac{1}{2}} = \begin{cases} h^n_i & u_{i+\frac{1}{2}} > 0 \\ h^n_{i+1} & u_{i+\frac{1}{2}} < 0. \end{cases} \tag{5}$$

The advection term uu_x in the NSWE is notoriously challenging to handle. To address this issue, we employ the approximation detailed in [15, 16, 17, 18].

$$uu_x = \frac{1}{h}\left(\frac{\partial qu}{\partial x} - u\frac{\partial q}{\partial x}\right). \tag{6}$$

Since we are using the upwind method, the discretization depends on the flow direction. When $\bar{h}_{i+\frac{1}{2}} > 0$, the discrete form of equation (6) can be expressed as

$$uu_{x_{i+\frac{1}{2}}} = \frac{1}{\bar{h}_{i+\frac{1}{2}}}\left(\frac{\bar{q}_{i+1}u^*_{i+1} - \bar{q}_i u^*_i}{\Delta x} - u_{i+\frac{1}{2}}\frac{\bar{q}_{i+1} - \bar{q}_i}{\Delta x}\right), \tag{7}$$

with $\bar{h}_{i+\frac{1}{2}} = \frac{1}{2}(h_i + h_{i+1})$ and $\bar{q}_i = \frac{1}{2}\left(q_{i+\frac{1}{2}} + q_{i-\frac{1}{2}}\right)$. In this case, the values of velocity u is required on the full grids. Consequently, we apply the upwind method, depending on the wave motion's direction. This upwind approximation is expressed as follows:

$$u^{*n}_i = \begin{cases} u^n_{i-\frac{1}{2}} & \bar{q}_i > 0 \\ u^n_{i+\frac{1}{2}} & \bar{q}_i < 0 \end{cases}. \tag{8}$$

To handle the dry area, we implement the wet-dry procedure. We calculate u where the water thickness is not zero.

4 Experimental Setup

This section briefly explains the experimental set-up used to collect the data which will be compared to the simulation results in the validation process. The experiments were conducted in the $50 \times 2 \times 1.6$ m wave flume, located in the Laboratory of Coastal Dynamics, National Research and Innovation Agency, Yogyakarta, Indonesia. Figure 3 illustrates the bathymetry used in the experiments, where a wave generator is positioned 1.5 m in front of the first slope, whilst a wave absorber is placed behind the dry area at the flume's right-end. In order to collect the data needed, several wave probes (WP) were placed within the observed flume. However, in this case, we will mostly use two of the provided wave probes, which are WP G and H. These two WPs are located at $x = 21.9$ m (after the sea dike) and at $x = 22.4$ m (after the trench), respectively.

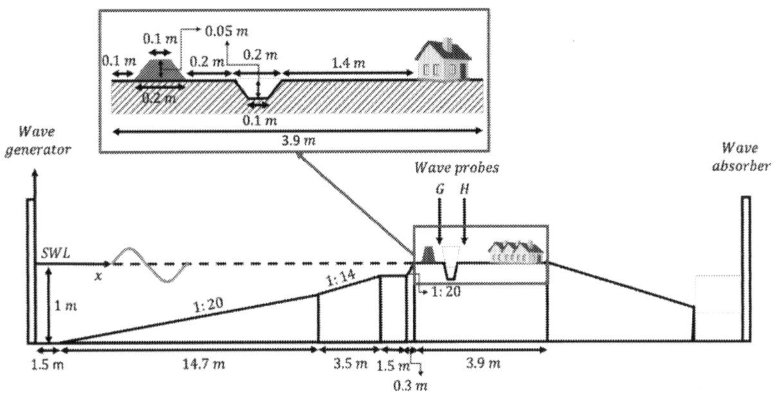

Figure 3: Description of the experimental set-up carried out at the Laboratory of Coastal Dynamics, National Research and Innovation Agency, Yogyakarta, Indonesia.

The bathymetry set-up illustrated in Figure 3 is a modification from the setup used in the study done by Chandler et al. [24]. In this case, three slopes of 1:20, 1:14, and 1:20, with different lengths were used in the submerged region, instead of one slope as used in [24]. In addition, the initial wave used in the experiments was reproduced by the tsunami wave generator at 1:50 scale based on the characteristics of the real recorded wave during the 2004 Indian Ocean Tsunami incident [24]. The wave characteristics mentioned are as follow: Wave height is set at 6.5 m, the wave crest' amplitude is 3.7 m, the wave trough amplitude is 2.8 m, and wave length is set at 14400 m [25]. The initial wave generated is depicted by Figure 4. Essentially, the experiments evaluate the propagation of the tsunamis waves and how they interact with structures. The wave elevation was measured and recorded before, during, and after it propagates over the structures. There are four scenarios considered for the set-up of the structures: (a) No structure; (b) Only trench or Only sea dike; and (c) A combination of trench and sea dike. Using both the data and the simulation results, the best scenario for the coastal protection structures between the mentioned four will be determined.

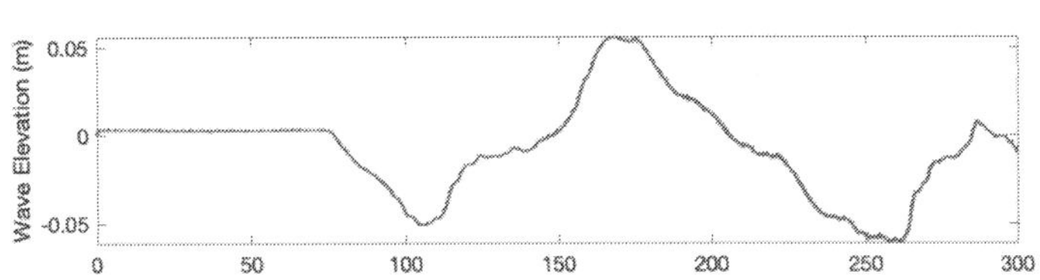

Figure 4: Initial wave used in the experiments, reproduced by the tsunami wave generator at 1:50 scale based on the characteristics of the real recorded 2004 Indian Ocean tsunami waves.

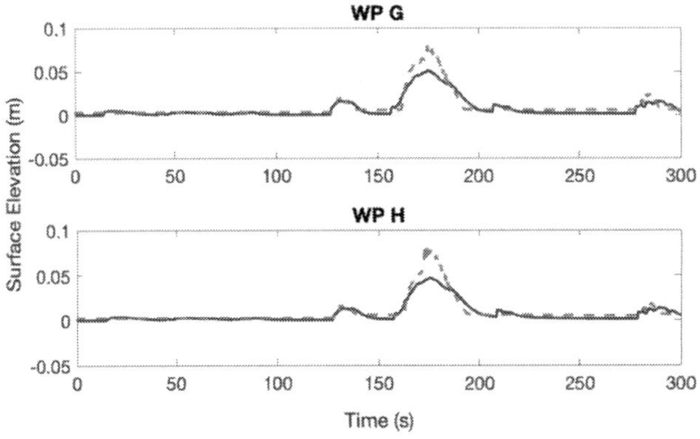

Figure 5: Comparisons between the numerical (blue solid line) and experimental (red dashed line) results at WP G and H for the case of no structure involved.

5 Results and Discussions

Here, the explained computational model is applied to examine the propagation of the tsunami waves generated based on the real 2004 Indian Ocean Tsunami waves and to evaluate how the implementation of a trench and/or a sea dike will affect its elevation. There are four scenarios presented, combining the structures. In all of the simulations, we used a 70 m long spatial domain, in which the 20 m extra domain is used to accommodate the wave absorbing area at the flume's right-end. However, the bathymetry set-up remains the same. All simulations were performed for 300 s, with $\Delta x = 0.05$ m and $\Delta t = 0.001$ s. In addition, we used $g = 9.81$ m/s^2 as the acceleration due to gravity. Moreover, the initial condition of each simulation is calm water, with the boundary condition on the left side follows the scaled-down Aceh's tsunami waves.

5.1 Case a: No structure

This scenario is presented to show the evolution of the tsunami waves when there is no structure implemented. In other words, this demonstrates the behaviour of the tsunami waves when it impacted the coastal area on the 2004 occurrence. The results of our developed numerical model are compared to the measured experimental data. As previously stated, we are only focusing on the approximation and the examination of the data collected at WP G and H, since these two locations are where the sea dike and the trench are supposedly located. Figure 5 shows the comparison for each of these two wave probes. The Root Mean Square Error (RMSE) formula has been used to compute the errors in each WP to quantify the comparisons. This resulted in an error of around 0.0057 and 0.0059 m for WP G and H, respectively. Considering that the maximum water elevation is around 0.77, these errors are relatively small. In addition, the wave transmission coefficient (K_t), defined as a ratio of the transmitted and the initial wave amplitude, is calculated to be around $K_t = 1.3925$ and $K_t = 1.4067$ for WP G and H, respectively. This indicate that the tsunami wave height on land was about 1.4 times the initial height. Depending on the initial tsunami wave height, damage is possible to occur on land due to this wave height multiplication, which is influenced by the shoaling phenomena as well.

5.2 Case b: Only trench or only sea dike

For this case, we simulated a scenario where we include only a trench or only a sea dike in the system. First structure to be addressed is a trench, which was located 0.5 m from the shoreline and has a length and a depth of 0.2 m and 0.05 m, respectively. This simulation was carried out to examine the effectivity of the trench alone in attenuating tsunami wave. Figure 6a shows wave evolution with respect to the observation time, measured at WP G and H, compared with the simulation results. In this case, the calculated RMSE is around 0.0068 m for WP G, while it is about 0.0070 m for WP H. Since we are examining wave reduction due to the trench only, we are looking into the wave transmission coefficient (K_t) measured at WP H, which is located after the trench. At the observed point, the value of K_t is about 1.3430, which is roughly 95% of the original K_t (the case of no structure involved). It means, the specific trench that is used here was able to reduce the tsunami waves by about 5%.

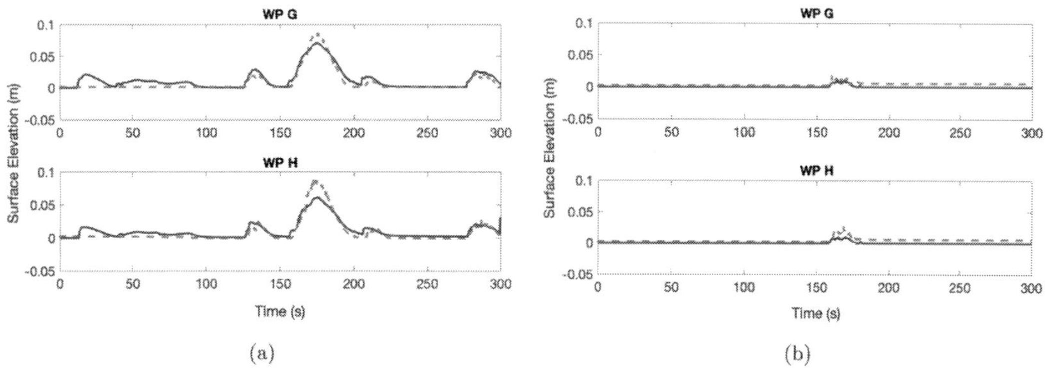

Figure 6: Comparisons between the numerical (blue solid line) and the experimental (red dashed line) results at WP G and H for the case when only (a) a trench or (b) a sea dike is included in the system.

The next structure to be discussed is a sea dike, which was located also on land, about 0.1 m away from the shoreline. The sea dike was set to have the height of 0.05 m and a length of 0.2 m. The crest of the sea dike has a length of 0.1 m, which means that the slope of the sea dike is 1 : 10 on both the front and back sides. This scenario is simulated using our developed numerical model, which results were then compared to the measured data. The comparisons are shown in Figure 6b. Visually, this comparison shows a really great agreement between the data and the simulation results for both WPs. This is supported by the calculated RMSE, which are about 0.0042 m and 0.0048 m for WP G and H, respectively. These errors are evidently smaller than previous cases. Compared to the previous trench-only scenario, this case, where we consider sea dike of the same size, generates significantly smaller wave transmission coefficient, which is about $K_t = 0.2882$, measured at WP G. This indicates that the structure of specific size was able to reduce the tsunami wave height by almost 80% (compared to the original K_t of WP G). This means, for the same size, the sea dike was more effective in reducing wave height than the trench was. In the following case, we are going to evaluate whether a combination between the two structures is even more effective or not.

5.3 Case d: A combination of trench and sea dike

This case combines the sea dike and the trench in a integrated coastal protection system, in which effectivity in reducing wave height is examined. The sizes and locations of both structures followed the ones mentioned previously, with the distance between the two structures is 0.2 m. Wave elevation recorded at WP G and H are presented in Figure 7, along with the measured data from the experiments. It was found that the errors between the data and numerical results are about 0.0029 m at WP G and 0.0020 m at WP H, calculated using RMSE formula. These errors are sufficiently small, indicating that our numerical model was also able to approximate wave propagation due to the presence of both structures at the same time. Furthermore, it was also discovered that the wave transmission coefficient at WP G and H are 0.3071 and 0.4687, respectively. From these values of K_t, it is acceptable to assume that whilst the sea dike was significantly efficient in reducing wave height (by around 70%), the trench gave an opposite effect. In this case, the trench was increasing the wave height by about 50% of the one induced by sea dike. This is interesting, considering that the implementation of the trench alone was able to reduce wave height, although only a slight reduction. More examination is needed on this matter to address the specific issue. However, for the used topography and wave characteristics, we can conclude that implementing the sea dike alone is better in attenuating wave height than applying the trench or even both structures at the same time. Although, it is noteworthy that the combined system was still able to reduce the generated tsunami wave by about 67%, which is significantly larger than applying the trench only.

6 Conclusion

In this study, we have demonstrated the evolution of the scaled-down 2004 tsunami wave centered at the Indian Ocean, using both experimental and numerical methods. The computational model employed the Staggered Finite Volume Method (FVM) to solve the Nonlinear Shallow Water Equations (NSWEs).

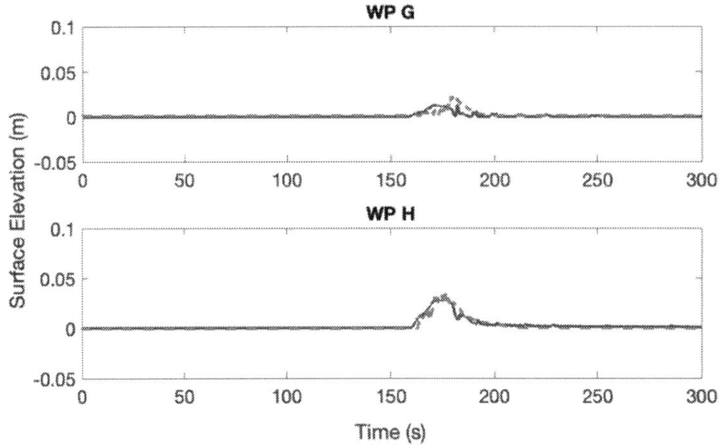

Figure 7: Comparisons between the numerical (blue solid line) and the experimental (red dashed line) results at WP G and H for the case when a combined system of sea dike and trench is implemented.

Three scenarios were analyzed: (a) no protective structures; (b) the presence of either a trench or a sea dike; and (c) a combination of both trench and sea dike. Comparing the computational and the experimental results revealed a relatively good agreement, which is supported by the resulted RMSE value of less than 0.007 m. The findings indicate that the most effective measure to reduce tsunami wave height is the use of a sea dike alone, which achieved a reduction of approximately 80%. The combination of both a trench and a sea dike resulted in a tsunami wave height reduction of about 67%, while the trench alone was the least effective, reducing wave height by only around 5%. It is crucial to note that these results are specific to the studied wave characteristics and coastal topography. Broader conclusions would require further investigation.

Author Contributions

H. Q. Rif'atin: Data Curation (equal), Formal Analysis (equal), Visualization (lead), Validation and Investigation (equal), Writing - Original Draft Preparation. **I. Magdalena:** Conceptualization (lead), Supervision (lead), Writing - Review & Editing (equal), Data Curation (equal), Formal Analysis (equal), Visualization (lead), Funding acquisition (lead), Project administration (lead), Methodology (lead), Validation and Investigation (equal). **W. Kongko:** Investigation (equal), Resources (lead), Data Curation (equal), Formal Analysis (equal). **M. Farid:** Validation and Investigation (equal), Formal Analysis (equal), Writing - Review & Editing (equal).

References

[1] M. A. Helal, M. S. Mehanna, Tsunamis from nature to physics, Chaos, Solitons Fractals 36 (4) (2008) 787–796. doi:10.1016/j.chaos.2007.08.044.

[2] R. Paris, Source mechanisms of volcanic tsunamis, Philosophical Transactions of the Royal Society A 373 (2015) 20140380. doi:10.1098/rsta.2014.0380.

[3] A. B. Rabinovich, E. L. Geist, H. M. Fritz, J. C. Borrero, Introduction to "tsunami science: ten years after the 2004 indian ocean tsunami. volume i, Pure and Applied Geophysics 172 (3-4) (2015) 615–619. doi:10.1007/s00024-015-1038-5.

[4] A. Suppasri, M. Kitamura, D. Alexander, S. Seto, F. Imamura, The 2024 noto peninsula earthquake: Preliminary observations and lessons to be learned, International Journal of Disaster Risk Reduction 110 (2024) 104611. doi:10.1016/j.ijdrr.2024.104611.

[5] When will it happen? a matter of time before the megathrust earthquakes strikes indonesia, https://seasia.co/2024/08/23/when-will-it-happen-understanding-indonesias-megathrust-earthquake-threat, accessed: 2024-08-27.

[6] G. Cuomo, A. Guercio, A. Dimakopoulos, Advanced numerical modeling of tsunami wave propagation, transformation and run-up, Engineering and Computational Mechanics 167 (3) (2014) 139–151. doi:10.1680/eacm.13.00029.

[7] V. Sriram, I. Didenkulova, A. Sergeeva, S. Schimmels, Tsunami evolution and run-up in a large scale experimental facility, Coastal Engineering 111 (2016) 1–12. doi:10.1016/j.coastaleng.2015.11.006.

[8] N. Shuto, Tsunami hazard mitigation, Proceedings of the Japan Academy. Series B, Physical and Biological Sciences 95 (2019) 151–164. doi:10.2183/pjab.95.012.

[9] A. Strusińska-Correia, Tsunami mitigation in japan after the 2011 tōhoku tsunami, International Journal of Disaster Risk Reduction 22 (2017) 397–411. doi:10.1016/j.ijdrr.2017.02.001.

[10] H. G. Guler, C. Baykal, T. Arikawa, A. C. Yalciner, Numerical assessment of tsunami attack on a rubble mound breakwater using openfoam, Applied Ocean Research 72 (2018) 76–91. doi:10.1016/j.apor.2018.01.005.

[11] N. M. Sirag, Mitigation of tsunami environmental implications on nuclear facilities through improvement of seawall design, International Research Journal Of Advanced Engineering And Science 4 (2019) 136–143.

[12] A. Silva, S. Araki, Investigating the behavior of an onshore wall and trench combination ahead of a tsunami-like wave, Geosciences. 10 (2020) 310–322. doi:10.3390/geosciences10080310.

[13] R. Gayathri, J.-Y. Chang, C.-C. Tsai, Wave power extraction from slanted oscillating water columns in the presence of an array of arbitrary trenches, Physics of Fluids 35 (9) (2023) 097105. doi:10.1063/5.0165164.

[14] Syamsidik, Tursina, A. Suppasri, M. Al'ala, M. Luthfi, L. K. Comfort, Assessing the tsunami mitigation effectiveness of the planned banda aceh outer ring road (borr), indonesia, Natural Hazards and Earth System Sciences 19 (1) (2015) 299–312. doi:10.5194/nhess-19-299-2019.

[15] S. Pudjaprasetya, I. Magdalena, Momentum conservative schemes for shallow water flows, East Asian J. Appl. Math. 4 (05 2014). doi:10.4208/eajam.290913.170314a.

[16] I. Magdalena, Non-hydrostatic model for solitary waves passing through a porous structure, J. Disaster Res. 11 (2016) 957–963. doi:10.20965/jdr.2016.p0957.

[17] I. Magdalena, N. Erwina, S. Pudjaprasetya, Staggered momentum conservative scheme for radial dam break simulation, J. Sci. Comput. 65 (2015) 867–874. doi:10.1007/s10915-015-9987-5.

[18] I. Magdalena, N. Erwina, An efficient two-layer non-hydrostatic model for investigating wave run-up phenomena, Comput. 8 (2020) 1. doi:10.3390/computation8010001.

[19] R. J. LeVeque, Finite Volume Methods for Hyperbolic Problems, Cambridge University Press, Cambridge, England, 2002.

[20] M. Zhang, Y. Ji, Y. Wang, H. Zhang, T. Xu, Numerical investigation on tsunami wave mitigation on forest sloping beach, Acta Oceanologica Sinica 39 (2020) 130–140. doi:10.1007/s13131-019-1527-y.

[21] I. Magdalena, K. Firdaus, D. Jayadi, Analytical and numerical studies for wave generated by submarine landslide, Alexandria Engineering Journal 61 (9) (2022) 7303–7313. doi:10.1016/j.aej.2021.12.069.

[22] K. Firdaus, A. M. A. Matin, N. Nurisman, I. Magdalena, Numerical study for sunda strait tsunami wave propagation and its mitigation by mangroves in lampung, indonesia, Results in Engineering 16 (2022) 100605. doi:10.1016/j.rineng.2022.100605.

[23] H. Q. Rif'atin, I. Magdalena, D. Dewata, C. Saengsupavanich, S. S. na Ayutthaya, Optimization of stepped revetment configuration on minimizing wave run-up and overtopping, Physics of Fluids 36 (2024) 046617. doi:10.1063/5.0195542.

[24] I. D. Chandler, W. H. Allsop, I. B. Granged, D. J. McGovern, Understanding wave generation in pneumatic tsunami simulators, in: Proceedings of the 6th International Conference on the Application of Physical Modelling in Coastal and Port Engineering and Science (Coastlab16), 2016, pp. 1–11.

[25] T. Rossetto, W. Allsop, I. Charvet, D. Robinson, Physical modelling of tsunami using a new pneumatic wave generator, Coastal Engineering 58 (6) (2011) 517–527.

IOP Conf. Series: Earth and Environmental Science **1464** (2025) 012009 doi:10.1088/1755-1315/1464/1/012009

Analytical and numerical studies of wave attenuation using mangroves and submerged breakwaters for coastal protection

Vinsensia Ferren[1], Ikha Magdalena[1,2,3,*]

[1]Faculty of Mathematics and Natural Sciences, Institut Teknologi Bandung, Indonesia
[2]Center for Coastal and Marine Development, Institut Teknologi Bandung, Indonesia
[3]Center for Mathematical Modelling and Simulations, Institut Teknologi Bandung, Indonesia

E-mail: ikha.magdalena@itb.ac.id

Abstract. This research investigates the effectiveness of mangroves and submerged breakwaters in mitigating wave energy and protecting coastal areas. Mangroves, with their dense root systems, and submerged breakwaters, which disrupt wave patterns, serve as natural and engineered solutions to reduce wave impact and erosion. The research highlights the synergistic benefits of combining these methods, demonstrating their potential to enhance coastal resilience against storm surges and rising sea levels. A mathematical model is employed to examine wave attenuation by mangroves and submerged breakwaters, solved analytically using the separation of variables method. Additionally, a staggered finite volume approach is implemented to numerically determine the wave transmission coefficient. For validation, numerical results are compared with analytical solutions. By evaluating different configurations and environmental conditions, the research provides insights into optimizing these strategies for sustainable coastal defense.

1 Introduction

Coastal regions are increasingly at risk from storm surges and rising sea levels, driving the need for effective mitigation strategies to counter erosion and flooding. Among various coastal protection methods, mangroves and submerged breakwaters have attracted significant attention due to their capability to attenuate wave energy and enhance coastal resilience. Mangroves, with their intricate root networks, serve as natural barriers by stabilizing sediments and dissipating wave energy [1, 2]. Conversely, submerged breakwaters are engineered structures specifically designed to disrupt wave patterns and reduce wave height [3, 4].

Integrating these natural and engineered approaches presents a promising solution for coastal defense. Previous studies have shown that mangroves can effectively lower wave heights and erosion rates, while submerged breakwaters further decrease wave energy by modifying wave propagation [5, 6]. However, the combined effects of these strategies, as well as their optimization under diverse environmental conditions, remain relatively unexplored.

This study investigates the effectiveness of combining mangroves with submerged breakwaters for coastal protection. A mathematical model is developed to analyze wave attenuation achieved through both mangrove ecosystems and submerged breakwater structures. The model is solved analytically using the separation of variables method, while a staggered finite volume approach is applied for numerical solutions to evaluate the wave transmission coefficient [7, 8, 9, 10]. A comparison between numerical results, experimental data, and analytical solutions is conducted to ensure accuracy.

By evaluating various configurations and environmental conditions, this research aims to provide insights into optimizing the combined use of mangroves and submerged breakwaters for sustainable coastal defense. The findings offer a deeper understanding of how these methods can be effectively integrated to strengthen coastal resilience against the adverse impacts of climate change and sea-level rise.

2 Mathematical Model

This section presents the mathematical model employed to analyze wave attenuation resulting from the interaction between mangroves and a submerged breakwater. The model is founded on the Shallow Water Equations, which are a recognized framework for addressing fluid dynamics issues, as previously documented in relevant literature [7, 8, 11, 12].

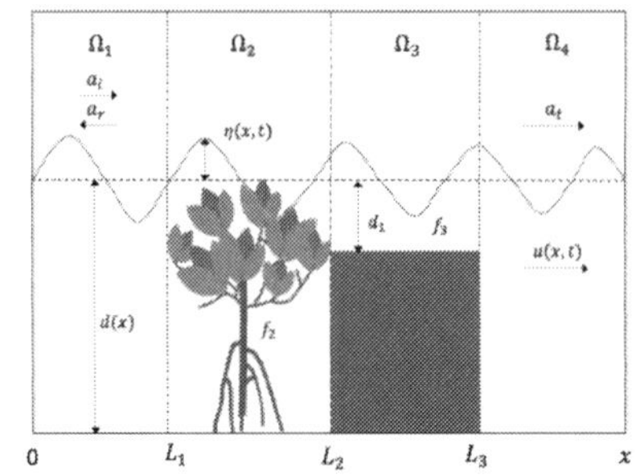

Figure 1: Illustration of Water Waves Interacting with Mangroves and a Submerged Breakwater

Figure 1 illustrates the key parameters within the model, which include the incident wave amplitude (a_i), transmitted wave amplitude (a_t), and reflected wave amplitude (a_r), all measured in meters. The water depth is denoted by d, while d_1 indicates the depth from the still water level to the top of the submerged breakwater. The friction coefficients relevant to the structures are represented as f_2 and f_3. The primary output of this model is the wave transmission coefficient (K_t), a metric that reflects the degree of wave amplitude reduction, computed as the ratio of the transmitted wave amplitude to that of the incident wave.

The Shallow Water Equations consist of two equations: the mass conservation equation and the momentum equation. For the analysis, two variables are introduced: η, representing water surface elevation, and u, denoting water wave velocity. Assuming linear wave behavior and flat bottom topography ($d(x) = d$), the model is formulated as follows:

$$\eta_t + du_x = 0, \quad (1)$$
$$u_t + g\eta_x = 0. \quad (2)$$

In this formulation, g signifies the gravitational acceleration.

To accommodate the specific context of this study, the observation domain is divided into four distinct sub-domains Ω_i, where $i = 1, 2, ..., 4$, as shown in Figure 1. In regions without structures, the original Equations (1) and (2) are applicable. However, in areas populated by mangroves, we introduce a friction term to account for fluid-structure interactions, leading to the following modified equations for sub-domain Ω_2:

$$\eta_t + du_x = 0, \quad (3)$$
$$u_t + g\eta_x + f_2\omega u = 0, \quad (4)$$

where f_2 is the mangrove's the friction coefficient, and ω is the wave frequency.

For the sub-domain encompassing the submerged breakwater, the bottom topography is adjusted to d_1, leading to the following equations for Ω_3:

$$\eta_t + (d_1u)_x = 0, \quad (5)$$
$$u_t + g\eta_x + f_3\omega u = 0. \quad (6)$$

In these equations, d_1 represents the water depth from the still water level to the submerged breakwater.

3 Analytical Solution

This section outlines the analytical approach employed to calculate the wave transmission coefficient, K_t. The process begins by establishing the water surface elevation, $\eta(x,t)$, and the horizontal water wave velocity, $u(x,t)$, for each respective domain. By applying continuity conditions to these definitions, we facilitate the derivation of K_t.

To derive $\eta(x,t)$ and $u(x,t)$, we utilize the method of separation of variables. For convenience, we introduce the parameter $\epsilon_i = \frac{1}{\sqrt{1-if_i}}$ for $i = 2, 3$. This parameter allows us to express the water surface elevation and horizontal water wave velocity in each domain as follows:

$$\eta(x,t) = e^{i\omega t}\begin{cases} a_i e^{-ik_1x} + a_r e^{ik_1x}, & x \in \Omega_1 \\ C_1 e^{-ik_2x} + C_2 e^{ik_2x}, & x \in \Omega_2 \\ C_3 e^{-ik_3x} + C_4 e^{ik_3x}, & x \in \Omega_3 \\ a_t e^{-ik_4x}, & x \in \Omega_4. \end{cases} \tag{7}$$

$$u(x,t) = e^{i\omega t}\begin{cases} (a_i e^{-ik_1x} - a_r e^{ik_1x})\sqrt{\frac{g}{d}}, & x \in \Omega_1 \\ (C_1 e^{-ik_2x} - C_2 e^{ik_2x})\sqrt{\frac{g}{d}}\epsilon_2, & x \in \Omega_2 \\ (C_3 e^{-ik_3x} - C_4 e^{ik_3x})\sqrt{\frac{g}{d_1}}\epsilon_3, & x \in \Omega_3 \\ (a_t e^{-ik_4x})\sqrt{\frac{g}{d}}, & x \in \Omega_4. \end{cases} \tag{8}$$

The wave numbers $k_1, k_2, \ldots, k_4$ are defined as $k_i = \frac{\omega}{\sqrt{gd}}$ for domains Ω_1 and Ω_4, and $k_i = \frac{\omega}{\sqrt{gd}}\sqrt{1-if_i}$ for domains Ω_2 and Ω_3. The coefficients $a_i, a_r, a_t, C_1, C_2, C_3, C_4$ represent the amplitudes of the waves interacting with the mangroves and submerged breakwaters.

To determine the wave transmission coefficient K_t, it is imperative to uphold continuity conditions at the points of discontinuity, specifically at $x = 0$, L_1, L_2, and L_3. These conditions ensure both wave height and horizontal momentum continuity, expressed by the following equations:

$$\eta(x^-,t) = \eta(x^+,t), \tag{9}$$

$$d(x^-)u(x^-,t) = d(x^+)u(x^+,t). \tag{10}$$

In these equations, x^- and x^+ denote the limits approaching the discontinuity from the left and right sides, respectively. The next step involves applying these continuity conditions to the previously derived expressions for $\eta(x,t)$ and $u(x,t)$, focusing on each interface.

To simplify the computation of wave amplitudes across the different domains, we introduce the following variables:

$$\begin{array}{lll} A = e^{-ik_4L_3} & B = e^{-ik_3L_3} & C = e^{ik_3L_3} \\ D = e^{-ik_3L_2} & E = e^{ik_3L_2} & F = e^{-ik_2L_2} \\ G = e^{ik_2L_2} & H = e^{-ik_2L_1} & I = e^{ik_2L_1} \\ J = e^{-ik_1L_1} & K = e^{ik_1L_1} & \end{array}$$

These variables assist in calculating wave amplitudes. Notably, the initial wave amplitude a_i can be derived using the following expression:

$$a_i = \frac{(1+R\epsilon_2)C_1H + (1-R\epsilon_2)C_2I}{2J}, \tag{11}$$

By solving this equation, we obtain the wave transmission coefficient K_t, expressed as follows:

$$K_t = \left|\frac{2J}{(1+R\epsilon_2)C_1H + (1-R\epsilon_2)C_2I}\right|, \tag{12}$$

where R is defined as $\sqrt{gd}$. The value of the wave transmission coefficient is influenced by several factors, including the friction coefficient, the widths of the mangroves and the submerged breakwater, as well as the height of the breakwater. The subsequent section on Sensitivity Analysis will explore how K_t varies with these parameters and dimensions.

4 A staggered finite volume method

To numerically solve the model, a staggered finite volume method is utilized. This technique involves dividing the spatial domain $\Omega_S = [0, L]$ into a combination of half and full grids with a grid spacing of Δx. The time domain $\Omega_T = [0, T]$ is discretized into finite time steps of duration Δt. Mass conservation is computed on the half-grid intervals $[x_{i-\frac{1}{2}}, x_{i+\frac{1}{2}}]$, while momentum conservation is addressed on the

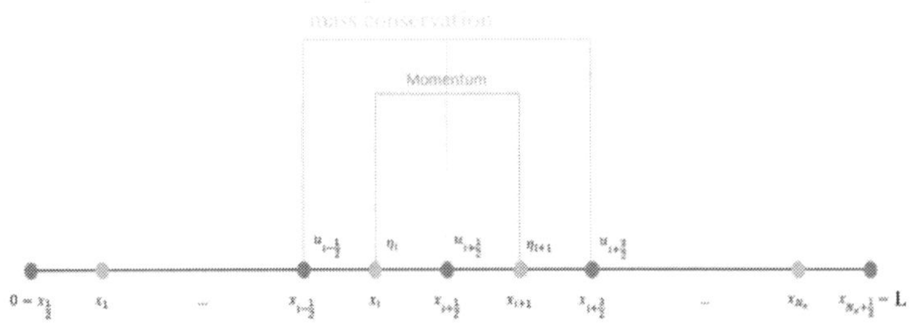

Figure 2: Staggered Grid Scheme

full-grid intervals $[x_i, x_{i+1}]$. An illustration of this grid scheme is shown in Figure 2. In this arrangement, the horizontal water wave velocity $u(x,t)$ is defined at the half-grid points, whereas the surface elevation $\eta(x,t)$ is defined at the full-grid points.

Using the staggered finite volume method in conjunction with a forward-time centered-space scheme, the discrete form of the mass conservation equation can be approximated as follows:

$$\frac{\eta_i^{n+1} - \eta_i^n}{\Delta t} + \frac{d^{*n}_{i+\frac{1}{2}} u^n_{i+\frac{1}{2}} - d^{*n}_{i-\frac{1}{2}} u^n_{i-\frac{1}{2}}}{\Delta x} = 0. \tag{13}$$

In a similar manner, the discrete version of the momentum equation is expressed as:

$$\frac{u^{n+1}_{i+\frac{1}{2}} - u^n_{i+\frac{1}{2}}}{\Delta t} + g\frac{\eta^{n+1}_{i+1} - \eta^{n+1}_i}{\Delta x} + (uu_x)^n_{i+\frac{1}{2}} + fu^{n+1}_{i+\frac{1}{2}} = 0. \tag{14}$$

To resolve the mass conservation equation calculated on the full grid, an approximation of d on the half grid is necessary. This approximation is obtained using the upwind method, which considers the fluid flow direction. The approximation is provided by:

$$d^{*n}_{i+\frac{1}{2}} = d^n_i \text{if } u_{i+\frac{1}{2}} > 0, d^n_{i+1} \text{if } u_{i+\frac{1}{2}} < 0. \tag{15}$$

5 Results and Discussions

5.1 *Waves Passing Through Mangroves*

This subsection focuses on the behavior of waves as they pass through a mangrove forest, known for its dense and intricate root systems that effectively reduce wave energy. The performance of the numerical model is evaluated under conditions without a breakwater and with a flat seabed at a water depth of d. The computational domain spans from $x = 0$ to $x = 25$ meters, with the water depth fixed at 10 meters. Within this domain, a 5-meter segment of mangroves is located between $x = 10$ and $x = 15$ meters.

The simulation begins with calm water, and a monochromatic wave with an amplitude of 0.5 meters and a wave frequency of $\omega = 3\pi$ is introduced. A fully absorbing condition is applied on the right boundary to prevent reflections. The initial conditions for the wave are given by:

$$\eta(x,0) = 0, u(x,0) = 0. \tag{16}$$

At the left boundary, the wave is introduced as:

$$\eta(0,t) = A\sin(\omega t), \tag{17}$$

While the right boundary condition is described by:

$$u(L,t) = \sqrt{\frac{g}{d}}\eta(L,t), \tag{18}$$

where $g = 9.81$ m$/s^2$.

In the mangrove region, the interaction between the fluid and vegetation is accounted for by incorporating a friction coefficient of $f = 0.3$. To ensure stability in the simulation, a spatial partition $\Delta x = 0.1$

is used, along with a time step $\Delta t = \Delta x/\sqrt{gd}$. To ensure comprehensive analysis, two simulations were conducted using different initial wave amplitudes: $a_i = 0.5$ m and $a_i = 1$ m. In the visual representation, green lines denote the mangrove locations, while the water surface elevation is illustrated by the blue line. The results shown in Figures 3 and 4 demonstrate a clear attenuation of wave energy as the waves

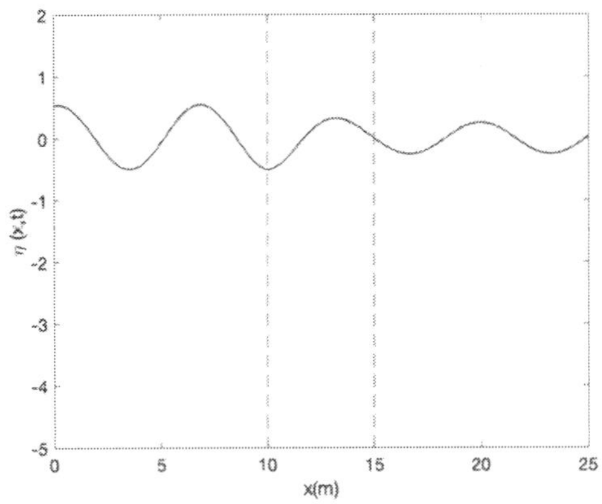

Figure 3: Evolution of wave amplitude through mangroves with an initial amplitude of $a_i = 0.5$ m.

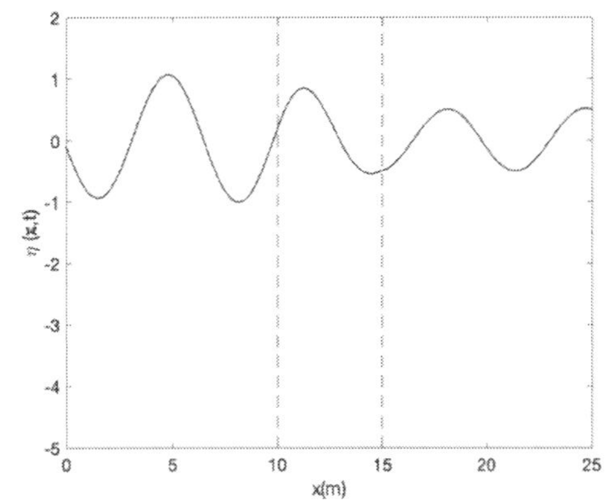

Figure 4: Evolution of wave amplitude through mangroves with an initial amplitude of $a_i = 1$ m.

pass through the mangrove area, evidenced by decreases in both wave amplitude and water velocity. This energy dissipation is primarily due to the drag force exerted by the mangrove roots, which effectively reduce wave energy. The impact of the mangroves on wave height is quantitatively assessed by comparing wave amplitudes before and after the mangrove zone. To further quantify this reduction, the numerical transmission coefficient (K_t) was calculated based on the maximum values of η at the initial and final points, analogous to the analytical K_t method. Although Figure 4 features a larger initial wave amplitude, it does not affect the K_t value since K_t is calculated as the ratio of transmitted wave amplitude to initial amplitude. Consequently, the K_t value remains constant regardless of the a_i value used in the simulation. The numerical K_t value of 0.5304 aligns closely with the analytical K_t of 0.5187, indicating an overall wave amplitude reduction of over 48% after passing through the mangroves.

5.2 Waves Propagation over A Submerged Breakwater

This section evaluates how waves behave as they pass through a submerged breakwater, known for its ability to reduce wave height due to its structural characteristics. The breakwater is positioned between $x = 10$ and $x = 15$ meters. For regions without structures, the water depth is set at $d = 10$ meters, while in areas with the breakwater, the depth from the still water level to the submerged breakwater is set at $d_1 = 7$ meters.

The same parameters as those described in the previous subsection were employed, except for the modified water depth resulting from the presence of submerged breakwaters. The initial and boundary conditions, as defined by Equations (16)-(18), remain unchanged. Additionally, the spatial domain partition Δx, the time step Δt, and the friction coefficient f were maintained at constant values. The results from the numerical simulation using different initial wave amplitudes are presented below. In the simulation, the red lines represent the locations of the submerged breakwaters, while the blue line shows the water's surface elevation. As seen in Figures 5 and 6, the wave amplitude decreases after passing through the submerged breakwater. The numerical transmission coefficient K_t, calculated from both simulations, quantifies this reduction. The simulation produces a K_t value of 0.5033, while the analytical K_t is 0.4972, confirming that the submerged breakwater reduces the wave amplitude by over 49% from its initial value.

5.3 Waves Passing Through a Combination of Mangrove and Breakwater

This section investigates the effectiveness of integrating mangroves with submerged breakwaters to improve wave attenuation. The breakwater alters wave patterns and decreases wave height, while the mangroves further dissipate wave energy through their dense root systems.

IOP Conf. Series: Earth and Environmental Science **1464** (2025) 012009 doi:10.1088/1755-1315/1464/1/012009

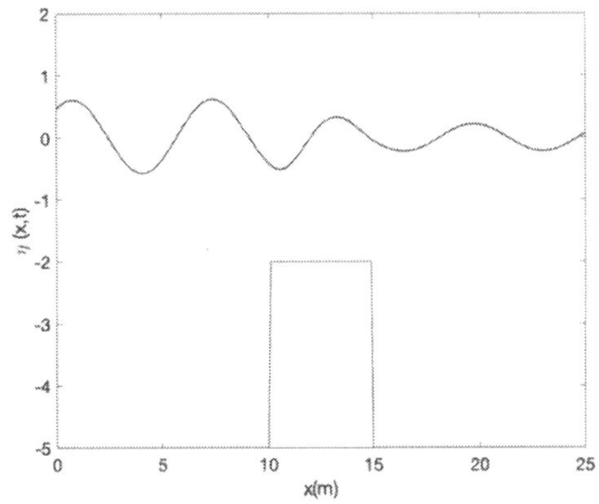

Figure 5: Evolution of wave amplitude through breakwater with an initial amplitude of $a_i = 0.5$ m.

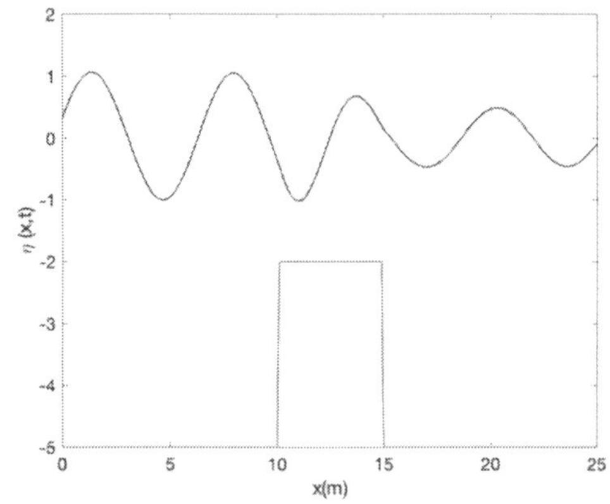

Figure 6: Evolution of wave amplitude through breakwater with an initial amplitude of $a_i = 1$ m.

The same parameters used in the previous analysis were applied to compare two scenarios: in the first, the mangroves are situated before the submerged breakwater, as depicted in Figure 7; in the second, the mangroves are positioned after the submerged breakwater, illustrated in Figure 8. The breakwater and mangrove structures are located in the intervals $5 \leq x \leq 10$ and $10 \leq x \leq 15$, respectively. The results from the numerical simulations for these two configurations are provided below.

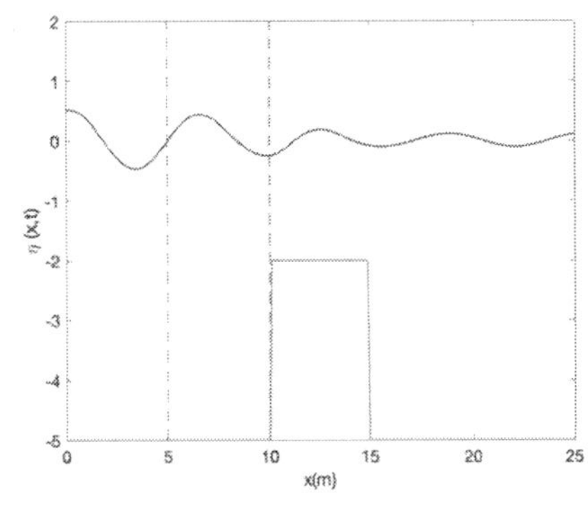

Figure 7: Wave Attenuation with Mangroves Positioned Before Submerged Breakwater.

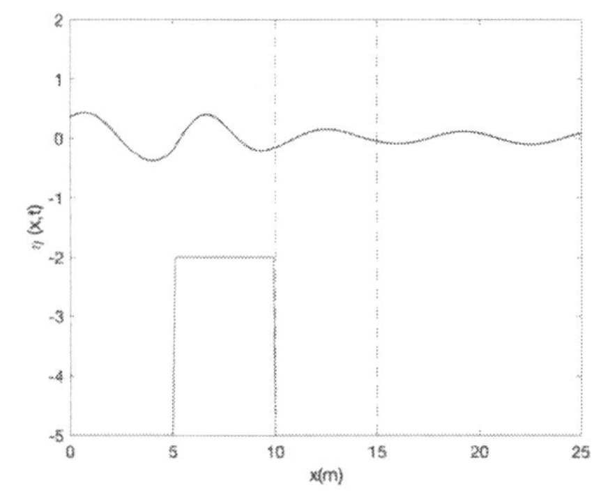

Figure 8: Wave Attenuation with Mangroves Positioned After Submerged Breakwater.

In the simulation, red lines indicate submerged breakwaters, green lines represent mangroves, and blue line displays the water surface elevation. Figures 7 and 8 demonstrate that the combination of these two elements significantly enhances wave attenuation compared to either structure used alone. The interaction between the breakwater and the mangroves creates a more effective barrier against wave energy, providing better coastal protection measured by the K_t values. At $x = 17$ meters, the numerical K_t for Figure 7 is 0.2374, while that for Figure 8 is 0.2371, demonstrating that placing mangroves after the submerged breakwater slightly increases the effectiveness of the combined setup. This configuration will be utilized in the subsequent sensitivity analysis.

For Figure 8, the analytical solution closely matches the numerical results, with a wave height of 0.2119 meters after passing through the combined structures. Overall, the numerical transmission coefficient (K_t) for the submerged breakwater-mangrove configuration reduces wave amplitude by over 76%, compared to a 48% reduction with mangroves alone and a 49% reduction with submerged breakwaters alone. This combined approach decreases K_t by approximately 27% more than the individual methods, highlighting the effectiveness of integrating both for coastal protection.

5.4 Validation with Experimental Data

In this section, we undertake a comparative analysis of the wave transmission coefficient values (K_t) derived from numerical simulations against empirical data obtained from physical models of submerged breakwaters. The primary goal of this analysis is to minimize discrepancies, thereby confirming that our numerical model effectively captures the wave attenuation phenomena associated with breakwater installations.

The experimental trials were carried out at the Coastal Research and Development Center, which operates under the Indonesian Ministry of Public Works, located in Buleleng, Bali. A comprehensive model was constructed to gather data across a range of configurations characterized by varying breakwater dimensions, including widths, heights, water depths, and wave parameters. The experimental setup featured a wave channel measuring 40 m in length, 0.6 m in width, and 1.2 m in height, with a designated 34 m testing section for manipulating structural configurations. An electric piston-type wave generator facilitated the production of regular waves, achieving heights of up to 25 cm and periods extending from 1 to 2.6 seconds. Wave probes were strategically placed both upstream and downstream of the breakwater to measure wave amplitudes before and after the interaction with the structure. At the end of the channel, a bed of hollow stones functioned as a wave-dampening mechanism. Figure 9 depicts the experimental setup, while Table 1 outlines the dimensions of the various breakwater types employed in the experiments across water depths of 35 cm, 40 cm, 45 cm, 50 cm, and 60 cm.

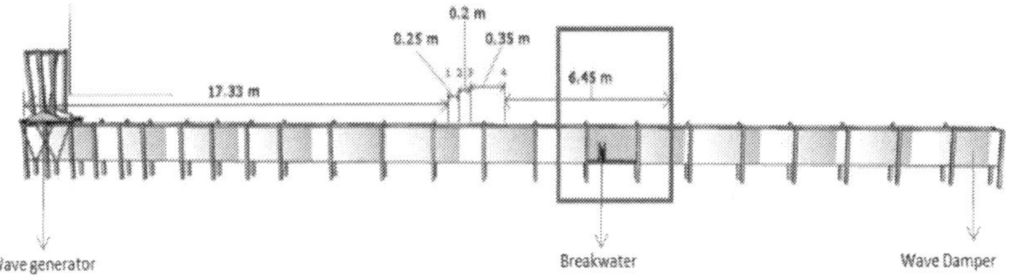

Figure 9: Experimental model of the breakwater setup.

Type No.	Height (H)	Length (L)
1	30 cm	45 cm
2	30 cm	30 cm
3	30 cm	15 cm
4	20 cm	45 cm
5	20 cm	30 cm
6	20 cm	15 cm

Table 1: Dimensions of breakwater types used in the experimental setup.

Utilizing these configurations, the wave transmission coefficients (K_t) were computed, aligning with the principles of the linear Shallow Water Equations (SWEs) for diverse breakwater sizes, water depths, wave heights, and periods. Table 2 presents a summary of the experimental data, numerical results, and the associated relative errors of the model. The table highlights the incoming wave height (H), the incoming wave amplitude (which is half of the wave height, denoted as A), the wave period in seconds (T), the undisturbed water depth (h_0), and the height of the breakwater (BW). Measurements of wave amplitude, expressed in meters, facilitate the evaluation of how wave characteristics impact the transmission coefficient K_t.

The results of this comparison reveal that the average error in the transmission coefficients produced by the numerical model is approximately 4.2% relative to the empirical data. This level of deviation is deemed acceptable within the context of numerical modeling, affirming that our model reliably simulates the wave attenuation effects induced by the breakwater structure.

No.	H (m)	A (m)	T (s)	h_0 (cm)	BW (cm)	L (cm)	Physical K_t	Numerical K_t	Error (%)
1	0.10879	0.05439	1.61	60	30	30	0.8435	0.8870	5.15708
2	0.07156	0.03578	2.04	60	30	15	0.8580	0.8578	0.02331
3	0.06793	0.03396	2.04	60	20	15	0.9483	0.9209	2.88938
4	0.10757	0.05378	1.61	60	20	30	0.9687	0.9392	3.04532
5	0.19330	0.09665	1.08	60	20	45	0.8311	0.8770	5.53380
6	0.09281	0.04641	1.61	50	20	30	0.8899	0.9295	4.44994
7	0.05919	0.02959	2.04	50	20	15	0.9254	0.9047	2.23687
8	0.09060	0.04530	1.61	50	30	30	0.8442	0.8652	2.48756
9	0.04082	0.02041	2.04	45	20	15	1.0860	0.9890	8.93186
10	0.08762	0.04381	1.61	45	20	30	0.9539	0.9249	3.04015
11	0.09100	0.04550	1.61	45	30	30	0.8492	0.8462	0.35317
12	0.06249	0.03125	1.61	40	20	30	1.0339	0.9209	10.92949
13	0.05665	0.02833	2.04	35	20	15	0.8240	0.8649	4.96359
14	0.06912	0.03456	1.61	35	20	30	0.8591	0.9179	6.84437
15	0.11318	0.05659	1.08	35	20	45	0.9158	0.9528	4.04018

Table 2: Comparison of wave transmission coefficients (K_t) from experimental observations and numerical simulations.

5.5 Comparison of Numerical Results vs. Analytical Solutions

Following the validation of the numerical scheme against experimental data, this subsection focuses on verifying the numerical model used in the simulations by comparing the results with analytical solutions. The accuracy of the numerical method is assessed by looking at how well the simulated wave behavior matches the theoretical predictions. This comparison includes an analysis of wave amplitudes, velocities, and attenuation effects across different scenarios. This validation confirms that the numerical model effectively represents physical processes and provides reliable predictions for wave attenuation and coastal protection.

The first numerical simulation tested different values of the friction coefficient f, ranging from 0 to 1. The observational domain remained unchanged, with the simulation time set to $t = 4.2$. The results from the numerical scheme closely match the analytical solutions regarding the relationship between K_t and f. As f increases, K_t gradually decreases, as shown in Figure 10 below.

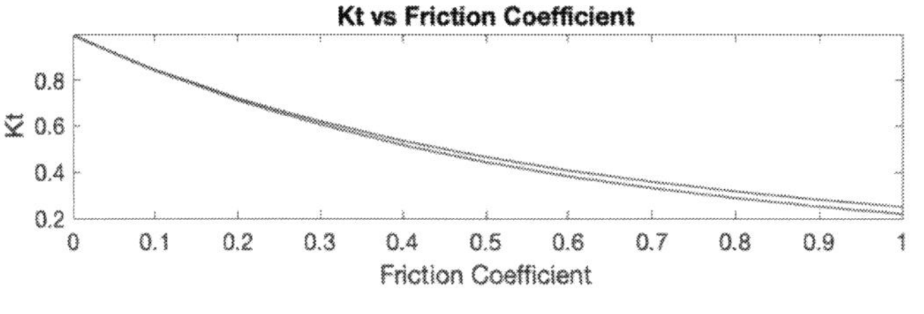

Figure 10: The effect of the friction coefficient on K_t.

Next, another numerical simulation was conducted using various mangrove widths from 0 to 10 meters. The observational domain remained the same and the results showed a good approximation of the analytical solutions when looking at the relationship between K_t and L_2. As L_2 increases, K_t also shows a gradual decrease, illustrated in Figure 11 below.

To examine the effect of submerged breakwater height on the wave transmission coefficient (K_t), the breakwater height, defined as $d - d_1$, was varied from 3 to 7 meters. Here, the total water depth d represents the distance from the seabed to the still water level, while d_1 denotes the depth from the still water level to the top of the breakwater. The results in Figure 12 show that K_t remains relatively constant across this range of breakwater heights, indicating that variations in breakwater height exert minimal impact on wave transmission. This consistent behavior of K_t with respect to $d - d_1$ is in agreement with the analytical solutions, which also reflect a limited influence of breakwater height on wave attenuation.

Additionally, the analytical and numerical values of K_t for different breakwater widths from 0 to 10 meters were explored. The results are illustrated in Figure 13 below. The numerical simulations indicate

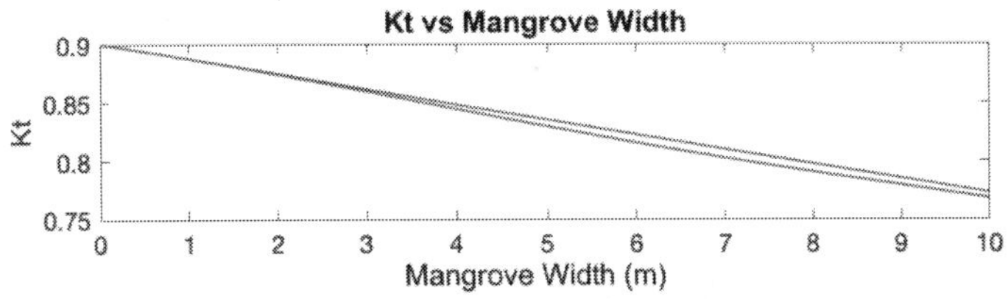

Figure 11: The effect of mangrove width on K_t.

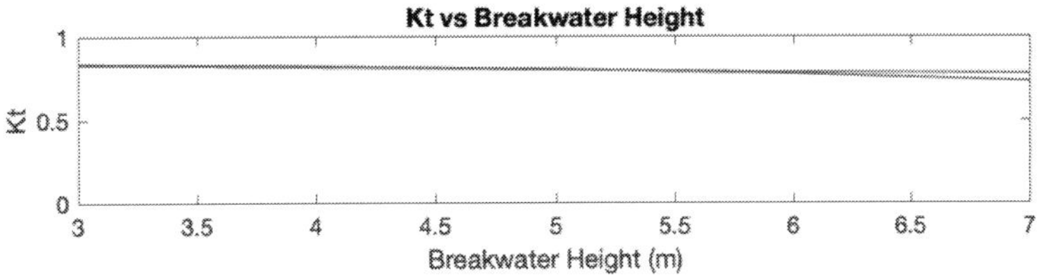

Figure 12: The effect of breakwater height on K_t.

that as the length of the submerged breakwater increases, K_t decreases, which aligns with the conclusions from the analytical solutions.

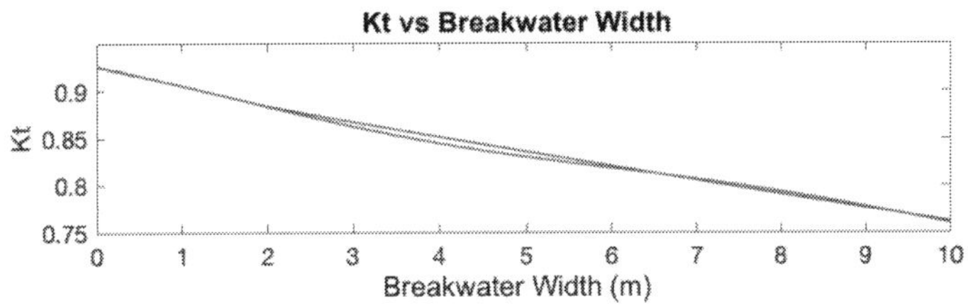

Figure 13: The effect of breakwater width on K_t.

Figures 10 to 13 show a close match between the numerical and analytical transmission coefficients (K_t), confirming the reliability of the numerical method. These figures also reveal a strong link between K_t and the parameters related to mangroves and submerged breakwaters. Notably, increasing the friction coefficient and the width of these structures leads to greater reductions in wave amplitude compared to changes in breakwater height.

6 Conclusion

A modified shallow water model has effectively captured the interactions between waves, mangroves, and submerged breakwaters. By employing the separation of variables method, an analytical solution for the wave transmission coefficient was derived, while the staggered finite volume method was used to simulate wave propagation with high accuracy. Both analytical and numerical methods produced a wave transmission coefficient, K_t, demonstrating a reduction of over 76% in initial wave amplitude when mangroves are combined with submerged breakwaters. This outcome highlights the considerable wave amplitude reduction achieved by these structures, with reductions influenced primarily by factors such as mangrove width, friction coefficient, and breakwater dimensions. The friction coefficients and structure widths proved to be key in wave amplitude reduction, as shown in Figures 10, 11, and 13. The significant slope of K_t along the x-axis in these figures indicates a strong sensitivity to changes in friction and width, with increases in these parameters leading to marked wave height dissipation. Conversely, Figure 12 shows a much flatter K_t slope along the x-axis for breakwater height, indicating a comparatively minor effect on wave attenuation. This result suggests that friction and width are more effective for reducing wave impact, with breakwater height playing a limited role in wave dissipation. The study further demonstrated that positioning the breakwater in front of the mangroves yields a more effective configuration, achieving a lower wave transmission coefficient compared to the reverse setup. These findings provide valuable guidance for designing efficient coastal protection strategies by pinpointing the parameters most influential for wave attenuation. Building on this work, future research could explore wave attenuation by mangrove trees and submerged breakwaters within a two-dimensional framework, allowing for the consideration of variations along the y-axis. This approach would facilitate a more detailed analysis of lateral heterogeneity effects, enhancing our understanding of wave dynamics and supporting further refinements in coastal protection strategies. Additionally, examining different wave types, such as irregular or multi-frequency waves, could offer a more comprehensive understanding of the protective role of mangrove and breakwater configurations in real-world, diverse wave environments.

Contribution Statement

V. Ferren: Data Curation (equal), Formal Analysis (equal), Visualization (lead), Validation and Investigation (equal), Writing – Original Draft Preparation. **I. Magdalena:** Conceptualization (lead), Supervision (lead), Writing – Review & Editing (equal), Data Curation (equal), Formal Analysis (equal), Visualization (lead), Funding acquisition (lead), Project administration (lead), Methodology (lead), Validation and Investigation (equal).

References

[1] A. Duran, The role of mangroves in coastal protection, Journal of Coastal Research 30 (1) (2014) 65–71.

[2] I. Möller, T. Spencer, M. Craig, The role of mangroves in coastal protection: A review, Global Environmental Change 29 (2014) 1–14.

[3] J. M. Graham, S. McCabe, A. Copping, The effectiveness of submerged breakwaters in coastal protection, Coastal Engineering 56 (5) (2009) 467–481.

[4] A. T. Williams, S. Pagliara, J. Bell, The efficacy of submerged breakwaters for coastal defense, Ocean & Coastal Management 80 (2013) 17–24.

[5] S. Narayan, M. W. Beck, M. K. L. Peyre, The value of coastal wetlands for flood damage reduction in the united states, Scientific Reports 6 (2016) 34009.

[6] A. S. Pacheco, A. Lima, Combined effects of mangroves and breakwaters in coastal protection, Journal of Coastal Research 33 (5) (2017) 1046–1058.

[7] I. Magdalena, M. F. Atras, L. Sembiring, M. A. Nugroho, R. S. B. Labay, M. P. Roque, Wave transmission by rectangular submerged breakwaters, Computation 8 (2) (2020) 56.

[8] I. Magdalena, J. L. E. Suhardi, M. B. Adityawan, Analytical and numerical studies for the reduction of wave run-up height by a submerged breakwater, GEOMATE Journal 20 (77) (2021) 1–9.

[9] R. J. LeVeque, Finite Volume Methods for Hyperbolic Problems, Cambridge University Press, 2002.

[10] W. Zhang, X. Li, Numerical simulations of tsunami waves, Computers & Fluids 39 (8) (2010) 1301–1314.

[11] I. Magdalena, R. La'lang, R. Mendoza, Quantification of wave attenuation in mangroves in manila bay using nonlinear shallow water equations, Results in Applied Mathematics 12 (2021) 100191. doi:https://doi.org/10.1016/j.rinam.2021.100191.
URL https://www.sciencedirect.com/science/article/pii/S2590037421000388

[12] I. Magdalena, H. Rif'atin, A. M. A. Matin, Analytical and numerical studies for seiches in a closed basin with bottom friction, Theoretical and Applied Mechanics Letters 10 (6) (2020) 429–437. doi:https://doi.org/10.1016/j.taml.2020.01.057.
URL https://www.sciencedirect.com/science/article/pii/S2095034920300738

IOP Conf. Series: Earth and Environmental Science **1464** (2025) 012010 doi:10.1088/1755-1315/1464/1/012010

A hybrid structure to reduce wave overtopping in urban coast of Jakarta

Octareza Siahaan[1*] and Hiroshi Takagi[2]

[1]Department of Water and Maritime Indonesia, Royal Haskoning DHV, Jakarta, Indonesia
[2]Department of Transdisciplinary Science and Engineering, Tokyo Institute of Technology, Tokyo, Japan

*E-mail: octareza.siahaan@rhdhv.com

Abstract. Jakarta is the Capital of Republic of Indonesia that expands in low lying coastal area. Frequency of coastal floods has been increasing these days and will be exacerbated by sea-level rise and land subsidence. Due to growing economic and population in this city, disaster risk would be further increased. In order to reduce the coastal flood risk, the Government of Indonesia plans to protect coastal area by constructing the tall coastal dike with a height of 4.8 m. Since design wave height of 100 years return period is 3.5 m, the height of the dike seems to be sufficient to prevent high waves. However, this new dike would cause many issues: for example, the dike appears too high for fisherman to access to the sea. In addition, a narrow tall dike may suddenly collapse due to a structural failure or unprecedented events such as a collision of ship or earthquake. In order to avoid such future risks, the authors propose a novel hybrid structure, which aims to lower the height of coastal dike without losing safety level by combining the elements of mangrove, landfill, and dike. Physical experimental was conducted with model scale 1:20. The authors also conducted site survey to adequately model mangroves of Jakarta. Based on the overtopping discharge, a non-dimensional graph is proposed to simply convert the quantity to real scale discharge. The cost of proposed hybrid structure is also estimated and compared with that of an existing coastal dike structure, demonstrating the proposed hybrid solution can be considered as an alternative solution of the planned tall dike by the government.

1. Introduction

Jakarta is the Capital of Republic of Indonesia that expands in low-lying coastal area. Jakarta suffers flood that caused by those from upstream, city drainage, and high tide on coastal area. Flood disaster frequently occurred in coastal area due to wave overtopping, which will be exacerbated by sea level rise and land subsidence. Subsidence rates along the coast of Jakarta ranged from 9.5 to 21.5 cm/year for the 2007 to 2009 period [1]. Muara Baru in Pluit Jakarta is one area that has large land subsidence rate. Between 1997-2011 the annual land subsidence rate was 0.20 m. Although the annual rate has slowed to 0.12 m between 2011-2018, significant subsidence is still in progress [2]. As a result of land subsidence, coastal dike has also been sinking, resulting in an increased risk of wave overtopping/ flood disaster.

In order to reduce the future disaster risk, the Government of Indonesia plans to protect coastal area by constructing the high coastal dike with a height of 4.8 m under Project National Capital Integrated Coastal Development (NCICD) [3]. However, this new dike would cause unforeseen issues in the near future. For example, the dike is too high for fisherman to access to the sea, and a narrow high dike may collapse due to a structural failure or unprecedented events such as a collision of ship or an earthquake [4].

Jakarta is the city where mangrove is able to grow naturally with a number of species of mangroves such as *Rhizophora sp.* and *Avicenna sp* [5]. Mangrove is expected to be used as an ecological solution for coastal protection to become more pronounced with the progress of land subsidence in Jakarta [6]. Muara Baru in North Jakarta where has experienced largest land subsidence rate is considered as target location in this research [2].

Currently, there appears to be a significant gap between engineering (structural) and ecological solutions. In order to harmonize these two solutions, an adaptive mangrove hybrid platform is proposed as shown in Figure 1 [7]. This research aims to elaborate utilization of a hybrid structure between mangrove and grey structure such as concrete dike and landfill. The authors propose a novel hybrid structure, which aims to lower the height of coastal dike with similar standard safety level with existing coastal dike plan. Mean overtopping discharge -a common design parameter that evaluates the performance of the dike- is used to measure the effectiveness of coastal defence.

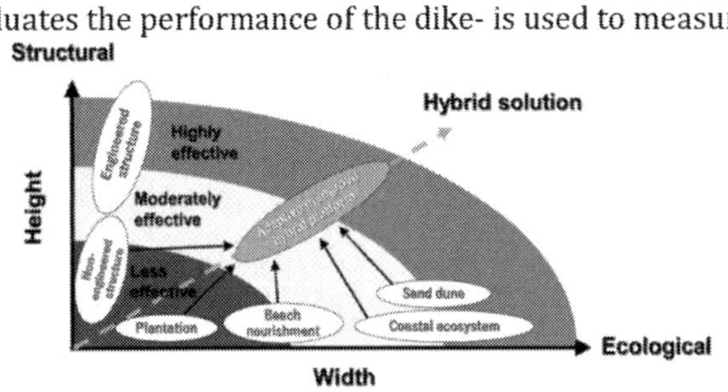

Figure 1 Illustration of gap between engineering (structural) and ecological solution (Takagi, 2019).

This research conducted survey for characterizing mangrove in Jakarta and laboratory experiment to examine wave overtopping for several scenarios with different dike height and mangrove width. The experiment results were compared with computational results. Wave overtopping non-dimensional graph was generated and compared with previous similar experimental. Furthermore, non-dimensional graph was created for usefulness in a basic design and cost estimation was presented for practical design of hybrid structure in Jakarta.

2. Data and Resources

This section summarizes data and resources used in this research, which include wave, tidal, bathymetry data, coastal dike design parameters in the NCICD Report [3], and mangrove observed in site visit.

2.1. Wave, tidal, and bathymetric conditions in Jakarta Bay

Wave condition in Jakarta Bay is collected from the Detail Design Coastal Protection [3]. The NCICD report uses the return period of 1/100 years, and the corresponding significant offshore wave height and period are 3.5 m and 10.06 s, respectively. The highest tidal amplitude in Jakarta Bay is 1.19 m from the lowest water level. In the bay near Muara Baru, water depth varies roughly from 3.0 m to 7.0 m below the lowest water level [3].

2.2. New coastal dike in NCICD project

New coastal dike constructed in Jakarta is a vertical wall. Spoon pile structure that has tight joint between piles, was adopted to reduce seepage from sea to the land. Sealing of seepage is one of the key technologies to mitigate coastal floods in low-lying area. Due to soft soil in Jakarta Bay, soil improvement is required. Soft soil is also disadvantage as it would promote scouring. Hence, toe protection was designed to be constructed in front of the new dike. The elevation of coastal dike was determined as high as 4.8 m from the lowest water level [3], which considered wave height, land subsidence, high tide, sea level rise and surplus elevation. On-going construction NCICD Project in Muara Baru considered the reclamation island in Jakarta Bay. Hence, wave height in research location could be reduced, resulted in slightly lower coastal dike with 3.5 m from the lowest water level. However, this research followed coastal dike design of NCICD [3] (i.e. the dike height is 4.8 m).

2.3. Mangroves in Jakarta
Our field survey observed characteristic of mature mangrove in coastal of Jakarta since it is resilient from wave, wind, and another environmental force. *Rhizophora sp.* and *Avicenna sp. are* dominated with the height of typical mature *Rhizophora sp.* is about 5 m with the root of 1.0 m high above the ground, while mature *Avicenna sp.* is 5 m high and has prop roots of 0.2-0.3 m above the ground. A critical factor affecting wave attenuation in mangroves is the density of obstacles that waves encounter as they pass through the mangrove, and the height of these obstacles relative to the water depth [8]. In the similar water level, the higher prop roots above the ground of *Rhizophora sp.* led to higher wave attenuation than *Avicenna sp.* Therefore, *Rhizophora sp.* is considered in this research model to expect effectiveness against wave attenuation and overtopping.

The observed mature *Rhizophora sp.* has a total of 15-20 attached and unattached roots with density 3.3 roots/ m^2. Attached root is root that grow from main trunk and penetrated into the soil, while unattached roots are hanged.

3. Experimental Set-up

Experimental laboratory is conducted to model the mangrove, landfill, dike, and environmental condition by reflecting the information obtained at the site. Properties used in the experimental are mangrove model made by metal and twisted as shown in Figure 2, dike model made by rubber and impermeable, landfill and slope are made by gravel and permeable with size around 10 mm.

3.1. Model properties and size
The experiment modelled 10 attached and 10 unattached roots. Canopy is not considered in the model for more conservative result (i.e. to avoid an overestimation of mangrove resistance). Scale of 1:20 was considered suitable to encompass the model mangrove in the experimental flume, and the Froude similarity law was used to decide model elements consisted of mangrove, dike, landfill, water reservoir to measure wave overtopping volume. Table 1 shows the size of each object and Figure 2 shows the dimension of mangrove model.

3.2. Experimental setup and procedures
The wave flume used in this research has the dimension of 30 cm high, 10 cm wide, and 300 cm long. Wave measurement gauge (KENEK CHT6-30) was used to measure the regular wave heights and periods at a location of 30 cm in front of the wave generator (Figure 2).

The wave flume was initially filled with water of 180 mm deep, and this water level was set to be zero as reference level in analysing water level variations. Prior to conducting the experiment, accuracy of wave gauge measurement was verified by comparing the measured water levels between wave gauge measurement and manual observation.

IOP Conf. Series: Earth and Environmental Science **1464** (2025) 012010 doi:10.1088/1755-1315/1464/1/012010

Table 1 Dimension of experiment properties.

Object	Prototype size	Model size
Dike height	1.4 m above the sea level	70 mm
Mangrove height	5 m above the ground	250 mm
Mangrove root height	1 m above the ground	50 mm
Mangrove root diameter	4-6 cm	2-3 mm
Landfill height	3 m	150 mm
Wave height	1.75 m	87.5 mm
Wave period	6.78 s	1.49 s
Water depth	3.6 m below the sea level	180 mm

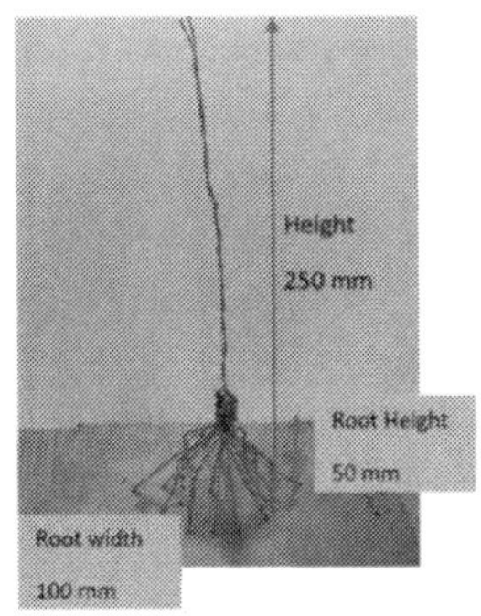

Figure 2 Mangrove model used in the experiment, composed of trunk and root system of *Rhizophora sp.*

The experiment aims to estimate the reduction in wave overtopping by placing the mangrove models in front of the vertical dike (Figure 3). If wave overtopped the dike, the overtopping discharge was measured by reading water accumulated in the container with the size of 400x100x100 mm. The discharge rate was calculated by dividing the volume by the duration of each experiment, resulted in wave overtopping discharge rate per width (l/s/mm). The main objective of experiment is to find the effectiveness of hybrid structure to the wave overtopping, and thereby wave overtopping between with and without mangrove was compared.

Owen and Allsop (1983) [9] suggest that experiment should be continued more than 20 minutes to produce reliable waves. However, due to limitation of the experiment flume with maximum case of 7 layers mangrove models and capacity of water container, the experiment was limited to be carried out within 4 minutes.

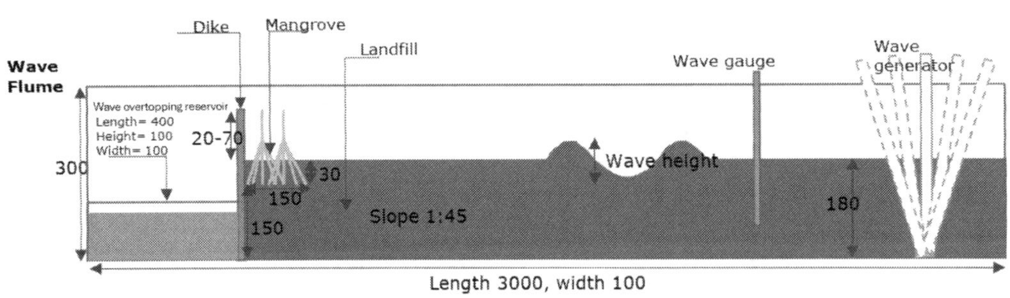

Figure 3 Illustration of experimental wave flume (unit: mm).

Figure 4 Photo of mangrove, landfill and dike (left) and wave flume set-up (right).

Ideally the experiment should be carried out with the model dike height size of 24 cm corresponding to the dike height of 4.8 m [3]. However, due to the wave flume height limitation, the dike elevation was limited to 15 cm above water level. When the experiment was conducted with this dike elevation, wave overtopping did not occur, and thus the elevation was gradually lowered to find the condition in which the overtopping starts. After the dike height was set lowered to 7 cm, 2 layers mangrove model were installed on the landfill, and continued the experiment for 4 minutes then water in container started to be accumulated. Several dike heights from 2 cm to 7 cm height from water level and the mangrove model varying from 2 layers (15 cm width) to 7 layers (40 cm width) were tested (Table 2).

Mangrove density is calculated by such as an illustration in Figure 4, which shows the case of 2 mangrove trees overlapped. Density of this case is 2 trees/ 6 m^2 or estimated equivalent as 33 trees/ 100m^2.

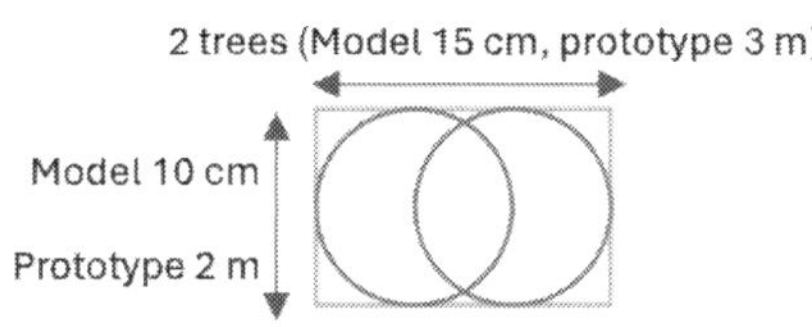

Figure 5 Illustration density of mangrove trees (Overlap in the direction of the dike).

With similar calculation, mangrove density for 5 m width (4 mangrove trees), 6 m width (5 mangrove trees) and 8 m width (7 mangrove trees) are estimated equivalent as 40 trees/ 100 m^2, 41.7 trees/ 100 m^2 and 43.8 trees/ 100 m^2, respectively.

3.3 Data validation

Wave overtopping is the main parameter in this research. Each experiment case repeated 4-10 times to obtain statistically reliable wave overtopping data. Data that may cause inconsistency to the result were eliminated, based on the standard deviation (those data less than 0.3 l/s/m were used).

The wave flume length is relatively short that cause reflective wave affects wave condition. Hence, wave height validation data also carried out. The authors used a wave separation analysis based on Goda and Suzuki [10] to determine both mean wave height H_{mean} and mean wave period T_{mean} of experiment. Upon completion of experiment, wave-related parameters such as wave height, wave period and wave overtopping rate are obtained.

Table 2. Case of experiment.

Case	Landfill width (cm)	Dike height (From water level in cm)	Mangrove width (cm)	Number of mangrove layer
1-1	15	7	-	-
1-2	15	2	15	2
1-3	15	7	15	2
2-1	25	7	-	-
2-2	25	2	25	4
2-3	25	4	25	4
3-1	30	7	-	-
3-2	30	4	30	5
3-3	30	7	30	5
4-1	40	2	40	7
4-2	40	5	40	7
4-3	40	7	40	7

4. Experimental Results

4.1 Wave overtopping discharge

Experiment result shows that addition of mangrove and dike height gradually (see the case number in Table 2) reduce the wave overtopping discharge (Figure 6). The error bar indicates the standard deviation in (l/s/m).

Absolute wave overtopping rate among all conditions may not be directly compared since incident waves were not necessarily same due to various wave height in the experimental. Hence, in order to normalize the wave heights, non-dimensional graph of wave overtopping and freeboard (dike) is generated.

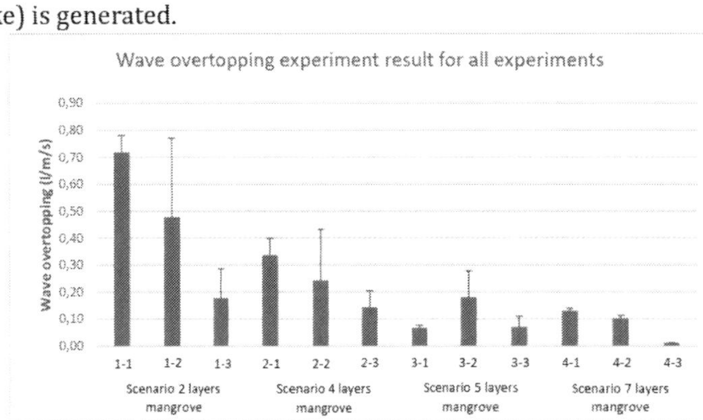

Figure 6 Result of wave overtopping (error bar is standard deviation in l/m/s).

The authors compared the non-dimensional wave overtopping graph with previous research as Ref. 1 cited from S. Dong [11] and Ref. 2 cited from Y.Q. Chim [12]. Both studies utilized the non-dimensional representation of $(q/gH_{m0}^3)^{0.5}$ and R_c / H_{m0} where q is wave overtopping and R_c is freeboard, and H_{m0} is significant wave height. Graph of the non-dimensional discharge is presented in the following section with purpose to compare result of this research with Ref. 1 and 2.

4.2 Comparison with previous studies and design manual

There is not much previous research that investigates basic design and construction cost of hybrid structure with mangrove forest, landfill, and low dike, which aim to lower coastal dike elevations in actual coasts. However, the following studies and manual are relevant with our study.

(1) Ref. 1 [11]

The research was carried out by utilizing four retrofit prototypes, including recurve wall, model vegetation, reef breakwater and diffraction pillars in front of a vertical wall to examine the performance in mitigating wave overtopping.

In the experiment of Ref.1, the mangrove density was 19 stems/ 100 m^2 (utilize pile), with dike and without landfill, while the present research considered the mangrove density 33-43 trees/ 100 m^2 with dike and landfill. Stem/ pile that performed in Ref. 1 is associated as tree. Therefore, the density with present research is similar and comparable. Ref. 1 produced a non-dimensional graph that consisted of dimensionless discharge $(q/gH_{m0}^3)^{0.5}$ and dimensionless freeboard (R_c/H_{m0}). The non-dimensional graph of Ref. 1 and the present research are shown on Figure 6 (Left) and (Right), respectively.

In Figure 7 (Left) the mangrove density 19 stems/ 100m^2, dimensionless freeboard range 0.8-1.1 shows non-dimensional discharge $1x10^{-3}$, while present research (Right) conducted with mangrove width 15 cm (the least mangrove density 33.3 trees/ 100 m^2) shows non-dimensional discharge $4x10^{-3}$ - $8x10^{-3}$. In addition, Figure 7 (Left) with only plain vertical wall, dimensionless freeboard shows non-dimensional discharge $2x10^{-3}$ while present research (Right) conducted with only landfill width 15 cm shows non-dimensional discharge $6x10^{-3}$ - $1x10^{-2}$. Both results gave not exactly same but similar order of overtopping discharge.

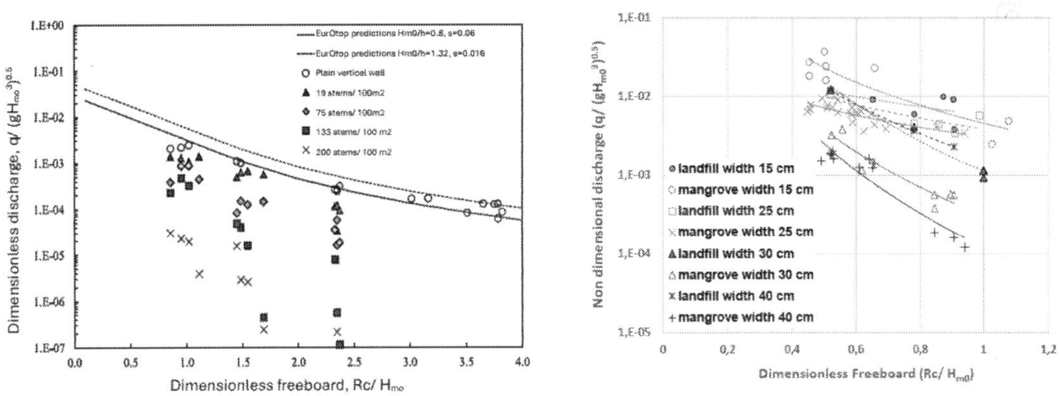

Figure 7 Non-dimensional graph of Ref.1[11] (Left) and this research (Right).

(2) Ref. 2 [12]

The objective of Ref. 2 research is to investigate the performance of hybrid coastal protection system consisting of nature system (mangrove) and hard structure (armoured revetment with berm). The revetment armour used is the Xbloc^{Plus}. Xbloc^{Plus} is a regularly placed armour which achieve its stability through block weight, friction, and interlocking between neighbouring blocks. Xbloc^{Plus} has circular hole in the centre which reduces the uplifting pressure and increase stability [12]. Ref. 2 produced non-dimensional graph that consisted of dimensionless discharge $(q/gH_{m0}^3)^{0.5}$ and relative crest height (R_c /H_{m0}). In this reference, the physical model of mangrove mimics *Rhizophora* species' trunk-prop root system. A total of 21 trees were placed with mangrove width of 18 m and density of 9.7 trees/ 100 m^2 in prototype scale. For the difference with this research, Ref.2 does not intend to lower the tall dike. Dike shape in this Ref. 2 is slope wall, whereas this research considered a vertical wall, mangrove root configuration is not overlap, mangrove less dense and utilized wider mangrove and water level is lower than this research. Even though the shape of the wall is different, configuration of wall and mangrove is comparable to see the similarity order of non-dimensional overtopping graph. Comparison non-dimensional graph is shown on Figure 8 and Table 3.

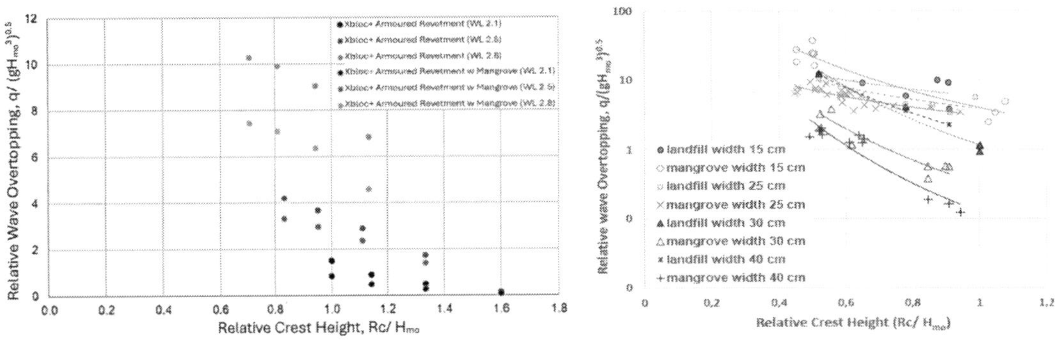

Figure 8 Non-dimensional graph of ref. 2 [12] (Left) and this research (Right).

Table 3 Comparison results of Fig 8 (Left) and (Right).

Configuration	Previous research, Figure 7 (Left)	Current research, Figure 7 (Right)	Comparison
With mangrove	• XblocPlus +Armoured revetment with Mangrove (WL 2.8 m), mangrove density 9.7 trees/ 100m2 • Non-dimensional graph 4-10	• Mangrove width 15 cm (density 33 trees/ 100 m2), WL 3.6 m • Non dimensional graph 4-40	Previous and current research with mangrove case shows similar order of non-dimensional graph
Without mangrove	• XblocPlus +Armoured revetment without Mangrove (WL 2.8 m) • Non-dimensional graph 6-10	• Vertical wall, WL 3.6 m • Non dimensional graph 6-10	Previous and current research without mangrove case shows similar order of non-dimensional graph (6-10)

(3) Design Manual (EurOtop) [13].

Design Manual that used for comparison is EurOtop application named XGB-overtopping. XGB-overtopping is a conceptual design tool to estimate wave overtopping discharges for a wide range of coastal structures. The model is based on about 10,000 physical model tests [14] at several institutes (Deltares, The Netherlands; Danish Hydraulic Institute, Denmark, etc. and other sources from countries including Iceland, Japan, Norway and U.S.A) [15].

Table 4. Comparison wave overtopping landfill case in this present experimental and EurOtop.

Case	Case	Present experimental Q (l/s/m)	EurOtop Q upper 97.5% (l/s/m)
Landfill with 3m and dike 1.4m	1-1	7,200 x 10^{-2}	2,500 x 10^{-2}
Landfill with 5m and dike 1.4m	2-1	3,300 x 10^{-2}	2,000 x 10^{-2}
Landfill with 6m and dike 1.4m	3-1	6,000 x 10^{-2}	2,000 x 10^{-2}

In the experiment under this research, vertical dike model with height 7 cm corresponding to prototype 1.4 m with landfill model width 15 cm, 25 cm and 30 cm corresponding to prototype 3 m, 5 m and 6 m respectively were carried out. Corresponding to these configurations, XGB-overtopping is performed to estimate the overtopping discharge with the similar geometry. As the result, the wave overtopping of experiment and EurOtop shows the same order of discharge as shown in Table 4, suggesting that the present experimental result without mangrove shows not exactly same but fair to good accuracy compared to EurOtop.

5. Discussion

5.1 Comparison of wave overtopping among different configurations

Wave overtopping measured in this experimental research was analysed to assess the difference in performance among three different configurations: (1) only vertical wall; (2) landfill + vertical wall (without mangrove) and (3) landfill + vertical wall+ mangrove (hybrid solution). The performance of each configuration was plotted as a non-dimensional graph (Figure 9 and Figure 10).

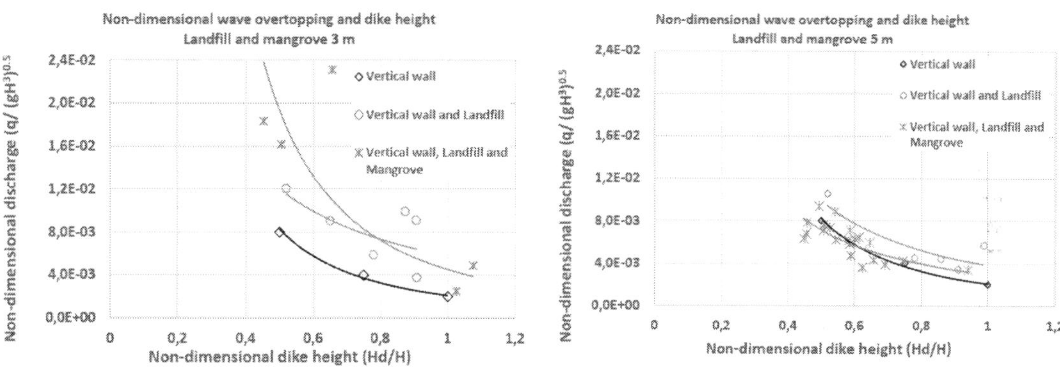

Figure 9 Comparison of wave overtopping rate among vertical wall, landfill, and hybrid solution with mangrove 3 m (2 layers) (Left) and mangrove 5 m (4 layers) (Right).

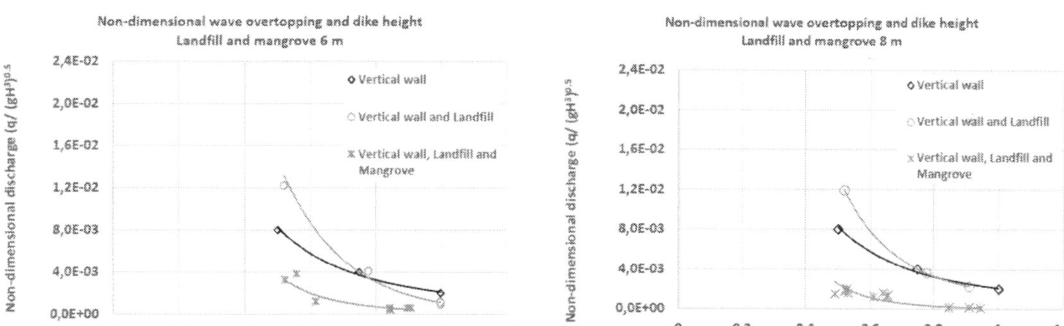

Figure 10 Comparison of wave overtopping rate among vertical wall, landfill, and hybrid solution with mangrove 6 m (5 layers) (Left) and mangrove 8 m (7 layers) (Right).

The cases of the addition of landfill to vertical wall in Figure 9 and Figure 10 are similar with the vertical wall graph. Only landfill case is less effective in reducing wave overtopping comparing to the hybrid system with mangrove. Landfill surface level with distance 3 cm below water level probably caused wave energy reduction due to turbulence and reflection was limited.

According to Magdalena and Azis [16], wave attenuation factor can be affected by friction coefficient, diffusion coefficient and vegetation length. Uncertainties in these fundamental factors seem to have caused underestimation of wave attenuation for the 2-layer mangrove case, resulting in larger wave overtopping. However, the non-dimensional graph of vertical wall and vertical wall and landfill for 4 cases shows relatively a similar order.

The hybrid systems with 2 mangrove layers and 4 layers (density 33.3 and 40 trees/ 100 m^2) show a close non-dimensional discharge to each other. It indicates that beside the mangrove length, the mangrove layers/ density in the rage of this study is not sufficient to significantly reduce wave overtopping. However, hybrid system with 5 mangrove layers and 7 layers (density 41.7 and 43.8 trees/ 100 m^2) showed that hybrid system in this mangrove layers/ density contributes to reduction in wave overtopping significantly probably due to wave energy reduction through turbulence and reflection. The authors found that mangrove density with 41.7 trees/ 100 m^2 or 5 layers as the least dimension for an effective mangrove and dike hybrid structure.

5.2 Basic design and cost estimate

According to EurOtop [13] and OCDI [17], tolerable wave overtopping for residence area is between 1-10 l/s/m. For this research, the tolerable wave overtopping is considered as 5 l/m/s.

In order to achieve wave overtopping being less than 5 l/m/s for the condition of H_{m0}: 3,5 m [3] and 5 layers mangrove of 5 m height (6 m width), according to Figure 10 (Left), the coastal dike height is required 1.8 m at least, such as shown in Figure 11.

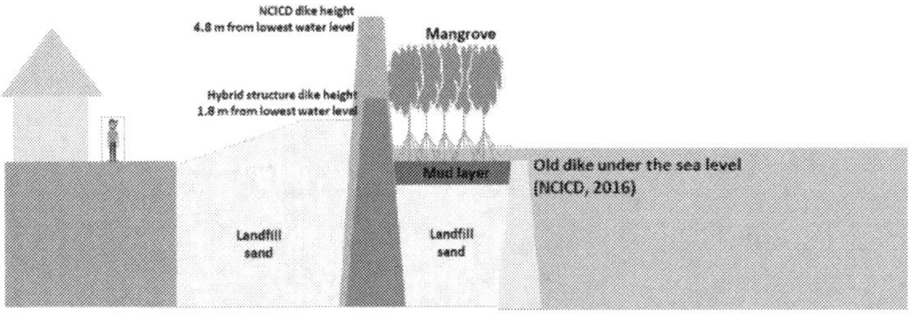

Figure 11 Proposed basic design dike height 1.8 m (Left) and old dike inundated (Right).

According to NCICD [3], typical coastal dikes with spoon pile, the construction cost is estimated USD 3,900/m^2 (including additional 10% for tax and operation and maintenance cost), which is similar with some major industrial port in Japan that costed approximately USD 4,000/m^2 to elevate the dyke by 1 m with a highly engineered solution [7]. Those references suggest that USD 3,900/m^2 is reasonable as reference unit construction cost. Cost estimation for our proposed hybrid solution is shown on Table 5.

Table 5. Comparison cost estimate of NCICD design and proposed hybrid solution [3].

Work item	Unit cost (USD/m^2)	Estimation cost for coastal dike 4.8 m (USD)	Estimation cost for coastal dike 1.8 m and 6 m width mangrove (USD)
Preparation	20	20	20
Pilling work	1,323	6,350	2,381
Caping beam work	610	610	610
Rubble mound installation	700	700	-
Landfill	542	2,600	975
Landfill for mangrove (sand)		-	540
Landfill for mangrove (mud)	18	-	72
Mangrove plantation	0.6	-	1.1
Sub-total	3,195	10,280	4,599
Total with maintenance cost	3,514	11,308	5,095
Total with tax	3,865	12,439	5,565
Total cost (round-up)	3,900	12,500	5,600

Cost estimate of construction 4.8 m coastal dike in Jakarta is USD 12,500/ m [3], while with the similar performance, construction cost for dike 1.8 m and additional 6.0 m mangrove is estimated USD 5,600/m.

6. Conclusion

The comparison between non-dimensional graph produced by this research and those in previous research indicates similar order of wave overtopping discharge. Although only addition landfill is less effective in reducing wave overtopping, a hybrid system with 5 layers mature mangrove (6 m width) on the landfill demonstrates remarkable reduction in wave overtopping.

For example, in order to achieve wave overtopping 5 l/m/s, for the condition of H_{m0}: 3.5 m and mature mangrove 6 m width, coastal dike height is required 1.8 m. The hybrid structure proposes a lower coastal dike and lower cost estimate than existing coastal dike plan NCICD [3].

References

[1] E. Chaussard, F. Amelung, H. Abidin, and S.-H. Hong, "Sinking cities in Indonesia: ALOS PALSAR detects rapid subsidence due to groundwater and gas extraction," Remote Sens. Environ., vol. 128, pp. 150–161, Jan. 2013, doi: 10.1016/j.rse.2012.10.015.

[2] Government of DKI Jakarta Water Resources Agency, 2021. Presentation Risk Coastal Area Under Sea Water Level: Webinar 02 Sept 2021

[3] Satuan Kerja BBWS Ciliwung Cisadane, "Detail Design Pembangunan Pengaman Pantai (NCICD) Lanjutan," Ministry of Public Work and Housing, Jakarta, Indonesia, Agustus. 2016.

[4] H. Takagi, D. Fujii, M. Esteban, and X. Yi, "Effectiveness and Limitation of Coastal Dykes in Jakarta: The Need for Prioritizing Actions against Land Subsidence," Sustainability, vol. 9, no. 4, Art. no. 4, Apr. 2017, doi: 10.3390/su9040619.

[5] A. Rumondang, C. KUSMANA, and S. W. Budi, "Species composition and structure of Angke Kapuk Mangrove Protected Forest, Jakarta, Indonesia," Biodiversitas J. Biol. Divers., vol. 22, Aug. 2021, doi: 10.13057/biodiv/d220932.

[6] H. Takagi, T. Mikami, D. Fujii, M. Esteban, and S. Kurobe, "Mangrove Forest against dyke-break-induced tsunami on rapidly subsiding coasts," Nat. Hazards Earth Syst. Sci., vol. 16, no. 7, pp. 1629–1638, Jul. 2016, doi: 10.5194/nhess-16-1629-2016.

[7] H. Takagi, "'Adapted mangrove on hybrid platform' – Coupling of ecological and engineering principles against coastal hazards," Results Eng., vol. 4, p. 100067, Dec. 2019, doi: 10.1016/j.rineng.2019.100067.

[8] A. L. McIvor, I. Möller, T. Spencer, and M. Spalding, "Reduction of wind and swell waves by mangroves. Natural coastal protection series: report 1. Cambridge Coastal Research Unit working paper 40," Reduct. Wind Swell Waves Mangroves Nat. Coast. Prot. Ser. Rep. 1, Jan. 2012.

[9] Owen M.W. & Allsop N.W.H., 1983, Hydraulic modelling of rubble mound breakwaters, Proceedings Conference on Breakwater: Design and Construction, ICE, London.

[10] Y. Goda and T. Suzuki, "Estimation of Incident and Reflected Waves in Random Wave Experiments," in Coastal Engineering Proceedings, Jan. 1976, p. 47. doi: 10.9753/icce.v15.47.

[11] S. Dong, S. Abolfathi, M. Salauddin, Z. H. Tan, and J. M. Pearson, "Enhancing climate resilience of vertical seawall with retrofitting - A physical modelling study," Appl. Ocean Res., vol. 103, p. 102331, Oct. 2020, doi: 10.1016/j.apor.2020.102331.

[12] Y. Q. Chim, M. Muilwijk, T. Ruwiel, B. Reedijk, and Z. Q. Yang, "Physical Model Investigation of Hybrid Coastal Protection, combining Nature-Based and Hard Infrastructure," Apr. 2022.

[13] J. van der Meer et al., EurOtop: Manual on wave overtopping of sea defences and related sturctures - An overtopping manual largely based on European research, but for worlwide application (2nd edition). 2016.

[14] J. Gosse, G. Steendam, J. van der Meer, D. Meer, P. Besley, and M. Gent, "The international database on wave overtopping," Jan. 2004, doi: 10.1142/9789812701916_0347.

[15] XGB Overtopping by Deltares https://dserie.deltares.nl/XGBOvertopping/default.aspx website was accessed on June 23, 2023

[16] I. Magdalena, V. Kusnowo, and M. Azis, "1D-2D Numerical Model for Wave Attenuation by Mangroves as a Porous Structure," Computation, vol. 9, Jun. 2021, doi: 10.3390/computation9060066.

[17] T. O. C. A. D. I. of Japan, Technical Standards and Commentaries for Port and Harbour Facilities in Japan. OCDI, 2002.

IOP Conf. Series: Earth and Environmental Science 1464 (2025) 012011 doi:10.1088/1755-1315/1464/1/012011

Morphodynamic modelling of Bandar Lampung Coast

Muhammad Fatkhurrozi[1*], Mustarakh Gelfi[1], and Aji Suraji[2]

[1] Ocean Engineering Department, Institut Teknologi Sumatera, Lampung Selatan, Indonesia
[2] Civil Engineering Department, Universitas Widyagama Malang, Malang, Indonesia

*E-mail: muhammad.fatkhurrozi@kl.itera.ac.id

Abstract. Modelling sediment transport is central in studying morpho dynamics of coastal areas where complex hydrodynamic features present. This study will investigate the sediment transport within the Pulau Pasaran Strait in Bandar Lampung coast. The problem involves sediment flow encountering rows of bridge piles in shallow water environment. The common hydrodynamic model based on non-linear shallow water equation is used. The results indicate a significant difference in sediment yield between the two sides of the bridge, highlighting the impact of the piles on disrupting the flow through the strait. Fully developed land formation around the bridge is projected to occur within 8.69 years.

1. Introduction

The coastline of Bandar Lampung features a complex geography, ranging from estuaries to the strait that separates Sumatra Island and Pasaran Island. The presence of the Pasaran Bridge, which connect the two islands, along with the bustling economic activities in the area, adds to the unique character of the coast (Figure 1). Way Kuripan River, that flows into the coast, further contributing to the complexity of the straits morphodynamics. The river's interaction with tidal forces, sediment transport, and currents creates dynamic changes in the shoreline and estuarine systems, influencing both the natural environment and human activities along the coast.

The issue of sediment accretion poses a significant challenge for the community, particularly for the sea transportation and fishing industry (Figure 2). This issue needs to be studied to gain a better understanding of the future coastal morphology. In this paper, a sediment transport model for the Bandar Lampung coast is developed using the widely available free hydrodynamics software, Delft3D. The coastal area features small islands (Pasaran Island) surrounded by waters and an estuary, making it an ideal case for studying complex sediment dynamics.

This case share some things in common with the morpho dynamics of the Muria Strait in Central Java, where intense sedimentation has driven land formation, transforming the coastal landscape significantly, a process believed to have materialized in the 17th century by De Graaf [1]. Such gradual reshaping of the coastal landscape underscores the complex interplay of sediment transport, deposition, and erosion, illustrating how sediment dynamics continuously reshape coastal morphology over time.

2. Coastal and sedimentation

2.1 Cases review

The investigation of morphological evolution of a tidal flat using numerical model Delft3D [2] was conducted in Jiangsu Province, China [3]. The study shows that the model demonstrates that the tidal flat's longitudinal profile adapts to a new equilibrium (becoming narrower and steeper) following multiple reclamation events, that is in line with existing field data and analytical formula. Fine sediments accumulate near the sea dike due to flood-dominated tidal currents, with peak deposition occurring when the dike is positioned at mean sea level. Over time and successive reclamations, sediment grain size becomes coarser offshore and across the tidal flat. This study highlights that human activity, e.g. land reclamation, significantly reshapes tidal flat profiles and leads to coarser sediment distribution, factors that should be carefully considered in future reclamation projects.

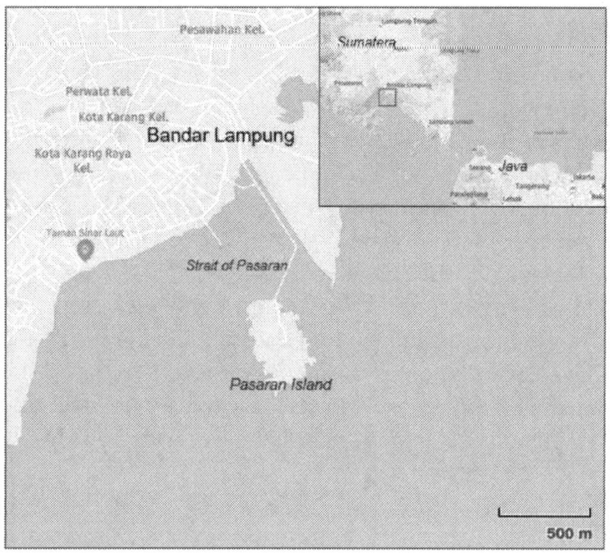

Figure 1. Geographical setting of Bandar Lampung coast.

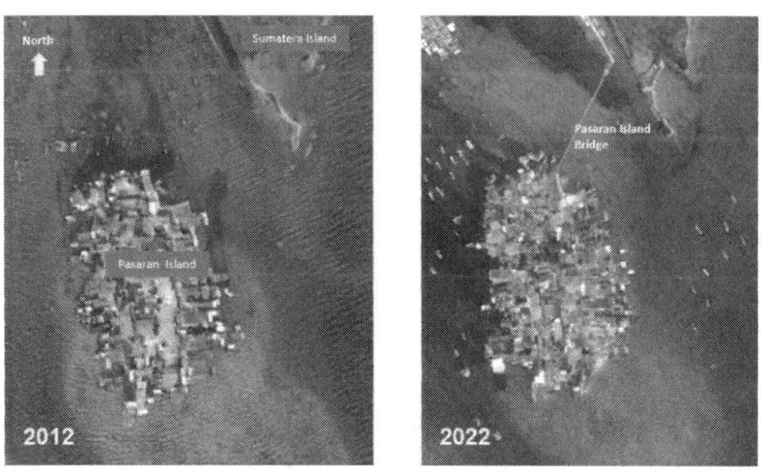

Figure 2. Bandar Lampung coast features strait that separate Sumatera Island and Pasaran Island. Satelite imagery (Google) suggests that sedimentation has been occuring in the strait.

One case studied by Hutley in Moreton Bay aimed to understand the behaviour of cohesive sediment dynamics in the region [4]. The area features relatively strong wind-driven waves with field data showing 0.5 to 1 m of significant wave height. A three-dimensional numerical model of Moreton Bay was built using Delft3D which focuses on cohesive sediment transport processes. The model underwent several calibration and validation processes. The study revealed that, according to the modelled scale of response in the study area, the total bed disturbance leading to sediment suspension was directly attributable to tidal forcing during 2015 and 2016. Hutley found that wind waves are more responsible for sediment dynamics in the area, accounting for 50% of the sediment disturbance, compared to tidal forces, which account for only 10%.

UNESCO-IHE [5] conducted a comprehensive numerical modeling study to evaluate the current longshore sediment transport regime along the northern coast of Sri Lanka and predict sediment behavior following the construction of planned harbors. The region's coastal dynamics are complicated by the presence of estuaries, and the harbors will include breakwaters that may disrupt sediment transport. The hydrodynamic model incorporated wind-wave mechanisms, as the coastal environment is predominantly wave-driven, using short time steps for accuracy. The study reveals sediment transport volumes of up to 150,000 m^3 per year, with wave behavior being the primary factor influencing sediment distribution.

The case of land forming is studied by Mirdan et. al. [6] where protective structures present in order to defend coastal form sand erosion. Mirdan et. al. used Delft3D to model the coastal environment with size 2100x790 meters with breakwaters of for 30 days simulation time. The model incorporated wave module to capture the wind wave effect and the time step of 0.05 minutes is used. It is found that tombolo and salient are formed due to the presence of the coastal structure which reduce the wave energy field.

2.2 Coastal sediment models

The Delft3D software consists of several modules that simulate flow dynamics, wave propagation, as well as sediment transport. Delft3D-FLOW is the "basic module" which provides the hydrodynamic foundation for other modules such as water quality, ecology, waves as well as sediment transport. The transport of suspended sediment in 3D is evaluated by solving the advection-diffusion equation for the suspended sediment. The governing equation is

$$\frac{\partial c_i}{\partial t}+\frac{\partial w c_i}{\partial x}+\frac{\partial v c_i}{\partial y}+\frac{\partial (w-w_{s,i})c_i}{\partial z}-\frac{\partial}{\partial x}\left(\varepsilon_{s,x,i}\frac{\partial c_i}{\partial x}\right)-\frac{\partial}{\partial y}\left(\varepsilon_{s,y,i}\frac{\partial c_i}{\partial y}\right)-\frac{\partial}{\partial z}\left(\varepsilon_{s,z,i}\frac{\partial c_i}{\partial z}\right)=S_i \quad (1)$$

where,

c_i = mass concentration of sediment fraction I [kg m^{-3}]

$\varepsilon_{s,x,i}$ = eddy diffusivities of sediment fraction I [$m^2\ s^{-1}$]

S_i = source and sink term representing the exchange with the bed [kg m^{-3}]

u, v, w = flow velocity [m^{-1}]

$w_{s,i}$ = sediment settling velocity of sediment fraction i [$m\ s^{-1}$]

The model first solve the local flow velocities and eddy diffusivities by Delft3D-Flow module. In the sediment transport calculation, the exchange of sediment between the bed and the flow, and the settling velocity of sediment under the action is incorporated in the model. For each type of sediment (mud or sand), formulation of several processes such as settling velocity, sediment deposition and pick-up are treated differently. The Delft3D also account the effect of sediment on fluid density.

Delft3D-FLOW utilizes staggered grid, i.e. the each of the quantities are not defined at the same location. Water level and concentration of constituents (sediment concentration) are defined in the center of cell while velocity components are defined on the interface of cell.

The choose of time step and grid size is governed by Courant number. Generally, the time integration of the hydrodynamic calculation utilizes explicit method. An implicit method for time integration is used instead when some potential instability occurs which utilizes Crank-Nicolson scheme. It is usually used when inversion of a large matrix present. After discretization, the spatial differential operators in both grid directions yield a system of difference equations forming a band matrix. Each row contains at least five non-zero entries, representing the surrounding grid cells associated with each water level point. The bandwidth of the matrix is relatively wide and all grid points are implicitly coupled [2].

3. Bandar Lampung Cases

This study involves numerical experiment on sediment transport at the vicinity of Pasaran Island, Bandar Lampung. Field data was collected include bathymetry, tide, and current. The initial flow model will be developed to validate key aspects of the setup, including the grid, bathymetry, and boundary conditions. This ensures that the model is accurately aligned with the objectives of the study. The sediment sample were collected from field survey to inspect the sediment properties used in the model. The morphology module, which simulate the temporal and spatial distribution of sediment is used.

For validation, flow model representing the estuary was used with the flow boundary condition at the river's upstream (north) and tidal boundary condition off Pasaran Island to the south, following the study of [7]. The river boundary condition was set based on the field current data. The tide boundary condition was extracted from the tide-model provided by Delft3D. The validation model was run for one day in Juli 24 2022 and it was found that the mean absolute percentage error of depth-averaged velocity for selected window time is 5%.

The model representing the sedimentation event run for 1 year start from March 2022 to March 2023. The river's flow boundary condition was determined by the analysis of river discharge by the method of FJ Mock [8] and the river catchment data was taken from [9]. The area of catchment is estimated to be 21.89 km^2. The one year time simulation was supposed to incorporate the river discharge variation due to hydrology and meteorology condition. The time step of 15 minutes and the grid size is of 25 m was used in the model. The bathymetry and grid of the model is shown in Figure 3.

The results of the model, i.e. velocity field and contour of sedimentation, are shown in Figure 4 and Figure 5 respectively. The velocity field is shown during the high discharge period. From Figure 4, it can be seen that the velocity inside the strait and at the east coast of the river is relatively slow ranging around 0-0.05 m/s. The highest velocity in the domain occurs at the 550 m to north of the river mouth reaching 0.75 m/s. The relatively low velocity of the river during high discharge confirms the effect of gentle slope of the riverbed which limits the development of the flow.

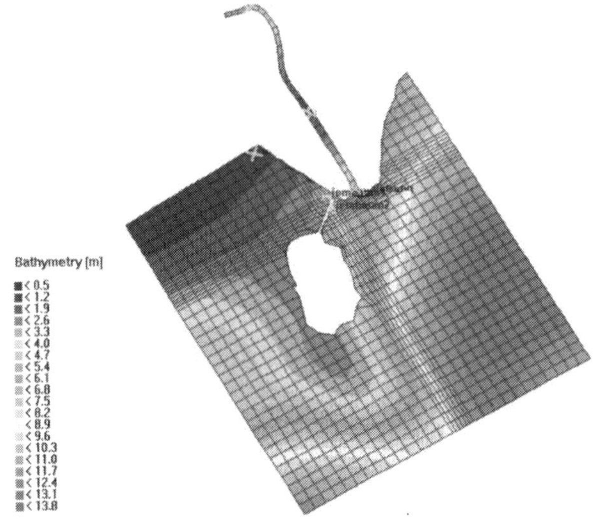

Figure 3. The model grid and bathymetry.

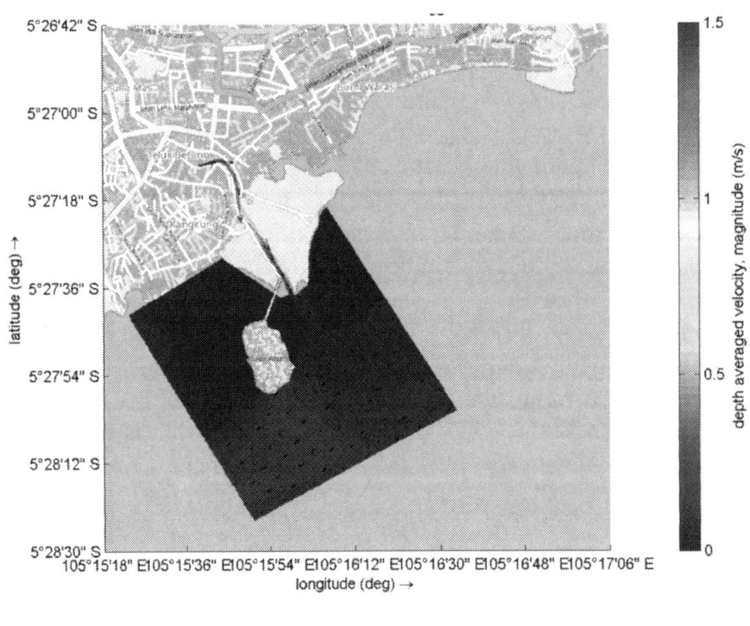

Figure 4. Velocity field during peak river discharge.

IOP Conf. Series: Earth and Environmental Science **1464** (2025) 012011 doi:10.1088/1755-1315/1464/1/012011

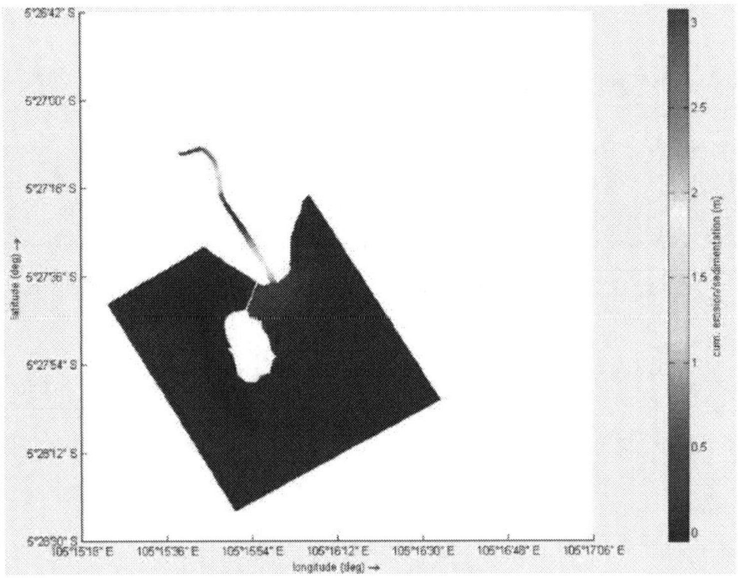

Figure 5. Contour of sedimentation after one year simulation.

The formation of sediment varies mildly around the estuary with the highest intensity occurring before the river mouth peaking at 1.6 m (Figure 5). At the east side of the bridge, cumulative sediment is estimated to be 0.1 to 0.3 m. The sedimentation is found to be distinctively small at the west side at just under 0.1 m confirming the effect of the bridge piles. Off the Pasaran Island to the south sedimentation occurs at negligibly small rate. While the abnormally high sediment formation at the vicinity of north upstream boundary that yield about 3 m maybe due to instability related to the boundary condition.

Sedimentation process over simulation time in selected monitoring points are then presented in Figure 6. The monitoring points, as shown previously in Figure 2 include: *jembatan*-1 (west side of the bridge) and *jembatan*-2 (east side of the bridge).

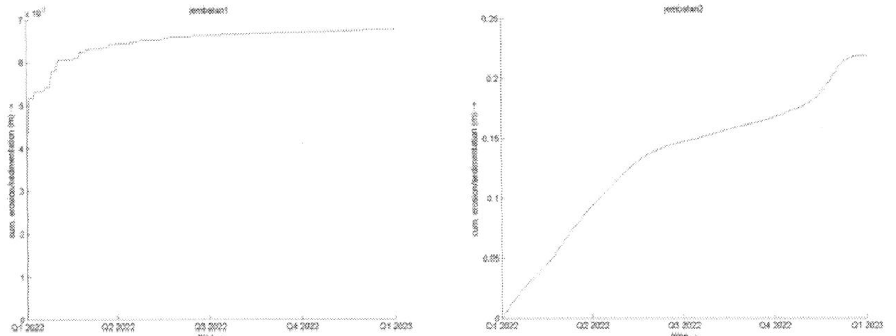

Figure 6. Sedimentation over time in each monitoring points, left: *jembatan*-1 (west side of the bridge), right: *jembatan*-2 (east side of the bridge).

At the east side of the bridge (*jembatan*-2), the sediment accumulation occurred gradually in relatively linear trend during the first quarter of the year, at the rate about 0.1 m per quarter (Figure 6). The rate of sedimentation then slowed in the next two quarter and then back to the pace at the last quarter. The variation of sedimentation rate maybe due to the variation of river discharge over the year. Periods of accelerated sedimentation showed correlation with times of high river discharge. It is shown that the sediment accumulation reached 0.23 m by the end of simulation. While at the west side of the bridge (*jembatan*-1), where the flow from the river is hindered by the bridge piles, the sediment formation is rather gentler than that of the east side. The sediment accumulation rose quickly during the first quarter and reaching about 0.0065 m at the second quarter. The accumulation then occurred slowly until the end of simulation with reaching just 0.0069 m.

The velocity distribution in the domain, as shown in the figure 3, indicates that the estuary shares a common hydraulic regime with other river-sea environments, characterized by a low Froude number dominating the stream conditions. This is associated with the gentle slope of the estuary, which limits rapid flow changes and supports a stable, low-energy hydraulic environment. Compounded by the presence of bridge piles, this condition results in sediment deposition concentrated predominantly near the east side of the bridge. The sediment yield difference between the east side and the west side of the bridge, with the ratio about 33.3 confirms the significance of the piles in disturbing the flow through the strait.

The case of "sediment dampening" occurred in the strait maybe correlated with the relatively uniform depth of the strait. In such bathymetry condition, the flow regime may be easily distracted by the presence of structure. The formation of sediment near the east side of the bridge could negatively impact sea-related activities around the island. If linear trend is assumed to materialize, the fully developed land formation in the east side of the bridge is expected to occur in 8.69 years. However, in managing the sedimentation, coastal structures can be installed as studied by Rakhmawati [7]. Her work demonstrates that the jetty reduced sedimentation volume by up to 91.7% at the selected monitoring point.

4. Conclusion

The sedimentation transport model in the area of Pasaran Island where complex features present was modelled using Delft3D. It was found that velocity in the domain is relatively low with the maximum value reaching just 0.75 m/s. This low Froude flow regime maybe due to the gentle slope of the domain. The formation of sediment varies mildly around the estuary with the highest intensity occurring before the river mouth peaking at 1.6 m. The sediment yield differs at both side of the bridge. In the east side the accumulation of sediment is 0.23 m by the end of simulation while at the west side is just 0.0069 m. This sharp difference confirms the significance of the piles in disturbing the flow through the strait. Assuming linear trend of sedimentation, the fully developed land formation in the east side of the bridge is expected to occur in 8.69 years.

References

[1] Sunarto, S. Geomorphological Development of The Muria Palaeostrait In Relation To The Morphodynamics Of The Wulan Delta, Central Java. Indonesian Journal of Geography, 40(2), 177-185.

[2] Deltares (2024) User Manual Deflt3D-FLOW, Deltares, Delft, The Netherlands. https://content.oss.deltares.nl/delft3d4/Delft3D-FLOW_User_Manual.pdf

[3] Chen, L., Zhou, Z., Xu, M., Xu, F., Tao, J., & Zhang, C. (2018). Exploring the influence of land reclamation on sediment grain size distribution on tidal flats: A numerical study. Coastal Engineering Proceedings, (36), 85-85.

IOP Conf. Series: Earth and Environmental Science **1464** (2025) 012011 doi:10.1088/1755-1315/1464/1/012011

[4] Hutley, N. (2017). Modelling cohesive sediment dynamics in Moreton Bay. School of Civil Engineering, University of Queensland
[5] UNESCO (2016) Final Report: Delft3D model based alongshore sediment transport rates at Pesalai, Gurunagar, Point Pedro and Mullaitivu, Sri Lanka.
[6] Mirdan, M., Tolba, E. R., Abdellaha, S. E., & Galal, E. M. (2024). Modeling Coastal Morphological Features in Proximity to Breakwaters Using Delft 3D. Journal of Maritime Research, 21(2), 196-203.
[7] Rakhmawati, A. L. (2021, March). Sedimentation and Longshore Current Simulation with Delft3D Software in Rejoso Shoreline. In IOP Conference Series: Earth and Environmental Science (Vol. 698, No. 1, p. 012021). IOP Publishing.
[8] Kiranaratri, A. H., Simarmata, N., & Hidayat, D. (2019). Analisis Potensi Tingkat Bencana Banjir Daerah Aliran Sungai Way Kuripan Kota Bandar Lampung. Rekayasa Sipil, 13(2), 147-152.
[9] Rasyid, I., & Afdhaliah, N. (2021). Analisis Debit Andalan Sungai Cisadane Dengan Metode FJ MOCK Bulanan dan Setengah Bulanan untuk Waduk Lepas Pantai. In ITB Graduate School Conference (Vol. 1, No. 1, pp. 549-562).

SWOT analysis of the new capital city based on the integrated coastal zone management approach

Destyariani Liana Putri[1*], Nurmawati[1], A Dianiswara[1], M K Wirawan[1]

[1] Ocean Engineering Study Program, Institut Teknologi Kalimantan, Balikpapan, Indonesia

*E-mail: putridestyariani@lecturer.itk.ac.id

Abstract. Coastal area as the buffered to the new capital city are including Samboja, Penajam Paser Utara and Muara Jawa. The criteria address the needs of coastal area which are stakeholder, expertise, resources, facility and regulation. These criteria are also known as the Integrated Coastal Zone Management (ICZM) indicators helped to mapping the condition of the area. Hence, the criteria need to be converting into SWOT (Strength, Weakness, Opportunity and Threat) questionnaire. Over 100 respondents participated in the survey which are coming from local government, fishermen, women community, agricultural community as well as tourism community. The coastal area in the new capital city has local resource potential that supports the main area related to fisheries, tourism and agriculture. However, there is still need to strengthening the coastal area alongside the development of the new capital city itself. Hence, the SWOT analysis provides the ideal condition for both internal and external perspectives. It also suggests the strategy that help the stakeholder on the developing the area based on their condition. As it indicates from the SWOT analysis that the coastal area should consider the implementation of progressive strategy to improve its readiness towards the development of new capital city. It means that the coastal area requires to use the strength components in order to mobilize the opportunities.

1. Introduction

The development and relocation of the New Capital City entitled Ibu Kota Nusantara (IKN) are running according to plan, as well as to assess its impact and sustainability. There are numbers of specific parameters has been addressed according to infrastructure and public facilities (e.g. roads, transportation, electricity, water, and other public facilities), environment (e.g. ecosystems, forests, and biodiversity, as well as mitigation of potential natural disasters), social and culture (e.g. local communities' empowerment), as well as regulation and policy (e.g. regarding spatial planning, licensing, and resource management).

The above-mentioned parameters were also known as the Integrated Coastal Zone Management (ICZM) indicators. The ICZM is usually used to evaluate the implementation of in several coastal areas of Indonesia, including in dealing with multisectoral stakeholders [1]. It commons to use ICZM combining with SWOT (Strength, Weakness, Opportunity and Threat) to analyse the condition of coastal area management in Indonesia and its contribution to climate change [2]. Furthermore, it is imperative to have comprehensive assessment of the IKN development, as it has implication into social aspects [3].

Hence, research related to the elaboration of SWOT in the context of Integrated Coastal Zone Management (ICZM) in the Capital City of the Republic of Indonesia (IKN), or more broadly in the coastal areas of Indonesia, is a relevant topic considering the government's grand plan to move the capital city. Although there may not be many specific studies that directly combine SWOT and ICZM specifically for IKN Nusantara, several previous studies related to ICZM in the coastal areas of Indonesia and SWOT analysis in similar contexts can provide useful insights.

By evaluating the readiness of the capital city comprehensively, the government and stakeholders can identify potential obstacles, and formulate the steps needed to overcome these challenges. This research can also help in formulating more integrated policy strategies to ensure that the development of the IKN is sustainable and pays attention to the sustainability of the coastal environment.

2. Method

This SWOT analysis highlights both the potential and challenges of implementing ICZM effectively, emphasizing the need for careful planning, resource management, and stakeholder engagement.

2.1 Issue Identification

SWOT Analysis requires the issue identification which includes the following key indicators.

1. Balancing Stakeholder Interests: The complexity of integrating various interests within ICZM frameworks can lead to conflicts and delays in decision-making.
2. Resource Allocation: Ensuring adequate resources (financial, technical, and human) is critical for the successful implementation of ICZM.
3. Governance and Policy Coordination: Effective ICZM requires strong governance structures and coordination across different policy domains and administrative levels.
4. Climate Resilience: Developing and implementing strategies to mitigate and adapt to the impacts of climate change on coastal zones is essential.
5. Data and Monitoring: Addressing data gaps and improving monitoring systems are vital for informed decision-making and adaptive management in ICZM.

2.2 Data and Information Collection

Sampling is a critical step in conducting a SWOT analysis, especially within the framework of Integrated Coastal Zone Management (ICZM). Effective sampling ensures that the data collected is representative, reliable, and comprehensive, enabling accurate identification of strengths, weaknesses, opportunities, and threats. The purposive sampling was used to select stakeholders who have specific knowledge, experience, or influence related to coastal management. This ensures that key perspectives are included (e.g. local fishermen, tourism operators, government officials, and environmental activists).

2.3 SWOT Matrix Formulation and Analysis

The assessment of SWOT matrix was identified from the external and internal factors (EFAS and IFAS matrices). It was carried out by using a scoring method, starting with very important (4) - not important (1) (Table 1).

The collected data from the SWOT matrix need to be categorized and examined the relationship as required in Integrated Coastal Zone Management (ICZM) perspectives. The alternative strategy might follow the EFAS and/or IFAS depending on the final score on each factor. Moreover, the pivotal step on the research procedure was developing tangible strategy which integrated the SWOT findings into ICZM. It was to ensure that the plans would be realistic and grounded enough based on the collected data.

Table 1. SWOT matrix diagram [4].

IFAS / EFAS	**STRENGTH (S)**	**WEAKNESSES (W)**
	Determine 5-10 internal strength factors	Determine 5-10 internal weakness factors
OPPORTUNITIES (O)	**STRATEGY SO**	**STRATEGY WO**
Determine 5-10 external opportunity factors	Create strategies that use strengths to take advantage of opportunities	Create strategies that minimize weaknesses to take advantage of opportunities.
THREATS (T)	**STRATEGY ST**	**STRATEGY WT**
Determine 5-10 external threat factors	Create strategies that use strengths to overcome threats.	Create strategies that minimize weaknesses and avoid threats.

3. Result and analysis

Analysis of community empowerment strategies and regional planning in the coastal area of the IKN using SWOT analysis. SWOT analysis refers to 5 attributes used, including, natural resources and tourism, human resources, supporting systems and facilities, rules and regulations, and relationships between stakeholders obtained from direct surveys, interview results, FGDs, questionnaire distribution, and secondary data.

The results of the SWOT analysis (Table 2-4) show that each coastal sub-district in the IKN area has a positive value for all conditions and almost all combinations of strategies (strengths, weaknesses, opportunities, and threats). Positive values illustrate that each coastal sub-district needs to implement a progressive strategy to develop the potential of existing resources. This means that the priority of the strategy must use strength to take advantage of opportunities. A progressive condition indicates that resources are feasible and ready to be expanded to achieve maximum progress.

Table 2. SWOT matrix on Sepaku.

Components	Score
Strength (S)	**3,94**
Weakness (W)	3,25
Opportunity (O)	**3,74**
Threat (T)	3,34

Table 3. SWOT matrix on Samboja and West Samboja.

Components	Score
Strength (S)	4,05
Weakness (W)	3,03
Opportunity (O)	3,80
Threat (T)	3,31

Table 4. SWOT matrix on Muara Jawa.

Components	Score
Strength (S)	3.98
Weakness (W)	3,33
Opportunity (O)	3,68
Threat (T)	3,20

The questionnaire was built based on the Likert-Scale. The scales consist of 5 scores which is representing the agreement of respondent [4]. The set of Likert-Scale used in the questionnaire are 1 (strongly disagree), 2 (disagree), 3 (neutral), 4 (agree) and 5 (strongly agree). Besides, the ICZM components (Chapter 2.1) are initially defined into internal conditions (human resources and facilities) and external conditions (regulation, natural resources and stakeholders). These conditions will build in the questionnaire in order to identify the S-W-O-T components after calculating Likert-Scale.

Each of area has particular SWOT values based on the questionnaire during the focus group discussions, respectively 66 people from Sepaku, 26 people from Samboja, and 21 people from Muara Jawa. The respondents are coming from local governments, local communities and leaders (including fisheries, tourism and palm oil) as well as the entrepreneurs. The Table 2-4 show that the areas need to consider their Strength, Weakness and Threat as it gives the higher weight values combinations.

Table 5. SWOT strategy based on integrated coastal zone management in IKN

No	**Alternative Strategy: Progressive (S-W and S-T)**		
	Sepaku	**Samboja & West Samboja**	**Muara Jawa**
1	Building public infrastructure and tourism support facilities	Developing formal and informal education related to the coast, technology, internet, and sustainable management of natural product processing (fisheries/agriculture)	Developing formal and informal education related to the coast, technology, internet, and sustainable management of natural product processing (fisheries/agriculture)
2	Developing formal and informal education related to the coast, technology, management, internet, and sustainable processing of	Improving the quality of human resources (HR) and community skills, as well as developing business capital assistance and government support	Improving the quality of human resources (HR) and community skills, as well as developing business capital assistance and government support

No	Alternative Strategy: Progressive (S-W and S-T)		
	Sepaku	**Samboja & West Samboja**	**Muara Jawa**
	natural products (fisheries/agriculture)		
3	Improving the quality of human resources (HR) and community skills and developing business capital assistance	Developing tourism support facilities such as accommodation, rest areas, and places to eat that are friendly to the disabled	Conducting data collection/inventory of local coastal resources and promoting tourism potential and home industries based on the internet/system
4	Developing tourist attractions, ecosystem rehabilitation, and promoting tourism potential	Conducting data collection/inventory of local coastal resources and promoting tourism potential	Spatial planning in coastal and marine areas and improving fisheries infrastructure
5	Sustainable utilization of coastal resource potential and development of other environmentally friendly businesses	Spatial planning in coastal and marine areas and improvement of fisheries infrastructure	Educating the community about coastal ecosystems and environments and carrying out ecosystem rehabilitation
6	Conducting data collection/inventory of local coastal resources	Utilizing coastal resource potential in a sustainable and environmentally friendly manner	Increasing regulations for industries that have the potential to damage the environment and increase carbon emissions
7	Spatial planning in coastal and marine areas	Accommodate activities that are local wisdom, especially related to women's empowerment	Develop tourism support facilities such as accommodation, rest areas, and places to eat that are friendly to the disabled

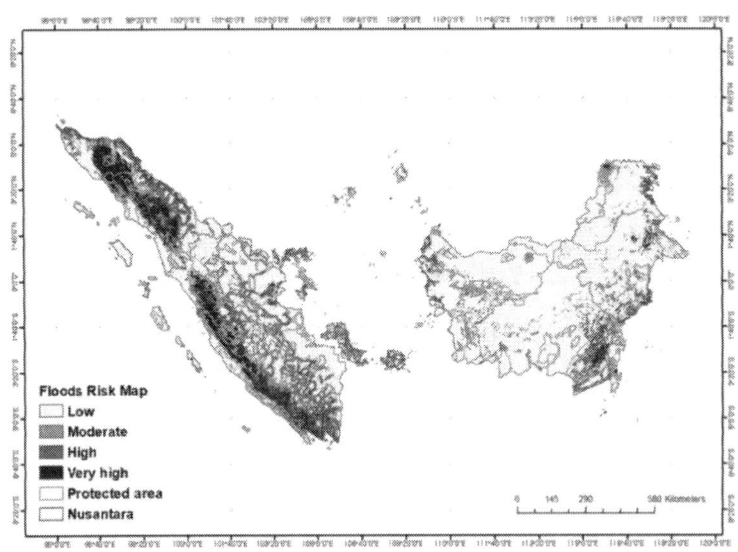

Figure 1. Flood Risk [5]

The higher weight values combinations is located in Strength - Opportunity (S-O) which means the areas have to provide progressive strategy in order to support the development of New Capital City based on the coastal local resources [6]. The Table 5 shows the set of strategies that should be considered by Sepaku, Samboja and Muara Jawa. Each of strategy represents the environmentally friendly suggestion as it is also known that the area is surrounding by natural hazard, particularly on flood and landslide (Figure 1-2). The IKN area has relatively moderate risk of flood rather than landslide which indicates low risk level [5].

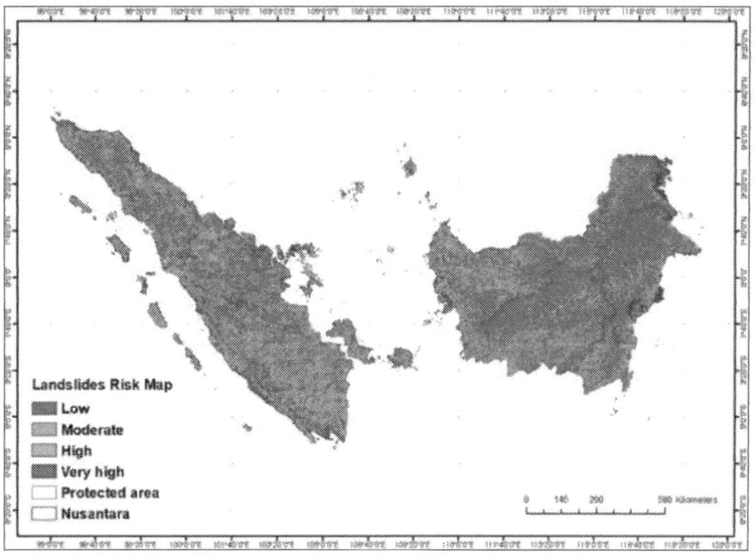

Figure 2. Landslides Risk [5]

In addition, there are requirement to improve the community capacity alongside the infrastructure development in IKN. Building both formal and informal education center will help the community to fulfil the needs in the region so it may reduce the social gap between local people and transmigrant. The gap of human capacity may increase the horizontal conflict (societal conflict) [7].

4. Conclusion

The IKN area particularly on coastal area need to provide progressive strategies (using Strength and Opportunity). The focus of Sepaku sub-district development is on the tourism and service sectors, the Samboja and West Samboja sub-district development is tourism and fisheries, while the Muara Jawa sub-district has a more diverse development sector ranging from tourism, fisheries, and agriculture. Hence, the coastal areas of the IKN have adequate local resource potential with regards to the role of the government, community, and investors to support the development of New Capital City (IKN).

Acknowledgment

This research was supported by the Regional Research and Development Agency, East Kalimantan Province and Community Service Institute (LPPM) Kalimantan Institute of Technology. Besides that, the author also appreciates the cooperation from the Government of Sepaku, Samboja, and Muara Jawa.

References

[1] E. Sofiyah, Challenges and opportunities for an integrated coastal management approach in Jakarta Bay, Indonesia., 2013.

[2] Basuki, T.M., Nugroho, H.Y.S.H., Indrajaya, Y., Pramono, I.B., Nugroho, N.P., Supangat, A.B., Indrawati, D.R., Savitri, E., Wahyuningrum, N., Purwanto and Cahyono, S.A., "Improvement of integrated watershed management in Indonesia for mitigation and adaptation to climate change: A review," *Sustainability,* pp. 14(16), p.9997, 2022.

[3] Aji, G., Arfani, Z., Sari, A.M. and Seprtiani, R., Dampak Pemindahan Ibukota Negara Baru terhadap Ekonomi dan Sosial di Provinsi Kalimantan Timur. Kultura: Jurnal Ilmu Hukum, Sosial, dan Humaniora, 1(5), pp.1-8, Vols. 1(5), pp.1-8., 2023.

[4] Nemoto, T. and Beglar, D., Likert-scale questionnaires. In JALT 2013 conference proceedings (Vol. 108, No. 1, pp. 1-6), 2014.

[5] Heo, S., Sohn, W., Park, S. and Lee, D.K., Multi-hazard assessment for flood and Landslide risk in Kalimantan and Sumatra: Implications for Nusantara, Indonesia's new capital. Heliyon, 10(18)., 2024.

[6] F. Rangkuti, Marketing Analysis Made Easy, Gramedia Pustaka Utama, 2005.

[7] Praditya, E., Suprapto, F.A., Ali, Y., Surjaatmadja, S. and Duarte, R., Nusantara Capital City (IKN): Threats and Defense Strategies for Indonesia's New Capital. The Journal of Indonesia Sustainable Development Planning, 4(1), pp.21-34., 2023.

IOP Conf. Series: Earth and Environmental Science **1464** (2025) 012013 doi:10.1088/1755-1315/1464/1/012013

Modeling of potential earthquake-generated tsunami wave hazard in the coastal area of Kelumbayan Subdistrict, Lampung, Indonesia

Davin Ridho Alfath[1,*], Mohammad Bagus Adityawan[2,3,*], Mohammad Farid[2,4], Arno Adi Kuntoro[2], Widyaningtias[2], M Gilang Indra Mardika[5,*], Dyah Kusumastuti[6]

[1] Undergraduate School of Civil Engineering, Faculty of Civil and Enviromental Engineering, Institut Teknologi Bandung, Bandung, Indonesia
[2] Center for Water Resources Development, Faculty of Civil and Environmental Engineering, Institut Teknologi Bandung, Bandung, Indonesia
[3] Undergraduate School of Water Resources Engineering and Management, Faculty of Civil and Environmental Engineering, Institut Teknologi Bandung, Bandung, Indonesia
[4] Center for Coastal and Marine Development, Faculty of Civil and Environmental Engineering, Institut Teknologi Bandung, Indonesia
[5] Center for Civil Engineering, Faculty of Infrastructure Technology and Regional Development, Institut Teknologi Sumatera, South Lampung, Indonesia
[6] Structural Engineering Research Group, Faculty of Civil and Environmental Engineering, Institut Teknologi Bandung, Indonesia

*E-mail: dvnalfath@gmail.com, mbagusadityawan@gmail.com

Abstract. Kelumbayan Subdistrict is one of the tourist destinations in Lampung Province, Sumatra Island, Indonesia, that holds the beauty of the sea and beaches as an attraction for tourists. However, the location of Kelumbayan Subdistrict around the Indian Ocean subduction zone makes this area prone to tsunamis due to submarine earthquakes generated by the meeting between the Indo-Australian Plate and the Euro-Asian Plate. Referring to the case of the Aceh tsunami in 2004, the potential damage generated by a tsunami disaster from the subduction zone in the south of Sumatra is enormous, with many fatalities and massive material losses. The case study of a potential tsunami in Kelumbayan Subdistrict was further investigated to determine the maximum water level of tsunami waves and the time when the waves reach land. The tsunami wave propagation is simulated using Delft3D. Historical tsunamis were reviewed from the tsunami catalog by the Agency for Meteorology, Climatology, and Geophysics (BMKG). To investigate this further, the potential maximum magnitude earthquake caused by the confluence of two tectonic plates in the vicinity of the study site was determined based on three scenarios of fault mechanisms that cause earthquakes with reference to the 2017 Indonesia Earthquake Hazard and Source Map (PUSGEN 2017). The research was conducted to determine the potential tsunami disaster that could occur in Kelumbayan Sub-district. The result of this research is the height of the tsunami wave and the arrival time. The data can be used as part of the formulation of mitigation measures in Kelumbayan Subdistrict.

1. Introduction

Indonesia is located at the confluence of several global tectonic plates, namely the Euro-Asian Plate in the north, the Indo-Australian Plate in the south, and the Pacific Plate in the east [1]. The meeting of these three plates causes collisions and the creation of faults on land or in the ocean, which makes Indonesia prone to earthquakes and tsunamis.

In 2004, Indonesia and the world were shocked by a natural disaster that occurred in Aceh Province, namely an earthquake accompanied by a tsunami. The source of the earthquake was the subduction of the Indo-Australian Plate into the Euro-Asian Plate in the western part of Sumatra Island. The subduction process resulted in an earthquake with an earthquake magnitude of 9.2 Mw that was accompanied by a tsunami event with a wave height of 5 to 30 meters. The sea receded for 30 minutes after the earthquake, then the first tsunami wave arrived at 08:42 UTC+7 (44 minutes after the earthquake). The second wave arrived at 08:53 UTC+7 (larger) and the third wave arrived at 09:15 UTC+7 (largest). The city of Banda Aceh was devastated by the magnitude 9 earthquake, which caused the infrastructure of the coastal areas to be hit by tsunami waves and carried far inland. At least 166,080 people were killed in Aceh and an estimated 6,245 people were missing. It is estimated that around 500,000 people have been left homeless [2].

One of the coastal areas, within the Sunda Strait region, that cannot be separated from the threat of earthquakes and tsunamis is the Kiluan Bay, Kelumbayan District, Tanggamus Regency, Lampung Province. The location of Kiluan Bay, which is close to the subduction zone and directly facing the Indian Ocean, makes this area prone to natural disasters caused by the meeting between the Indo-Australian Plate and the Euro-Asian Plate. Kelumbayan sub-district is located not far from the subduction zone in the Sunda Strait and the southwestern part of Sumatra Island [3]. Tsunami disasters caused by earthquakes in this subduction zone, namely in 1931 in Southwest Sumatra with Mw 7.9; in 1958 in Bengkulu, Southwest Sumatra with Mw 6.5; in 2007 in Bengkulu, Southwest Sumatra with Mw 8.4; in 2011 in Lampung with Mw 6.1; in 2019 in Labuan, Banten with Mw 6.9; and in 2022 in Labuan, Banten with Mw 6.6. With this earthquake history [4], Kiluan Bay is not free from the threat of earthquakes with the same magnitude or even greater.

The relationship between earthquake magnitude and fault plane area is a causal relationship. Therefore, many of the previous studies addressed the topic of the relationship between these two parameters in the form of empirical equations of moment magnitude (Mw) as a logarithmic function of fault plane area obtained from earthquake statistics [5]. One commonly used modeling of the relationship between magnitude and fault plane area is an empirical model that uses 244 earthquake data from around the world, including earthquake data with depths of less than 40 km by [6].

In modeling tsunami waves, numerical models are commonly used to predict tsunami wave inundation. A numerical model commonly used to predict tsunami waves is the Shallow Water Equation [7]. In addition to modeling tsunamis, numerical models can also be used to determine the damage to structures affected by tsunamis and the number of potential casualties [8].

In this study, tsunami wave modeling will be conducted to determine the potential tsunami wave height and wave arrival time on the coast around Kiluan Bay. Tsunami wave inundation analysis is conducted based on available spatial data around Kiluan Bay, fault segments that have the potential to generate tsunamis in the coastal area of Kiluan Bay, and modeling using the Delft3D-Flow numerical tool. The use of the Delft3D model in this study is in line with previous studies that have shown the accuracy of this model in predicting tsunami inundation [9], [10] .

The results of this study can be used to plan tsunami disaster mitigation measures in Kiluan Bay. One example of mitigation efforts is the development of tsunami early warning systems. Previous research has developed innovative approaches by utilizing maritime wireless communication technology to detect and transmit tsunami early warnings quickly to the community [10].

2. Methodology

2.1 Study Area

Teluk Kiluan is located in Kiluan Negeri Village, Kelumbayan Sub-district, Tanggamus Regency, Lampung Province. Kiluan Bay is located at a latitude of -5.774° and a longitude of 105.10°. Based on data obtained from the official website of Teluk Kiluan Tourism Village, the village has 1619 inhabitants with a total female population of 776 and a male population of 843 [11]. The study Area is shown in Figure 1.

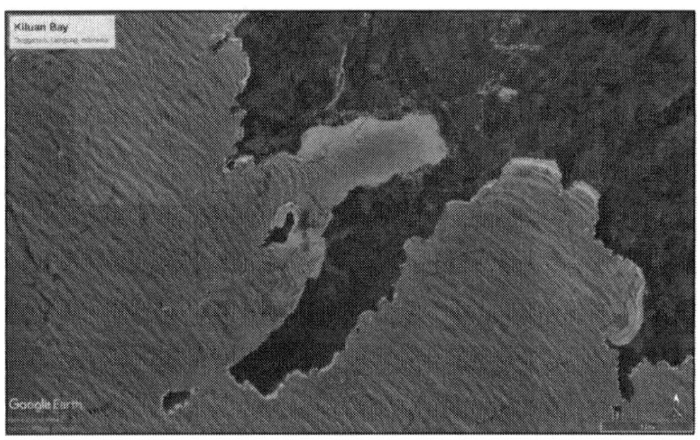

Figure 1. Kiluan bay [12].

Kiluan Bay is famous for its stunning natural beauty, especially its beaches and sea. In this village, there are clean white sandy beaches, clear and calm sea water, and beautiful views of small islands. Kiluan Bay is also a natural habitat for various marine life, including dolphins and whales. Locals say around 300 visitors come to Kiluan Bay every weekend [13].

2.2 Tsunami modeling tool

The tsunami modeling tool used is Delft3D-FLOW. Delft3Flow is a hydrodynamic simulation program developed by Deltares that is used to simulate tsunamis, tides and the interaction between the two cases. The model can be used for two-dimensional calculations with the scale of fault length and time interval used. Based on the Delft3D-FLOW manual procedure [14], the theories used in this modeling are shallow wave theory. Then, in modeling the tsunami, the theory used is the Okada Model [15].

The Delft3D-FLOW tool can simulate the movement of tsunami waves towards the coast. In addition, the tool can also output graphical data of sea level during the tsunami process from predetermined observation points.

2.3 Spatial data

The spatial data used in this research are bathymetry and topography data that will be used as initial modeling input and detailed modeling input. The use of bathymetry and topography data is based on the resolution and grid size of each model. In conducting the initial modeling input, the spatial data used are bathymetry data from GEBCO with a data resolution of 15 arc-second and BATNAS data with a resolution of 6 arc-second. Furthermore, the input for detailed modeling uses BATNAS data with the same resolution and DEMNAS with a resolution of 0.27 arc-second [16].

2.4 Fault models for tsunami simulation

The tsunami waves generated are derived from the relationship between the earthquake that occurred and the identified fault parameters [9], [17]. In this study, tsunamis were generated by three scenarios of megathrust segment fault models. Each fault scenario has a different earthquake magnitude based on the fault location, fault dimensions, and the resulting earthquake magnitude. The first scenario is due to the Enggano Segment megathrust fault, the second scenario is due to the Banten Sunda Strait Segment megathrust fault, and the scenario is due to the combined megathrust fault between The Enggano Segment and The Selat Sunda Banten Segment. The fault locations are shown in Figure 2 and the parameters of each fault scenario are shown in Table 1.

The fault dimension as the source of the earthquake that caused the tsunami was obtained using several data sources, namely the Indonesia Earthquake Hazard and Source Map 2017, USGS, and calculations performed by Delft3D-FLOW using the Okada Model (1985) [15]. The following are the input parameters used for each tsunami-generating earthquake scenario. In the modeling, a fault strike angle of 90 degrees was used to obtain a tsunami model with extreme fault generation.

The earthquake moment magnitude used as a reference is the maximum magnitude predicted in the Indonesian Earthquake Hazard and Source Map 2017 and the combined magnitude formed from scenario 3 using the correlation equation between the dimensions of the fault and the magnitude of the resulting magnitude. Furthermore, moment magnitude predictions were made for the combined Enggano Segment and Banten Sunda Strait Segment using several empirical equations that have been developed in previous studies. Prediction of the combined magnitude value between the two fault segments using the equation previously used in research by Wells dan Coppersmith [6].

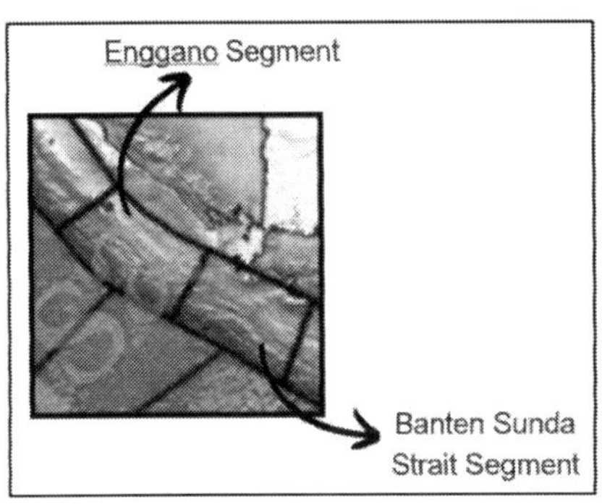

Figure 2. Fault Segment around Kiluan Bay [1].

IOP Conf. Series: Earth and Environmental Science **1464** (2025) 012013 doi:10.1088/1755-1315/1464/1/012013

Table 1. Fault parameters to generate tsunami.

Scenario	L (Fault Length)	W (Fault Width)	Earthquake Magnitude	Focal Mechanism			Depth of Epicenter
	km	km	Mw	Strike (°)	Dip (°)	Slip (°)	km
1st Scenario	250	131.78	8.8	107	11	7.9	30.5
2nd Scenario	250	131.78	8.8	107	11	7.9	30.5
3rd Scenario	530	131.78	9.1	90	11	11.2	30.5

2.5 Tidal wave

Tidal wave data processing is carried out to determine the highest water level value when the tide occurs in Kiluan Bay. Tidal data processing was carried out using the least square method for the time span of June 24, 2024 to July 8, 2024 so that the following tidal characteristics were obtained in Figure 3. Tidal data analysis is also carried out to determine the important tidal water level data in Kiluan Bay which can then be used as a reference wave height in modeling. Important tidal water level data is shown in Table 2.

The HHWL value with an elevation of 1.2 meters will be used as the initial elevation for detailed tsunami wave modeling.

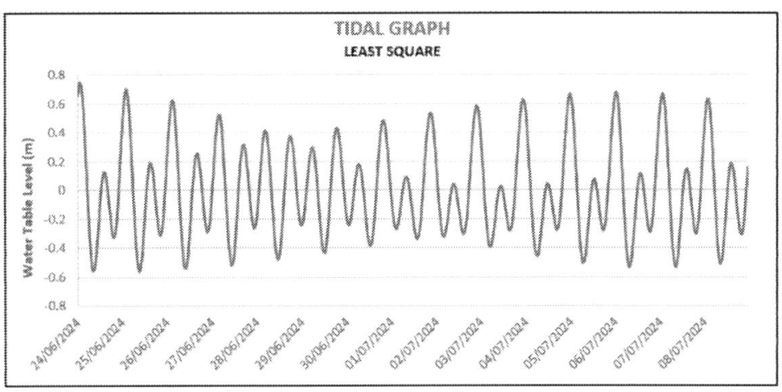

Figure 3. Tidal graph.

Table 2. Water Table Elevation Parameter

Location	Water Table Elevation	Least Square - Water Level (m)
Kiluan Bay	HHWL	1.2
	MHWL	0.5
	MSL	0.0
	MLWL	-0.5
	LLWL	-1.2

2.6 Detailed model

Detailed modeling was carried out on the scenario with the largest earthquake magnitude, which is the earthquake originating from the 3rd scenario or the combination of two fault segments, namely the Enggano Segment and the Banten Sunda Strait Segment. The modeling is performed in stages using the Grid Nesting procedure which involves the use of small grids within a larger coarser grid. Modeling using a nesting grid system can provide accurate simulation results in predicting tsunami waves, arrival times, and affected areas. In this study, the grid used for wave modeling in Kiluan Bay is shown in Figure 4.

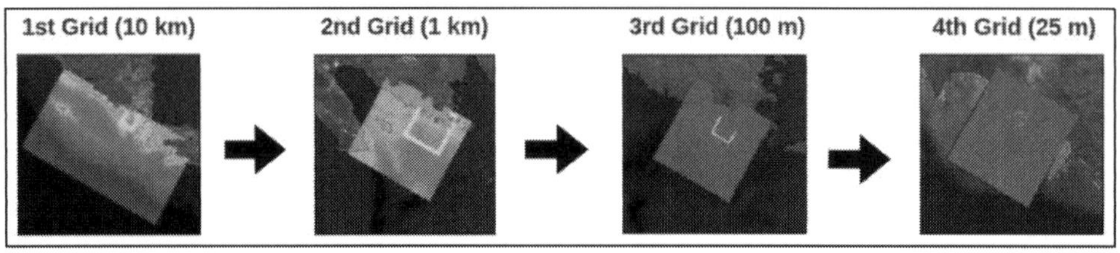

Figure 4. Grid model.

The nesting process is also done by varying the alpha value of each grid boundary. The alpha value is the value of the wave absorption ability of the previous grid to be passed on to the grid being used. The alpha value of each grid is varied to produce stable tsunami wave propagation and avoid significant changes in wave speed.

3. Result of Tsunami Simulation

3.1 Modeling results of three tsunami scenarios

Modelling was conducted for three scenarios of earthquake generation by fault segmentation. The results of modeling the potential tsunami inundation height with 3 different scenarios in Kelumbayan sub-district is shown in Figure 5, Figure 6, and Figure 7. A graph of the relationship between tsunami wave height and wave arrival time is shown in Figure 8. It is found that the wave height in the first scenario occurred at the 38th minute with a height of 0.62 meters. In the second scenario, the maximum wave height occurred in the 29th minute with a height of 1.8 meters. Then, in the third scenario, the maximum wave height occurred in the 32nd minute with a height of 2.34 meters.

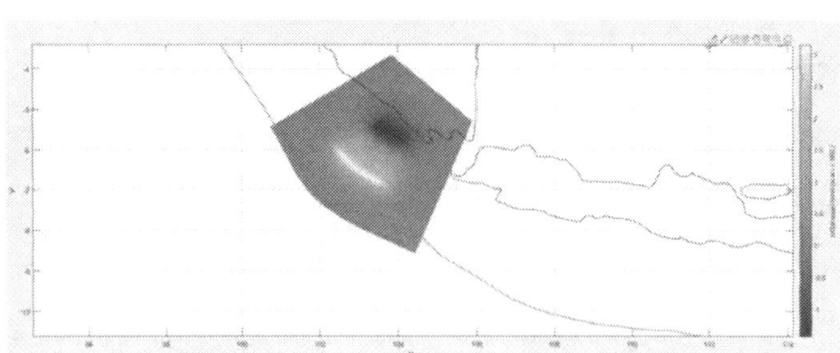

Figure 5. Initial wave for 1st scenario.

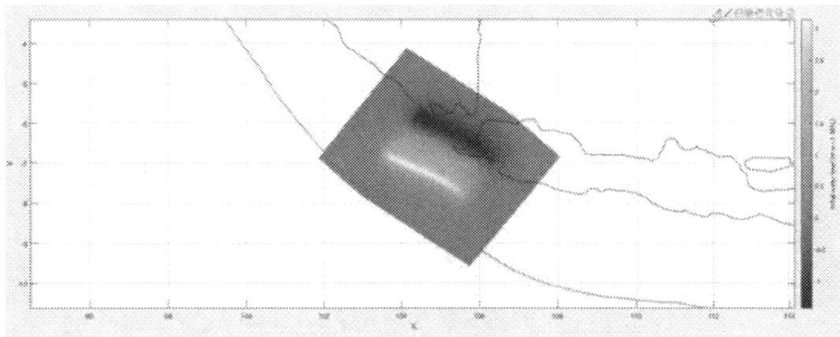

Figure 6. Initial wave for 2nd scenario.

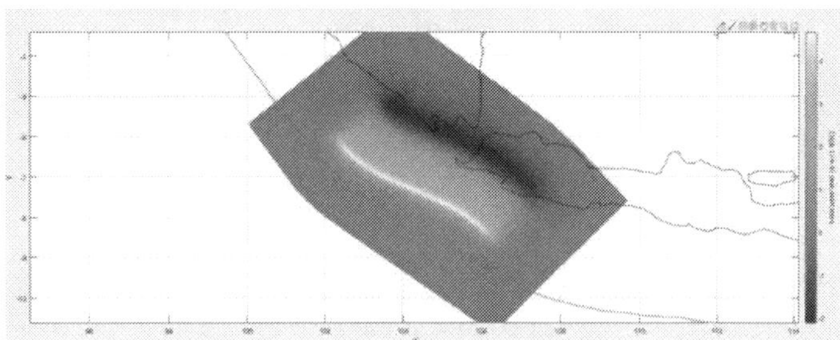

Figure 7. Initial wave for 3rd scenario.

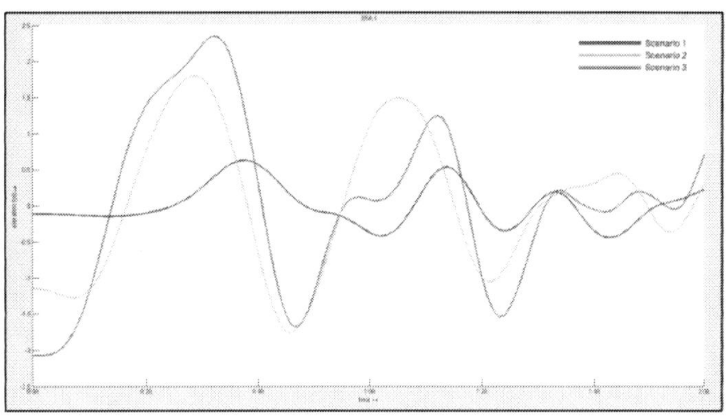

Figure 8. Water level of tsunami for all scenarios.

It is known that the third scenario is a tsunami propagation scenario that can produce the highest waves. Therefore, a more detailed modeling of the third scenario with faults originating from the combination of the Enggano Fault and the Banten Sunda Strait Fault was conducted for the study site.

3.2 Modeling results of the worst tsunami scenario (The Third Scenario)

Tsunami modeling results with the worst-case scenario are displayed based on the values recorded in the most detailed model. In observing the elevation values in Kiluan Bay, the placement of the observation location points is the closest area to each residential as can be seen in Figure 9.

Figure 9. Observation point.

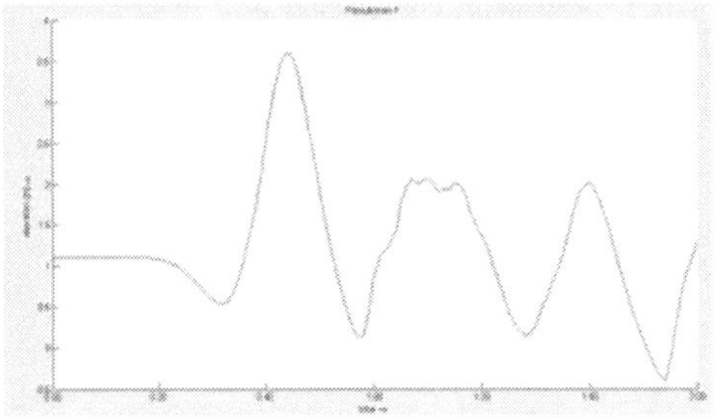

Figure 10. Water level on observation point A.

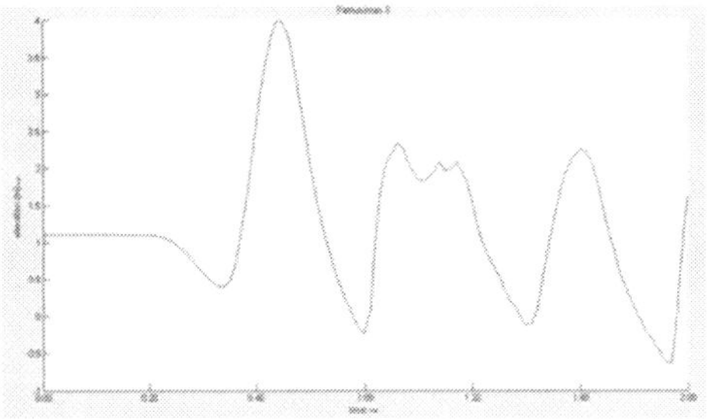

Figure 11. Water level on observation point B.

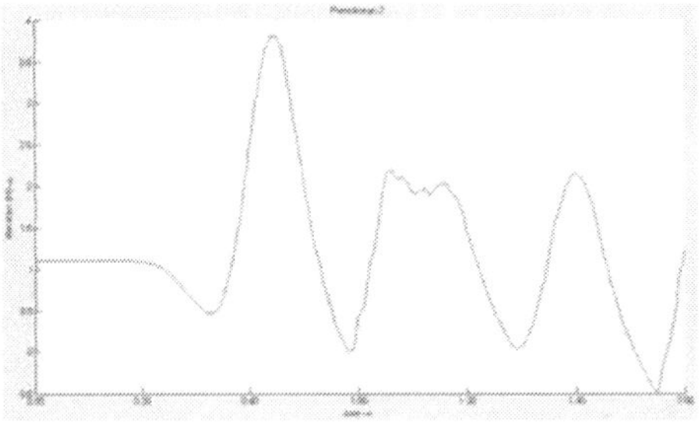

Figure 12. Water level on observation point C.

Based on Figure 10, it can be seen that in the area around residential location near observation point A, the peak wave elevation reached 3.59 m with a wave arrival time of 44th minute. It is also known that the tsunami wave exceeded the HHWL elevation at the 36th minute. Then, based on Figure 11, in the area around residential location near observation point B, the peak wave elevation reached 3.81 m with a wave arrival time of 44th minute. It can also be seen that the tsunami wave exceeded the HHWL elevation in the 37th minute. Then, as can bee seen in Figure 12, in the area around residential location near observation point C, the peak wave elevation reaches 3.98 m with the wave arrival time at the 44th minute, and the tsunami wave will exceed the HHWL elevation at the 38th minute.

In the wave height graph of the modeling results, it is known that tsunami waves occurred several times at 20 minutes and 40 minutes after the highest tsunami wave reached the observation location. The characteristics of the area in the form of a bay also make tsunami wave energy concentrated at certain points and produce resonance of incoming tsunami waves. The interaction of tsunami waves can produce complex wave patterns. The tsunami wave modeling results provide valuable information for the development of tsunami disaster information and mitigation plans in Kiluan Bay.

4. Conclusion

This study has shown the potential hazard caused by earthquake-generated tsunami in Kiluan Bay, Lampung. Based on historical data, the area suffered several tsunami incidents. It is also shown that there are 2 major faults, Banten Sunda Strait Segment and Enggano Segment, play a significant role in the tsunami generation.

This study has shown that a combination of both faults with a magnitude of 9.1 Mw may generate a major tsunami. The tsunami propagation was investigated using Delft3D. The initial condition for sea level is set at highest tide level of 1.2 m. It is found that the generated tsunami may reach the shoreline of Kiluan Bay within 44 minutes with a wave height around 3.59 meter to 3.98 meter.

It is hoped that this study may provide information for decision maker to develop Kiluan Bay area and to plan a mitigation plan. Future study will investigate the tsunami run-up overland and the required infrastructure for vertical evacuation building in Kiluan Bay.

Acknowledgements

This study was supported by Institut Teknologi Bandung Research (Riset ITB), P2MI ITB, and Riset Unggulan 3P. The authors would also like to express gratitude for the support of this research from the SATREPS Building Sustainable System for Resilience and Innovation in Coastal Community (BRICC) Project.

References

[1] PUSGEN 2017 Peta Sumber dan Bahaya Gempa Indonesia 2017 (Jakarta: Kementrian Pekerjaan Umum dan Perumahan Rakyat)
[2] Pusat Gempa Bumi dan Tsunami Kedeputian Bidang Geofisika 2019 Indonesia tsunami catalog (Jakarta: Badan Klimatologi, Metereologi, dan Geofisika)
[3] Supendi P et al 2022 On the potential form megathrust earthquakes and tsunamis off The Southern Coast of West Java and Southeast Sumatra, Indonesia Springer Natural Hazard 116 1316

IOP Conf. Series: Earth and Environmental Science **1464** (2025) 012013 doi:10.1088/1755-1315/1464/1/012013

[4] United States Geological Survey 2024 Earthquake catalog United States Department of The Interior [Online]. Available: https://earthquake.usgs.gov/earthquakes/. [Accessed March 2024].
[5] Ayudia G P and Prastowo T 2020 Analisis relasi antara magnitudo momen gempa dan luas bidang patahan sumber gempa Universitas Negeri Surabaya Inovasi Fisika Indonesia 09(3) 7-9
[6] Wells D L and Coppersmith K J 1994 New empirical relationships among magnitude, rupture length, rupture width, rupture area, and surface displacement Bulletin of The Seismological Society of America 84 975-1002
[7] Bousaha A, Laouar A, Guerziz A, and Hassan H S 2014 A new modified scheme for linear shallow-water equations with distant propagation of irregular wave trains tsunami dispersion type for inviscid and weakly viscous fluids Global Journal of Pure and Apllied Mathematics 10 793-818
[8] Koshimura S, Oie T, Yanagisawa H, and Imamura F 2009 Developing fragility functions for tsunami damage estimation using numerical model and post-tsunami data from Banda Aceh, Indonesia Coastal Eng. J. 51 243-273
[9] Purnama M R et al 2021 Development of tsunami inundation map for thecoast of Palu City IOP Conf. Ser.: Earth Environ. Sci. 737
[10] Adityawan M B et al 2023 Development of a tsunami early warning system on the coast of Palu Based On maritime wireless communication Progress in Disaster Science 19
[11] Desa Wisata Kiluan 2023 Profil Desa Wisata Kiluan, Kabupaten Tanggamus. [Online]. Available: https://desakiluan.com/profile. [Accessed 2024].
[12] Image of Kiluan Bay on Google Earth (Online). Available: https://earth.google.com/web. [Accessed 2024].
[13] Kompas 2013 Teluk Kiluan yang Berkilau [Online]. Available: https://nasional.kompas.com/read/2013/11/23/1042399/Teluk.Kiluan.yang.Berkilau?page=all. [Accessed 2024].
[14] Deltares 2024 Delft3D Flow Manual Version 4.05 [Online]. Available: https://content.oss.deltares.nl/delft3d4/Delft3D-FLOW_User_Manual.pdf. [Accessed 2024].
[15] Okada Y 1985 Surface deformation due to shear and tensile faults Bulletin of the Seismological Society of America 75 1135-1154.
[16] Badan Informasi Geospasial 2024 Information of seamless digital elevation model (DEM) and national bathimetry [Online]. Available: https://tanahair.indonesia.go.id/demnas/#/. [Accessed 2024].
[17] Alam R R R 2021 Tsunami-induced inundation on the coast of Palu City IOP Conf. Ser.: Earth Environ. Sci. 708

IOP Conf. Series: Earth and Environmental Science **1464** (2025) 012014 doi:10.1088/1755-1315/1464/1/012014

Near-bed flow dynamics and bed shear stress in a mangrove-vegetated estuary

Faruq Khadami[1,2*], Totok Suprijo[1,2], Alfiansyah Rizky Hidayat[2], Ivonne Milichristi Radjawane[1,2], Ayi Tarya[1,2], Gandhi Napitupulu[1,2]

[1] Research Group of Environmental and Applied Oceanography, Faculty of Earth Sciences and Technology, Institut Teknologi Bandung, Bandung 40132, Indonesia
[2] Oceanography Study Program, Faculty of Earth Sciences and Technology, Institut Teknologi Bandung, Bandung 40132, Indonesia

*E-mail: faruq@itb.ac.id

Abstract. Understanding sedimentation in riverbanks requires investigating near-bed flow velocity and bed shear stress. Studying flow in vegetated rivers presents complexities due to vegetation acting as a drag force that slows down water flow, especially near the bed. In tidal rivers, aside from river discharge influence, flow patterns are also affected by tides. This study investigates near-bed flow velocity and bed shear stress in the Ender River estuary, Cirebon, West Java, Indonesia. The Ender River estuary is a micro-tidal estuary bordered by mangrove vegetation on its riverbanks with a dominant clay sediment substrate. High-frequency measurement of near-bed flow velocity were conducted using an Acoustic Doppler Velocimeter (ADV) from October 4 to 10, 2023. The data were analyzed to determine the mean near-bed flow velocity ($\bar{u}$) and flow fluctuation, which was then used to estimate bed shear stress using the Log Profile (LP) method. The result revealed that the average near-bed flow velocity pattern is dominated by river flow and varies with tides. Strong seaward currents occur during ebb tides and weaken during flood tides. Maximum near-bed flow velocity occurs during ebb and flood conditions ($4x10^{-3}$ to $6x10^{-3}$ m/s) and minimum at peak flood ($1x10^{-3}$ m/s). Bed shear stress ranged from 0 to 0.07 N/m^2, with maximum values coinciding with ebb tide conditions. Given the calculated shear stress values, theoretically, the settled sediment is unlikely to be eroded or resuspended. This study provides insights into near-bed flow patterns in mangrove-vegetated tidal rivers, essential for understanding sedimentation and erosion processes in riverbanks with clay sediment substrates. However, the specific implications of these findings may vary depending on site-specific conditions.

1. Introduction

The critical conditions for mud erosion are highly site-specific, influenced by factors such as flow history, clay content, consolidation, and liquefaction potential [1-3]. To accurately assess sediment transport in muddy environments, precise measurements of near-bed current velocity and bed shear stress are essential. This study aims to characterize these parameters in the mud-dominated Ender River estuary, Cirebon.

The estuary is characterized by extensive mangrove vegetation. While mangroves reduce mean flow through increased friction, they also enhance turbulence by converting mean kinetic energy to turbulent kinetic energy (TKE). The mangrove vegetation also significantly influences turbulent kinetic energy dissipation. These complex interactions impact sediment transport, settling rates, and deposition.

Given the scarcity of studies on near-bed velocity in muddy riverbanks and the site-specific nature of mud erosion, this research is crucial for understanding shear stress in mangrove-dominated areas, particularly along riverbanks. By analyzing near-bed current velocity and shear stress variations in response to tides and river flow, this study seeks to discuss their impact on sediment transport within the Ender River estuary.

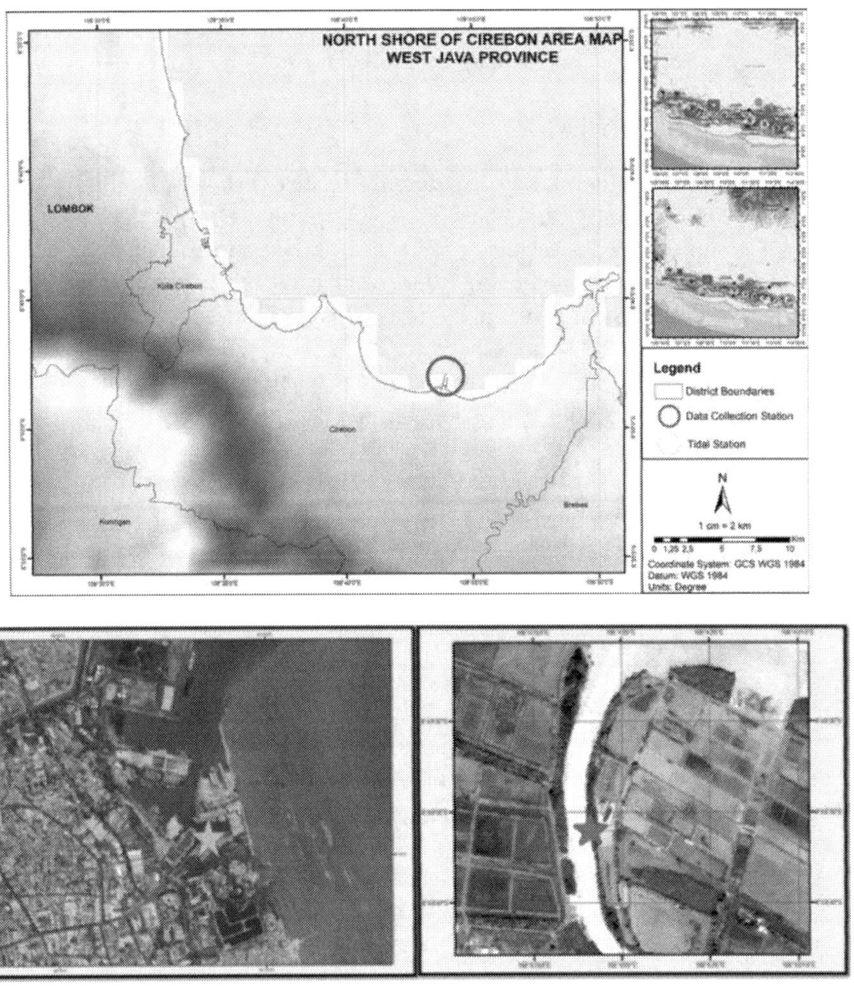

Figure 1. Study area in the northern part of Cirebon. The map shows the locations of the ADV (Acoustic Doppler Velocimeter) and water elevation observation stations. The red circle with a star indicates the ADV station, while the yellow circle with a star marks the water elevation station.

2. Data and method

2.1. Study site

The Ender River Estuary, located on the northern coast of Java, specifically in the Cirebon Regency of West Java, was chosen as the study site. This estuary is typical of those found along the northern coast of Java, characterized by native mangrove habitats and a dominant clay sediment substrate [4–6]. The Ender River Estuary is a micro-tidal estuary with tidal ranges between 0.8m and 1m [7], featuring a mixed semi-diurnal tidal pattern. Wave conditions in the Java Sea typically have heights ranging from 0.4 m - 0.8 m [4].

The study site under mangrove vegetation is located at the apex of the river mouth and extends along the riverbank in the area of the river mouth. This vegetated area is relatively a tidal flat. The location of the study area is shown in Figure 1.

2.2. Survey methods

Field observations were conducted during the transitional season, from 4 to 10 October 2023. These observations were aimed at covering one full period of spring and neap tides. The selected observation site was on a riverbank, as shown in Figure 1. Due to the limited availability of Acoustic Doppler Velocimeters (ADV), a single ADV was placed at this location, representing a sample of a riverbank bordered by mangrove vegetation.

The ADV used in this study measured the three velocity components—x, y, and z—which correspond to the eastward, northward, and upward directions, respectively. The ADV was installed vertically, with the sensor positioned 20 cm above the bed. The installation scheme of the ADV is illustrated in Figure 2. The instrument was programmed to collect data at a frequency of 8 Hz, using a burst mode that recorded data every 5 minutes. The ADV was also equipped with a calibrated compass to correct the measured current directions. Additionally, the ADV featured a pressure sensor. However, due to the limited duration of submersion for the pressure sensor, the tide gauge from the MTCRC station at Pelindo Harbor was used as a supplementary data. Although this station is somewhat distant from the study site, the phase of the water level recorded at the MTCRC station aligns with the pressure data collected by the ADV. The location of the MTCRC tidal station is shown in Figure 1. These elevation data were used as a reference for the tidal periods.

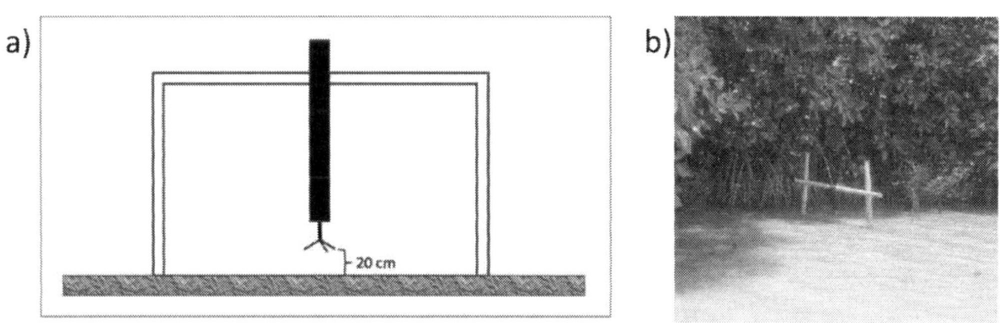

Figure 2. a) Schematic of instrument installation. b) Photo of ADV installation at the study site.

2.3. Data processing methods

Using the high-frequency velocity data obtained from the ADV, the data were processed to calculate the near-bed mean velocity and bed shear stress. The calculations for the near-bed mean velocity and bed shear stress were performed using the following equations:

$$u = \bar{u} + u' \quad (1)$$

$$v = \bar{v} + v' \quad (2)$$
$$w = \bar{w} + w' \quad (3)$$

Where u, v and w are the eastward, northward, and upward velocity components recorded by the ADV, respectively, and $\bar{u}, \bar{v}$, and $\bar{w}$ represent the mean currents. The terms u', v' and w' are the velocity fluctuations associated with the mean velocities.

Next, the values of the velocity fluctuations u', v', and w' were used to calculate the shear stress. In this study, the Log Profile (LP) method was employed. The following equation was used to compute the bed shear stress using the Log Profile method:
The friction velocity u_* was defined by Stull (1988) [8] as:

$$u_* = \left(\overline{u'w'}^2 + \overline{v'w'}^2\right)^{\frac{1}{4}} \quad (4)$$

The bed shear stress τ_B (N/m^2) was then calculated using:

$$\tau_B = \rho u_*^{\ 2} \quad (5)$$

Or equivalently,

$$\tau_B = \rho\left(\overline{u'w'}^2 + \overline{v'w'}^2\right)^{\frac{1}{2}} \quad (6)$$

3. Result and Discussion

3.1. Near bed velocity

The current velocity data obtained using the Acoustic Doppler Velocimeter (ADV) includes high-frequency current velocities in the u (west-east), v (north-south), and w (vertical up-down) directions. The hourly averaged velocities, denoted as ū, v̄, and w̄, were calculated from the measurement data and are shown in Figure 3a. Figure 3a shows that the ū velocity tends to be negative, and the v̄ velocity tends to be positive, indicating that the horizontal current tends to move northwestward. Meanwhile, the w̄ velocity is relatively small and shows little fluctuation, indicating minimal vertical current movement. The current is dominated by the north-south direction (v̄) with oscillations towards the north and south within the range of -5×10^{-3} to -1.5×10^{-2}. In contrast, the west-east current is dominated by eastward flow with a lower magnitude compared to the north-south current, within the range of -1×10^{-2} to 2×10^{-3}. This indicates that the near-bed current at the study site is primarily influenced by river flow, modulated by tidal effects. The current rose data shown in Figure 3b also clarifies that the near-bed velocity has a dominant direction towards the northeast, further indicating that the bottom flow is dominated by northward river flow."

To analyze the tidal variation's effect on the near bed velocity magnitude, daily averaging was performed (Figure 3c). This averaging makes the daily pattern of near bed velocity more distinct. During the first high tide, the near bed velocity increases from 2×10^{-3} m/s as it approaches high tide, then rises and peaks as it transitions to low tide with a value of 5×10^{-3} m/s. During the second high tide, the velocity peak also occurs as it transitions to low tide, but with a higher magnitude, reaching 7×10^{-3} m/s.

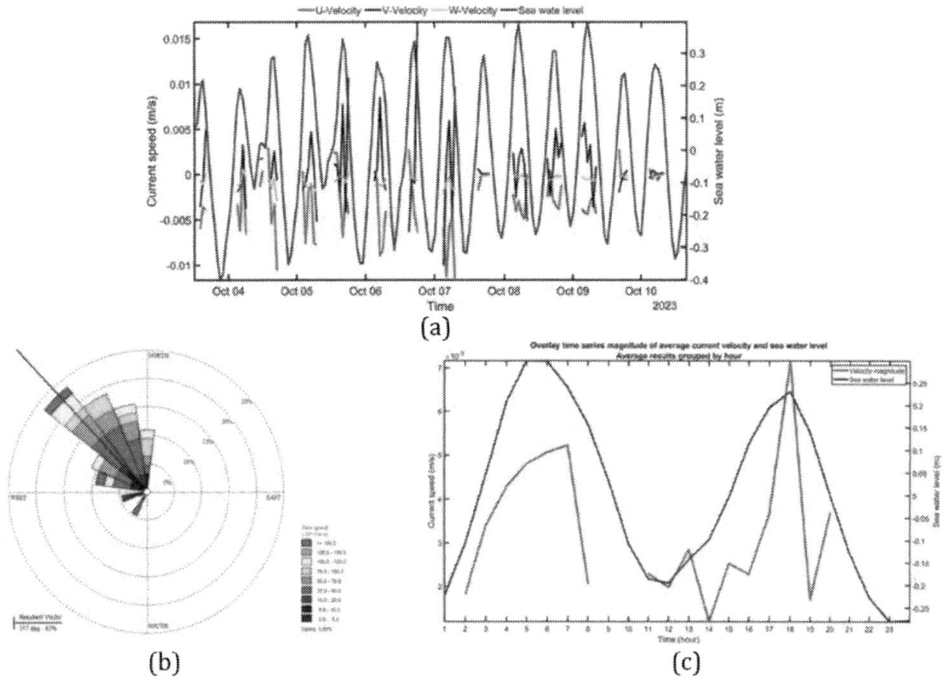

Figure 3. Time series of sea water level with (a) current velocity and (b) Current rose of near bed current velocity and (c) 25 hours-averaging of water level and current velocity magnitude.

3.2. Shear stress variation

The shear stress calculated using the Log Profile method (Equation 6) is presented in Figure 4. Similar to the near-bed current, the shear stress also exhibits tidal variations, generally increasing during high tide and decreasing as it approaches low tide. The calculation of shear stress using the log profile method shows a maximum value of 0.07 N/m^2 at the peak of high tide on October 4, 2024. Additionally, during October 6-7,2024 the current and shear stress values are also the highest compared to other days, which is associated with neap tide conditions occurring during that period, resulting in the dominance of river flow.

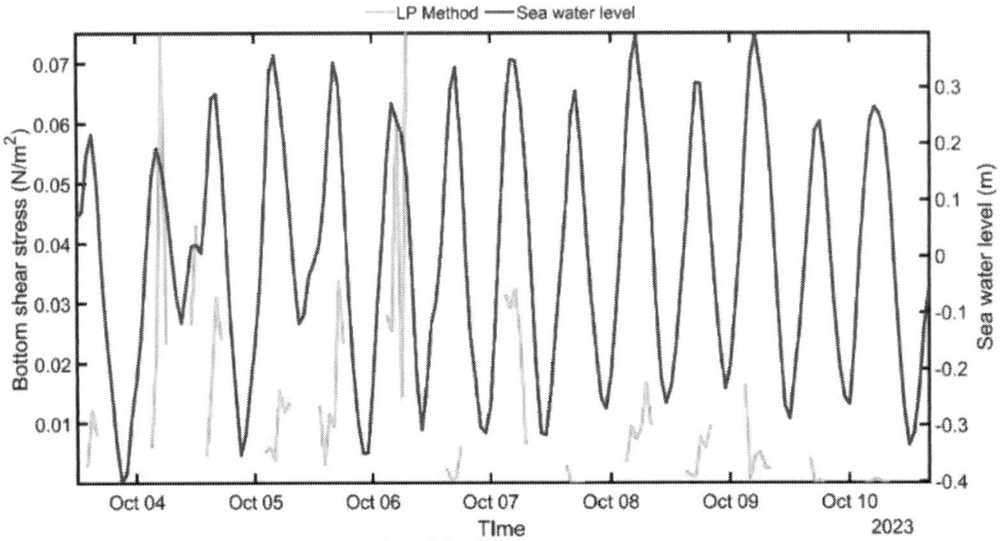

Figure 4. Time series of sea water level with bed shear stress.

3.3. Implication on Sediment Transport

Sediment in Cirebon waters is predominantly composed of silt and mud [4–6]. Given the silty nature of the sediment, the critical shear stress required for erosion falls within the range of 0.06-2.018 N/m^2 [9]. Based on our measured shear stress values of 0-0.07 N/m^2, it appears that settled sediment is unlikely to be resuspended. However, further investigation is warranted to confirm these findings, as shear stress calculations can vary depending on the method employed. These methods may show similar patterns but yield different shear stress values. Additionally, each method has its advantages and disadvantages; for example, the log profile method excels in estimating shear stress under conditions with a logarithmic velocity profile, while the max log profile method is better suited for measuring shear stress in zones with strong velocity fluctuations [3]. To definitively determine the resuspension potential of bottom sediment, future studies should focus on near-bed sediment dynamics. This involves measuring near-bed turbidity and correlating it with sediment concentration, current velocity, and bed shear stress.

4. Summary

The ADV observations conducted in this study revealed that the near-bed flow velocity was dominated by river flow, with variations influenced by the tidal cycle. The shear stress estimated using the Log Profile method exhibited a similar pattern. The maximum flow velocity was recorded during ebb and flood tides ($4x10^{-3}$ to $6x10^{-3}$ m/s), while the minimum value occurred at peak flood ($1x10^{-3}$ m/s). The bed shear stress ranged from 0 to 0.07 N/m^2, with the highest values during ebb tides. Based on these shear stress values, theoretically, the settled sediment would not be eroded or suspended. However, this study did not directly measure sediment transport dynamics. Therefore, further research is needed to confirm the sedimentation processes in the Ender estuary, particularly regarding direct measurements of sediment transport.

Acknowledgement

We'd like to thank the ITB Faculty of Earth Sciences and Technology for providing us with the resources, facilities, and funding we needed through the PPMI FITB 2024 program. We're also grateful for the financial support we received from ITB's Research and Community Service Institute (LPPM) through the Young Lecturer Capacity Enhancement Program.

References

[1] Andersen T J, Fredsoe J and Pejrup M 2007 In situ estimation of erosion and deposition thresholds by Acoustic Doppler Velocimeter (ADV) *Estuar Coast Shelf Sci* 75 327–36

[2] Myrhaug D, Holmedal L E and Rue H 2006 Erosion and deposition of mud beneath random waves *Coastal Engineering* 53 793–7

[3] Salehi M and Strom K 2012 Measurement of critical shear stress for mud mixtures in the San Jacinto estuary under different wave and current combinations *Cont Shelf Res* 47 78–92

[4] Triana K, Solihuddin T, Husrin S, Risandi J, Mustikasari E, Kepel T L, Salim H L, Sudirman N, Prasetyo A T and Helmi M 2023 An integrated satellite characterization and hydrodynamic study in assessing coastal dynamics in Cirebon, West Java *Reg Stud Mar Sci* 65 103107

[5] Nurkhasanah D, Ambariyanto, Suprijanto J, Pusparini N and Prasetyo B 2019 The sediment distribution based on types of grain size in the Cirebon and Demak waters *IOP Conf Ser Earth Environ Sci* 246 012024

[6] Agusta V C and Putri M R 2023 Grain size trend analysis for sediment characteristic of Cirebon Waters *IOP Conf Ser Earth Environ Sci* 1163 012016

[7] Badriana M R, Abdurrahman U, Arifin Nur A, Jeon C, Miliachristi Radjawane I and Park H 2024 Tidal Characteristic Analysis Utilizing Radar Tide Gauge in Cirebon Seawater *Jurnal Ilmu dan Teknologi Kelautan Tropis* 16 105–15

[8] Stull R B 1988 *An Introduction to Boundary Layer Meteorology* ed R B Stull (Dordrecht: Springer Netherlands)

[9] Chen D, Zheng J, Zhang C, Guan D, Li Y and Wang Y 2021 Critical Shear Stress for Erosion of Sand-Mud Mixtures and Pure Mud *Front Mar Sci* 8

IOP Conf. Series: Earth and Environmental Science **1464** (2025) 012015 doi:10.1088/1755-1315/1464/1/012015

The effect of jetty layout on waves at TPI Beach, Lumajang Regency with Delft3D modelling

Afifa Nurchalim[1], Retno Utami Agung Wiyono[2*], Wiwik Yunarni Widiarti[2], and Entin Hidayah[2]

[1] Undergraduate Program of Civil Engineering, Jember University, Jember, Indonesia
[2] Civil Engineering Departement, Jember University, Jember, Indonesia

*E-mail: retnoutami@unej.ac.id

Abstract. TPI beach is adjacent to the river mouth and is directly opposite the south coast which has big waves. The condition of TPI Beach raises a problem, namely the decline of the coastline due to abrasion by high waves and river currents when flooding. A jetty structure is modelled in the study area using Delft3D to determine the effect on current velocity and wave height. The simulation is carried out with three scenarios, the first scenario is the addition of one jetty on the southwest side of the river mouth with a length of 400 m, the second scenario is the addition of one jetty on the northeast side of the river mouth with a length of 347 m, and the third scenario is two jetties with a height of 400 m on the southwest the mouth of the river and 347 m to the northeast of the estuary. The research data includes wind data for 10 years, tidal data, and bathymetry maps. In the first scenario, the reduction in wave height is 9%, in the second scenario the reduction in wave height is 5%, and with two jetties, it is 9%. The reduction of current velocity in the first scenario is 61%, the reduction of current velocity in the second scenario is 7%, and in the third scenario is 56%. The most effective scenario to reduce both wave height and current velocity is Scenario 1.

1. Introduction

TPI Beach is located in Lumajang Regency which is directly opposite the south coast. TPI Beach is adjacent to the river mouth and is located in the south coast of Java Island which has big waves. The location of the beach raises a problem, namely the change in the coastline. This shoreline change is in the form of a coastline retreat caused by abrasion by high waves from the southern sea and abrasion by river currents during flooding[1].

High waves that have abraded TPI Beach have occurred several times since 2016 [2]. Based on previous studies and news, it is known that coastal abrasion causes the coastline to retreat at least 1 km [3] and affected at least 60 houses belonging to fishermen in two neighborhoods (RT). Associations have become victims of abrasion [4]. The fish auction place (TPI), which is usually used as a place to trade fish, was also destroyed by the high waves on the south coast [4]. River estuaries that are often covered by sediment can cause siltation and reduce the appearance of the river estuary flow, thus blocking the discharge of river flow into the sea [5]. The estuary of this river is used as a shipping lane by the TPI community, but the presence of this deposition has disrupted ship traffic [6]. This situation gets worse in the rainy season because it makes the river mouth discharge near TPI Beach increase and will erode the surrounding area as a new route to the sea.

With such coastal conditions and the TPI beach which until now still does not have any coastal protection structures at all, the same thing may happen again. Therefore, it is necessary to plan a coastal protective building. In this study, a jetty structure was chosen because it was considered more suitable for the site conditions. Many engineers use jetty buildings to stabilize and maintain the shape of river mouths [7]. Based on several studies [8] jetty is suitable for use in cases at river mouths because they can withstand sediment transport and improve coastal erosion.

This study will model the jetty layout using Delft3D so the effect of the jetty on the wave height and current velocity can be analyzed. Previous research [1] used Delft3D to determine changes in coastal morphology due to jetty construction. [9] used Delft3D as a tool to study the effect of the placement of breakwaters and groins on tidal and current heights. This study is different from previous researches because it modelled three scenarios of jetties placement to show the difference between the effect scenario to wave height and current velocity.

2. Methods

Figure 1 shows the research location in TPI Beach, Tempursari District, Lumajang Regency together with the fetch line which has latitude coordinates of -8.333129° and longitude 112.968903°.

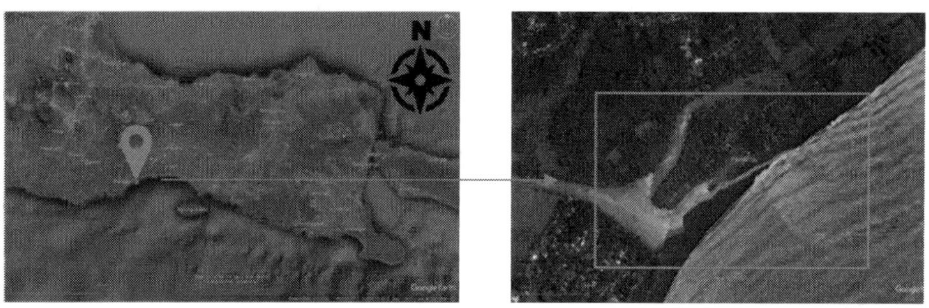

Figure 1. The study area in TPI Beach, Lumajang.

In this study, several data are necessary: wind data, bathymetry, and tidal data. The wind data include 10 years obtained from Malang BMKG Station. Malang BMKG Station is located at coordinates -7.90080 ° and longitude 112.59790 ° with an elevation of 285 m. Wind data is processed into windrose with the help of a website (windrose.xyz) to obtain the dominant wind direction. Wind data is processed into significant wave heights and significant periods using hindcasting methods[10][11]. Tidal data was obtained from the ina-sea water level BIG website (https://www.big.go.id/). The data used is data for 3 days starting from March 13-15, 2022. This data is utilized based on previous interviews conducted with local residents that the tidal wave occurred in March. The last one is bathymetric data obtained from the official website of Tanah Air Indonesia, owned by BIG. This data has a resolution of 6 arc-seconds or 183.6 meters using the MSL datum (Mean Sea Level). All of these data are included in the model simulation process. Ona way ANOVA and Tukey HSD tests were also utilized for analysis in the study.

2.1 Wave Forecast

This wave forecasting is done to get the significant wave height and significant wave period. Later the significant wave height and significant wave period will be included in the simulation process. The analysis was carried out using the hindcasting method[11].

The fetch length is shown in Figure 2. The effective fetch is 98.9 km to the southeast and the wind stress factor value is 15.09 m/s. Based on calculations using hindcasting, it is obtained the significant wave height is 2.43 meters and wave period is 7.13 seconds.

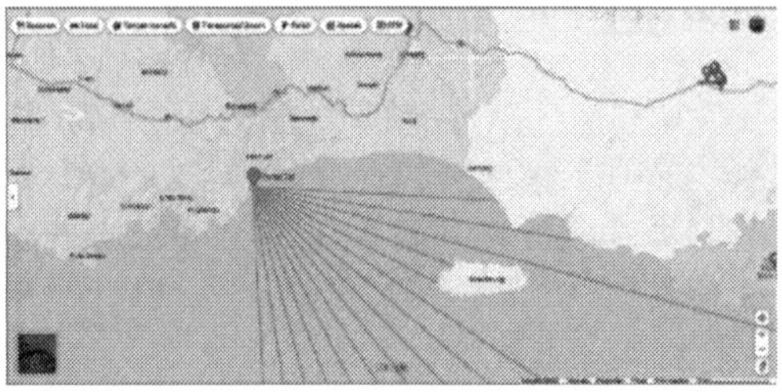

Figure 2. Fetch line at TPI Tempursari Beach.

2.2 Delft3D Modeling

Delft3D program was utilized to model one existing model (initial conditions) and three scenarios with the additional jetty building. This model utilized a curvilinear grid with a minimum grid size of 10 m.

By hindcasting method, significant wave height and wave period is calculated for simulation. In Delft3D-Wave, the significant wave height is 2.43 meters from Southeast with wave period 7.13 second.

The first scenario is the addition of a jetty with a length of 400 meters on the southwest side of the river mouth, the second scenario is the addition of a jetty building to the northeast of the river mouth 347 meters long, and the third scenario is to add two jetties on the southwest and northeast side of the jetty with the length of each jetty is 400 meters on the southwest of the river mouth and 347 meters on the northeast side of the river mouth.

The selection of length is based on the location of the beach which is in the south coast area which has relatively large waves, so the length chosen is expected to be able to protect the coastal area and river estuary. Simulation carried out for 3 days starting from March 13, 2022, to March 15, 2022, and takes four (4) observation points as shown in Fig. 3a.

2.2.1 Delft3D-Flow

Delft3D-FLOW is a system in the Delft3D section that is used to calculate SWAE (Shallow Water Equation) or equations during shallow water conditions by calculating velocity and altitude variables in two or three dimensions on a grid or auxiliary line[1]. The following are some of the equations used in the Delft3D software [12].

2.2.2 Delf3D-Wave

Delft3D-WAVE is one of the features in Delft3D which is used to simulate the propagation [1]. Delft3D-WAVE uses a SWAN (Simulating Waves Nearshore)[13] model that utilized action balance equation [14]. In SWAN the evolution of the wave spectrum is described by the spectral action balance equation

3. Result

Results of tidal verification, and simulation results contains one existing simulation, and three other scenarios are shown in this section.

3.1 Tidal Verification

Verification of tidal simulation is conducted between the simulation results and Badan Informasi Geospasial (BIG) data. The results showed a standard deviation of 0.0031 and the graph can be seen in Figure 3.

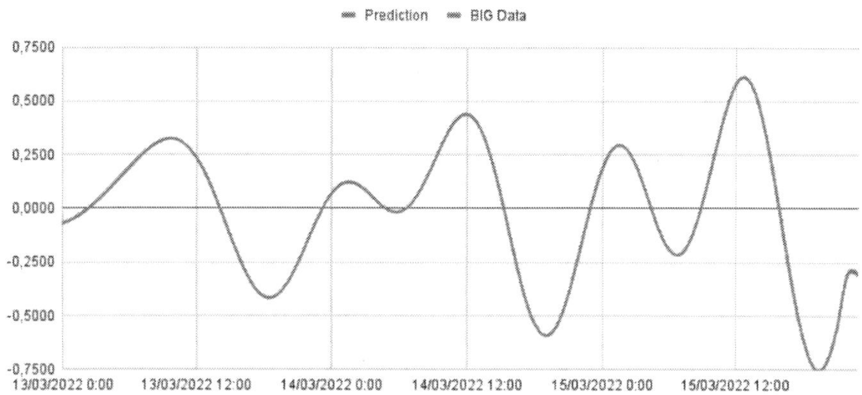

Figure 3. Tidal verification.

3.2 Simulation Result

The simulation results of the Delft3D show the maximum wave height (Fig.4) and current speed occurred on March 15, 2022, at 00.00. In Scenario 1, the modeled jetty building has a length of 400 meters and is placed on the southwest side of the river mouth. Scenario 2 is building a jetty on the northeast side of the river mouth with a length of 347 meters, and Scenario 3 adding a jetty building on both sides of the river mouth with a length of 400 meters each the southwest and 347 meters on the northeast side of the river mouth. The simulation results at each observation point on 3/15/2022 at 00:00 are as follows.

Table 1. Wave height reduction.

Observation point	Significant Wave Height (meters)				Wave Height Reduction (%)		
	Existing	Scenario 1	Scenario 2	Scenario 3	Scenario 1	Scenario 2	Scenario 3
A	2,355	2,029	2,208	2,040	14%	6%	13%
B	2,354	2,103	2,210	2,112	11%	6%	10%
C	2,343	2,180	2,220	2,186	7%	5%	7%
D	2,328	2,224	2,241	2,227	4%	4%	4%

Based on the calculations in Table 1, the average reduction in wave height is obtained. In the first scenario, it is 9% in the second scenario it is 5%, and in the third scenario, it is 9%.

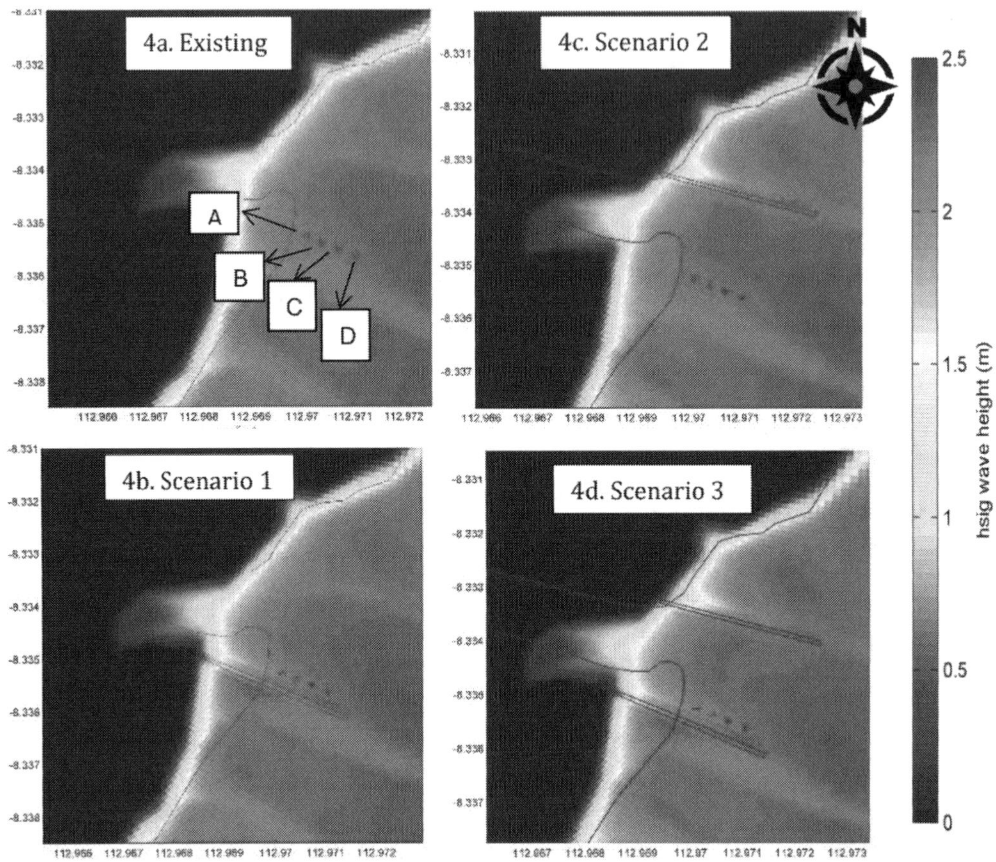

Figure 4. Simulation results of wave height in TPI Beach .

Figure 4a shows the wave height in the existing state, with the height in the area at the observation point ranging from 2 to 2,5 meters. The other three figures show a reduction in wave height in the observation area behind the jetty building. The change in wave height in the observation point area for Scenario 1 and 3 ranges from 1 to 1,5 meters. The reduction in wave height is due to the jetty building protecting the area behind it from waves from the southeast. In the case of Scenario 2, the wave height does not change significantly from existing case. This is because the location of the observation point is far from the jetty building, thus making the area less protected from the jetty building.

The addition of the jetty building also has an influence on the speed of the current in the wave. The results of the simulation obtained a reduction in current velocity as shown in Figure 5. Figure 5 shows a change in current velocity around the jetty building with the same water level conditions, namely in the range of 3 meters.

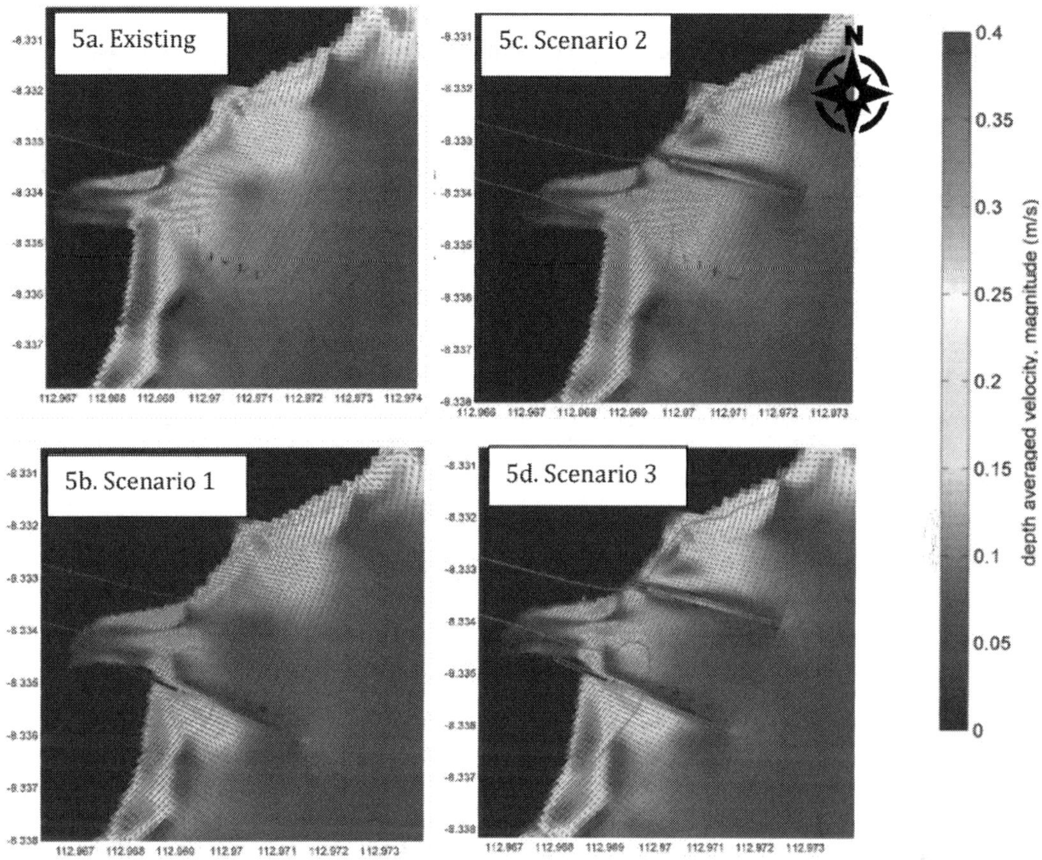

Figure 5. Simulation results of current velocity in TPI Beach.

In Scenario 2, the observation area does not get protection from the jetty because the observation area is located in front of the building, so the reduction in the observation area is relatively small. As in scenario 1, the observation area in scenario 3 is also protected by the southwest jetty but is also affected by currents bouncing off from the jetty in the northeast. The reduction value due to the addition of jetty buildings on 3/15/2022 at 00:00 can be seen in Table 2.

Table 2. Current speed reduction.

observation point	current speed (m/s)				current speed reduction (%)		
	Existing	Scenario 1	Scenario 2	Scenario 3	Scenario 1	Scenario 2	Scenario 3
A	0,142	0,040	0,129	0,054	72%	9%	62%
B	0,127	0,036	0,114	0,044	72%	11%	65%
C	0,103	0,042	0,096	0,045	60%	7%	57%
D	0,090	0,054	0,088	0,054	39%	1%	40%

IOP Conf. Series: Earth and Environmental Science **1464** (2025) 012015 doi:10.1088/1755-1315/1464/1/012015

4. Analysis

4.1 The Result of One-Way ANOVA Test

ANOV test is used to determine whether the hypothesis that has been made meets or not. One type of Anova test is the one-way test. That matters because the analysis process uses variations observational data is one-way and the analysis involves only one independent change. Anova test which is called the F score and subsequently in the journal will be referred to as F_{count}, is named after R.A. Fisher [15][16].

The hypotheses made in this study are:

1. The ANOVA test for the wave height reduction

There are two assumptions: H_a = Scenario 1, Scenario 2, and Scenario 3 will provide a reduction in wave height; and H_0 = Scenario 1, Scenario 2, and Scenario 3 will not provide a reduction in wave height.

2. The ANOVA test for the current speed reduction

There are also two assumptions: H_a = that Scenario 1, Scenario 2, and Scenario 3 will provide a reduction in current speed; and H_0 = Scenario 1, Scenario 2, and Scenario 3, will not reduce the current speed.

To calculate the Anova or F(F_{count}), calculate using the formula:

$$F_{count} = \frac{V_A}{V_D} = \frac{KR_A}{KR_D} = \frac{JK_A:db_A}{JK_D:db_D} \quad (1)$$

which V_A represents the variance between groups, V_D represents the variance between groups. KR_A or MSB is mean sum of squares between the groups and KR_D or MSWis mean sum of squares within the groups. Variants in a group can also be called Error Variants. It is formulated:

$$JK_A = \sum \frac{(\sum X_{Ai})^2}{n_{Ai}} - \frac{(\sum X_\tau)^2}{N} \text{ for } db_A\text{=A-1} \quad (2)$$

$$JK_D = (\sum X_\tau)^2 - \sum \frac{(\sum X_{Ai})^2}{n_{Ai}} \text{ for } db_D\text{=N-A} \quad (3)$$

JK_A represents the sum of squares within the groups, JK_D is sum of squares between the groups, db_A and db_D is degrees of freedom, A is total number of populations and N is total number of samples in a population. After knowing the value of the F distribution in this case F_{count} [17], this value will be used to find out the significant probability value or in this case the F_{table}. The F_{table} the significance 95% or α = 0.05.

Table 3. Calculation of wave height reduction based on ANOVA test.

Source of Diversity	degrees of freedom	sum of squares	mean square	F count
Between Groups	3	0.11529	0.03843	10.52
In Group (error)	12	0.04382	0.00365	
Total		0.15911		

In this study, the ANOVA test uses degrees of freedom 1 is 3 and degrees of freedom 2 is 12 with probability (α)= 0.05, so the F_{Table} is 4.256. The calculation of the ANOVA test of wave height reduction is shown in Table 1. Based on the calculations, it is known that F_{count} is 10.52, while F_{Table} is 4.256. Because F_{count} more than F_{Table}, the hypothesis Ha is rejected. Thus, the addition jetty on several scenarios will provide a reduction in wave height.

Table 4 shows the calculation of the ANOVA test for the reduction of the current velocity. Based on the calculations, it is known that F_{count} is 23.41, while F_{Table} is 4.256. F_{count} is more than F_{Table}, then the hypothesis H_a is accepted. Thus, the addition of jetty on several scenarios will provide a reduction in current speed.

Table 4. Calculation of the current speed reduction based on ANOVA test.

Source of Diversity	degrees of freedom	sum of squares	mean square	F count
Between Groups	2	0.017131	0.005710	23.41
In Group (error)	9	0.002927	0.000244	
Total		0.020058		

4.2 The result of Tukey HSD Test

When an analysis of variance (ANOVA) gives a significant result, this indicates that at least one group differs from the other groups, and in order to analyze the pattern of difference between means, the ANOVA is often followed by specific comparisons, and the most commonly used involves comparing two means (the so-called pairwise comparisons)[18]. Tukey developed an easy and frequently used pairwise comparison technique under the name of the Honestly Significant Difference (HSD) Test. The main idea of the HSD Test is to test whether there was any similarity between the mean of the comparison groups [19].

Analysis of Variance shows that there is significant difference in wave height among the scenarios. The subsequent Tukey HSD post-hoc test shows that the significant differences in wave height are found between the existing simulation and Scenario 1 (p value = 0.001) and between the existing

Table 5. The result of Tukey HSD Test for the wave height reduction.

rownames(tukey$skenario)	p adj
Scenario.1-Existing	**0.001683241**
Scenario.2-Existing	0.053367898
Scenario.3-Existing	**0.002229869**
Scenario.2-Scenario.1	0.238789472
Scenario.3-Scenario.1	0.998163652
Scenario.3-Scenario.2	0.304024813

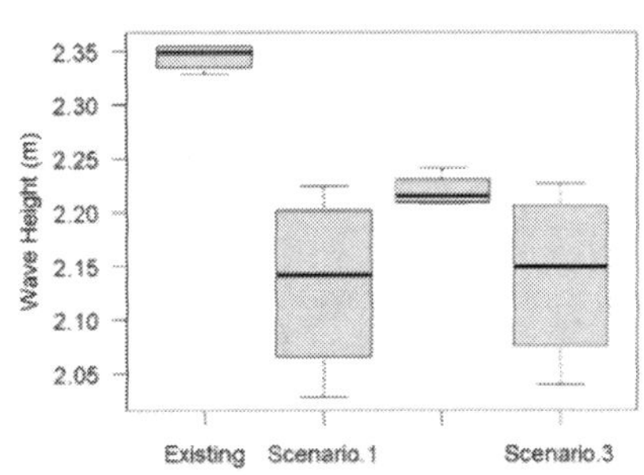

Figure 6. Boxplot of simulation wave height reduction.

simulation and Scenario 3 (p value = 0.002). The significant difference can also be confirmed by looking at boxplot in Figure 6.

Table 6. The result of Tukey HSD Test for the curent velocity reduction.

rownames(tukey$skenario)	p adj
Scenario.1-Existing	**0.0001359464**
Scenario.2-Existing	0.8565434946
Scenario.3-Existing	**0.0003137996**
Scenario.2-Scenario.1	**0.0004435914**
Scenario.3-Scenario.1	0.9402909891
Scenario.3-Scenario.2	**0.0010846536**

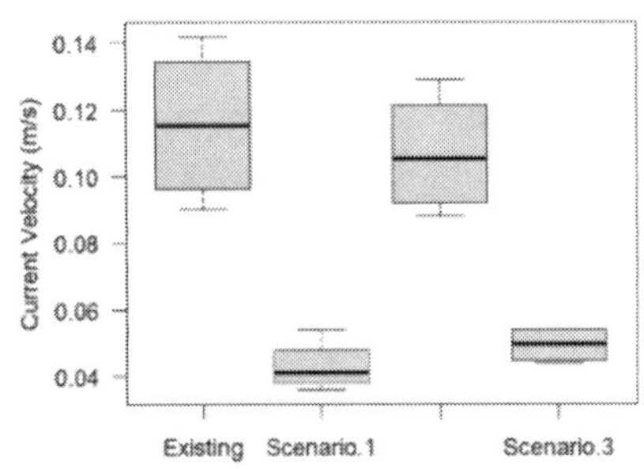

Figure 7. Boxplot of Simulation current velocity reduction.

Analysis of Variance shows that there is significant difference in wave height among the scenarios. The subsequent Tukey HSD post-hoc test shows that the significant differences in wave height are found between the existing simulation and Scenario 1 (p value = 0.0001) and between the existing simulation and Scenario 3 (p value =0.0003). The significant difference can also be confirmed by looking at the sboxplot in Figure 7.

5. Discussion

Based on the modeling results, it is shown that one jetty on the southwest side (Scenario 1) has the greatest reduction in wave height and current velocity than the other two scenarios. This is caused by the direction of the waves and wind at the research site.

The direction of the waves in this study comes from the southeast. The reduction of wave height and velocity in Scenario 1 and Scenario 3 has greater values than in Scenario 2 with one jetty in the northeastern part of the river mouth. This condition occurs because in Scenario 2, the location of observation points is in front of the jetty, thus the observation points unprotected by jetty. This protection of the jetty occurs because of the wave reflections. This reflection occurs because the waves hit the jetty[20]. Water waves may be partially or totally reflected by both natural and manmade barriers [10].

In the simulation, the results of the current velocity reduction in Scenario 3 with two jetties on both sides of the river estuary are quite large but not greater than Scenario 1. This may occur because the observation area is located between the two jetties. The observation area is protected by the jetty building on the southwest but also gets an increase in current that is reflected from the jetty in the northeast, so the flow in Scenario 3 will increase because of the reflection. This condition shows the reduction of current velocity with two jetties (Scenario 3) is smaller than just one jetty in the southwest (Scenario 1).

According to the simulations and ANOVA tests that have been carried out, it is known that the jetty could reduce both the wave height and the current velocity. This is in line with the function of the jetty, which is to protect rivesr mouths and beaches from sediment transport along the coast[20]. Longshore transport waves are caused by the turbulence of sediments when the waves break, then move along the currents and wave components parallel to the coast [5]. A study explains that the inclusion of tidal and wind-driven currents not only increases the advection of sediment stirred up by the waves but also leads to an amplification of the bed shear stress, consequently increasing the mixture and the sediment concentration in the water column[11]. Therefore, the jetty is made perpendicular to the beach to withstand sediment transport. Based on the simulation results, the jetty may reduce the current velocity and wave height so it can be estimated that the addition of the jetty building on the TPI beach may reduce sedimentation.

6. Conclusion

Based on the simulation results, the reduction in wave height in Scenario 1 with one jetty on the Southwest side of the estuary is 9%, and in Scenario 2 with one jetty on the Northeast side of the river mouth is 5%, and in Scenario 3 with two jetties on the Northeast and Southwest side of the river is 9%. In addition, the simulation results also show a reduction in current velocity, scenario 1 reduced the current velocity by 61%, Scenario 2 by 7%, and Scenario 56%. It is show that one jetty on the southwest side (Scenario 1) has the greatest reduction in wave height and current velocity than the other two scenarios. This is caused by the direction of the waves and wind at the research site. According to the simulation and ANOVA test that have been carried out, it is known that the jetty is able to reduce both wave height and current velocity. This is in line with the function of the jetty, which is to protect river mouths and beaches from sediment transport along the coast. And the subsequent Tukey HSD post-hoc test show that the significant differences in wave height are found between the existing simulation and Scenario 1 and between the existing simulation and Scenario 3. This may occur due to the location of jetty in Scenario 2 which could not reduce wave coming from the predominant wave direction.

References

[1] Fahmi M, Halfi TM. 2019. Simulasi Numerik Perubahan Morfologi Pantai Akibat Konstruksi Jetty Pada Muara Lambada Lhok Aceh Besar Menggunakan Software Delft3D. Jurnal Teknik Sipil **8** 50–9.

[2] https://lumajangsatu.com/baca-12008-wisata-pantai-tpi-tempursari-lumajang-akan-jadi- kenangan. Accessed February 1, 2020.

[3] https://www.wartabromo.com/2020/05/15/bupati-lumajang-berencana-relokasi-rumah-warga- korban-abrasi-tpi-tempursari/ (accessed February 1, 2022).

[4] https://surabaya.tribunnews.com/2020/09/01/solusi-bupati-lumajangterkait-penanganan- warga-tpi-tempursari-terdampak-abrasi. accessed February 1, 2022.

[5] Pokaton KY, Tawas HJ, Jasin MI, Mamoto JD. 2013. Perencanaan Jetty Di Muara Sungai Ranoyapo Amurang. Jurnal Sipil Statik **1** 434–43.

[6] Irawan Y, Dharmita S. 2019. Jetty Sebagai Alternatif Pengaman Muara Sungai Air Bengkulu. Majalah Teknik Simes. **13** 25–33.

[7] Flor-Blanco G, Pando L, Morales JA, Flor G. 2015. Evolution of beach–dune fields systems following the construction of jetties in estuarine mouths (Cantabrian coast, NW Spain). Environ Earth Sci. **73** 1317–30.

[8] Pietrafesa LJ. 2012. On the continued cost of upkeep related to groins and jetties. J Coast Res. **28** iii–ix.

IOP Conf. Series: Earth and Environmental Science **1464** (2025) 012015 doi:10.1088/1755-1315/1464/1/012015

[9] Royyan MAA, Utami Agung Wiyono R, Fery Irawan J, Bintang Pratama M, Halik G, Yunarni Widiarti W. 2020. Numerical Simulation of Breakwater Layout in Puger Beach Jember Due to Tidal Wave. IOP Conf Ser Earth Environ Sci. **437** 1–9.
[10] Coastal Engineering Research Center. Shore Protection Manual. 1984. Shore Protection Manual, vol. I. 4th ed., Vicksburg, Mississippi: U.S. Government Printing Office Washington, D.C.. p. 1–337.
[11] Uğurlu A, Balas C. 2024. Integrative Probabilistic Design of River Jetties by 3D Numerical Models of Transport Phenomena: The Case Study of Kabakoz River Jetties. Mar Sci Technol Bull. **13** 151–67.
[12] Deltares. 2021. Delft3D-FLOW Simulation of multi-dimensional hydrodynamic flows and transport phenomena, including sediments. The Netherlands: Deltares.
[13] Kaji A, Luijendijk A, Van Thiel de Vries J, De Schipper M, Stive M. 2014 . Effect of Different Forcing Processes on the Longshore Sediment Transport At the Sand Motor, the Netherlands. Coastal Engineering Proceedings, **vol. 1p**. 11.
[14] Deltares. 2014. Delft3D-WAVE Simulation of short-crested waves with SWAN. The Netherlands: Deltares.
[15] Kim TK. 2017. Understanding one-way ANOVA using conceptual figures. Koren Journal of Anesthesiology. **70** (1).
[16] Sawyer SF. 2009. Analysis of Variance: The Fundamental Concepts. Journal of Manual & Manipulative Therapy. **17** 27E-38E.
[17] Woods NN, Zinnert JC. 2024. Shrub encroachment of coastal ecosystems depends on dune elevation. Plant Ecol **225** 1047–57.
[18] Abdi H, Williams LJ. 2010. Encyclopedia of Research Design. In Neil Salkind (Ed.), Encyclopedia of Research Design., Dallas, Richardson, USA.
[19] Mardhotillah B, Rozi S, Rodgiyah Z. 2021. Tukey HSD Post Hoc Test untukPerbandingan Karakteristik lIngkungan dan Sumber Daya Provinsi-Provinsi di Indonesia. Jurnal Engineering. **3**8191.
[20] Triatmodjo B. 2020. Perencanaan Bangunan Pantai. 2nd ed. Yogyakarta: Beta Offset Yogyakarta.

Flood inundation hazard mapping and vulnerability analysis in the downstream area of the Cipunagara River Basin of Indonesia using Google Earth Engine

Hendra Achiari [1*], **Gurusu** [1], **Ravindra Jayaratne** [2] **and Reynaldi J. Purba** [1]

[1] Ocean Engineering, Faculty of Civil and Environmental Engineering, Bandung Institute of Technology, Indonesia
[2] School of Architecture, Computing and Engineering, University of East London, United Kingdom

*E-mail: h_achiari@itb.ac.id

Abstract. Tidal flooding is a natural phenomenon where seawater enters land areas with high sea levels. Tidal flooding is caused by factors, namely forest area decline, land degradation, hydrological conditions disturbances, household waste, improper drainage, and rising tides. The North Coast of Java that has the potential to be affected by tidal flooding is the downstream area of the Cipunagara River Basin, Subang Regency. Factors causing disasters include runoff from the Cipunagara river, sea water intrusion, changes in coastal land use, sea level rise, and the presence of river water overflows every year, which play a role in the expansion of the tidal area of flood inundation. The purpose of this study is to map the risk of tidal flooding based on the hazard index, vulnerability index. The determination of the risk index is carried out using a spatial method to identify the hazard index and vulnerability index, by scoring or overlaying through the GEE and ArcGIS 10.8 applications, as well as qualitative descriptive methods in determining disaster capacity. The high-vulnerability category for tidal flooding is found in the villages of Legon Kulon, Patimban, Pusakaratu, Mundusari, Rancasari, Pamanukan Sebrang, Mulyasari, Pamanukan Hilir, Pamanukan, Lengkong Jaya, and Pusakajaya with a tidal flood vulnerability index scoring range, which is 3.2 – 3.8. The moderate vulnerability category was found in the villages of Bobos, Karangmulya, Rancadaka, and Gempol with a scoring range of 2.6 – 3.2. While the category of low tidal flood vulnerability index was in Tegalurung, Mayangan, Legon Wetan, Pengarengan, and Ranca Hilir with a tidal flood vulnerability index score range of 2.0 – 2.6. This vulnerability index will be expanded to determine the Capacity index and Risk index for the same location of study.

1. Introduction

The ongoing phenomenon of climate change is leading to rising sea levels, primarily due to the melting of polar ice caps. This presents a significant challenge for archipelagic nations around the globe. It is predicted that by 2100 sea level will rise due to global warming as high as 440 mm to 740 mm [1]. The consequence is the rise of issues in coastal regions, including tidal flooding or coastal inundation. Tidal flooding is caused by several factors, namely a decrease in forest area, land degradation, disturbance of hydrological conditions, household waste, improper drainage,

IOP Conf. Series: Earth and Environmental Science **1464** (2025) 012016 doi:10.1088/1755-1315/1464/1/012016

and rising tides [2]. This disaster will have a greater impact and intensity along with the phenomena of global warming and climate change [3]. The North Coast of Java, the East Coast of Sumatra, Kalimantan, the South Coast of Sulawesi, and other small islands are lowland coastal areas in Indonesia [4]. The downstream region of the Cipunagara River Basin, located on the North Coast of Java, is potentially vulnerable to tidal flooding, particularly in Subang Regency. Disasters are influenced by factors such as runoff from the Cipunagara River, seawater intrusion, alterations in coastal land use, rising sea levels, and annual river water overflows, which contribute to the expansion of the tidal flood inundation area. Over time, the impact of tidal flooding is expected to intensify, with a growing area experiencing inundation from these floods year after year. Developing tidal flood risk mapping for the downstream Cipunagara watershed is crucial. Key factors in disaster risk mapping encompass hazard mapping and vulnerability analysis. Disaster risk mapping is designed to reduce the impacts and losses from disasters by employing disaster risk management strategies. The use of map visualization techniques is anticipated to expedite the management of tidal flood inundation issues in the downstream region of the Cipunagara watershed.

2. Methodology

2.1 Research data and equipment

The study utilizes both secondary and primary data. Secondary data consists of supporting information gathered from relevant agencies and institutions, whereas primary data is collected directly from the research site using the methods outlined in Table 1. The types and sources of data used are shown in Table 1 and Table 2.

Table 1. The type of tools and devices utilized.

Tools and Devices	Utility
Field survey device:	
1. GPS	Positioning and tracking the coastline
2. Camera	Documentation of conditions in the field
Data analysis tools:	
1. Computer Hardware and Software (ArcGIS 10.8 and Google Earth Engine)	Data analysis tools

Table 2. Data types and their sources

No	Data Type	Data Properties		Source
		Primary	Secondary	
1	Subang Regency Administration Map		√	https://tanahair.indonesia.go.id/
2	Data Digital Elevation Model (DEM)		√	https://developers.google.com/earth-engine/datasets/catalog/USGS SRTM
3	*Global Surface Water* (GSW) 1996-2020		√	https://global-surface-water.appspot.com/
4	SHP Cipunagara Watershed Administration		√	https://tanahair.indonesia.go.id/

No	Data Type	Data Properties		Source
		Primary	Secondary	
5	Subang in Figures 2020		√	https://subangkab.bps.go.id/
6	Coordinate Point of Each Land Cover Class	√		Direct collection on the spot
7	Inundation Height Data Measurement	√		Direct collection on the spot

2.2 Data processing stage

2.2.1 Rob flood hazard index

The assessment of tidal flood hazards is conducted using data from areas susceptible to flooding, taking into account factors such as river flow distance, proximity to the shoreline, inundation height, and the extent of river and tidal flooding, in accordance with Head of BNPB Regulation No. 2 BNPB of 2012 [5]. The scoring of the Hazard Index is presented in Table 3 below.

Table 3. Composing parameters and scoring of the hazard index.

No	Hazard Index Parameters	Weight (%)	Hazard Index		
			Low	Medium	High
1.	River Flow Distance	15	>500 m	201-500 m	0-200 m
2.	Coastline Distance	15	>500 m	201-500 m	0-200 m
3.	Flood Height	20	0.0-0.2 m	0.2-0.5 m	> 0.5 m
4.	River Flood Area	25	0-4 ha	4-8 ha	>8 ha
5.	Tidal Inundation Area	25	0-4 ha	4-8 ha	>8 ha

Hazard Index = (0.15 * River Flow Distance score) + (0.15 * Coastline Distance score) + (0.2 * Flood Height score) + (0.25 * River Flood Area score) +(0.25 * Tidal Inundation Area score) (1)

Determine the classification of scoring results with the following formula:

$$\frac{\text{Highest Score} - \text{Lowest Score}}{3\ (\text{Number of Classes})} \qquad (2)$$

Value Score each parameter based on Regulation of the Minister of Public Works and Housing 12/PRT/M/2014 [6, with modification]:
- In the LOW hazard class has a score of 1
- In the MEDIUM hazard class has a score of 3
- In the HIGH hazard class has a score of 5

2.2.2 Rob flood vulnerability index

- Social vulnerability

Social vulnerability consists of parameters of population density and vulnerable groups. Vulnerable groups consist of the ratio of sex, the ratio of the vulnerable age group, the ratio of the poor, and the ratio of the population with disabilities. The preparation parameters are presented in Table 4.

IOP Conf. Series: Earth and Environmental Science **1464** (2025) 012016 doi:10.1088/1755-1315/1464/1/012016

Table 4. Compiling parameters and scoring of social vulnerability. [5]

Parameters	Weight (%)	Class		
		Low	Medium	High
Population density	60	<5 soul/ha	5-10 soul/ha	>10 soul/ha
Vulnerable Group - Gender Ratio (10%) - Ratio of Vulnerable Age Groups (10%)	40	>40	20-40	<20
- Poor Population Ratio (10%) - Disabled Population Ratio (10%)		<20	20-40	>40

- Physical vulnerability

Physical vulnerability encompasses the parameters of residential properties, public amenities, and essential facilities. The cumulative value in rupiah of these properties, amenities, and facilities is determined according to the hazard classification of the impacted region.

Table 5. Table of physical vulnerability parameters. [5]

Parameters	Weight (%)	Class		
		Low	Medium	High
House	40	<400 million	400 - 800 million	>800 million
Public facilities	30	<500 million	500 million -1 billion	> 1 billion
Critical Facilities	30	<500 million	500 million -1 billion	> 1 billion

Physical Vulnerability = (0.4 * House score) + (0.3 * Public facilities score) + (0.3 * Crisis Facility score) (3)

Calculation of the value of each parameter (except House) is carried out based on:
- In the LOW hazard class has 0% effect
- In the MEDIUM hazard class has a 50% effect
- In the HIGH hazard class has 100% effect

The calculation of the House parameter value is carried out based on:
- In the LOW hazard class, the number of houses affected is multiplied by 5 million
- In the MEDIUM hazard class, the number of houses affected is multiplied by 10 million
- In the HIGH hazard class, the number of houses affected is multiplied by 15 million

- Economic vulnerability

Economic vulnerability is characterized by parameters such as the contribution of Gross Regional Domestic Product (GRDP) and productive land. These parameters are detailed in Table 6.

- Environmental vulnerabilities

Environmental vulnerability is determined by parameters such as protected forests, natural forests, mangroves, shrubs, and swamps. Each of these parameters can be identified through land cover data. The parameters prepared are displayed in the following Table 7.

Table 6. Table of economic vulnerability parameters. [5]

Parameters	Weight (%)	Class		
		Low	Medium	High
Productive Land	60	<50 million	50 - 200 million	>200 million
GDP	40	<100 million	100 – 300 million	> 300 million

Economic Vulnerability = (0.6 * Productive Land score) + (0.4 * GDP score) (4)

Calculation of the value of each parameter is carried out based on:
- In the LOW hazard class has 0% effect
- In the MEDIUM hazard class has a 50% effect
- In the HIGH hazard class has 100% effect

Table 7. Table of environmental vulnerability parameters. [5]

Parameters	Weight (%)	Class		
		Low	Medium	High
Protected forest	30	<20 ha	20 - 50 ha	>50 ha
Natural Forest	30	<25 ha	25 - 75 ha	>75 ha
Mangrove/Mangrove	20	<10 ha	10 - 30 ha	>30 ha
Shrubs	10	<10 ha	10 - 30 ha	>30 ha
Swamp	10	<5 ha	5 - 20 ha	>20 ha

Environmental Vulnerability = (0.3 * Protected Forest Score) + (0.3 * Natural Forest Score) +(0.2 * Mangrove Score) + (0.1 * Shrubs Score) + (0.1 * Swamp Score) (5)

Calculation of the value of each parameter is carried out based on:
- In the LOW hazard class has 0% effect
- In the MEDIUM hazard class has a 50% effect
- In the HIGH hazard class has 100% effect

- Vulnerability Index

The vulnerability index for each threat is derived by integrating the scores of social, physical, and economic vulnerabilities, applying the respective weights to each component as outlined in [5].

$$FVI = (SVI x\, 40\%) + (PVI\, x\, 25\%) + (EVI\, x\, 25\%) + (ENVI\, x\, 10\%) \quad (6)$$

where:
FVI : Flood Vulnerability Index
SVI : Social Vulnerability Index
PVI : Physical Vulnerability Index
EVI : Economic Vulnerability Index
ENVI : Environmental Vulnerability Index

2.2.3 Rob flood capacity index

Regional capacity in disaster management is a crucial parameter for determining the success of disaster risk reduction. It is imperative that regional capacity aligns with the National Disaster Management System as stipulated in Law Number 24 of 2007 on Disaster Management and its subsequent regulations. The capacity index values are categorized into five levels of regional achievement in disaster management. These levels are detailed in Table 8.

Table 8. Levels of regional achievements in disaster management. [5]

Levels	Information
1	Regions have had minor achievements in disaster risk reduction efforts by implementing some advanced actions in plans or policies.
2	The regions have implemented several disaster risk reduction actions with achievements that are still sporadic due to the absence of systematic institutional and/or policy commitments.
3	The commitment of the government and several communities regarding disaster risk reduction in an area has been achieved and supported by systematic policies, but the achievements obtained with these commitments and policies are considered not comprehensive, so they are still not significant enough to reduce the negative impacts of disasters.
4	With the support of commitments and comprehensive policies in disaster risk reduction in an area, successful achievements have been obtained, but it is recognized that there are still limitations in commitment, financial resources, or operational capacity in implementing disaster risk reduction efforts in the area.
5	Comprehensive achievements have been achieved with adequate commitment and capacity at all levels of the community and levels of government

The determination of flood capacity index will be completed in the next stage of research.

2.2.4 Mapping of rob flood risk index

Determination of the risk index is carried out using a spatial method to identify the hazard index and vulnerability index, by scoring and overlaying through the ArcGIS 10.8 application and qualitative descriptive methods in determining disaster capacity, and to recommend structural and non-structural mitigation based on the description of the index. risk seen from the results of the table and visual form in the form of a map. The formula for determining the risk index is as follows [5]:

$$\text{Risk (R)} = \text{Hazard (H)}\,\frac{\text{Vulnerability (V)}}{\text{Capacity (C)}} \tag{7}$$

Determine the classification/class of the Risk Index scoring results with the following formula [6]:

$$\frac{Highest\ Score\ -\ Lowest\ Score}{3\ (Number\ of\ Classes)} \tag{8}$$

3. Results and discussion

3.1. Rob flood hazard index

The classification of the tidal flood hazard index for each village, derived from overlaying various parameters such as distance from river flow, distance from shoreline, inundation height, area of river flooding, and area of tidal flooding per village in the downstream region of the Cipunagara watershed, is depicted in Table 9.

Table 9. Class of rob flood hazard index per village.

Nu	Village	Score					Weight (%)					Score Flood Rob	Rob Flood Class
		RD	CD	FH	RF	IT	RD	CD	FH	RF	IT		
1	Tegalurung	1	3	3	5	5	15	15	20	25	25	3.7	High
2	Mayangan	5	5	5	5	5	15	15	20	25	25	5.0	High
3	Legon Wetan	3	5	5	5	5	15	15	20	25	25	4.7	High
4	Legon Kulon	3	1	3	5	5	15	15	20	25	25	3.7	High
5	Pangarengan	5	3	5	5	5	15	15	20	25	25	4.7	High
6	Bobos	5	1	3	5	1	15	15	20	25	25	3.0	Medium
7	Karangmulya	5	1	3	5	1	15	15	20	25	25	3.0	Medium
8	Patimban	5	3	5	5	5	15	15	20	25	25	4.7	High
9	Rancadaka	1	1	3	5	1	15	15	20	25	25	2.4	Medium
10	Gempol	1	1	1	1	1	15	15	20	25	25	1.0	Low
11	Pusakaratu	1	1	1	5	1	15	15	20	25	25	2.0	Low
12	Mundusari	3	1	1	1	1	15	15	20	25	25	1.3	Low
13	Rancasari	1	1	1	1	1	15	15	20	25	25	1.0	Low
14	Pamanukan S	5	1	3	5	1	15	15	20	25	25	3.0	Medium
15	Mulyasari	3	1	1	3	1	15	15	20	25	25	1.8	Low
16	Pamanukan H	3	1	1	1	1	15	15	20	25	25	1.3	Low
17	Pamanukan	3	1	1	1	1	15	15	20	25	25	1.3	Low
18	Ranca Hilir	3	1	3	5	1	15	15	20	25	25	2.7	Medium
19	Lengkong Jaya	3	1	1	1	1	15	15	20	25	25	1.3	Low
20	Pusakajaya	1	1	1	3	1	15	15	20	25	25	1.5	Low

Information:
RD : River Distance
CD : Coastline Distance
FH : Flood Height
RF : River Flood
IT : Inundation Tidal

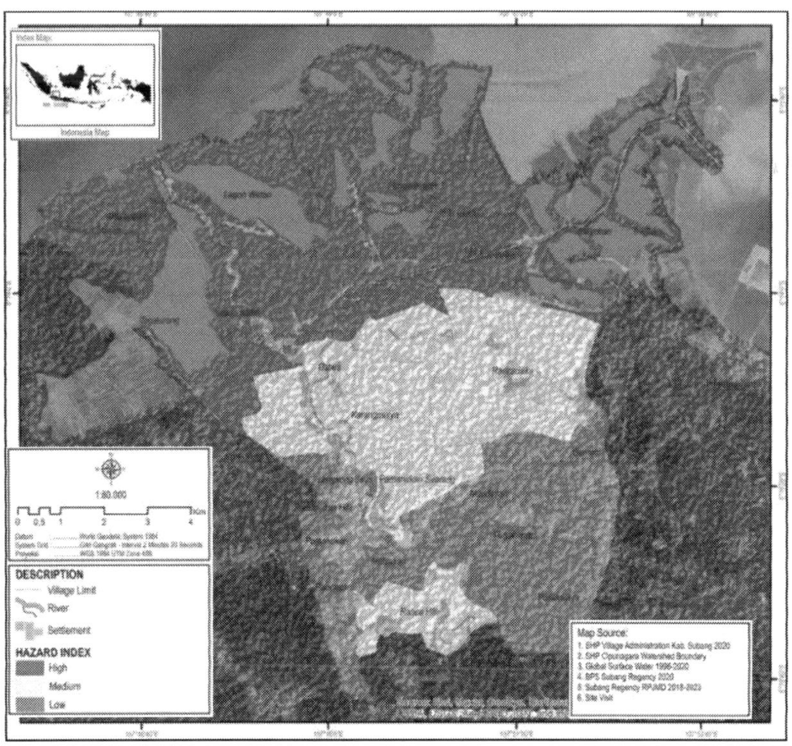

Figure 1. Map of flood hazard index per village.

According to Table 9, the villages with the highest tidal flood hazard index are Tegalurung, Mayangan, Legon Wetan, Legon Kulon, Pangarengan, and Patimban. The villages with a medium tidal flood hazard index include Bobos, Karangmulya, Rancadaka, Pamanukan Sebrang, and Ranca Hilir. The villages with a low-category tidal flood hazard index include Gempol, Pusakaratu, Mundusari, Rancasari, Mulyasari, Pamanukan Hilir, Pamanukan, Lengkong Jaya, and Pusakajaya. The map depicting the tidal flood hazard index for each village in the downstream region of the Cipunagara watershed is illustrated in Figure 1.

3.2. Rob flood vulnerability index

The assessment of vulnerability to flood disasters is conducted according to disaster risk assessment standards. The assessment categorizes vulnerability into four indices: social vulnerability index, physical vulnerability index, economic vulnerability index, and environmental vulnerability index. The assessment of the index allows for the determination of the potential number of populations exposed, environmental damage, and potential losses due to tidal floods. The analysis of the tidal flood vulnerability index for each village in the downstream area of the Cipunagara watershed is detailed in Table 10.

Table 10. Rob flood vulnerability index per village.

Nu	Village	Score				Weight (%)				Score Flood Rob	Rob Flood Class
		SVI	PVI	EVI	ENVI	SVI	PVI	EVI	ENVI		
1	Tegalurung	1	1	5	5	40	25	25	10	2.4	Low
2	Mayangan	1	1	5	5	40	25	25	10	2.4	Low
3	Legon Wetan	1	1	5	5	40	25	25	10	2.4	Low
4	Legon Kulon	3	5	5	1	40	25	25	10	3.8	High
5	Pangarengan	1	1	5	5	40	25	25	10	2.4	Low
6	Bobos	3	1	5	1	40	25	25	10	2.8	Medium
7	Karangmulya	3	1	5	1	40	25	25	10	2.8	Medium
8	Patimban	1	5	5	5	40	25	25	10	3.4	High
9	Rancadaka	3	1	5	1	40	25	25	10	2.8	Medium
10	Gempol	3	1	5	1	40	25	25	10	2.8	Medium
11	Pusakaratu	5	1	5	1	40	25	25	10	3.6	High
12	Mundusari	5	1	5	1	40	25	25	10	3.6	High
13	Rancasari	5	1	5	1	40	25	25	10	3.6	High
14	Pamanukan S.	5	1	5	1	40	25	25	10	3.6	High
15	Mulyasari	5	1	5	1	40	25	25	10	3.6	High
16	Pamanukan H.	5	1	5	1	40	25	25	10	3.6	High
17	Pamanukan	5	1	5	1	40	25	25	10	3.6	High
18	Ranca Hilir	1	1	5	1	40	25	25	10	2.0	Low
19	Lengkong Jaya	5	1	5	1	40	25	25	10	3.6	High
20	Pusakajaya	5	1	5	1	40	25	25	10	3.6	High

Information:
SVI : Social Vulnerability Index
PVI : Physical Vulnerability Index
EVI : Economic Vulnerability Index
ENVI : Environmental Vulnerability Index
FVI : Flood Vulnerability Index

Results Based on the scoring and weighting of all parameters the highest tidal flood vulnerability index was found in Legon Kulon Village, Patimban Village, Pusakaratu Village, Mundusari Village, Rancasari Village, Pamanukan Sebrang Village, Mulyasari Village, Pamanukan Hilir Village, Pamanukan Village, Lengkong Jaya Village, and Village Pusakajaya with a tidal flood vulnerability index scoring range, which is 3.2 – 3.8. The category of moderate tidal flood vulnerability was found in Bobos Village, Karangmulya Village, Rancadaka Village, and Gempol Village with a scoring range of 2.6 – 3.2. While the tidal flood vulnerability index with a low-class category is found in Tegalurung Village, Mayangan Village, Legon Wetan Village, Pengarengan

Village, and Ranca Hilir Village with a tidal flood vulnerability index score range of 2.0 - 2.6. The tidal flood vulnerability index map is presented in Figure 2.

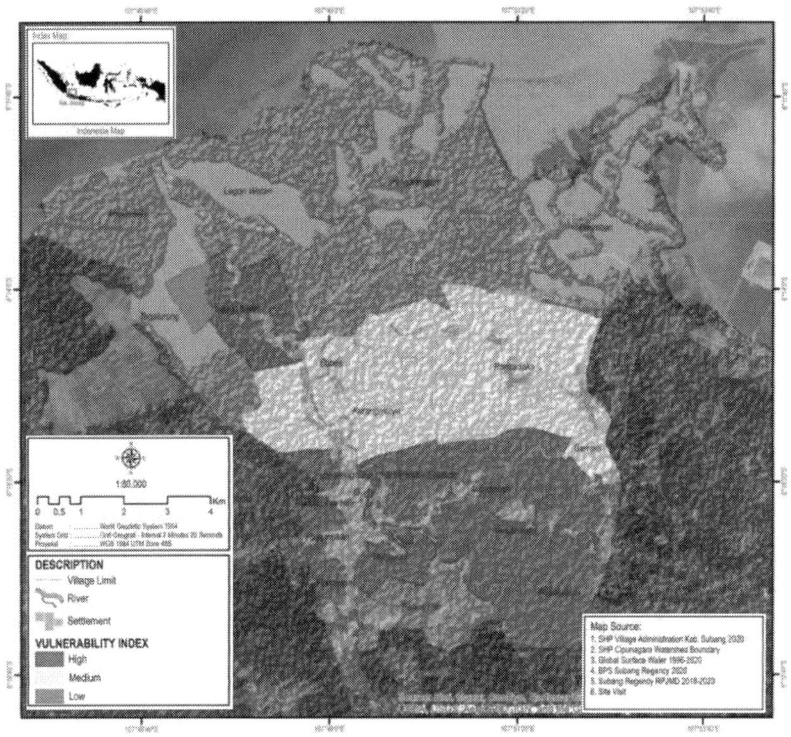

Figure 2. Map of rob flood vulnerability index per village.

4. Conclusion

The tidal flood vulnerability index per village in the downstream area of the Cipunagara watershed is divided into three classes. Based on the result analysis and compared each vulnerability index output, the high-vulnerability category for tidal flooding is found in Legon Kulon Village, Patimban Village, Pusakaratu Village, Mundusari Village, Rancasari Village, Pamanukan Sebrang Village, Mulyasari Village, Pamanukan Hilir Village, Pamanukan Village, Lengkong Jaya Village, and Village Pusakajaya with a tidal flood vulnerability index scoring range, which is 3.2 – 3.8. The moderate vulnerability category was found in Bobos Village, Karangmulya Village, Rancadaka Village, and Gempol Village with a scoring range of 2.6 – 3.2. While the category of low tidal flood vulnerability index was in Tegalurung Village, Mayangan Village, Legon Wetan Village, Pengarengan Village, and Ranca Hilir Village with a tidal flood vulnerability index score range of 2.0 – 2.6. Once the vulnerability index is determined, the next stage is to set up to which the region has the critical vulnerability and more consideration. Futher more, the administrative policy will recommend for villages that have low tidal flood vulnerability risk focus on environmental management, especially on the management of river border areas and coastlines.

Reference

[1] Intergovernmental Panel on Climate Change (IPCC), Sixth Assessment Report (2023). Climate Change: Mitigation of Climate Change. Cambridge: Cambridge University Pr.

[2] Marwah, S. and Alwi, L.O. (2014). Analysis of the Impact of Land Use Change on Tidal Flood in Kendari City. International Journal of Applied Science and Technology, Vol. 4, No. 7, December 2014.

[3] Marfai, M.A., Mardiatno, D., Cahyad, A., and Nucifera, F. (2013). Spatial Modeling of Rob Flood Hazard Based on Climate Change Scenarios and Its Impacts in Coastal Pekalongan. Jurnal Bumi Lestari, Vol. 13, No. 2, August 2013: 244-256.

[4] Numberi, F. (2009). Climate Change, its Implications for Marine, Coastal and Small Island Life. Jakarta: Citrakreasi Indonesia

[5] National Disaster Management Agency (BNPB). (2012). Head of BNPB Regulation Number 2 of 2012 on General Guidelines for Disaster Risk Assessment. Jakarta: National Disaster Management Agency.

[6] Ministry of Public Works and Housing. (2014). Regulation of the Minister of Public Works and Housing on the Implementation of Urban Drainage Systems No. 12/PRT/M/2014. Jakarta: Ministry of Public Works and Housing.

IOP Conf. Series: Earth and Environmental Science **1464** (2025) 012017 doi:10.1088/1755-1315/1464/1/012017

Estimation of wave transmission coefficient for natural coastal shore protection using data driven predictive model "genetic programming"

Suciana[1*], Nita Yuanita[2] and Alamsyah Kurniawan[2]

[1] Department of Ocean Engineering, Institut Teknologi Sumatera, Indonesia
[2] Ocean Engineering Program, Institut Teknologi Bandung, Indonesia

*E-mail: suciana@kl.itera.ac.id

Abstract: Erosion is one of the most common problems in Indonesia's coastal areas. Mangrove forests are a natural form of coastal protection, however planting and establishing mangroves requires temporary protection from waves and currents, allowing mangroves to grow and function optimally. Before conducting field construction, physical model testing is necessary to know the effectiveness of temporary protection systems. The physical model provides water elevation data, with the advancement of technology, which is used to analyze and predict the wave transmission coefficient (Kt). The transmission coefficient is a key factor in evaluating the effectiveness of a breakwater structure. This research focuses on deriving an equation for Kt by analyzing relationships between variables from the physical model of a coastal protection system that includes variations in the weight and slope of geobag structures and different wave data. Data-driven models are increasingly being developed and used, with one such approach being genetic programming (GP). GPTIPS serves as an offline data-driven modeling tool to catch the trend of the wave transmission coefficient and then use it to predict the Kt equation. The results suggest that combining data analysis with GP models enhances forecasting accuracy by identifying significant predictive parameters. It was found that the Genetic programming for the Kt prediction error model can lead to substantial improvements when used as a data assimilation technique to update Kt predictions derived from primary physical models. The results show that the prediction equation of Kt was successfully obtained, with the error prediction formula Kt being less than 5%.

1. Introduction

Coastal erosion is a significant hazard that poses a major challenge to coastal areas, affecting both natural ecosystems and human communities on a global scale [1]. To address this issue, many regions around the world are increasingly favoring natural coastal protection methods that utilize vegetation, such as mangroves [2]. However, developing of coastal protection with natural systems presents challenges. Mangrove seedlings may be destroyed by waves or currents before they become sufficiently established, necessitating suitable protection for two years post-planting. To address this issue, a proposed solution involves a coastal protection system that integrates primary natural defenses with temporary man-made structures [3], [4]. To assess the effectiveness of a temporary structure for a natural coastal protection system in reducing wave height, preliminary research was presented through a physical laboratory experiment [5]. Water elevation data from physical laboratory experiments used to calculate the value of the wave transmission coefficient (Kt). The wave transmission coefficient (Kt) quantifies how much of an incoming wave's energy or height is transmitted past a structure like a breakwater or seawall. It is often defined as the ratio of the transmitted wave height (Ht) to the incident wave height (Hi), or mathematically, Kt=Ht/Hi. This coefficient, which ranges between 0 (indicating full reflection) and 1 (indicating complete transmission), is critical in coastal and ocean engineering to evaluate how structures affect wave energy dissipation and protect shorelines or marine facilities [6]. The transmission coefficient is a key factor in evaluating the effectiveness of a breakwater structure to reduce waves. Sensitivity experiments were conducted using a third-generation numerical wave model to analyze wave attenuation through mangroves while varying the sea bottom slope. The results showed that waves were reduced by 93 to 98% on mild slopes (1:80, 1:40), around 84% on a 1:20 slope, and around 67% on a sharp slope (1:10). The study concluded that wave height decreases exponentially for relatively in mild slopes, while wave reduction becomes more gradual as bottom steepness increases, depend on wave transformation characteristics and water depth variations with different slopes [7].

In conducting a physical model, there are several interrelated parameters that affect the resulting value of Kt such as transmitted and Incident wave height, wave steepness, wavelength, width and slope structure, and water depth. The analytical process is made significantly more difficult by the intricate governing mechanisms, which are multi-scale, multi-dimensional, time-varying, and exhibit highly non-linear dynamics. Conventional numerical and analytical models offer the main approach for addressing the complex task of characterizing and predicting the transmission coefficient by framing the underlying physics through solvable equations. Yet, capturing the wave dynamics in totality, accounting for many physical parameters in one equation calls for rigorous tuning of the models for further improvement.

The aim of this research is to assess the feasibility of applying the cross correlation function (CCF), autocorrelation function (ACF), and average mutual information (AMI) theory to evaluate the information content in observed and predicted equations for wave transmission coefficients with Genetic Programming. This involves assuming that the physical model by Yuanita et al. [4] accurately represents the physics of the domain based on the available data, and relating this to variables reflecting the state of predictions, such as input data, state variables, and model outputs. Additionally, the research investigates the potential use of genetic programming (GP) with the GPTIPS tool for use with MATLAB as an offline data-driven modeling tool to capture and update wave transmission coefficient predictions in real-time applications. To the authors' knowledge, this is the first study exploring the combination of CCF, ACF, AMI, and GP for updating equations of wave transmission coefficient predictions.

2. Methodology

2.1 Data availability

The present work is based on observations data used by Yuanita et al. [4] consisting of water elevation data from the physical model. The physical model was conducted in the 2D wave flume with dimensions of length 40m, width 1.2 m, and height 1.5 m. Four wave gauges were positioned at various points along the flume to measure wave parameters. The wave gauges were positioned at CH0, CH1, CH2, and CH3. The water depth (d) is 0.65 m near the wavemaker and more shallow in the beach area a depth 0.4 m with consistent across all scenarios. The setup of the physical model experiment is illustrated in Figure 1.

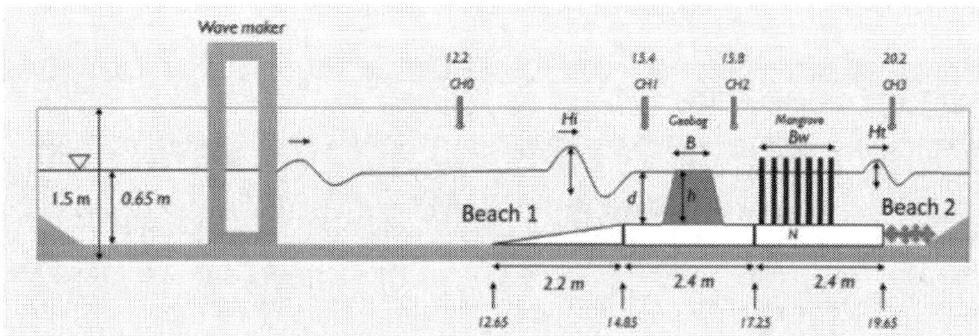

Figure 1. Configuration of physical model experiment on wave flume [4].

Table 1. Structure configuration scenario.

Code	Structure Scenario
K	System configuration without structure
P	Beach-only configuration
M	Beach with mangroves, without geobag structure
MG1	mangrove system and geobag structure with a 1:1.5 slope and 0.5 kg weight
MG2	mangrove system and geobag structure with a 1:2 slope and 0.5 kg weight
MG3	mangrove system and geobag structure with a 1:1.5 slope and 1 kg weight
MG4	mangrove system and geobag structure with a 1:2 slope and 1 kg weight
MG5	mangrove system and geobag structure with a 1:1.5 slope and 2 kg weight
MG6	mangrove system and geobag structure with a 1:2 slope and 2 kg weight

Table 2. Wave configuration scenario.

Code		1	2	3	4	5	6	7	8	9
Wave Scenario	Wave steepness		0.006			0.003			0.0005	
	Significant wave height Hs (m)	0.15	0.1	0.08	0.15	0.1	0.08	0.15	0.1	0.08
	Wave crest period Tp (s)	1.59	1.3	1.16	2.25	1.84	1.64	5.53	4.51	4.03

This study employed three different geobag units, each weigh 0.5 kg, 1 kg, and 2 kg with two types of geobag configurations with different front slope variations of 1:1.5 and 1:2.0. [4]. Structure configuration becomes 9 scenarios, without structure, beach only, and different combinations of mangrove system and geobag weight that can be seen in Table 1.

The wave variations used involve significant wave height (Hs) and peak period (Tp). The range of wave steepness is adjusted to match wave conditions in Indonesia, ranging from 0.0005 to 0.006. The significant wave heights used in this study are 0.15 m, 0.1 m, and 0.08 m, with peak periods ranging from 1.2 seconds to 5.5 seconds. The wave configuration scenario can be seen in Table 2. The next sections explore a data model integration approach consisting of two parts to improve the Kt prediction (Figure 2).

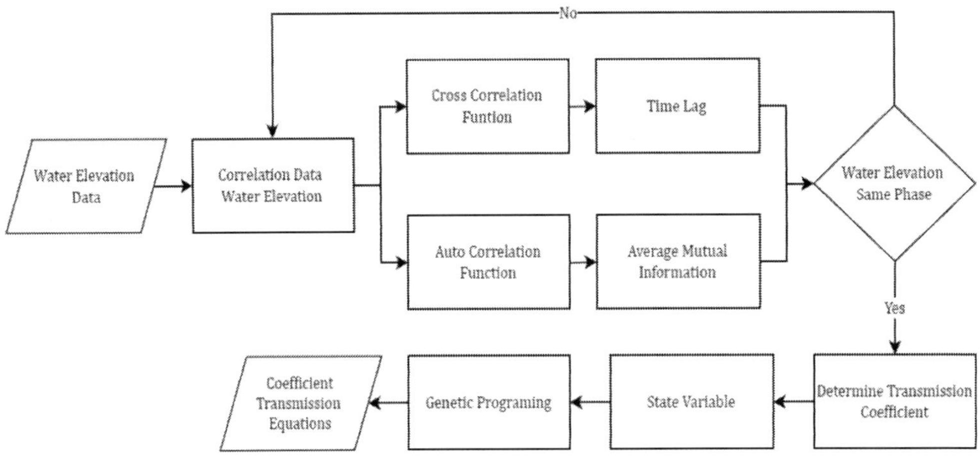

Figure 2. Schematic flowchart of improving the equations of transmission coefficient prediction.

2.2. Correlation data

The incident wave heights (Hi) were collected from the CH1 wave gauge while the transmitted wave heights (Ht) were measured using the CH2 wave gauge. The data must be same phase to get the transmission coefficient value, with the process correlation data. Before entering into genetic programming modeling, surface water level (SWL) data from physical models in the laboratory are processed first and correlations are sought between the data using the crosscorrelation function (CCF), autocorrelation function (ACF), and average mutual information (AMI). This CCF can perform time series data analysis, see the size of the data similarity, pattern recognition, predict data series relationships and generate time lag values. In the process of computing CCF data using the xcorr function. In the water elevation data there is information that can be analysis from one data set that has a relationship to another data set or what is called serial correlation.

In determining the category of serial correlation (high or low or periodic) it can be seen by looking at the value of AMI. Average mutual information (AMI) measures information that can be learnt from one set of data having knowledge of another set of data. The AMI does not depend on any particular function and therefore can help to detect both linear and non-linear correlations [8]. The AutoCorrelation Function is used in the application of the AMI program. This ACF is one of the tools used to find patterns in the data. To calculate the AMI between a time series s(t) and its time-delayed counterpart s(t+T), the average mutual information concerning the time delay is defined as shown in Equation [1].

$$AMI(T) = \sum_{s(t),s(t+T)} P\big(s(t), s(t+T)\big)\log\left[\frac{P\big(s(t), s(t+T)\big)}{P\big(s(t)\big)P\big(s(t+T)\big)}\right] \tag{1}$$

Here, *P(s(t))* and *P(s(t), s(t+T))* represent probability density functions (PDFs), with P(s(t), s(t+T)) being the joint PDF. This formulation shows is evident that as *T* increases, the chaotic nature of the signals causes s(t) and *s(t+T)* to become practically independent, leading AMI(T) to approach zero [9]. The AMI between the original signal and its time-lagged version can be calculated for each time lag value. Interpretation of the AMI values, if the AMI values between *s(t)* and *s(t+T)* are computed for different values of lag time *T*, the following observation can be made as the following. High serial correlation occurs when the AMI at smaller values of *T* is comparable to that at *T=0*. Low serial correlation when increasing *T* the AMI values drop abruptly. Periodicity is when at nearly regular time intervals the AMI values rise periodically [10], [11].

2.3. Data driven model, genetic programming and GPTIPS

Genetic programming (GP) is an evolutionary computing approach driven by data, designed to automatically generate input-output relationships [12]. GP is data-driven modeling techniques like artificial neural networks, fuzzy logic, or regression trees. GP offers mathematically interpretable structures that optimally relate the system's input and output variables. GP has been used in conducting research related to the prediction of sea-level rise anomalies with a combination of visual data and simulation data [13]. Starting with a set of observed input-output data (training data), GP creates a population of models with random structures, without requiring prior knowledge of the process mechanisms. These models are then assessed using validation data (a portion of the original data not used for training) based on an appropriate fitness measure. A model's likelihood of survival in the evolutionary process depends on how well it predicts the validation data [14], [15]. Successful model components are recombined to generate new models. With each generation, GP refines the model structure, using a nonlinear least-squares algorithm to estimate model parameters. GP has been widely and successfully applied in modeling various hydrological processes [16], [17], [18], [19].

GPTIPS is a GP tool designed for MATLAB. GP, inspired by biological evolution, creates computer programs to solve tasks by randomly generating a population of tree-structured programs. The best-performing programs are then mutated and crossed to form new populations, repeating the process until effective solutions emerge. When applied to building empirical mathematical models from data, this approach is called symbolic regression. Unlike traditional regression, where the model's structure must be predefined, GP evolves both the model's structure and parameters automatically. GPTIPS offers tools for exploring evolved models, simplifying them post-run, and exporting them in various formats such as graphics files, LaTeX, or MATLAB code. A key feature of GPTIPS is its ability to evolve multigene individuals, enabling the creation of simpler and more accurate models. A multigene individual is made up of one or more genes, where each gene is a conventional GP tree. These genes are added progressively by individuals to enhance fitness, such as by minimizing a model's sum of squared errors on a dataset. The complete model is a weighted linear combination of the individual genes. In GPTIPS, the optimal weights for these genes are determined automatically by applying ordinary least squares regression on the genes with respect to the output data. This pseudo-linear model is capable of representing non-linear behaviors (Figure 3). The result of the multiple-gene model is

$$y = do + d1(0.41\,x1 + tanh(x1\,x3)) + d2(0.45x3 + sqrt(x2)).$$

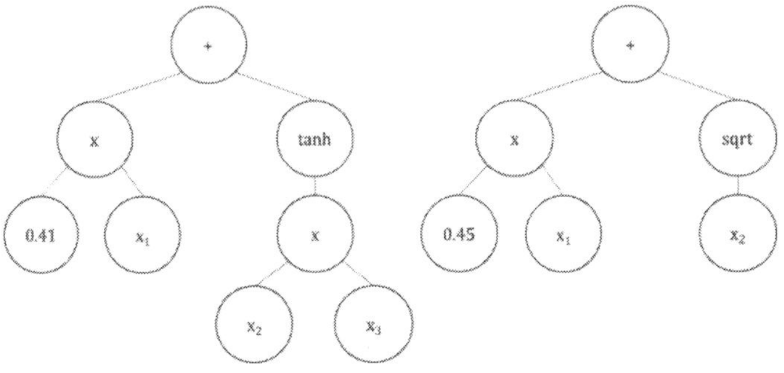

Figure 3. A pseudo-linear multigene model of output y with inputs, X_2 and X_3. The weights d_0, d_1 and d_2 are automatically obtained by least squares [20].

There are 4 symbolic regression demos included with GPTIPS. First is gpdemo1: which runs ordinary GP symbolic regression on Koza's 1992 quartic polynomial problem. Then gpdemo2: runs multigene GP symbolic regression on a dataset generated from a non-linear mathematical function comprising one output (y) and 4 inputs (x_1, ..., x_4). A separate test data set is used to evaluate the evolved models. The function is $y = \exp(2x_1\sin(\pi\, x_4)) + \sin(x_2x_3)$, Both testing & training data is noise-free. Third is gpdemo3: which runs multigene GP symbolic regression on a dataset generated from a simulation of a non linear pH neutralization process. This also comprises one output (pH) and 4 inputs. A test data set is used to evaluate the evolved models. Both testing & training data is noise-free. Last is gpdemo4: which runs multigene GP symbolic regression on a "real-life" dataset comprising one output (concrete compressive strength) and 8 inputs. A holdout validation data set and a testing data set are used to evaluate the evolved models. In GPTIPS, the fitness function implements multigene symbolic regression and is used by gpdemo2, gpdemo3 and gpdemo4. Using this fitness function GPTIPS attempts to minimise the root mean squared error (RMSE) between the output y and the predicted output y'. The gppretty function can also return other outputs, for instance to extract a string containing the LaTeX version of the simplified genes of a multigene expression. [20]

2.4. State parameter

In this research there are five parameters with one as output (y) transmission coefficient as shown in Equation 2 and four for input x. The input parameters are the slope of geobag as shown in Equation 3, wave steepness as shown in Equation 4, relative top width of the geobag structure as shown in Equation 5, and relative wave height as shown in Equation 6.

$$y = Kt = \frac{H_t}{H_i} = \frac{\overline{H_4}}{\overline{H_2}} \quad (2)$$

$$x = tan\,\boldsymbol{\alpha} \quad (3)$$

$$x = \frac{H_i}{gT^2} \quad (4)$$

$$x = \frac{B}{L}; \quad (5)$$

$$x = \frac{H}{d} \quad (6)$$

Where H_t is transmission wave height, H_i is incident wave height, $\tan\alpha$ is slope geobag, B geobag top width, L geobag top Lenght, H is wave height and, d in depth.

3. Result and discussion

3.1. Data relationship analysis

Water level data is recorded when the wave maker is activated and pushed forward the water and creates wave height, as the next wave recorder captures the water level data, there is a delay in recording the first wave. The cross-correlation function finds the time delay inter-channel and creates water level data starting at the same time. Water level data is obtained in the same phase and the data wave individual can be compared to get Kt observation data (See Figure 4).

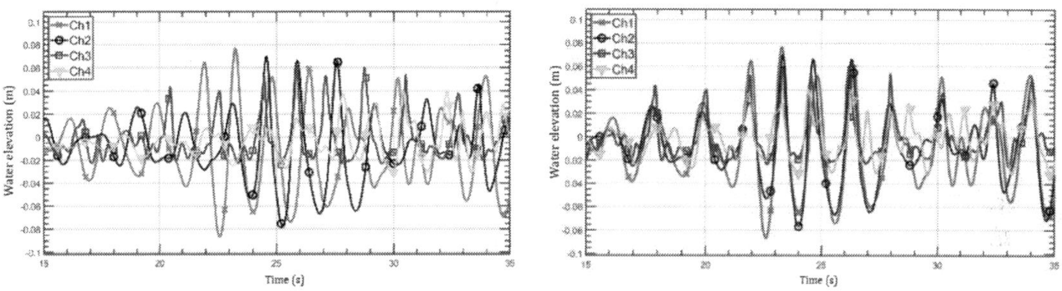

Figure 4. water level before (left) and after use Cross Correlation Function (right).

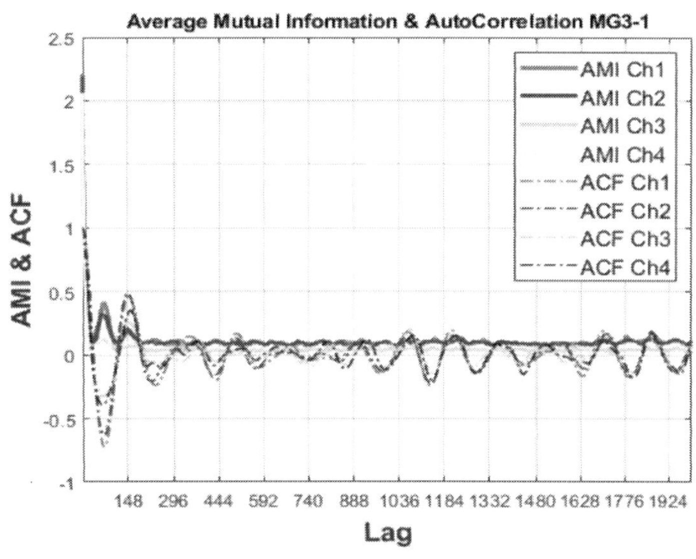

Figure 5. AMI and ACF output.

Figure 5 shows that the Average Mutual Information (AMI) value drops suddenly with increasing time lag. This suggests that the mutual information between data is low. The low level of mutual information between the data may occur because the waves reflection after hitting a structure, causing the water to become highly non-linear.

In addition, the peak value of the AutoCorrelation Function (ACF) and Average Mutual Information (AMI) in Figure 5 is the same as the value of Hs output from the CrossCorrelation Function (CCF) in Figure 4. Therefore, the data correlation process is deemed to be accurate.

P (Beach only configuration)

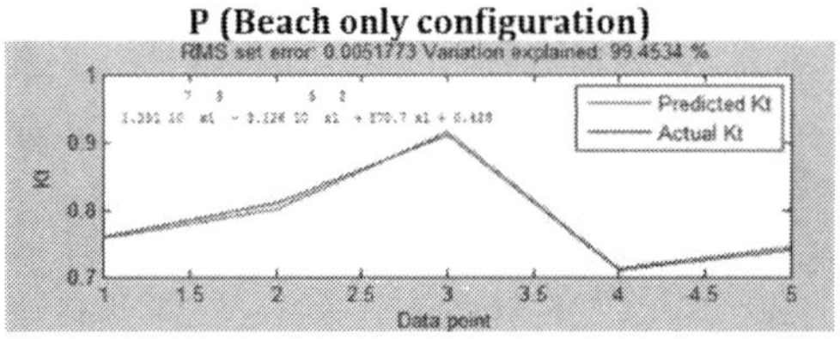

$$Kt = 1.351\,10^7 \left(\frac{H_i}{gT^2}\right)^3 - 3.126\,10^5 \left(\frac{H_i}{gT^2}\right)^2 + 870.7 \left(\frac{H_i}{gT^2}\right) + 0.428$$

M (Beach, mangroves,without geobag)

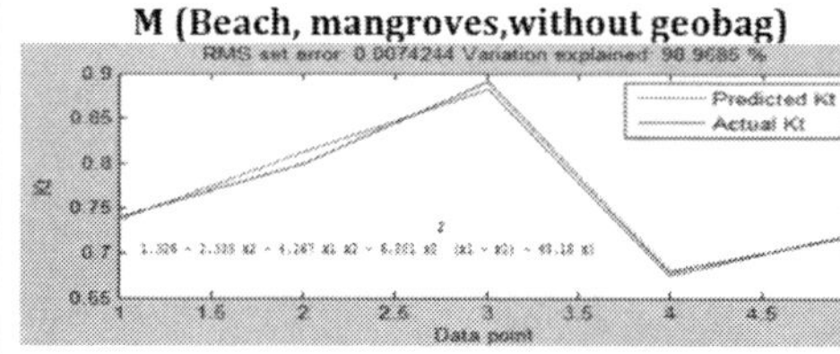

$$Kt = 1.326 - 2.333\left(\frac{H}{d}\right) - 4.267\left(\frac{H_i}{gT^2}\right)\left(\frac{H}{d}\right) - 6.201\left(\frac{H}{d}\right)^2\left(\frac{H_i}{gT^2} - \frac{H}{d}\right) - 43.18\left(\frac{H_i}{gT^2}\right)$$

MG1 (mangrove model and geobag structure, weight 0.5 kg, slope 1:1.5

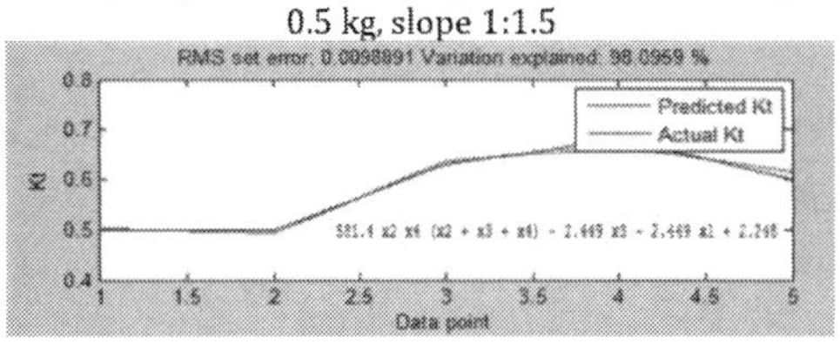

$$Kt = 581.4\left(\frac{H_i}{gT^2}\right)\left(\frac{H}{d}\right)\left(\left(\frac{H_i}{gT^2}\right) + \left(\frac{B}{L}\right) + \left(\frac{H}{d}\right)\right) - 2.449\left(\frac{B}{L}\right) - 2.499\,\tan\boldsymbol{\alpha} + 2.248$$

MG2 (mangrove model and geobag structure, weight 0.5 kg, slope of 1:2

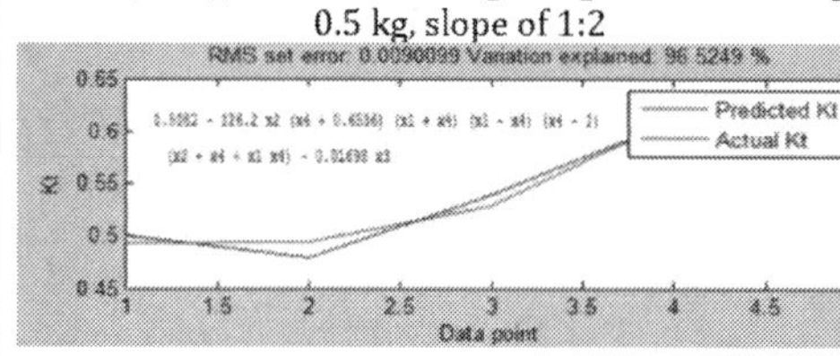

$$Kt = 0.5082 - 226.2\left(\frac{H_i}{gT^2}\right)\left(\left(\frac{H}{d}\right) + 0.6536\right)\left(\tan\boldsymbol{\alpha} + \left(\frac{H}{d}\right)\right)\left(\tan\boldsymbol{\alpha} - \left(\frac{H}{d}\right)\right)\left(\left(\frac{B}{L}\right) - 2\right)\left(\left(\frac{H_i}{gT^2}\right) + \left(\frac{B}{L}\right) + \left(\frac{H}{d}\right)\tan\boldsymbol{\alpha}\right) - 0.01698\left(\frac{B}{L}\right)$$

MG3 (mangrove model and geobag structure, weight 1 kg, slope 1:1.5

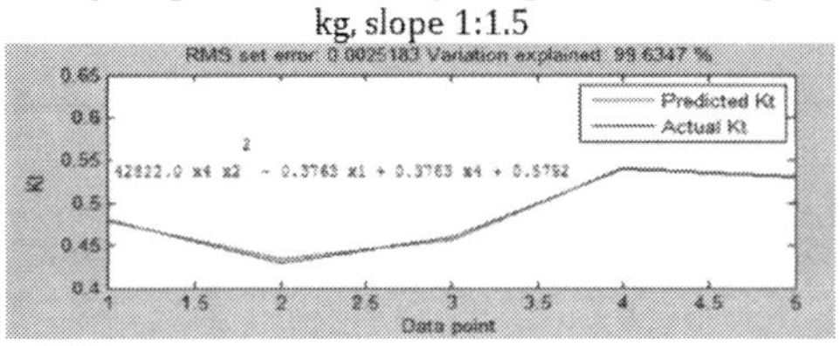

$$Kt = 42822\left(\frac{H_i}{gT^2}\right)\left(\frac{H}{d}\right)^2 - 0.3763\,\tan\boldsymbol{\alpha} + 0.3763\left(\frac{H}{d}\right) + 0.5792$$

MG4 (mangrove model and geobag structure, weight 1 kg, slope of 1:2

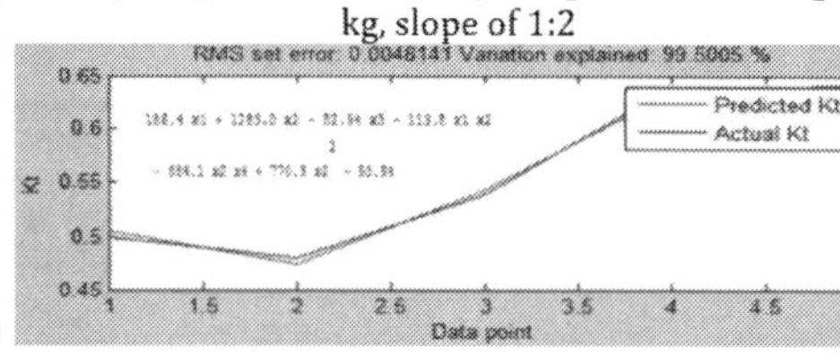

$$Kt = 186.4\,\tan\boldsymbol{\alpha} + 1283\left(\frac{H_i}{gT^2}\right) - 52.94\left(\frac{B}{L}\right) - 113.8\,\tan\boldsymbol{\alpha}\left(\frac{H_i}{gT^2}\right) - 884.1\left(\frac{H_i}{gT^2}\right)\left(\frac{H}{d}\right) + 770.3\left(\frac{H_i}{gT^2}\right)^2 - 93.94$$

MG3 (mangrove model and geobag structure, weight 2 kg, slope 1:1.5

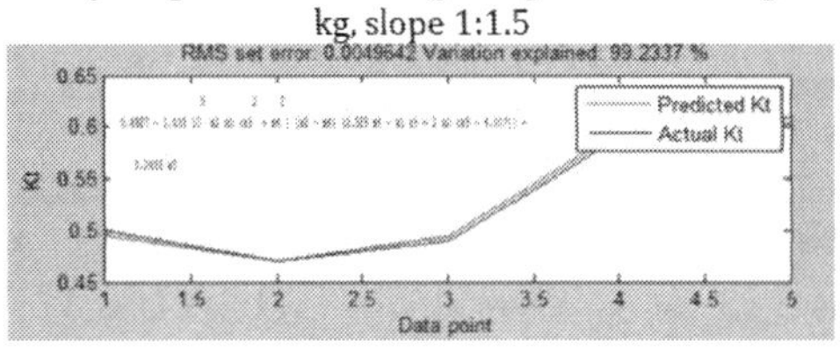

$$Kt = 0.4927 - 1.639\,10^5\left(\frac{H_i}{gT^2}\right)\left(\frac{B}{L}\right)\left(\left(\frac{B}{L}\right)^2 + \left(\frac{H}{d}\right)^2\right)\left(\left(\frac{H_i}{gT^2}\right) - \left(\frac{H}{d}\right)\right)\left(2.529\left(\frac{H}{d}\right) - \tan\boldsymbol{\alpha}\left(\frac{H}{d}\right) + 2\left(\frac{H_i}{gT^2}\right)\left(\left(\frac{B}{L}\right) - 4.937\right)\right) - 0.2435\left(\frac{B}{L}\right)$$

MG4 (mangrove model and geobag structure, weight 2 kg, slope of 1:2

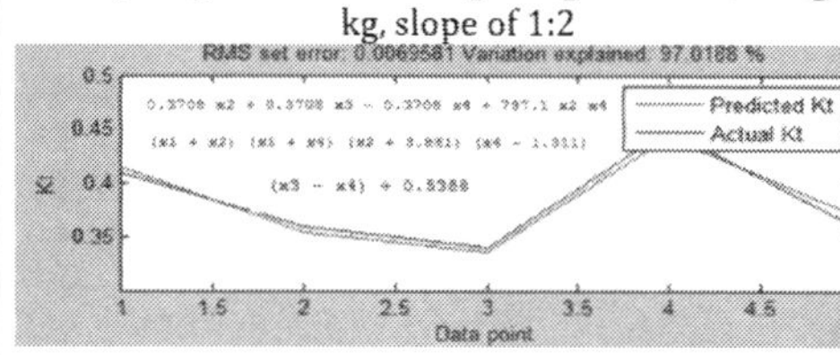

$$Kt = 0.3708\left(\frac{H_i}{gT^2}\right) + 0.3708\left(\frac{B}{L}\right) - 0.3708\left(\frac{H}{d}\right) + 797.1\left(\frac{H_i}{gT^2}\right)\left(\frac{H}{d}\right)\left(\tan\boldsymbol{\alpha} + \left(\frac{H_i}{gT^2}\right)\right)\left(\tan\boldsymbol{\alpha} + \left(\frac{H}{d}\right)\right)\left(\left(\frac{H_i}{gT^2}\right) + 3.881\right)\left(\left(\frac{H}{d}\right) 1.311\right)\left(\left(\frac{B}{L}\right) - \left(\frac{H}{d}\right) + 0.5388\right)$$

Figure 6. GPTIPS output for equation of coefficient transmission (Kt) prediction scenario P, M, MG1, MG2, MG3, MG4, MG5 and MG6.

3.2. Equation of coefficient transmission (Kt)

State parameters input for GPTIPS before generating the equation prediction of coefficient transmission (Kt). In this research there are five parameters with one as output (y) transmission coefficient and four for input x_1 as slope of geobag, x_2 as wave steepness, x_3 as the relative top width of the geobag structure, and x_4 as relative wave height. gpdemo2 modul from GPTIPS was chosen to generate the Equation of coefficient transmission (Kt). The graphic and formula output from the GPTIPS are shown in Figure 6. The Equation of scenarios P and M is different from other scenarios because the input parameter is just y as coefficient transmission, x_1 as wave steepness, for scenario P, and added x_2 as relative wave height for scenario M. All the scenario successes represent Kt prediction. The graphic shows the formula of the Kt model GPTIPS is capable of representing non-linear behaviors with the multiple parameter input and output.

4. Conclusion

A data relationship analysis has been conducted to evaluate the time dynamics of information regarding the observed water level inter-channels. AMI provides initial information indicating that the water elevation data is non-linear. Predicting the transmission coefficient helps determine the extent to which physical and environmental parameters influence the effectiveness of coastal protection, represented by Kt. The significance test was performed using GPTIPS, with the predicted data from the generated formula showing over 95% representativeness. Based on the results of the Kt equation obtained from GPTIPS, it shows that each parameter contributes to the Kt value. This means that if this physical model is to be applied in the field, physical parameters such as the slope of the geobag, wave steepness, the relative top width of the geobag structure, and relative wave height need to be considered.

References

[1] Y. Chang, K. Chu, and L. Z. H. Chuang, "Sustainable coastal zone planning based on historical coastline changes: A model from case study in Tainan, Taiwan," *Landscape and Urban Planning*, vol. 174, pp. 24–32, 2018. DOI: 10.1016/j.landurbplan.2018.02.012.

[2] S. Narayan, M. W. Beck, B. G. Reguero, I. J. Losada, B. van Wesenbeeck, N. Pontee, *et al.*, "The effectiveness, costs and coastal protection benefits of natural and nature-based defences," *PLoS ONE*, vol. 11, no. 5, p. e0154735, 2016. DOI: 10.1371/journal.pone.0154735.

[3] N. Yuanita, A. Kurniawan, P. Paramashanti, and A. A. Laksmi, "Natural coastal protection system preliminary design," in *Proceeding of Ocean, Mechanical and Aerospace-Science and Engineering*, vol. 4, Padang, Indonesia, 2017.

[4] N. Yuanita, A. Kurniawan, Setiawan, F. Hasan, and M. Khasanah. "Physical model of natural coastal protection system: Wave transmission over mangrove seedling trees". *The 3rd International Water Safety Symposium. Journal of Coastal Research*, Special Issue No. 91, pp. 176-180. 2019

[5] N. Yuanita, A. Kurniawan, M. L. Al Hakim, K.R. Irawan, N.W. Saputra. "Mapping of waves parameter in Indonesian water for application of natural coastal protection," Jou*rnal of Sustainability Science and Management. Volume 15 Number 6, August 2020: 85-99.*

[6] R. G. Dean and R. A. Dalrymple, *Water Wave Mechanics for Engineers and Scientists*. Singapore: World Scientific, 1991. DOI: 10.1142/1232.

[7] K. G. Parvathy and P. K. Bhaskaran, "Wave attenuation in presence of mangroves: A sensitivity study for varying bottom slopes," *The International Journal of Ocean and Climate Systems*, vol. 8, no. 1, pp. 1–9, 2017. DOI: 10.1177/1759313116679076.

[8] A. Kurniawan, S. K. Ooi, and V. Babovic, " Understanding tidal and non-tidal representation of numerical model using data relationship analysis in singapore regional waters," *E-proceedings of the 36th IAHR World Congress*, 2015.

[9] H. D. I. Abarbanel, *Analysis of Observed Chaotic Data*. New York: Springer-Verlag, 1996, p. 272.

[10] A. J. Abebe and R. Price, "Information theory and neural networks for managing uncertainty in flood routing," *J. Comput. Civ. Eng.*, vol. 18, no. 4, pp. 373–380, 2004.

[11] V. Babovic and M. Keijzer, "Forecasting of river discharges in the presence of chaos and noise," in *Coping with Floods: Lessons Learned from Recent Experiences*, J. Marsalek, Ed., NATO ARW Series, Dordrecht: Kluwer, 1999, pp. 405–420.

[12] J. R. Koza, *Genetic Programming: On the Programming of Computers by Means of Natural Selection.* Cambridge, MA: The MIT Press, 1992, p. 835.

[13] A. Kurniawan, S. K. Ooi, and V. Babovic, "Improved sea level anomaly prediction through combination of data relationship analysis and genetic programming in Singapore regional waters," *Computers & Geosciences*, vol. 62, pp. 94–104, 2014.

[14] V. Babovic, D. Rodriguez Aguilera, M. Keijzer, and J. Harrington, "An evolutionary approach to knowledge induction: Genetic programming in hydraulic engineering," *ResearchGate*, pp. 1–10, 2001.

[15] V. Babovic, "Introducing knowledge into learning based on genetic programming," *Journal of Hydroinformatics*, vol. 11, no. 3, pp. 181–193, 2009

[16] V. Babovic and M. Keijzer, "Genetic programming as a model induction engine," J. Hydroinformatics, vol. 2, pp. 35–60, 2000.

[17] J. Shiri and O. Kisi, "Comparison of genetic programming with neuro-fuzzy systems for predicting short-term water table depth fluctuations," Comput. Geosci., vol. 37, no. 10, pp. 1692–1701, 2011.

[18] O. Kisi, J. Shiri, and B. Nikoofar, "Forecasting daily lake levels using artificial intelligence approaches," Comput. Geosci., vol. 41, pp. 169–180, 2012.

[19] S. Karimi, O. Kisi, J. Shiri, and O. Makarynskyy, "Neuro-fuzzy and neural network techniques for forecasting sea level in Darwin Harbor, Australia," Comput. Geosci., vol. 52, pp. 50–59, 2013.

[20] D. P. Searson, D. E. Leahy, and M. J. Willis, "GPTIPS: An open-source genetic programming toolbox for multigene symbolic regression," in *Proc. Int. Multiconference of Engineers and Computer Scientists (IMECS 2010)*, vol. 1, Hong Kong, 2010, pp. 77–80.

IOP Conf. Series: Earth and Environmental Science **1464** (2025) 012018 doi:10.1088/1755-1315/1464/1/012018

Coastal flood modelling of Jakarta using the SFINCS, a reduced-physics solver with high computational efficiency

Rafi Ahmadi Yusuf[1*], Munawir Bintang Pratama[2], Entin A. Karjadi[1], Hiroshi Takagi[3]

[1] Ocean Engineering, Institut Teknologi Bandung, Bandung, Indonesia
[2] School of Engineering, The University of Edinburgh, Scotland, United Kingdom
[3] School of Environment and Society, Tokyo Institute of Technology, Japan

*E-mail: rafi2805@gmail.com

Jakarta, the largest city in Indonesia has been on the receiving end of many floods in the past three decades, with flooding driven by tides, sea-level rise, wind, surges, and land subsidence. The risk of coastal flooding in Jakarta has been steadily increasing due to several factors, the first being sea level rise due to global warming and land subsidence as the consequence of rapid urban development. The present work uses the Super-Fast Inundation of CoastS (SFINCS), a computationally efficient hydrodynamic modelling software to project the extent of coastal floods on Jakarta. The model shows a good agreement when validated against field data in Tanjung Priok tide station and few other key areas while also being compared to hydrodynamic modelling results from Delft3D-FLOW (Takagi *et al.*, 2016). Four modelling scenarios are investigated, each scenario consists of the sea-level rise variable and two scenarios with land subsidence combined. The modelling results indicate that land subsidence has the most significant impact on the scale of inundation over the coast and inland. Scenarios that only consider sea level rise only show inundation reaching a few hundred meters from the coastline. The significance of land subsidence on inland areas near the coast of Jakarta can be seen with regards to the scale of areas affected by inundation compared to scenarios using the same tidal data but only considering sea level rise. The present work demonstrates the benefit of using the SFINCS, where a faster, higher-resolution result can be achieved compared to the Delft3D-FLOW. The SFINCS model have proven that it can provide similar results during water level validation and coastal flooding inundation extent compared to the Delft3D-FLOW model with a significant decrease in time required.

1. Introduction

Jakarta, the capital of Indonesia has been one of the fastest growing cities with an estimated population of over 12.5 million people in 2005 [1], with an estimation of 30 million in the year 2030 [2]. The corresponding growth of population pushes growth in other key areas such as infrastructure and economy, such rapid development has caused significant environmental changes. Increased frequency of riverine and coastal flooding can be observed dating back to the year 1992, with the worst major coastal flood occurring during November 2007.

Jakarta's coastal areas are marked as vulnerable to coastal flooding [3], with various cases occurring almost every year. While not directly affected by tropical storms and its immediate effects, Jakarta's coasts are susceptible to residual effects of storms from the Philippines or Vietnam [3], [4], alongside high waves (storm surges), heavy winds and rain in the regions that storms generate frequently in [3]. Risk of flooding has been increasing due to climate change raising the sea level [5], in combination with land subsidence due to urbanization and freshwater extraction [6], [7].

The constant increase of mean sea level elevation and the subsiding land along the coastline, and urban areas increases the flooding frequency and extent of flooded areas into urban Jakarta [8], thus increasing the potential damage. During the National Seminar for Construction of the Great Sea Wall, Coordinating Minister of Economy explained that the economic damage caused by coastal flooding has reached 2.1 trillion Rupiah each year. The economic impact sparked development of countermeasures to mitigate infrastructure damage. One such method is running an efficient but accurate coastal flooding simulation to predict flooding extent on urban and coastal areas.

Enter SFINCS, a reduced-physics hydrodynamic model capable of running accurate coastal flooding simulations efficiently, using less computing power and system requirements while also reducing the time needed for hydrodynamic computations. These feats make SFINCS a suitable candidate for early warning systems and to an extent, much more accessible to lesser powerful or mobile computational systems.

The purpose of the present research is to validate the capabilities of SFINCS to compute coastal flooding scenarios based on the research titled "Projection of coastal floods in 2050 Jakarta" [9]. There will be a total of four scenarios that are investigated, using a combination of sea-level rise and with land subsidence considered. The author of this work attempts to present SFINCS as an alternative to full process-based physics hydrodynamic models while maintaining the accuracy of such models, with the added benefits of faster and efficient computation.

2. Methods

2.1 SFINCS description

SFINCS or Super-Fast Inundation of CoastS is a newly developed reduced-physics solver by Deltares, with the capability to efficiently simulate compound flooding events with limited computational power and good accuracy [10]. SFINCS uses simplified governing equations for continuity and momentum based on Local Inertial Equations (LIE) used on the LISFLOOD-FP model [11].

SFINCS by default runs LIE mode, which excludes processes necessary for calculating wave dynamics. Running SFINCS with advection turns it into SFINCS-SSWE (Simplified Shallow Water Equations), while neglecting atmospheric forcing, Coriolis, and viscosity terms present on full-physics process-based models [10]. SFINCS has the option to include spatially varying friction, infiltration and precipitation as additional forcings aside from tides, wind, and waves.

The present research uses SFINCS-SSWE to simulate flooding on coastal systems. SFINCS has been validated on various cases such as compound flooding by Hurricane Irma, wave-driven flooding during Typhoon Haiyan [10], Tohoku Tsunami offshore propagation [12], coastal flooding in Houston, Texas during Hurricane Harvey [13] with accurate and fast results.

2.2 Model domain and bathymetry

To replicate the Delft3D-FLOW model as close as possible, the SFINCS domain is located at the same domain extent on the north of Jakarta Bay. The exact computational domain is extended towards the offshore and urban areas of Jakarta as seen on Figure 1.

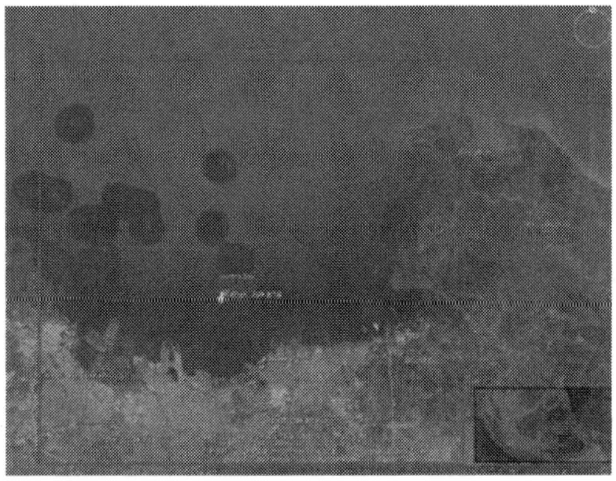

Figure 1. Map of Jakarta Bay, the red continuous box depicts the computational domain for the SFINCS model.

The bathymetry data is obtained from GEBCO whilst the topography data is taken from the SRTM 90 m Digital Elevation Database. The computational domain spans from 5.898°S–106.68°E to 6.201°S–107.03°E. Another part of this research is to validate SFINCS towards field data, utilizing the latest bathymetry and topography acquired from Indonesia's DEMNAS and BATNAS.

2.3 Boundary conditions
As of SFINCS v2.0.3, the model only accepts boundary condition inputs time-series format. SFINCS requires at least one type of forcing, the options being water levels, waves, discharges, and meteorological forcing set on the desired boundary condition.

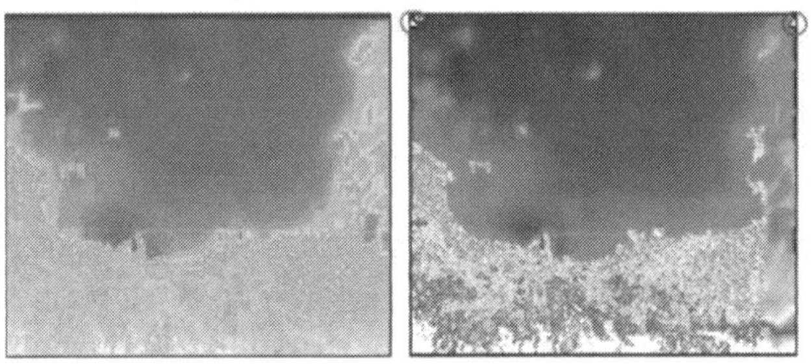

Figure 2. Northern offshore open boundary condition as shown in the Delft3D-FLOW model (Left) and the two water level boundary points on the northern domain of the SFINCS model (Right).

Boundary conditions on the SFINCS model are assigned on the top right and left corner to simulate a northern boundary present on the Delft3D-FLOW model as shown on Figure 2. Tidal driven forcings are defined onto boundary points, obtained from the eight main tidal constituents assigned to the offshore open boundary on the Delft3D-FLOW model derived from the TPXO7.1 Global Inverse Tide Model.

2.4 SFINCS inputs and setting
SFINCS can only use cartesian coordinates (UTM Zone) and a uniform grid system. The grid characteristics for the SFINCS model takes after the smallest grid size of the Delft3D-FLOW reference model, as it has gradually increased grid sizes further offshore.

Table 1. Grid characteristics comparison between Delft3D-FLOW and SFINCS models.

Model	Model Resolution (Easting × Northing)	Vertical × Horizontal Cells	Total Cells
Delft3D-FLOW (Reference Model)	50 × 120 m – 200 × 440 m	434 × 164	69.984 Cells
Delft3D-FLOW (1)	50 × 120 m	779 × 671	219.900 Cells
Delft3D-FLOW (2)	50 × 50 m	779 × 281	522.709 Cells
SFINCS (1)	50 × 120 m	777 × 278	216.006 Cells
SFINCS (2)	50 × 50 m	777 × 667	517.482 Cells

The present research tests SFINCS with a uniform grid size of 50 m × 50 m for further analysis of the model's capabilities. Additional Delft3D-FLOW tests will be run at the same resolution as SFINCS for model runtime comparison as seen on Table 1.

Table 2. Investigated coastal flooding scenarios and the variables for each scenario.

Scenario	Abnormal Tides	Sea Level Rise	Land Subsidence
Jakarta Bay 2000 (Scenario 1)	✓	X	X
Jakarta Bay 2025 (Scenario 2)	✓	✓	✓
Jakarta Bay 2050 (Scenario 3)	✓	✓	X
Jakarta Bay 2050 (Scenario 4)	✓	✓	✓

Four coastal flooding scenarios are investigated, each will be run for 6 plus 2 days to accommodate the spin-up process present on the Delft3D-FLOW model. Reference time for both models are set at the 20th until 28th November 2007, including the extensive coastal flooding event that occurred on the 26th. The first scenario will be the base model, only with abnormal tides variable present during November 2007. Subsequent scenarios as seen in Table 2 will include sea level rise and land subsidence variables to present the effect of future extreme coastal flooding events. Input of both variables for future scenarios will be added as linear increases or decreases in the corresponding variable's elevation per year [9].

3. Results and discussions

3.1 Validation: SFINCS vs Delft3D

SFINCS is validated towards results from the Delft3D-FLOW reference model. Water level validation methods used are the Root Mean Squared (RMS) error method and Pearson's Correlation Coefficient. Corresponding validation results can be seen on Table 3.

Validations are conducted on six areas, including Tanjung Priok tide station used for validation of observed tides on the Delft3D reference model [9] as seen on Figure 3. Additional points include ports (Sunda Kelapa and Northern Jakarta Container Industrial Area) and coastal areas (Ancol and Mutiara Coast) with estuaries that are interconnected with Jakarta's main rivers.

Table 3. SFINCS water level validation towards Delft3D-FLOW results, (A) is SFINCS with the 50 × 120 m grid and (B) is SFINCS with the 50 × 50 m grid, the final row being the average on all areas.

Validation Location	Validation Method	Scenario 1	Scenario 2	Scenario 3	Scenario 4
Tanjung Priok (A)	RMSE (m)	0.006643	0.001008	0.000960	0.001061
	Pearson's Correlation Coefficient	0.999797	0.999802	0.999820	0.999793
Tanjung Priok (B)	RMSE (m)	0.006616	0.001012	0.0009641	0.001032
	Pearson's Correlation Coefficient	0.999782	0.999794	0.999800	0.999778
Average	RMSE (m)	0.009359	0.001355	0.001008	0.001054
	Mean Pearson's Correlation Coefficient	0.999389	0.999548	0.999803	0.999785

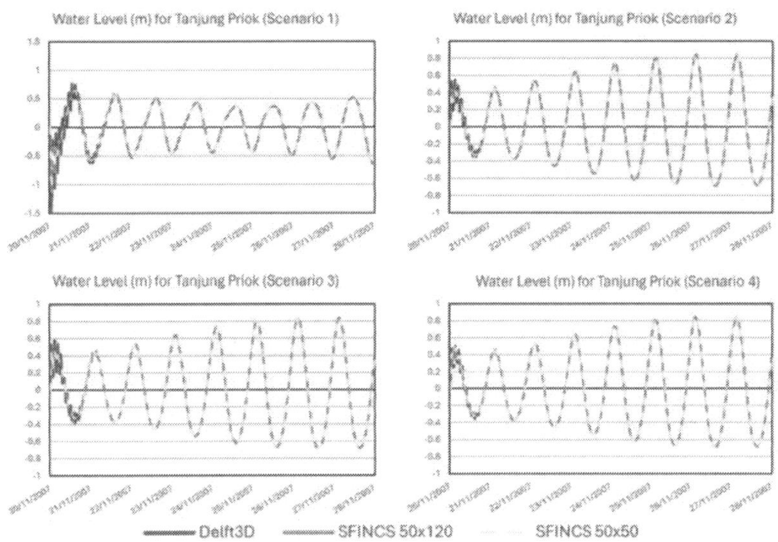

Figure 3. Water level results for Tanjung Priok tide station, the overlaid data shows both SFINCS (Blue for 50x120 m grid and Yellow for 50x50 m grid) models and the Delft3D (Red) water level results for every scenario investigated in this research.

SFINCS simulates water levels at Tanjung Priok tide station with Root Mean Squared (RMS) error values ranging from 0.0066 m to 0.001 m as seen on the fourth scenario. All scenarios have a near 1 correlation coefficient with Delft3D's results. The near identical water levels can be seen in Figure 4. Outliers during validation due to bathymetry utilization from smaller grids resulted in error values reaching 0.037 m for the port validation areas on the first scenario.

Simulated water levels at Tanjung Priok tide station are similar for SFINCS models. These results are consistent throughout every validation point investigated. Though fluctuations can be seen during the first two days of simulation, specifically the spin-up process. The first two days of

IOP Conf. Series: Earth and Environmental Science **1464** (2025) 012018 doi:10.1088/1755-1315/1464/1/012018

simulation (88 out of 385 data points) were cut from the validation process. Differences seen during the spin-up period on both models can be seen in Figure 4.

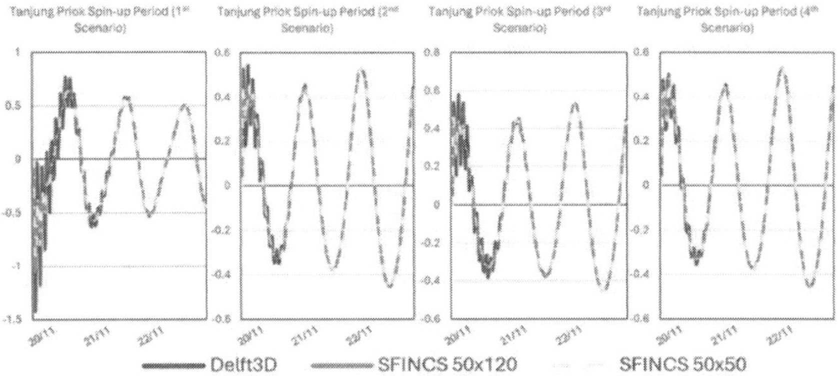

Figure 4. Spin-up behaviour comparison for both models at every investigated scenario.

Delft3D-FLOW has substantially higher fluctuations in simulated water levels compared to SFINCS, with the highest average gap on all six validation points being 0.354 m on the first scenario. Scenarios two until four have less fluctuations on the Delft3D model, but still visible. SFINCS can go through the spin-up period faster than Delft3D-FLOW, as seen visually that it takes less than one day for the SFINCS model to be fully spun-up.

3.2 Comparisons between rough and fine grid resolution

Simulation times for SFINCS differ greatly compared to Delft3D-FLOW results, even when utilizing less than ideal computational setups like laptops and personal computers as seen on Table 4. A speed up factor of 2.59 on the first scenario, 2.15 on the second, 2.24 on the third scenario, and 2.34 on the fourth scenario. Utilizing uniform grid on Delft3D-FLOW models increased computational times significantly, with SFINCS achieving speed up factors of 10.1 for the 50 × 120 m grid resolution models and 9.58 for the 50 × 50 m grid resolution models. All models and scenarios have increasing runtime following every scenario, the longest time needed being on the fourth scenario on every model investigated.

Table 4. Runtime comparison for all models and scenarios. Models with (1) uses 50 × 120 m grid, and (2) uses 50 × 50 m grid resolution. Model runtimes are presented in the HH:MM: SS format.

Model	Total Grid Cells	Simulation Times			
		Scenario 1	Scenario 2	Scenario 3	Scenario 4
Delft3D-FLOW (Reference Model)	69.984	00:27: 39	00:28: 41	00:29: 32	00:32: 12
Delft3D-FLOW (1)	219.900	01:48: --	01:55: --	02:06: --	02:13: --
Delft3D-FLOW (2)	522.709	04:36: --	04:43: --	04:52: --	05:06: --
SFINCS (1)	216.006	00:10: 41	00:13: 19	00:13: 09	00:13: 46
SFINCS (2)	517.482	00:28: 47	00:32: 07	00:32: 01	00:33: 20

[a] All simulations were run from a personal laptop with the following specifications: Lenovo Ideapad Gaming 3i with 16GB 3200MHz RAM, Ryzen 5th Gen 4600H Laptop CPU, and Geforce GTX 1650Ti Laptop GPU.

3.3 Sea level rise and land subsidence scenarios

This section presents the inundation results utilizing SFINCS, to show its capabilities in simulating coastal flooding and sea water inundation. Visual results for inundation modelled by SFINCS and Delft3D is shown in Figure 5.

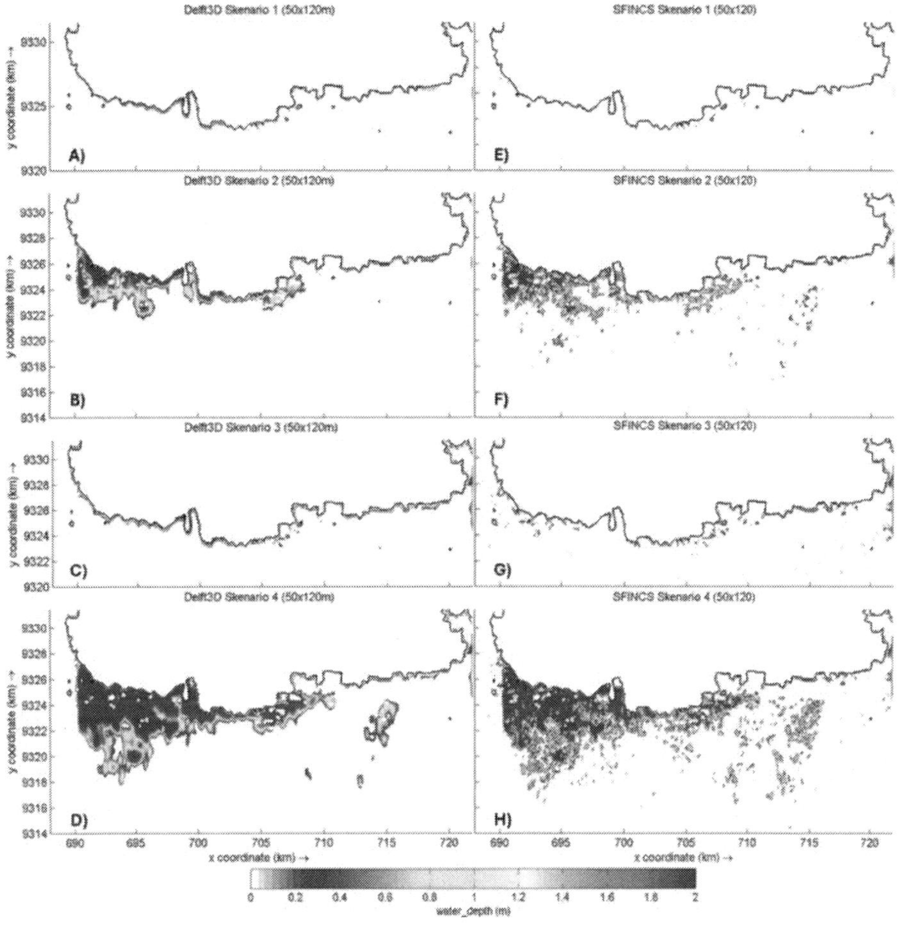

Figure 5. Sea water inundation comparison at the model domain, the left row is the Delft3D model followed by SFINCS on the right. A) through D) presents scenarios 1 to 4 (in descending order) of Delft3D-FLOW's inundation results while E) through H) presents the SFINCS inundation results.

On all presented scenarios, SFINCS overestimates inundation points deeper into Jakarta's urban areas. Areas in eastern Jakarta can be seen with significantly less isolated inundation points, especially on the second and fourth scenario, though SFINCS manages to capture inundation quite accurately on the western part of Jakarta.

Factors such as interpolation method play a role in simulating inundated coastal and urban areas since it affects the utilization of bathymetry and topography. Utilizing uniform grids with the smallest size taken from Delft3D-FLOW's model, a more detailed topography can be utilized

IOP Conf. Series: Earth and Environmental Science **1464** (2025) 012018 doi:10.1088/1755-1315/1464/1/012018

by SFINCS to simulate a more detailed flooding path. Due to this, differences in elevation can be observed during validation using Delft3D's QUICKPLOT.

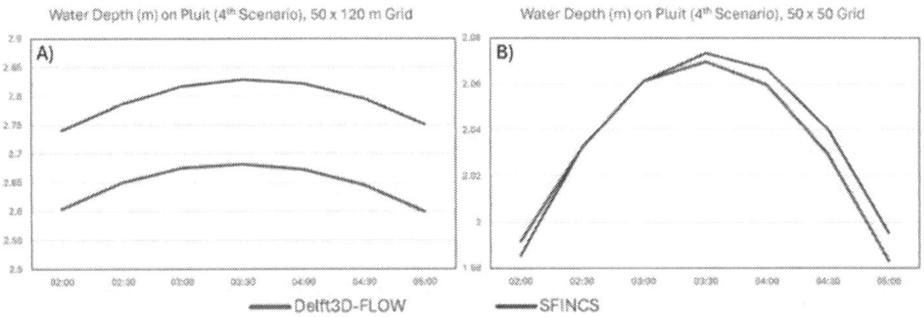

Figure 6. Flooding water depth comparison on the 4th scenario between 50x120 m and 50x50 m grid resolutions.

Heavily inundated areas such as the City of Pluit in North Jakarta can be seen with water levels and flooding depths that are identical to Delft3D-FLOW's results. SFINCS models water levels correctly, but due to the difference in depth interpolation as the direct result of different grid systems, SFINCS overestimates the inundation levels deeper into Jakarta's urban areas. On the SFINCS models, bed level elevation is underestimated by 0.01-0.15 m compared to Delft3D models of the same resolution.

4. Conclusions

Previous sections show that SFINCS can provide accurate and fast results for coastal flooding systems in comparison to full-physics process-based models such as Delft3D-FLOW, even when utilizing much finer grid resolutions and subpar computing systems. As seen during the validation process, SFINCS recreated a near perfect water level simulation at two to three times faster than Delft3D-FLOW, with RMS error values less than 1 cm and near perfect correlation, this behaviour is consistent throughout every validation point as the average RMS error value for each scenario still manages to be below 1 cm.

SFINCS took less time to normalize water levels during spin-up. This means that simulation time frames can be cut shorter to save computational resources on time-sensitive scenarios, such as the utilization of SFINCS as early warning and flood/tsunami prediction systems. Less spin-up time also means that with the same timeframe, SFINCS can provide more quantities of usable data for validation purposes or predictions.

Though the use of different grid systems can affect the resulting inundation levels, SFINCS still manages to show excellent results in comparison to Delft3D-FLOW. SFINCS presented more isolated inundation areas overall, but intensely inundated areas were presented very well. The use of external interpolation methods on the SFINCS model could be the direct cause of overestimation of inundated areas on the eastern part of Jakarta.

Inundation on scenarios without land subsidence presents very little flooding even in coastal areas, such results could be due to the different interpolation method used on the SFINCS model utilizing the provided bathymetry and elevation data to create much more accurate flood paths. Scenarios with land subsidence show lesser differences on coastal areas largely due to the implementation of the variable being a linear decrease in elevation, and as such, the effect of interpolation are seen much deeper into urban Jakarta on the SFINCS model results.

Overall, SFINCS can be utilized as an alternative to full-physics hydrodynamic models. A combination of accuracy, computational efficiency, and speed makes SFINCS suitable for use as a base for early warning systems while having the ability to be utilized by lesser powerful computers.

References

[1] Djaja, R., Rais, J., Abidin, Z.H., Wedyanto, K., (2004): Land subsidence of Jakarta Metropolitan Area. 3rd FIG Regional Conference, Jakarta, Indonesia, October 3–7, 2004, 14.

[2] Sagala, S., Lassa, J., Yasaditama, H., Hudalah, D., (2013): The Evolution of Risk and Vulnerability in Greater Jakarta: Contesting Government Policy in Dealing with a Megacity's Exposure to Flooding. An Academic Response to Jakarta Floods in January 2013. IRGSC Working Paper No.2, 18 pp.

[3] Ningsih, N.S., Suryo, W., Anugrah, S.D., (2011): Study on characteristics of residual water level in Jakarta, Semarang, and Surabaya waters Indonesia and its relation to storm events in November 2007. Int. J. Basic Appl. Sci. 11 (5), 19–25.

[4] Takagi, H., Esteban, M., Shibayama, T., Mikami, T., Matsumaru, R., De Leon, M. Thao, N.D., Oyama, T., Nakamura, R., (2014): Track analysis, simulation, and field survey of the 2013 Typhoon Haiyan storm surge. J. Flood Risk Management 10. 42-52.

[5] Nerem, R.S., Chambers, D.P., Leuliette, E.W., (1999): Variations in global mean sea level associated with the 1997–1998 ENSO event: impli cations for measuring long-term sea-level change. Geophys. Res. Lett. 26 (19), 3005–3008.

[6] Hutabarat, L.E., Ilyas, T., (2017): Mapping of Land Subsidence Induced by Groundwater Extraction in Urban Areas as Basic Data for Sustainability Countermeasures. International Journal of Technology. Volume 8(6), 1001-1011.

[7] Hutasoit, L.M., (2001): Kemungkinan Hubungan antara Kompaksi Alamiah dengan Daerah Genangan Air di DKI Jakarta. Buletin Geologi, Jilid/Volume 33, No. 1, 2001. Bandung: Departemen Teknik Geologi ITB.

[8] Abidin, H.Z., Andreas, H., Gumilar, I., Gamal, M., Fukuda, Y., Deguchi, T., (2009): Land Subsidence and Urban Development in Jakarta (Indonesia). 7th FIG Regional Conference, Hanoi, Vietnam, 16.

[9] Takagi, H., Esteban, M., Mikami, T., Fujii, D. (2016): Projection of Coastal Floods in 2050 Jakarta. Urban Climate Journal, 17:135-145.

[10] Lejinse, T., Ormondt, M., Nederhoff, K., Dongeren. A. (2021): Modeling Compound Flooding in Coastal Systems Using a Computationally Efficient Reduced-physics solver: Including Fluvial, Pluvial, Tidal, Wind, and Wave Driven Processes.Coastal Engineering Journal 163:103796.

[11] Bates, P.D., Dawson, R.J., Hall, J.W., Horrit, M.S., Nicholls, R.J., Wicks, J., Mohamed Hassan, Mohamed Ahmed Ali (2005). Simplified two-dimensional numerical modelling of coastal flooding and example applications. Coast. Eng. 52, 793-810.

[12] Röbke, B.R., Leijnse, T., Winter, G., van Ormondt, M., van Nieuwkoop, J., de Graaff, R. (2021): Rapid Assessment of Tsunami Offshore Propagation and Inundation with D-FLOW Flexible Mesh and SFINCS for the 2011 Tohoku Tsunami in Japan. J. Mar. Sci. Eng. 9, 453

[13] Sebastian, A., Bader, D.J., Nederhoff, C.M., Lejinse, T.W.B., Bricker, J.D., Aarnikhof, S.G.J. (2017): Hindcast of Pluvial, Fluvial, and Coastal Flood Damage in Houseton, Texas during Hurricane Harvey using SFINCS. Natural Hazards Journal, 109:2343-2362.

[14] Egbert, Gary D., and Svetlana Y. Erofeeva. "Efficient inverse modeling of barotropic ocean tides." Journal of Atmospheric and Oceanic Technology 19.2 (2002): 183-204.

IOP Conf. Series: Earth and Environmental Science **1464** (2025) 012019 doi:10.1088/1755-1315/1464/1/012019

The effect of the construction method on the tombolo process at Palabuhanratu seaport

Eka Oktariyanto Nugroho[1,2*], Cipta Riyana[3,5], Dantje Kardana Natakusumah[1,2], Dedi Nurpadilah[3], Widyaningtias[1], Fatma Nurkhaerani[4], Siska Wulandari[5], Hana Syakira[6], Ken Wiralino Taredja[5], and Benedictus Arie Moniaga[5]

[1] Water Resources Engineering Research Group, Faculty of Civil and Environmental Engineering, Bandung Institute of Technology, Jl. Ganesha 10, Bandung, 40132, Indonesia
[2] Fluid Mechanics and Hydrodynamics Laboratory, Center for Industrial Technology, Institut Teknologi Bandung, Bandung, Jl. Ganesha 10, Bandung, 40132, Indonesia
[3] PT. New Atlas Consultant, Jl. PHH Mustopa No. 143, Bandung, 40619, Indonesia
[4] Karawang Singaperbangsa University, HS. Ronggo Waluyo St., Puseurjaya, Telukjambe Timur, Karawang, West Java, 41361 Indonesia
[5] Master Program of Water Resources Engineering, Faculty of Civil and Environmental Engineering, Bandung Institute of Technology, Jl. Ganesha 10, Bandung, 40132, Indonesia
[6] Master Program of Water Resources Management, Faculty of Civil and Environmental Engineering, Bandung Institute of Technology, Jl. Ganesha 10, Bandung, 40132, Indonesia

*E-mail: nugrohoeka@itb.ac.id

Abstract. Palabuhanratu seaport is not yet operational due to several land facilities still to be constructed and a poorly performing port basin. The offshore facilities consist of a jetty and a breakwater. During the construction of the breakwater, a tombolo formed; a tombolo being a natural tongue of land or a spit formed by deposition. In this case, sedimentation in the tombolo became much worse during the construction of the jetty, continuing up to the end of the project. Nowadays, the port basin is nearly parallel to the pier, while the elevation values in front of the pier range between +0.8 and +2.5 meters LWS. The objective of the study is to explain the sediment transport processes that lead to a tombolo formation, regarding construction sequences. A simple simulation of sediment movement will be developed using the backtracking analysis. The spatial analysis indicates the changes of the shoreline during and after the construction periods; up to a remarkable 65.69% area of change from June 6 to September 16, 2018. Hydro-oceanographic simulations showed significant bed level changes around the site, with bed elevations up to 3.4 meters of sedimentation on the left bank and 3.0 meters of erosion on the right bank. These findings have shown shallow conditions in the seaport basin of Palabuhanratu, indicating that the sediment transport is cross-shore, which is perpendicular to the coast. These have disrupted the pre-existing balance of longshore sediment transport at the ports, which then resulted in significant sediment deposition compromising the depth and functionality of the port basin.

1. Introduction

Coastal zones support industries like tourism, shipping, and fishery [1], offering significant ecological, social, and economic benefits such as fish habitats and transportation hubs [2], [3]. Ports handle 90% of global trade by sea but face environmental challenges from urbanization and industrial activities, necessitating sustainable practices [4]. Growing competition among shipping ports is driven by offshoring, technological advances, and market demands [5]. In Indonesia, coastal and marine resources significantly contribute to the national economy through fisheries and maritime activities [6].

Palabuhanratu Seaport in Sukabumi Regency, West Java, Indonesia, is under development and designated as a regional feeder in the national port master plan by Decree KP 432 of 2017 [7]. The 2019–2024 master plan focuses on completing facilities like docks and breakwaters, but construction faces challenges from sedimentation due to tombolo formation during breakwater construction.

Sedimentation from coastal construction threatens sandy shorelines, causing erosion or deposition [8], [9]. Creating sediment models is crucial to understanding coastal processes and managing sediment movement [10]. Harbors with extensive breakwaters are prone to sedimentation, disrupting shipping and requiring dredging, as seen in the ports of IJmuiden and Ulee Lheue [11], [12], [13].

Submerged breakwaters dissipate wave energy and affect sediment transport, leading to accretion and erosion [14]. To avoid forming tombolos or salients, detached breakwaters should be placed over six times their length offshore [9], [15], [16]. These landforms are shaped by wave transformation, currents, and sediment transport [2]; tombolos form when sediment connects the salient to the breakwater, depending on spacing, length, and gap width [1], [17]. If breakwaters are too close or long, they hinder wave energy needed for sediment transport, causing tombolo formation within 50 to 75 days after installation [18], [19].

This research investigates sediment transport leading to tombolo formation at Palabuhanratu Seaport. Using backtracking analysis and bathymetric data, it examines the link between construction sequences and sediment patterns. Understanding these changes is crucial for designing marine structures like jetties and breakwaters [20], [21], addressing current challenges, and guiding future projects.

2. Study Area

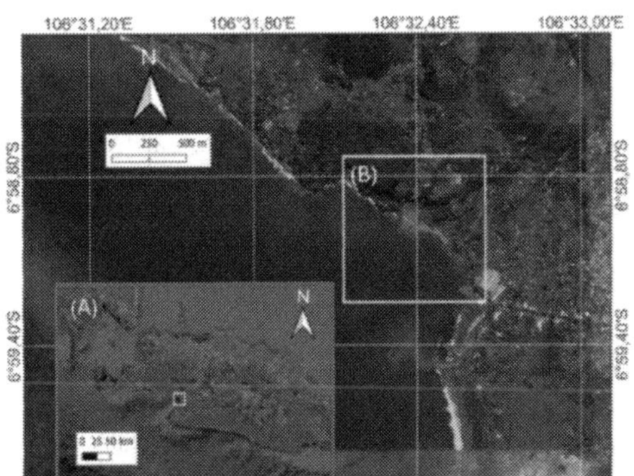

Figure 1. Palabuhanratu Seaport is located on Java Island, Indonesia (A). The study area is highlighted in more detail at the location marked in yellow (B).

This study is located on Palabuhanratu Seaport, Palabuhanratu Subdistrict, Sukabumi Regency, West Java, Indonesia. The coordinate is on 6° 58' 56.40" N and 106° 32' 24.49" E. Administratively, Palabuhanratu Seaport is bordered by Cikakak District to the north, the Indian Ocean to the south, Simpenan District to the west, and Bantargadung District to the east. In Pelabuhan Ratu Bay, there are four types of coastal landforms: erosional coast, volcanic coast, fluvial coast, and depositional coast, each with different beach characteristics based on these landforms [22]. The study area is presented in Figure 1.

3. Methodology

3.1 Conceptual work

The construction of a breakwater and jetty at the Palabuhanratu Seaport aimed to stabilize the port basin and protect it from strong waves. While the breakwater reduced wave energy, it also decreased the water flow velocity behind the structure. This reduction led to sediment accumulation. Over time, the ongoing deposition of sediments formed a tombolo—a landform that connects the breakwater to the shore. This unintended sedimentation affects port operations, requiring frequent dredging, and may alter local marine ecosystems. Understanding the hydrodynamic changes caused by the breakwater, and their effect on sediment transport, is crucial for finding construction methods that reduce long-term impacts (Figure 2).

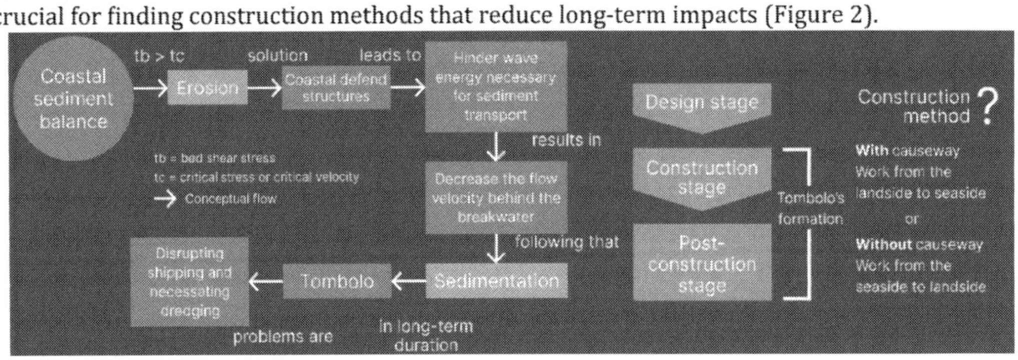

Figure 2. Conceptual work of the study.

3.2 Data collection and preparation

This study utilizes both primary and secondary data collected through surveys and literature reviews. The primary data comprises sedimentation data, tidal data, and bathymetry obtained during a 2022 survey. The secondary data includes Google Earth spatial imagery from 2017 to 2022, Ocean Map Pushidrosal No. 93, wind data sourced from the European Centre for Medium-Range Weather Forecasts (ECMWF) for 2018–2019, and global tidal data [24]. This data was used to create a model designed to accurately simulate real-world conditions, serving as a dependable resource for coastal planning and management at Palabuhanratu Seaport.

3.3 Workflow

The workflow consists of data collection (primary and secondary), data processing (spatial and numerical using MIKE 21) and conclusion, which is shown in Figure 3.

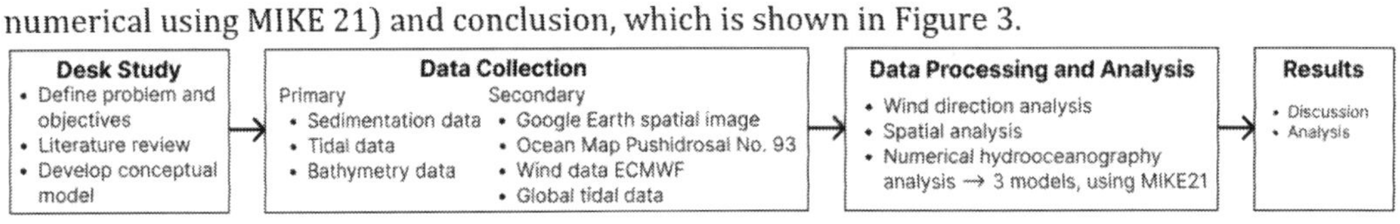

Figure 3. Workflow diagram of the study.

The Sediment Transport module equation is derived from the Advection-Dispersion (AD) module. The MIKE 21 AD equation as written in Equation 1:

$$\frac{\partial \bar{c}}{\partial t} + u\frac{\partial \bar{c}}{\partial x} + v\frac{\partial \bar{c}}{\partial y} = \frac{1}{h}\frac{\partial}{\partial x}\left(hD_x\frac{\partial \bar{c}}{\partial x}\right) + \frac{1}{h}\frac{\partial}{\partial y}\left(hD_y\frac{\partial \bar{c}}{\partial y}\right) + Q_L C_L \frac{1}{h} - S \quad (1)$$

Where:

$\bar{c}$ = average mass concentration [kg/m³]
u, v = average flow velocity [m/s],
D_x, D_y = dispersion coefficients,
h = water depth,
S = deposition/erosion [kg/m³/s]
Q_L = discharge in the horizontal area [m³/s/m²]
C_L = discharge concentration [kg/m³]

3.4. Coastal sedimentation trends: spatial data overview for input data

Spatial analysis is conducted by comparing the image retrieved from the Google Earth Spatial Image from time to time during the construction (Figure 4) and after the construction stages (Figure 5) to determine the change in the shoreline that occurred at the location of the study. The result will be later to be confirmed by the sediment transport model.

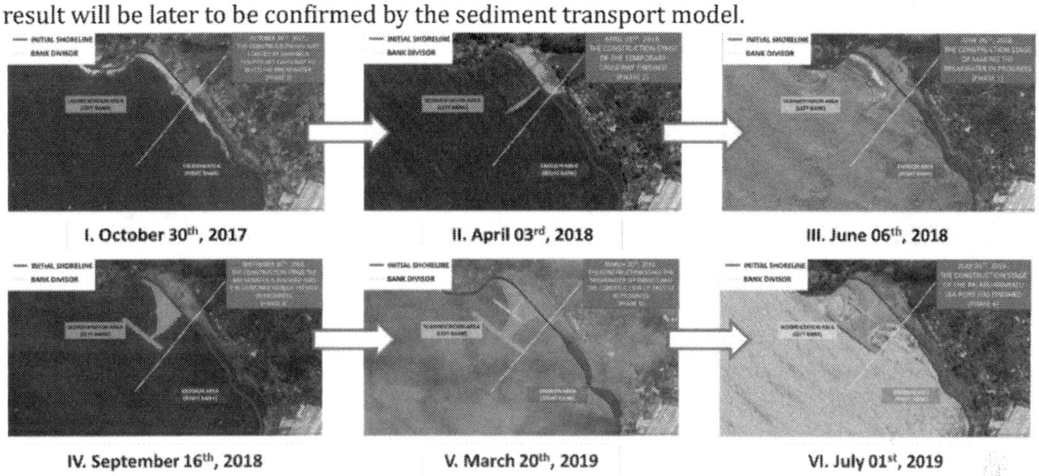

Figure 4. The change of the shoreline at the location of study from Day 1 to Day 610 (October 30th, 2017 – July 1st, 2019).

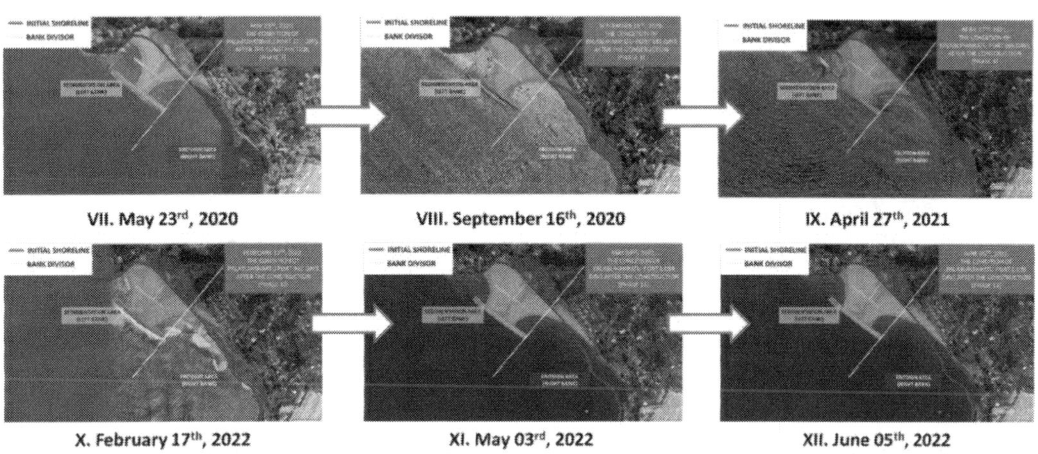

Figure 5. The change of the shoreline at the location of study from Day 937 to Day 1680 (May 23rd, 2020 – June 05th, 2022).

3.5. Comparison of hydrodynamic model outputs
Calibrating a 2D MIKE21 model to water surface elevations involves adjusting Manning's roughness coefficients to align simulated and observed water levels. Step 1 sets a baseline for roughness values, while Step 2 refines Manning's roughness distribution for tidal analysis. Root Mean Square Error (RMSE) measures the error magnitude between predicted and observed values; a smaller RMSE indicates a better fit. Classification: Near 0 (minimal error), moderate values (acceptable), and high values (large error) [25], [26], [27], [28], [29]. RMSE-observation Standard Deviation Ratio (RSR) evaluates model performance by combining model error with observation standard deviation. Classification: 0 (perfect), 0 < RSR ≤ 0.5 (very good), 0.5 < RSR ≤ 0.6 (good), 0.6 < RSR ≤ 0.7 (satisfactory), and RSR > 0.7 (poor) [26], [29], [30], [31]. Nash-Sutcliffe Efficiency (NSE) compares predictions to observations, with values close to 1 indicating high accuracy. Classification: 1.0 (perfect), 0.7–1.0 (very good), 0.5–0.7 (good), 0.0–0.5 (acceptable), and negative values (<0) indicating performance worse than the mean. Positive NSE reflects better model accuracy, while negative NSE suggests otherwise [26], [27], [29], [30], [31], [32].

4. Result and Discussion

4.1. Spatial analysis
Spatial analysis used Google Earth satellite images from before, during, and after construction stages to assess shoreline changes at the study location. The initial shoreline is marked by a red line, while each subsequent shoreline position is shown with a yellow line, indicating substantial shifts over time. This visual assessment confirmed ongoing shoreline changes.
To quantify these shifts, the area was divided into two parts: the left bank (indicating sedimentation) and the right bank (indicating erosion). Measurements showed significant sedimentation increase after construction, with erosion initially rising alongside sedimentation, then stabilizing over time. The largest area change, a 65.69% increase, occurred between June 6 and September 16, 2018 (Figure 6).

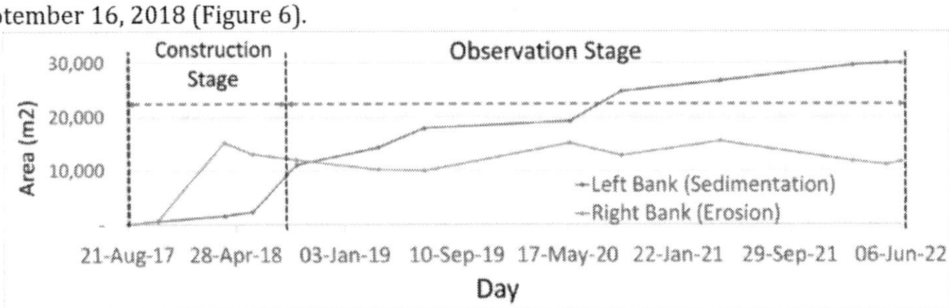

Figure 6. The change in the sedimented and eroded area.

4.2. Windrose diagram for wind direction analysis and numerical input data
The analyzed data includes hourly wind speeds from 2018 and 2019, sourced from ECMWF at coordinates 7°17'24.00" S, 106°04'48.00" E. This data was used for hindcasting to generate wave data [24]. The wind predominantly originated from the Northwest, Southeast, and East, with wind speed visualized in a windrose diagram (Figure 7). Waves from June to November 2018 were primarily driven by northwest winds, from December 2018 to March 2019 by southeast winds, and from April to July 2019 by northwest winds again.

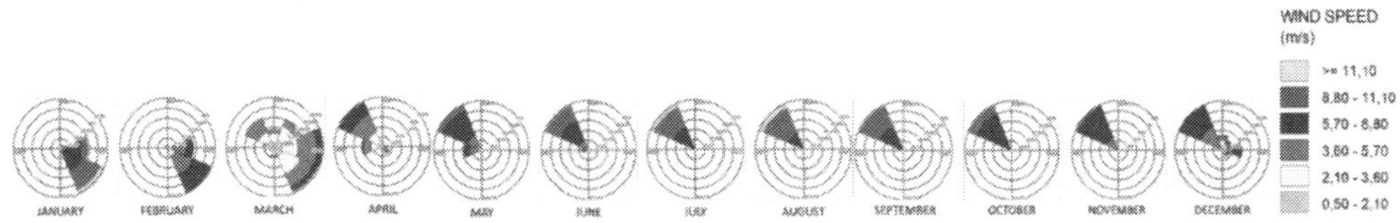

Figure 7. Windrose for the location of study and the wave generated at the face line of the jetty.

4.3. Hydrooceanography analysis

The hydro-oceanographic analysis involves: (1) creating a mesh with a generator, (2) generating waves using a spectral wave module, (3) modeling tides and sedimentation with a flow model for each case, and (4) analyzing results to understand tombolo formation due to the causeway. The model includes three conditions: during causeway construction (first model), with the breakwater built completely with the causeway (second model), and with the breakwater built without the causeway (third model). Validation was performed on tidal data and the first model to ensure accuracy for the second and third models.

The validation results of the tidal modeling show an RMSE value of 0.11, which is close to 0; the closer the RMSE value is to 0, the more the model aligns with real conditions. The RSR value is 0.37, where a range of $0 < RSR \leq 0.5$ indicates a very good model, and the NSE value is 0.86, where an NSE range of 0.7 - 1.0 denotes a very good model. The calibration results of the tidal modeling are shown in Figure 8.

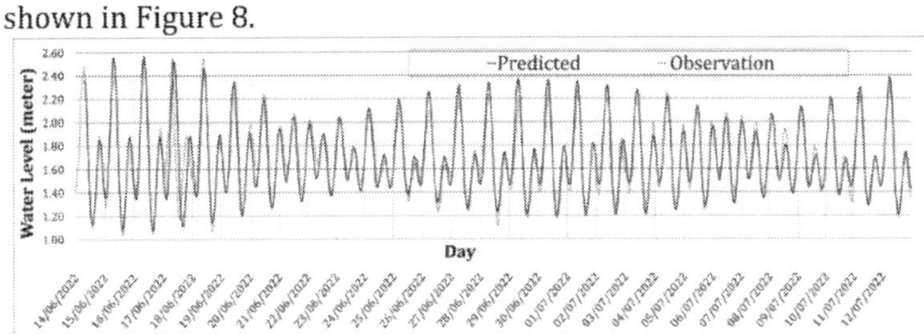

Figure 8. Tidal calibration.

The first condition modeled in this study is when the causeway was under construction from June 2018 to August 2018, when the waves dominantly occurred from the Northwest direction. The activity resulted in the bed level on the northwest side of the location of the study being elevated to 3.42 meters and the southeast side to 3.00 meters. This indicates that the bed level changes caused the Northwest side of the location of study has become shallowed and created a tombolo (Figure 9).

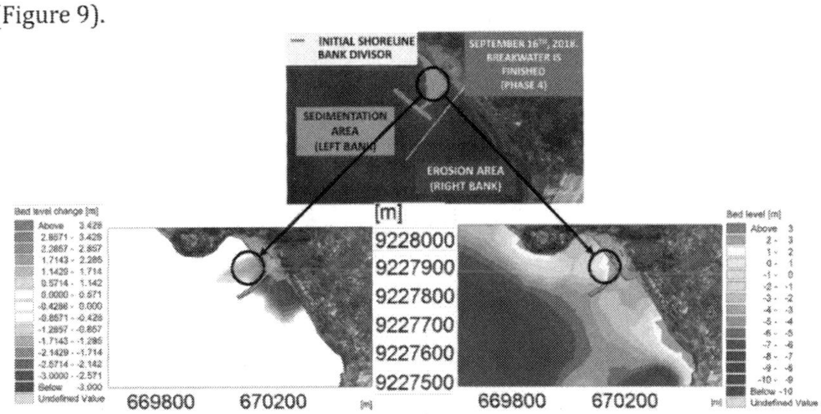

Figure 9. Bed level change under the condition of the built causeway.

The second model in this study represents conditions from September 2018 to March 2019, after the breakwater and causeway were built, with waves mainly from the Southeast. The model showed bed levels rising to 3.43 meters on the Southeast side and dropping up to 3.00 meters on the Southwest side. This suggests that the bed on the Southwest side stabilized and sediment shifted to the Southeast, causing shallowing and the formation of a tombolo. (Figure 10).

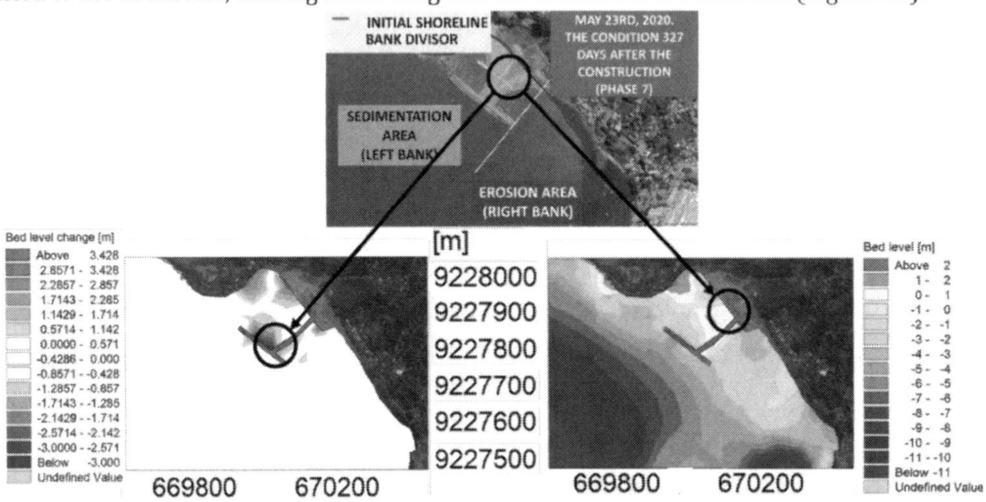

Figure 10. Bed level changes under the condition when the breakwater is built with the causeway.

The 1st third condition that was modeled in this study is a condition when the breakwater was built without the causeway. The model showed that there was no significant change in the bed level of the location of the study. The highest change of the bed level is 0.2067 meters. This showed minimal bed level changes, suggesting that the causeway had a major role in promoting tombolo formation. (Figure 11).

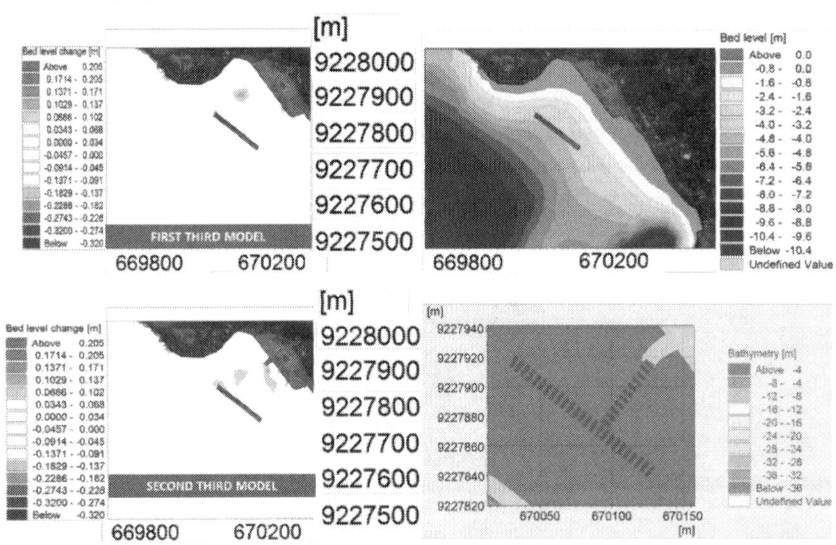

Figure 11. Bed level changes with a breakwater but no causeway or wharf (top). Bed level changes with a breakwater, no causeway, and a wharf (bottom).

Table 1. The change in sedimentation and erosion area.

Date	Sedimentation Area (m^2)	Area Change (%)	Erosion Area (m^2)	Area Change (%)	Remarks
21-Aug-17	-	-	-	-	Initial Condition
30-Oct-17	553	0,00%	598	0,00%	Causeway construction stage
03-Apr-18	1.576	48,05%	15.131	92,40%	Causeway construction stage
06-Jun-18	2.281	18,28%	13.017	7,51%	The Causeway construction stage was finished, and started to build the breakwater
16-Sep-18	11.017	65,69%	11.967	4,20%	The breakwater construction stage is finished, and the trestle construction stage under progress
20-Mar-19	14.285	12,92%	10.163	8,15%	The construction stage of the breakwater, trestle and jetty finished
01-Jul-19	17.859	11,12%	9.962	1,00%	The observation stage after all the construction stages had been done
23-May-20	19.181	3,57%	15.090	20,47%	Observation stage
16-Sep-20	24.742	12,66%	12.763	8,35%	Observation stage
27-Apr-21	26.651	3,71%	15.444	9,50%	Observation stage
17-Feb-22	29.571	5,19%	11.737	13,64%	Observation stage
03-May-22	29.938	0,62%	11.103	2,78%	Observation stage
05-Jun-22	29.977	0,07%	11.599	2,18%	Observation stage

The 2nd third condition model is a breakwater with no causeway and with a wharf, which is a proposed construction method for future projects. Without a causeway and with the presence of a wharf, the bed level changes slightly more significantly but does not have an impact as much as the causeway, which plays a major role in promoting tombolo formation (Figure 12). The construction of a 210-meter-long breakwater, positioned 140 meters from the shoreline, creates a salient under these conditions without forming a tombolo (a Theoretical approach). The spatial analysis reveals a 65.69% increase in sedimentation over approximately three months (June 6 to September 16, 2018) (Table 1), coinciding with the breakwater's completion and ongoing trestle construction. This surge indicates their impact on sediment buildup. Two initial models examined how constructing the breakwater from land using a causeway affected sediment transport and tombolo formation, showing bed level changes of up to 3.4 meters on the left bank and erosion with bed level changes of -3.0 meters on the right (Figure 9 and Figure 10). A third model tested breakwater construction from the ocean, which resulted in minimal bed level changes (up to 0.12 meters) (Figure 11), underscoring the causeway's role in sedimentation and tombolo formation. Coastal sediment transport involves both onshore-offshore and longshore movements. This finding is supported by research that reports port structures, such as those on Dadap Beach in Indramayu Regency, induce wave diffraction and current turbulence, resulting in reduced longshore current velocity. It further underlines the critical role that structure type and placement play in influencing sediment transport and deposition patterns [33].

5. Conclusion

This study examines the sediment transport processes that led to tombolo formation at Palabuhanratu Seaport. Using backtracking analysis and time-series bathymetric data, it explored how construction sequences influenced sediment movement. Spatial analysis showed shoreline changes over time due to construction. The causeway for the breakwater facilitated one tombolo, while the breakwater itself caused another in front of the jetty. Post-construction, sedimentation, and shallow waters persisted until stability was achieved. The first model showed sedimentation on the northwest side of the causeway, forming a tombolo around 3.4-meter sedimented. The second model indicated sediment stabilization on the northwest side and seabed shallowing on the southeast, forming a second tombolo. The third model demonstrated 0.12 meters of sedimentation due to limited barriers. The first and second model shows that the construction method using a causeway did a massive change to the sea bed level compared to the third model where the construction method did not use a causeway. It means that the construction method from the landside to the seaside has given the consequences of the form of tombolo. Meanwhile, the construction method from the seaside to the landside without using a causeway had the advantage of less bed level change that lessened the chance of the form of tombolo. The recommendation for future research should explore the dredging method, the sedimentation and alternate the breakwater designs and placements to reduce sediment buildup and assess tombolo impacts on port operations and marine ecosystems.

Acknowledgments

Deepest gratitude to MIKE DHI for the manuals, guides, and modules. Especially for generously providing credentials to the Student Labkit license for 25024058@mahasiswa.itb.ac.id. The support has been invaluable in enabling the study to be completed. The tools and resources offered by MIKE DHI have significantly contributed to the successful development and simulation of the model. The research team is also indebted to the Water Resources Engineering and Management Laboratory and Water Resources Engineering Research Group, Institut Teknologi Bandung.

References

[1] Saengsupavanich, C., Ariffin, E.H., Yun, L.S. & Pereira, D.A., 2022. Environmental impact of submerged and emerged breakwaters. Heliyon, Dec., doi: 10.1016/j.heliyon.2022.e12626.

[2] Tsai, C.P., Chen, Y.C. & Ko, C.H., 2023. Prediction of Bay-Shaped Shorelines between Detached Breakwaters with Various Gap Spacings. Sustainability (Switzerland), 15(7), doi: 10.3390/su15076218.

[3] Vona, I., Gray, M.W. & Nardin, W., 2020. The impact of submerged breakwaters on sediment distribution along marsh boundaries. Water (Switzerland), 12(4), doi: 10.3390/W12041016.

[4] Johnston, E.L., Hedge, L.H. & Mayer-Pinto, M., 2015. The urgent global need to understand port and harbour ecosystems. CSIRO, doi: 10.1071/MFv66n12_ED.

[5] Chapapría, V.E. & Peris, J.S., 2021. Vulnerability of coastal areas due to infrastructure: The case of Valencia port (Spain). Land (Basel), 10(12), doi: 10.3390/land10121344.

[6] Arini, D., 2016. Shoreline Change Analysis for Aia Bangih Beach Mapping, West Pasaman. Proceeding of the First International Conference on Technology, Innovation and Society, ITP Press, pp. 627–635, doi: 10.21063/ICTIS.2016.1055.

[7] Pasaribu, D., 2023. Building Sustainability Through Port Integration: A Case Study PT Pelabuhan Indonesia. IOP Conference Series: Earth and Environmental Science, Institute of Physics, doi: 10.1088/1755-1315/1188/1/012023.

[8] Cáceres, R.A., Zyserman, J.A. & Perillo, G.M.E., 2016. Analysis of Sedimentation Problems at the Entrance to Mar del Plata Harbor. Journal of Coastal Research, 32(2), pp. 301–314, doi: 10.2112/JCOASTRES-D-14-00056.1.

[9] Noble, R.M., 1978. Coastal Structures' Effects on Shorelines. Coastal Engineering, pp. 2068–2085, doi: 10.1061/9780872621909.127.

[10] Yola et al., 2022. Coastal protection system design at the Indonesian Mangrove Center in Pekalongan, Central Java – Indonesia. IOP Conference Series: Earth and Environmental Science, 1065(1), p. 012062, doi: 10.1088/1755-1315/1065/1/012062.

[11] Elnabwy, M.T. et al., 2024. Harbor Sedimentation Management Using Numerical Modeling and Exploratory Data Analysis. Advances in Civil Engineering, doi: 10.1155/2024/1209460.

[12] Winterwerp, J.C., 2005. Reducing Harbor Siltation. I: Methodology. Journal of Waterway, Port, Coastal, and Ocean Engineering, 131(6), pp. 258–266, doi: 10.1061/(ASCE)0733-950X(2005)131:6(258).
[13] Ananda, C.R., Syamsidik & Fauzi, A., 2020. Analisis Sedimentasi Kolam Pelabuhan Ulee Lheue dengan Menggunakan Piranti Lunak Delft3D. Journal of The Civil Engineering Student, 2(3), pp. 232–238. Available at: https://jim.usk.ac.id/CES/article/view/10327
[14] Ahmadian, A.S., 2016. Chapter 1 - Introduction. In Numerical Models for Submerged Breakwaters. Boston: Butterworth-Heinemann, pp. 1–15. Available at: https://www.sciencedirect.com/science/article/pii/B9780128024133000018
[15] Inman, D.L. & Frautschy, J.D., 1965. Littoral processes and the development of shorelines. In Santa Barbara Specialty Conference. American Society of Civil Engineers, p. 1006. Available at: https://repository.tudelft.nl/file/File_6f971425-c15d-48f4-9c3e-a34425710ae8?preview=1
[16] Sauvage de Saint Marc, M.G. & Vincent, M.G., 1955. Transport littoral formation de filches et de tombolos. In 5th Conference Coastal Engineering, pp. 296–328.
[17] Fairley, I., Davidson, M., Kingston, K., Dolphin, T. & Phillips, R., 2009. Empirical orthogonal function analysis of shoreline changes behind two different designs of detached breakwaters. Coastal Engineering, 56(11–12), pp. 1097–1108, doi: 10.1016/j.coastaleng.2009.08.001.
[18] Walker, J.R., Clark, D. & Pope, J., 1980. A Detached Breakwater System for Beach Protection. Coastal Engineering, pp. 1968–1987, doi: 10.1061/9780872622647.120.
[19] Razak, M.S.A. & Nor, N.A.Z.M., 2018. XBeach Process-Based Modelling of Coastal Morphological Features Near Breakwater. MATEC Web of Conferences, 203, p. 01007, doi: 10.1051/MATECCONF/201820301007.
[20] Sakhaee, F. & Khalili, F., 2021. Sediment pattern & rate of bathymetric changes due to construction of breakwater extension at Nowshahr port. Journal of Ocean Engineering and Science, 6(1), pp. 70–84, doi: 10.1016/j.joes.2020.04.002.
[21] Taredja, K.W. & Nugroho, E.O., 2018. Penanganan Sedimentasi di Pelabuhan Tanjung Api-Api pada Sungai Banyuasin Provinsi Sumatera Selatan. RekaRacana: Jurnal Teknik Sipil, 4(3), p. 117, doi: 10.26760/rekaracana.v4i3.117.
[22] Giovani, C., Damayanti, A. & Susiloningtyas, D., 2018. Coastal Typology of Landform in Pelabuhan Ratu Bay, Sukabumi Regency, Jawa Barat Province. E3S Web of Conferences, 73, p. 04012, doi: 10.1051/e3sconf/20187304012.
[23] Oktariyanto Nugroho, E., Moniaga, B.A., Suciaty, F., Chrysanti, A., Harlan, D. & Syahrial Badri Kusuma, M., 2019. Comparative study on wave hindcasting using wind downscaling data at Bojong Salawe Beach. MATEC Web of Conferences, 270, p. 04024, doi: 10.1051/matecconf/201927004024.
[24] Copernicus Climate Change Service (C3S), 2018. ERA5 hourly data on single levels from 1940 to present. Climate Data Store (CDS), doi: 10.24381/CDS.ADBB2D47.
[25] Immerzeel, W.W. & Droogers, P., 2008. Calibration of a distributed hydrological model based on satellite evapotranspiration. Journal of Hydrology, 349(3–4), pp. 411–424, doi: 10.1016/j.jhydrol.2007.11.017.
[26] Garg, A. et al., 2022. Machine learning models for predicting the compressive strength of concrete containing nano silica. Computers and Concrete, 30(1), pp. 33–42, doi: 10.12989/CAC.2022.30.1.033.
[27] Mediero, L., Garrote, L. & Martín-Carrasco, F.J., 2011. Probabilistic calibration of a distributed hydrological model for flood forecasting. Hydrological Sciences Journal, 56(7), pp. 1129–1149, doi: 10.1080/02626667.2011.610322.
[28] Chai, T. & Draxler, R.R., 2014. Root mean square error (RMSE) or mean absolute error (MAE)? Geoscientific Model Development Discussions, doi: 10.5194/gmdd-7-1525-2014.
[29] Kumar, N., Singh, S.K., Srivastava, P.K. & Narsimlu, B., 2017. SWAT Model calibration and uncertainty analysis for streamflow prediction of the Tons River Basin, India, using Sequential Uncertainty Fitting (SUFI-2) algorithm. Modeling Earth Systems and Environment, 3(1), p. 30, doi: 10.1007/s40808-017-0306-z.
[30] Chen, H., Xu, C.-Y. & Guo, S., 2012. Comparison and evaluation of multiple GCMs, statistical downscaling and hydrological models in the study of climate change impacts on runoff. Journal of Hydrology, 434–435, pp. 36–45, doi: 10.1016/j.jhydrol.2012.02.040.
[31] Setegn, S.G., Dargahi, B., Srinivasan, R. & Melesse, A.M., 2010. Modeling of Sediment Yield From Anjeni-Gauged Watershed, Ethiopia Using SWAT Model1. Journal of the American Water Resources Association, 46(3), pp. 514–526, doi: 10.1111/j.1752-1688.2010.00431.x.
[32] Mizukami, N. et al., 2019. On the choice of calibration metrics for 'high-flow' estimation using hydrologic models. Hydrology and Earth System Sciences, 23(6), pp. 2601–2614, doi: 10.5194/hess-23-2601-2019.
[33] Supriyanto, M.S., Cahyono, M., Oktariyanto Nugroho, E., Monarulhuda Sulaiman, D., 2023. The Effect of Dadap Port Structure in Indramayu on Coastal Morphology. JTS ITB 30. https://doi.org/10.5614/jts.2023.30.1.4

IOP Conf. Series: Earth and Environmental Science **1464** (2025) 012020 doi:10.1088/1755-1315/1464/1/012020

Understanding the sedimentation process in Tanjung Carat, South Sumatra, Indonesia

Ahmad M Firdaus[1*], Rajin Sihombing[2] and Arya P Iwantoro[3,4]

[1] Ocean Engineering Program, Coastal Engineering Research Group, Faculty of Civil and Environmental Engineering (FCEE), Institut Teknologi Bandung, Bandung, Indonesia
[2] Civil Engineering Doctoral Program, Faculty of Civil and Environmental Engineering (FCEE), Institut Teknologi Bandung, Bandung, Indonesia.
[3] Department of Geography, University of Exeter, Exeter, United Kingdom.
[4] Lingkari Institute, Jakarta, Indonesia

*E-mail: ahmad.firdaus@itb.ac.id

Abstract. South Sumatra Province is one of the provinces in Indonesia with high revenue compared to other provinces. However, they have faced a significant challenge with their logistics and transportation efficiency. Their main ports are within the river area which limits the size of the vessels it can accommodate. One of the major ports is Boom Baru Port, which is situated in Palembang, the capital of South Sumatra Province. It lies on the Musi River and has a draft of approximately -6 meters from the lowest water spring (LWS). To address this challenge, plans are underway to develop a new major port in the coastal area. Tanjung Carat has emerged as a promising candidate for port development because of its strategic location at the estuary of major rivers in South Sumatra, which includes the Musi Banyuasin and Musi River. The upstream area from these rivers is heavily utilised by mining and agriculture. However, the geographic location and upstream human activities lead to important challenges due to sedimentation, which will be a major constraint for the port development project. The limited availability of published studies on sedimentation in this area further hinders effective planning for the proposed port development. Additionally, the tidal forces in this estuary are complex. The complexity lies in the presence of tidal asymmetry driven by the interaction between diurnal and semidiurnal tidal components. The estuary's connection to the Bangka Strait adds another layer of complexity to the tidal regime. Therefore, this research aims to improve our understanding of the key factors influencing sedimentation and develop strategies to mitigate its impact on the proposed port development. Delft3D numerical model is employed to simulate and analyse the hydrodynamic processes in the Tanjung Carat Estuary. This analysis provides a comprehensive picture of the main drivers of sediment dynamics in this critical location.

1. Introduction

Tanjung Carat is planned as a location for the New Palembang port, situated in the estuary of the Musi Banyu Asin River. Therefore, sedimentation that mainly occurs in the estuary could be a problem for this plan. The sedimentation or siltation problem occurs in the navigational channel of Palembang Port in Musi River. This river has the same estuary as the Musi Banyu Asin River (see Figure 1). This sedimentation problem in Musi's Navigational Channel has caused the government to plan to build a new port in the Musi River estuary area. The new port plan is expected to increase the capacity of logistics services in the South Sumatra region. However, the planned location of the port is also in an estuary area, so a study is needed to see the sedimentation conditions in the area.

This study aims to understand the sedimentation process in the Musi River delta area, which will be one of the alternative plans for the Palembang New Port. So, it is expected to provide input regarding sediment problems that may arise at the location of the port plan. A hydro-morphodynamic modelling studies have been undertaken by the developer or government during the planning process. However, studies have yet to be conducted by the academic community or other independent entities available in the public domain regarding sedimentation conditions in the Musi River estuary area.

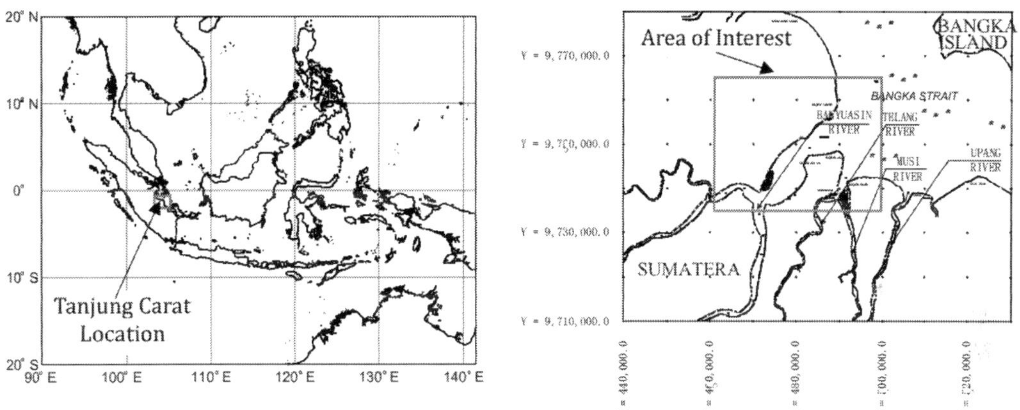

a. Location of Tanjung Carat b. Area of Tanjung Carat

Figure 1. Area of study.

2. Tidal Asymmetry

Sedimentation within estuarine environments presents a compelling area of study for researchers in coastal engineering and geomorphology. Typically, sediment deposition in estuaries is concentrated in the intertidal zones, where the interaction between freshwater inflows from rivers and saline oceanic waters promotes the flocculation of cohesive sediments. This behaviour, however, may differ for non-cohesive sediments, which are less susceptible to flocculation processes. The estuarine setting is generally characterised by lower flow velocities relative to upstream river sections, which encourages sediment deposition.

Residual sediment transport in estuaries is significantly influenced by tidal asymmetry—variations in the duration and strength of ebb and flood tidal currents. The convergence of riverine and tidal flows generates a distinct pattern in both tidal elevations and tidal stream velocities. Tidal asymmetry arises as the tidal wave undergoes deformation while propagating across the continental shelf and into estuarine systems. This study explores the role of tidal asymmetry in sedimentation processes within estuaries. Through theoretical analysis, general principles governing this relationship are derived and subsequently assessed using field observations.

Tidal asymmetry is commonly understood to result from the interaction between the primary semidiurnal tidal constituents and shallow water harmonics, specifically the M4 and M6 components (see [1], [2], [3], [4], and [5]). However, asymmetries can also arise from interactions among the astronomical tidal components. Previous studies by Hoitink et al. [6] and Nidzieko [7] have illustrated that tidal asymmetry may manifest in mixed diurnal tidal regimes. However, their investigations focused on asymmetry in terms of tidal elevation and head differences, which may not fully represent the tidal regime observed in this study.

Globally, 37 primary constituents are recognised as having a significant impact on tidal patterns, serving as the basis for tidal harmonic analysis [8]. Nevertheless, data from DISHIDROS (2004) indicate that only nine constituents exert a meaningful influence on tidal processes within Indonesian waters. These include the four principal constituents (M2, S2, K1 and O1), two semidiurnal lunar-elliptic components (N2 and L2), a diurnal elliptic component (P1), and two shallow-water constituents (M4 and M6). Analyzing tidal asymmetry thus necessitates a detailed examination of interactions among these key components.

In many Indonesian locations, tidal elevations exhibit a predominantly semidiurnal pattern. However, the Tanjung Carat area, located near the Bangka Strait, provides an interesting case for examining sedimentation in relation to tidal asymmetry. Notably, the Bangka Strait experiences a predominantly diurnal tidal regime [9], making it one of the few such regions in Indonesia. This study aims to investigate the asymmetry processes characteristic of diurnal tidal environments, as observed in the Bangka Strait and the nearby Tanjung Carat area.

3. Materials and Methods

3.1 Model description

Hydro-morphodynamic processes, including sedimentation and erosion processes, were simulated using a well-established open-source model, Delft3D (version 4.01.00). Delft3D employs the finite difference method to simulate complex hydro-morphodynamic processes [10]. The model employs a semi-implicit numerical scheme, known as the Alternating Direction Implicit (ADI) scheme, on a staggered grid to solve unsteady shallow water equations, as well as sediment dynamic and morphodynamic equations. A detailed description of the model can be found in Lesser et al. [10].

3.2 Model setup

Figure 2 illustrates the model domain that covers the study area. This includes two rivers that merge into the estuary and debouch in Bangka Strait. The model domain was spatially discretised into rectangular grid cells, with cell sizes (dx) ranging from 9 to 1000 meters in both the x and y directions. Finer grid cells are implemented in the river and estuary regions to capture the detailed morphodynamic processes, while coarser grid cells are used offshore in the Bangka Strait. The temporal discretization (dt) was set at 0.2 minutes to ensure numerical accuracy and stability. Meanwhile, the horizontal eddy viscosity was set to 10 m^2/s. This value was applied since applying a smaller value will cause a numerical instability in the systems with complex coastlines. This is because flow magnitude and direction rapidly change in these locations and applying a larger value will not significantly affect the results [11]. Regarding the bed roughness, a Chézy coefficient was applied with the value of 60 $m^{0.5}/s$ as typically applied in the tide-dominated systems [11], [12], [13].

The initial bathymetry for each grid cell was established using data from field surveys in the river and estuary areas, complemented by secondary data from the Indonesian Hydrographic Data Center/Pushidrosal [14] for the offshore region. Bathymetric data collection was conducted in March and April 2022. A survey vessel, equipped with echo-sounding instruments and GPS, navigated along a pre-planned sounding route prepared in HYDRO Navigation software. The recorded depth data was corrected for tidal fluctuations, transducer depth, bar check corrections, and map datum adjustments to provide accurate depth measurements at each point along the route relative to the Chart Datum Level. The final depth values for each point represented depths during Low Water Springs (LWS). This bathymetric data was then plotted on a track plot map, with coordinates in UTM (Universal Traverse Mercator) at a scale of 1:50,000.

For sediment composition in riverbed and seabed, we consider a single sediment type of cohesive sediment (silt). This assumption is reasonable because, according to our field data collection, the sediment deposit in the study area is predominantly composes of silt, as also typically observed in tide-dominated estuaries and deltas [15], [16]. Based on model calibration, the silt settling velocity (w_s) was set to 0.2 mm/s while the critical bed shear stress for erosion (τ_{ce}) and deposition (τ_{cd}) was set to 0.5 and 0.4 N/m^2, respectively. In this study, a Morfac of 4 was applied over a simulation period of three months, which is acceptable for the tide-influenced systems as in our study area. This is because this value remains below the typical Morfac used in tide-dominated systems [11], [17], [18]. Consequently, the model effectively represents one year of morphodynamic development.

3.3 Boundary conditions

The model includes two types of open boundaries, namely offshore and river boundaries. Offshore boundaries are located at the northern and eastern ends of the offshore area (Figure 2). Tidal boundaries were defined using time series of tidal water levels for January to March 2022 obtained from NAOTIDE, a program that predicts tidal water level at a given time and location based on a global ocean tide model developed using TOPEX/POSEIDON altimeter data [19]. Two river boundaries were defined at the upstream end of the two rivers within the model domain (Figure 2). The river flows at these boundaries were derived from current measurements conducted in March 2022. The flow fluctuates with the spring-neap tidal cycle, as the river discharge is significantly influenced by tidal forces at these locations. Additional suspended sediment concentration (SSC) of silt was also imposed from the northern and southern river boundary. SSC values were defined based on the time-averaged concentration from the field observations at the same points as the current data collection, with the value of 100 mg/l.

4. Results

4.1 Model validation

Model validation constitutes an essential phase in hydrodynamic modeling, serving to verify that model outputs provide an accurate representation of field conditions. In this study, validation was conducted through comparisons at five designated observation points, selected to enable comprehensive analysis of hydrodynamic behavior (refer to Figure 2). Validation focused on two primary outputs from the Delft3D-FLOW model: water surface elevation and current velocity.

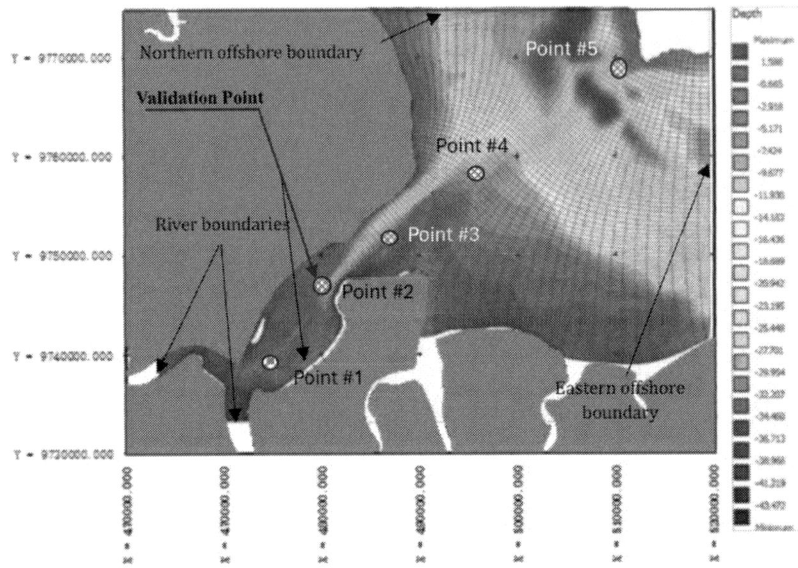

Figure 2. Model mesh and bathymetry with boundary conditions.

Model outputs were systematically compared with survey data, as illustrated in Figure 3 and Figure 4. The validation process demonstrates that the modeled patterns of water surface elevation and current velocity are sufficiently consistent with those observed in the field. Notably, while the general alignment is robust, the model tends to slightly overestimate current velocity magnitude. Specifically, the model predicts an average current velocity of 0.61 m/s, in contrast to the observed field average of 0.52 m/s.

Despite this minor discrepancy in magnitude, the strong concordance in both pattern and average values indicates that the model reliably captures the in-situ conditions relevant to this study. Thus, the Delft3D-FLOW model is considered suitably validated for further analysis within the scope of this work.

4.2 Hydrodynamic of shallow water modelling

Building upon the model validation findings, current velocity emerges as a crucial parameter for interpreting potential sedimentation dynamics within the coastal area. Analysis of the hydrodynamic model outputs indicates that currents in the Tanjung Carat region predominantly flow northeast during neap tides and shift to a southwest orientation during spring tides. Current velocities are observed to vary between 0.0 m/s and 1.2 m/s, with spatial variation in magnitude.

Based on these flow patterns, it can be hypothesised that sediment deposition is most likely in zones where current velocities are minimal, as lower flow speeds promote settling. Conversely, areas experiencing higher current velocities may be predisposed to erosion due to the increased potential for sediment mobilisation. Figure 5 depicts the modelled current conditions during spring and neap tidal phases, respectively, and supports these hypotheses on sediment dynamics. The current patterns from the hydrodynamic model in the estuary area of Tanjung Carat are illustrated in Figure 5.a for the spring tide condition and Figure 5.b for the neap tide condition.

The water surface elevation and velocity magnitude are then extracted in five observation points (see Figure 2 for the location of observation points in the model domain). The water surface elevation (WSE) for all observation points is presented in Figure 6, while the velocity magnitude is shown in Figure 7. Based on the water surface elevation in the observation points, it shown that the WSE all points have a similar pattern of diurnal type (Figure 6a).

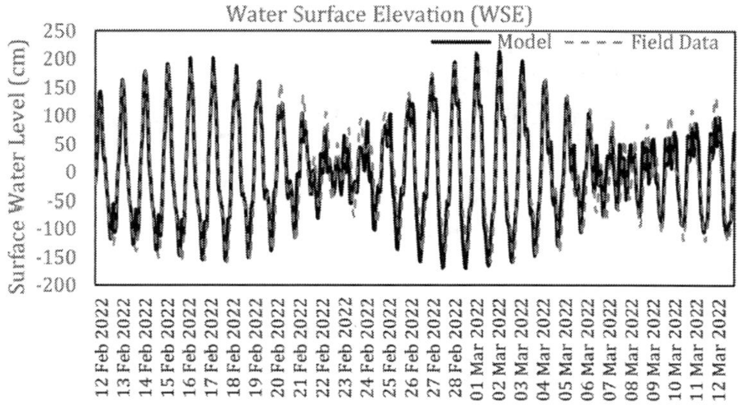

Figure 3. Model validation to the water surface elevation.

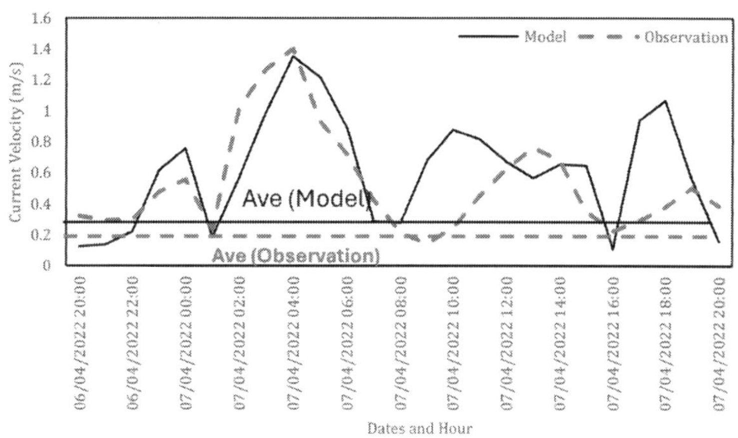

Figure 4. Model validation to the current velocity.

The water level in the long section of the Tanjung Carat estuary shows that the water level varies depending on ebbs and flood conditions. As shown in the water level observation from points #1 to #6, the water level tends to have the same mean sea level (as shown in the average from Figure 6b for obs#2 obs#3 and Average). The water level showed higher in the river channel side (Point#1) than in the sea area (Point#5) on flood conditions (obs#1). Meanwhile, on ebbs condition (obs#4) the river side is lower than in the ocean side. This condition shows that the there are phase differences between the river side and the ocean side. This condition shows that the tidal forcing dominates the movement of the water body in the Tanjung Carat estuary.

The velocity of in the observation points also shows an interesting pattern. The time series of all points are varied (as shown in Figure 7a). The average velocity magnitude is lower when the points is nearer to the seaside (see Figure 7b). Only for obs#2 the velocity of the river side (point 1) is at the same magnitude to the ocean side (point 5). This condition shows that at the ocean side area the velocity magnitude tends to be lower than in the river side area. Therefore, the sedimentation is expected to occurs in the ocean side.

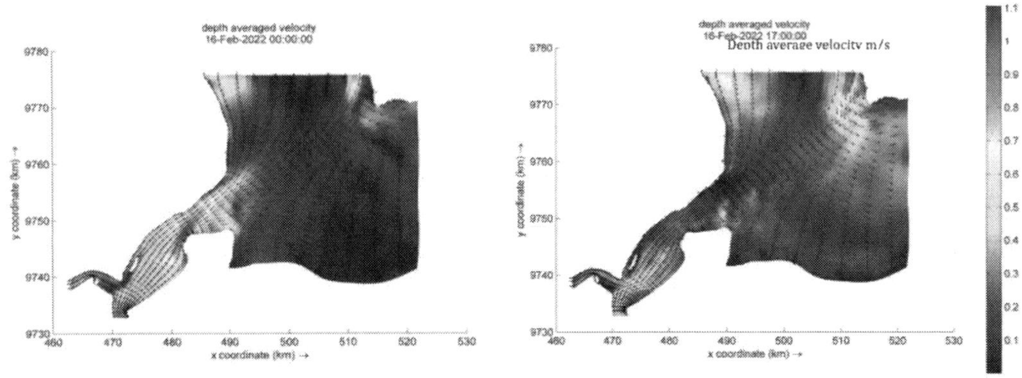

a. Spring tide

b. Neap tide

Figure 5. Illustration on the flow pattern.

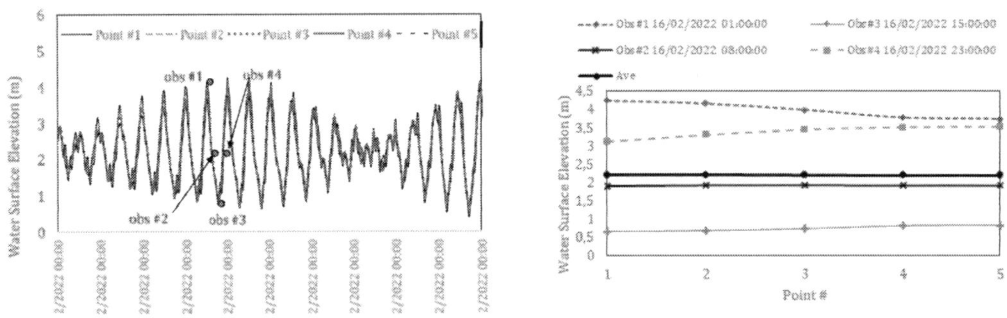

a. Time series of WSE in the observation point

b. Snapshot of WSE in the observation points for the particular time in the spring tide.

Figure 6. Water Surface Elevation (WSE) of the observation points.

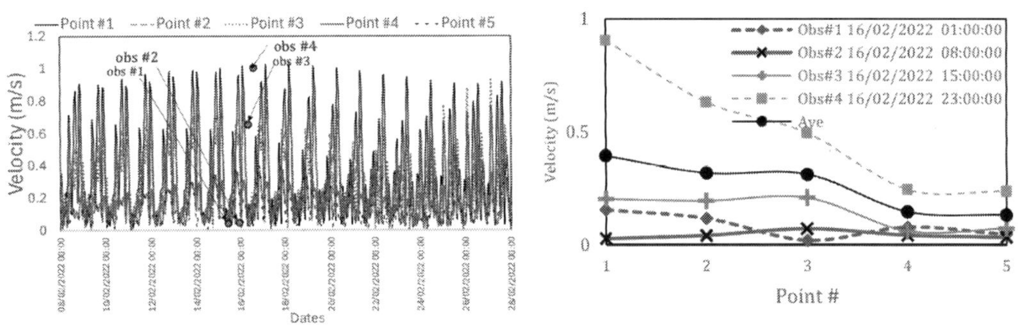

a. Time series of Velocity Magnitude on the observation points.

b. Velocity Magnitude on the Observation points for the particular time in the spring tide.

Figure 7. Water Surface Elevation (WSE) of the observation points.

4.3 Sedimentation

After obtaining results for water surface elevation and current velocity, the analysis can proceed to sedimentation conditions. Sedimentation conditions will be verified by the sedimentation rate around the island in of Tanjung Carat area using Google Earth satellite imagery. Figure 8 and Table 1 show a comparison of the island's area in satellite images from 1985 and 2020, revealing a sedimentation rate of approximately 0.19 meters per year. In the same area, observation points for the model were selected around the island to examine sedimentation conditions, and the model results (Figure 9) indicated a sedimentation rate of about 0.2 meters per year around the island. Furthermore, when considering the current velocity results, this island area has low current speeds with the typical flow below 0.5 m/s, which aligns with the initial hypothesis that areas with low current speeds are prone to sedimentation. These results demonstrate that the sedimentation model used in this study is sufficiently representative to illustrate sedimentation conditions in the Tanjung Carat area.

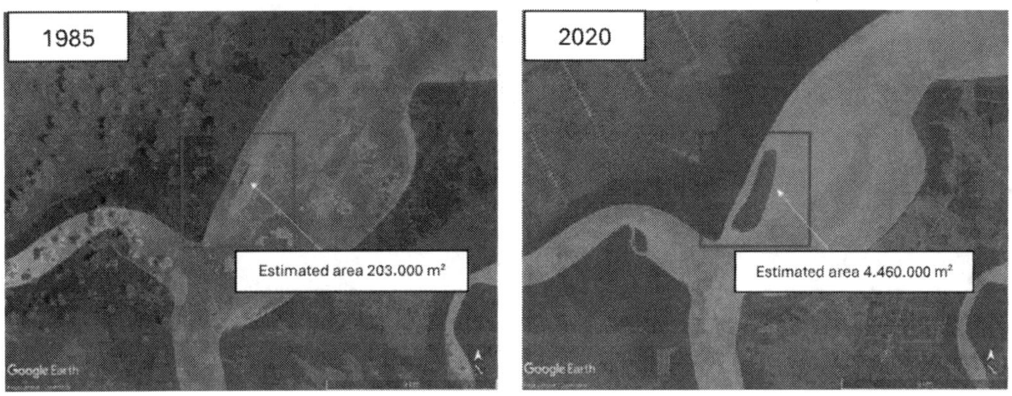

Figure 8. Satellite images comparison near island area.

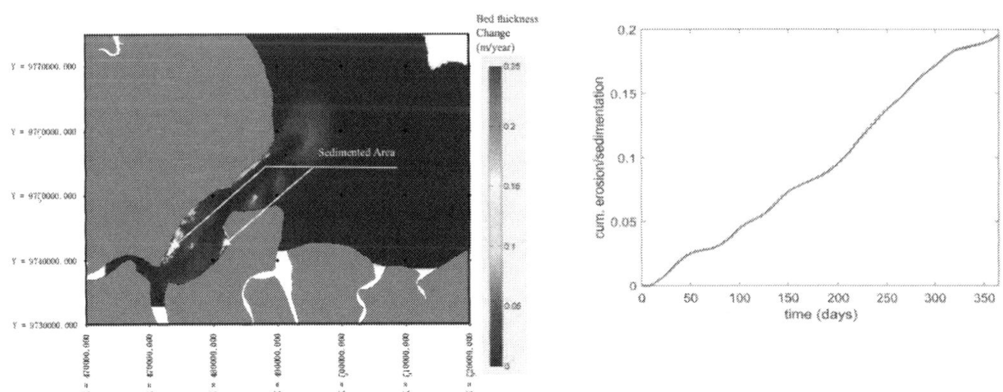

a. Bed level change b. Sediment Rate

Figure 9. Sedimentation model result.

5. Analysis of Tidal Asymmetry and Its Impact to the Sedimentation Process

As mentioned in Section 2, the sedimentation process in the area is correlated to the tidal asymmetry conditions in the area. Therefore, to understand the phenomenon, co-tidal chart analysis is conducted in this study. A co-tidal chart was developed by separating the tidal constituents in the area. Using the Fast Fourier Transform (FFT), the amplitude and phase of each tidal constituent are obtained.

5.1 Co-tidal chart of water elevation

The first analysis of tidal asymmetry is for the tidal elevation. The model results in water surface elevation is extracted in each node. The water level in the time domain is then transformed into the frequency domain. The distribution of the constituent's amplitude is presented in Figure 10. It shows that the principal semidiurnal components (M_2 and S_2) are not the dominant components in this area. The amplitude of M_2 ranges from 0.1 to 0.4 m, and the amplitude of S_2 is only from 0.12 to 0.24. Whilst both diurnal components (K_1 and O_1) have higher amplitude. The amplitude of K_1 is the most dominant of another component, with amplitude ranging from 1.2 to 1.4 m. The O_1 is the second highest constituent with amplitude ranging from 0.55 to 0.75 m, or higher than both of the semidiurnal components.

This condition is also reflected in the amplitudes of over-tides component components of semidiurnal (M_4, M_6 S_4 and MS_4). Those components are only observable in the intertidal area or estuary zone of Tanjung Carat. Theoretically, two shallow-water constituents (M_4 and M_6) are the reflection of the interaction between M_2 and the geometry and bathymetry of the area. These two components are unobservable at the Bangka Strait.

The phase difference between components is presented in Figure 11. It shows that the phase difference between M2 and S2 is small while the diurnal components (K1 and O1) tend to be in phase with each other. At the same time, the overtides components tend to have the same phase to each other as well.

5.2 Co-tidal chart of velocity

The harmonic analysis approach is also applicable to the tidal current. The time variation of velocity can be written in the same manner as water surface elevation (WSE), as shown in eq. (1).

$$v(t) = \sum_{i}^{n} A_i \cos(\omega_i t - \emptyset_i), \qquad (1)$$

where the amplitude (A_i) is the velocity magnitude for each tidal constituent and the angular speed (ω_i) for each constituent is the same as the angular speed of tidal elevation constituents, whilst the phase lag ($\emptyset_i$) of velocity will differ from the phase lag of each constituent in the tidal elevation.

Tidal elevation uses mean sea level (MSL) as a reference, the elevations above MSL are positive and below MSL are negative. In a similar way, tidal velocity also could be distinguished by its direction. As the tidal flow moves bi-directionally, the direction changing of the tidal flow could be considered similar with the rise and fall in the tidal elevation. A typical channel has a tidal flow that moves back and forth in opposite directions, while headland types usually have more complicated patterns since the directions are affected by the geometry of the headland. Therefore, the direction of a channel could be represented as positive and negative in the Cartesian system.

However, currents are more complicated than water elevation in term of analysis, since currents are vector quantities on each component (e.g., for the northing and easting direction). There are many more difficulties with implementing harmonic analysis for the current, such as the influence on the current of non-tidal factors. For instance, Ekman transport (wind-induced current) may interact with the tidal current. The current measurements usually show a greater proportion of non-tidal energy compared to tidal elevation. However, since astronomical forces predominantly drive the ocean current, the analysis for the current still can be done.

A similar approach to the WSE, an a FFT analysis is conducted for the velocity. The co-tidal chart for the velocity magnitude is presented in Figure 12 and the phase is in Figure 13. The velocity magnitude co-tidal chart in Figure 12 shows that the K_1 and O_1 still plays an important role in the tidal velocity. Meanwhile, the S_2 is the only semidiurnal component with significant magnitude, while M_2 is relatively insignificant in the Bangka Strait. Although, both constituents show the significant magnitude in the estuary area of Tanjung Carat. The magnitude of other is also small in this area.

The co-tidal chart for the velocity phase is also shows that the semidiurnal M_2 and S_2 and diurnal (K_1 and O_1) are in approximate of 180° phase differences to each other. The M_2 and S_2 have 180° phase differences to each other as well as K_1 and O_1. These results shows that the sedimentation process in Tanjung Carat estuary might lead to the interaction of diurnal component rather than the semidiurnal.

6. Conclusion, Discussion and Future Work

The results of sedimentation modelling carried out for a period of one year indicate that sedimentation occurs in Tanjung Carat waters with a seabed growth rate of around 0.2 m per year. The simulation and morphological data from Google Earth show that sedimentation and morphological changes occur in this area. The theory discussed in this article argues that sedimentation in the estuary can be identified by tidal asymmetry.

This article studied the characteristics of sedimentation in areas with pronounced diurnal components like Tanjung Carat. As shown in the analysis, the form factor or *formzahl* number in this area ranges from 3.5 to 8.5. This *from all numbers indicates* that this area is diurnal type tidal variation.

Most of study on the sedimentation in the intertidal variation is in the semidiurnal variation. The interaction in the estuary area is mostly due to the interaction of the principal Semidiurnal component (M2) and amplitudes of over-tides component components of semidiurnal (M4, M6 S4 and MS4). Many studies have shown the asymmetry in the area that could lead to the sedimentation process in the area. Meanwhile, the Tanjung Carat area shows the dominant components as the diurnal constituents (K_1 and O_1). Therefore, it requires a deeper understanding of the interaction of K_1 and O_1 and other over-tides components or events with the principal diurnal (M_2 and S_2).

The co-tidal chart that is shown in the simulation result shows that this area is an example of an asymmetry problem. Furthermore, this case is an example of the interaction of strong diurnal components in creating asymmetric. However, further analysis is required to determine the correlation between asymmetry phenomena with the transport sediments and estuary morphology in future research.

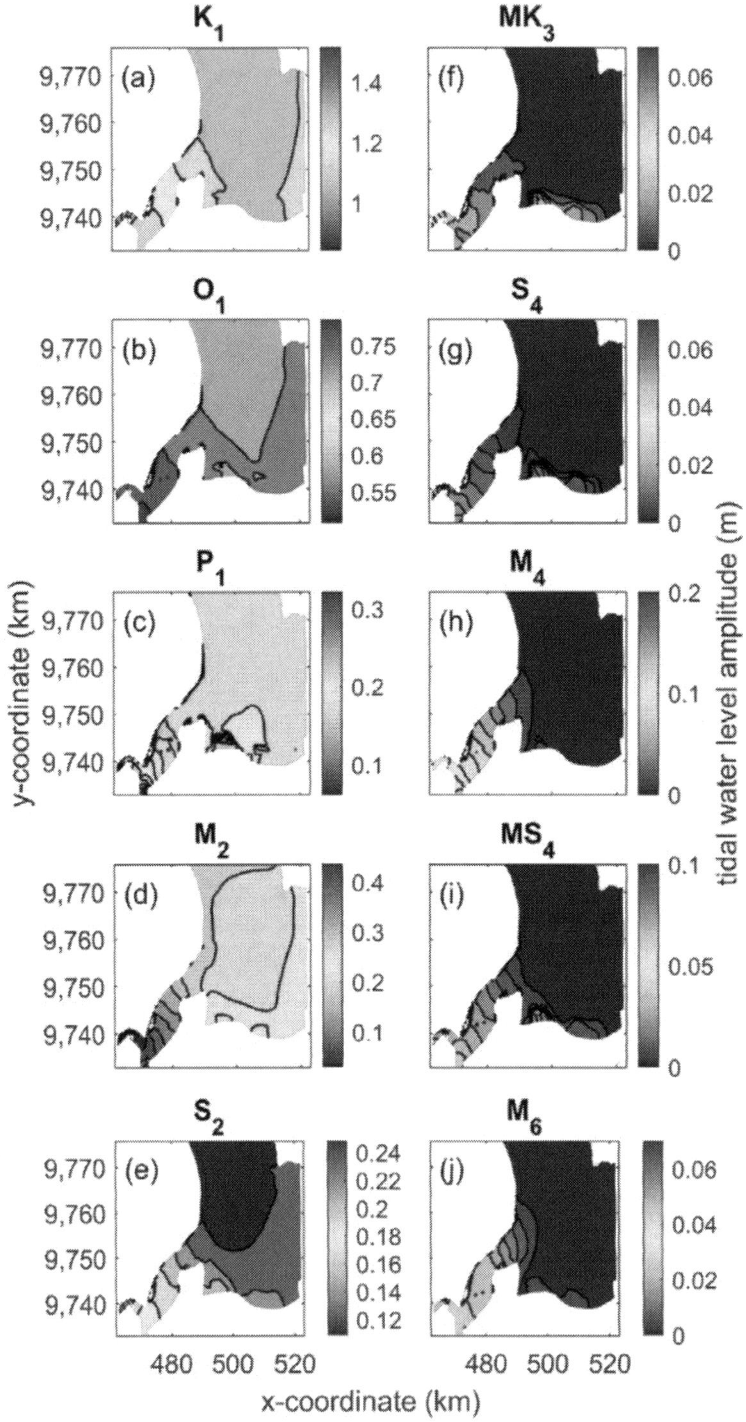

Figure 10. Distribution of water level amplitudes for different tidal components across the study area, including diurnal components (a-c), semidiurnal components (d, e), diurnal-semidurnal overtides (f), and semidiurnal overtides (g-j).

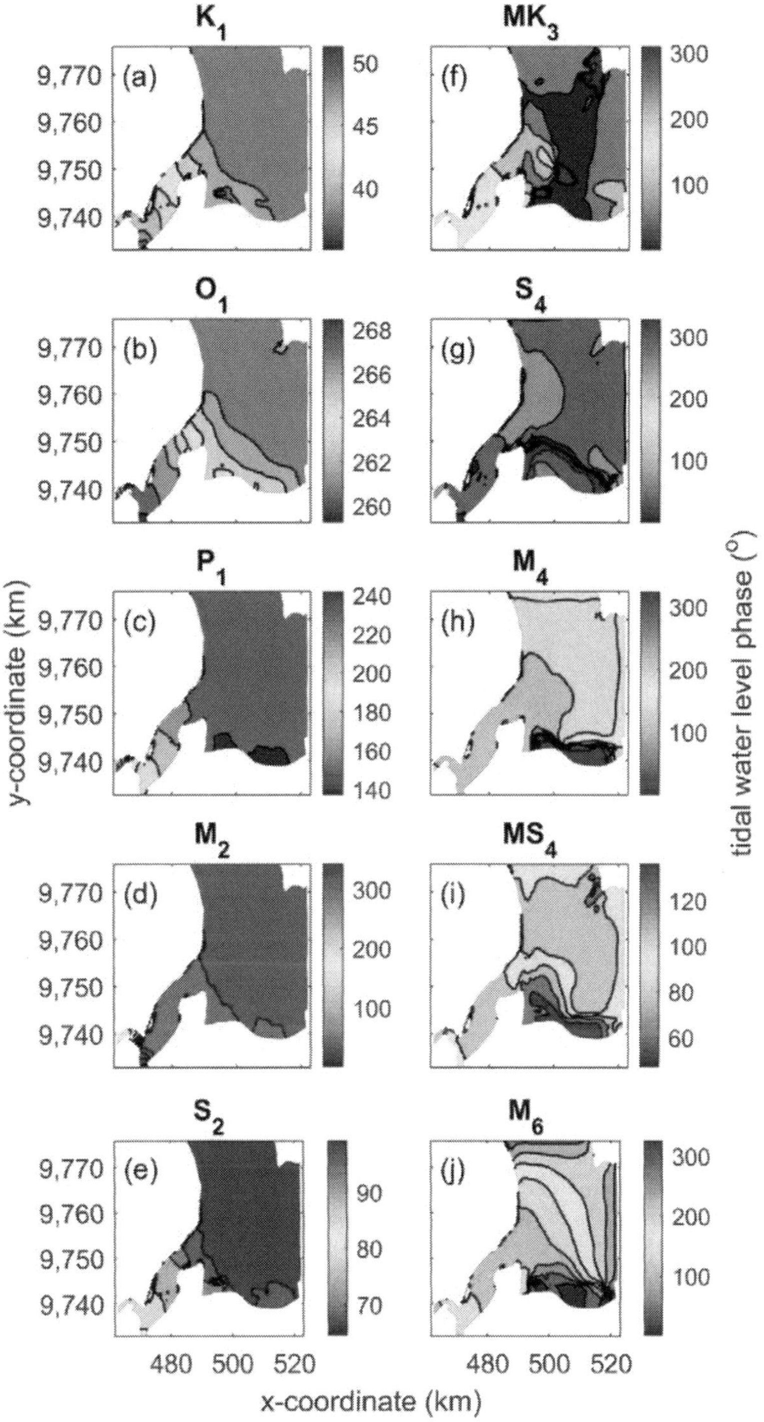

Figure 11. Distribution of water level phases for different tidal components across the study area, including diurnal components (a-c), semidiurnal components (d, e), diurnal-semidurnal overtides (f), and semidiurnal overtides (g-j).

IOP Conf. Series: Earth and Environmental Science **1464** (2025) 012020 doi:10.1088/1755-1315/1464/1/012020

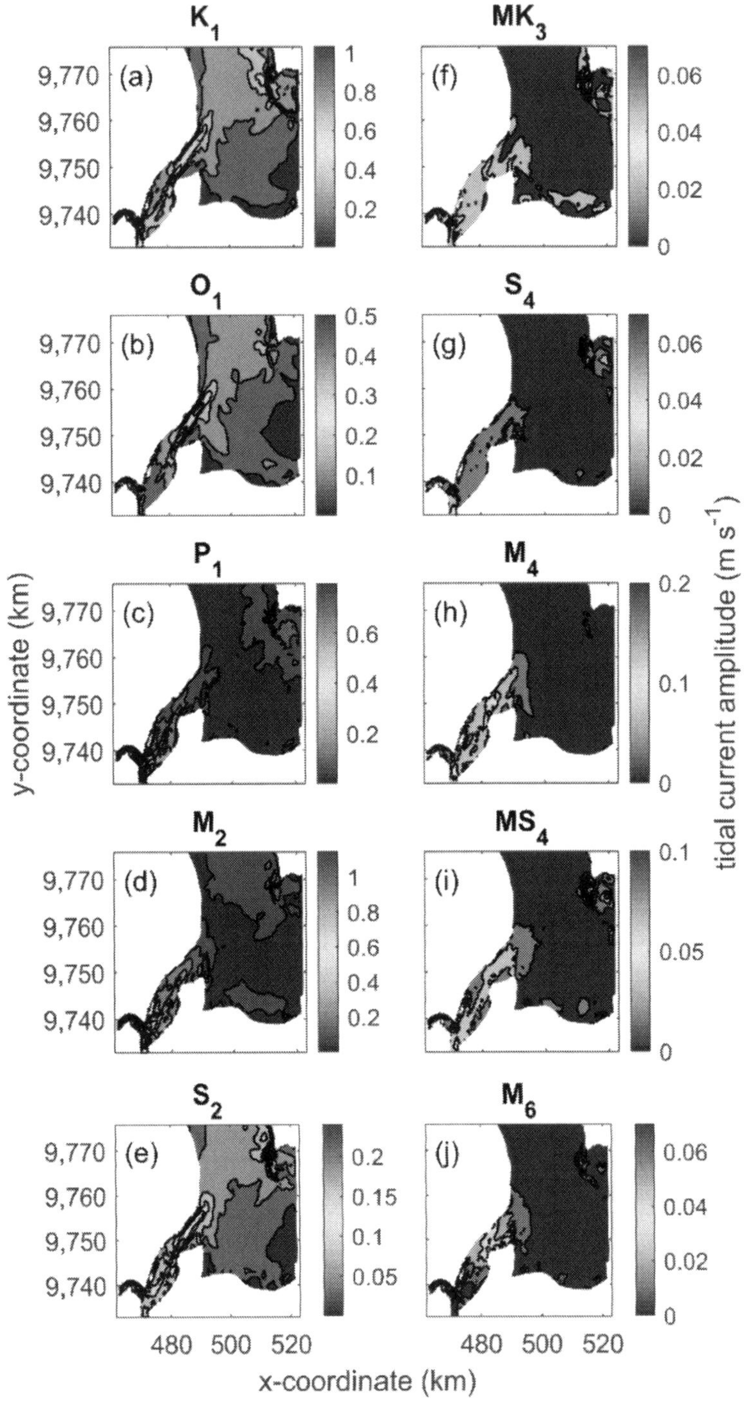

Figure 12. Distribution of tidal current amplitude for different tidal components across the study area, including diurnal components (a-c), semidiurnal components (d, e), diurnal-semidurnal overtides (f), and semidiurnal overtides (g-j).

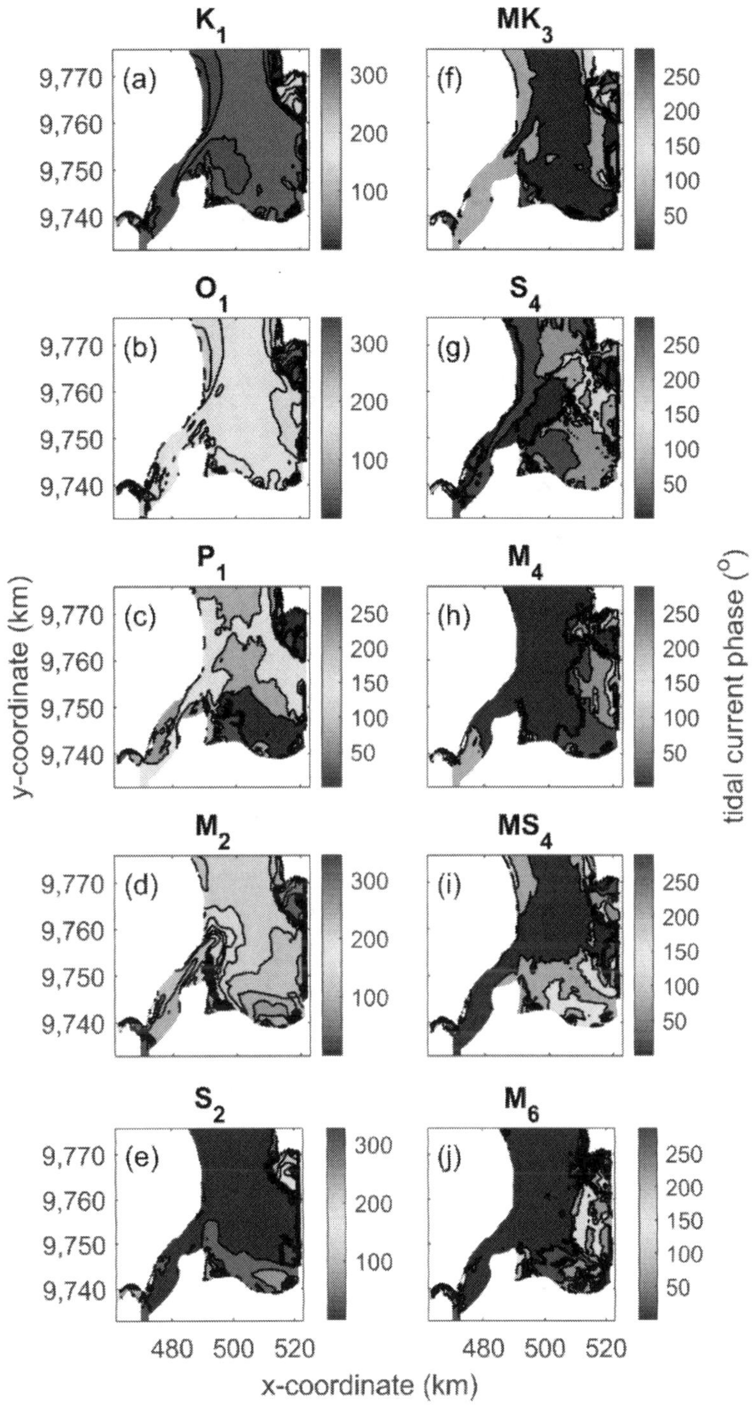

Figure 13. Distribution of tidal current phase for different tidal components across the study area, including diurnal components (a-c), semidiurnal components (d, e), diurnal-semidurnal over-tides (f), and semidiurnal overtides (g-j).

Acknowledgements

This research was funded by the P2MI program of ITB in 2022.

References

[1] J. Dronkers, "Tidal asymmetry and estuarine morphology," Netherlands Journal of Sea Research, vol. 20, no. 2–3, 1986, doi: 10.1016/0077-7579(86)90036-0.

[2] J. O. Blanton, G. Lin, and S. A. Elston, "Tidal current asymmetry in shallow estuaries and tidal creeks," in Continental Shelf Research, 2002. doi: 10.1016/S0278-4343(02)00035-3.

[3] Y. Mazda, N. Kanazawa, and E. Wolanski, "Tidal asymmetry in mangrove creeks," Hydrobiologia, vol. 295, no. 1–3, 1995, doi: 10.1007/BF00029110.

[4] A. S. Chernetsky, H. M. Schuttelaars, and S. A. Talke, "The effect of tidal asymmetry and temporal settling lag on sediment trapping in tidal estuaries," Ocean Dyn, vol. 60, no. 5, 2010, doi: 10.1007/s10236-010-0329-8.

[5] M. E. Scully and C. T. Friedrichs, "Sediment pumping by tidal asymmetry in a partially mixed estuary," J Geophys Res Oceans, vol. 112, no. 7, 2007, doi: 10.1029/2006JC003784.

[6] A. J. F. Hoitink, P. Hoekstra, and D. S. Van Maren, "Flow asymmetry associated with astronomical tides: Implications for the residual transport of sediment," J Geophys Res Oceans, vol. 108, no. 10, 2003, doi: 10.1029/2002jc001539.

[7] N. J. Nidzieko, "Tidal asymmetry in estuaries with mixed semidiurnal/diurnal tides," J Geophys Res Oceans, vol. 115, no. 8, 2010, doi: 10.1029/2009JC005864.

[8] D. Pugh and P. Woodworth, Sea-level science: Understanding tides, surges, tsunamis and mean sea-level changes. 2012. doi: 10.1017/CBO9781139235778.

[9] K. Wyrtki, "Physical Oceanography of the Southeast Asian Waters. Naga Report Volume 2. Scientific Results of Marine Investigation of the South China Sea and the Gulf of Thailand 1959-1961," Scientific Results of Marine Investigation of the South China Sea and the Gulf of Thailand 1959-1961, vol. 2, no. Naga Report, 1961.

[10] G. R. Lesser, J. a. Roelvink, J. a T. M. van Kester, and G. S. Stelling, "Development and validation of a three-dimensional morphological model," Coastal Engineering, vol. 51, no. 8–9, pp. 883–915, 2004, doi: 10.1016/j.coastaleng.2004.07.014.

[11] A. P. Iwantoro, M. Van Der Vegt, and M. G. Kleinhans, "Morphological evolution of bifurcations in tide-influenced deltas," Earth Surface Dynamics, vol. 8, no. 2, pp. 413–429, Jun. 2020, doi: 10.5194/esurf-8-413-2020.

[12] L. Guo, M. van der Wegen, D. J. A. Roelvink, Z. B. Wang, and Q. He, "Long-term, process-based morphodynamic modeling of a fluvio-deltaic system, part I: The role of river discharge," Cont Shelf Res, 2015, doi: 10.1016/j.csr.2015.09.002.

[13] A. P. Iwantoro, M. van der Vegt, and M. G. Kleinhans, "Stability and Asymmetry of Tide-Influenced River Bifurcations," J Geophys Res Earth Surf, vol. 127, no. 6, p. e2021JF006282, 2022, doi: https://doi.org/10.1029/2021JF006282.

[14] IHDC, "Peta Batimetri Pantai Timur Selat Bangka," 2013.

[15] W. E. Galloway, "Process framework for describing the morphologic and stratigraphic evolution of deltaic depositional systems," in Deltas: Models for Exploration, M. L. Broussard, Ed., Houston: Houston Geological Society, 1975, pp. 87–98.

[16] J. T. Wells, "Chapter 6 Tide-dominated estuaries and tidal rivers," Developments in Sedimentology, vol. 53, no. C, 1995, doi: 10.1016/S0070-4571(05)80026-3.

[17] L. Braat, T. Van Kessel, J. R. F. W. Leuven, and M. G. Kleinhans, "Effects of mud supply on large-scale estuary morphology and development over centuries to millennia," Earth Surface Dynamics, vol. 5, no. 4, 2017, doi: 10.5194/esurf-5-617-2017.

[18] D. Xie et al., "Mangrove removal exacerbates estuarine infilling through landscape-scale bio-morphodynamic feedbacks," Nat Commun, vol. 14, no. 1, 2023, doi: 10.1038/s41467-023-42733-1.

[19] K. Matsumoto, T. Takanezawa, and M. Ooe, "Ocean tide models developed by assimilating TOPEX/POSEIDON altimeter data into hydrodynamical model: A global model and a regional model around Japan," J Oceanogr, vol. 56, no. 5, 2000, doi: 10.1023/A:1011157212596.

IOP Conf. Series: Earth and Environmental Science **1464** (2025) 012021 doi:10.1088/1755-1315/1464/1/012021

Study of the seasonal and interannual shoreline changes in Pesaren Beach, North Bangka, Indonesia

Munawir Pratama[1*], Entin A. Karjadi[2], Billy Y. Anugrah[2], and Devi U. Tias[2]

[1] Institute for Energy Systems, School of Engineering, The University of Edinburgh, Edinburgh, Scotland, United Kingdom

[2] Ocean Engineering Department, Faculty of Civil and Environmental Engineering, Institut Teknologi Bandung, Bandung, Indonesia

*E-mail: munawir.pratama@ed.ac.uk

Abstract. Coastal villages at Bangka Island have seen severe erosion that endangers their livelihood and infrastructure, likely to be induced by climate changes and human activities. Only a few studies have attempted to investigate the shoreline dynamics at Bangka Island, making this issue is still poorly understood. This paper presents a case study at Pesaren Beach, located in the Northeast of Bangka Island. The seasonal and interannual shoreline dynamics of Pesaren Beach are investigated by combining numerical hydrodynamic modelling using Delft3D and long-term satellite images and offshore wave analysis. Shoreline dynamics obtained using CoastSat reveal a cyclical pattern influenced by seasonal variations of the wave. During the northeast monsoon period (DJF), the strong and northern wave induces a receding shoreline. Meanwhile, during the west monsoon period (JAS), weaker and southeastern waves form a beach recovery period, advancing the shoreline. Tidal currents are less likely to contribute significantly to this seasonality, with spring peak flow speed < 0.5 m/s. The yearly fluctuations of shoreline position appear to be attributed to the interannual variability of wave climate. These results provide valuable insights into Pesaren Beach and lend a baseline understanding of other embayed sandy beaches on Bangka Island. Furthermore, this study demonstrates the potential of using publicly available data and tools to investigate shoreline dynamics in data-scarce regions like Indonesia. Built on this study and supported by field measurement, the potential impact of offshore tin mining on the beach could be explored.

1. Introduction

Currently, coastal areas around the world are under various pressures caused by the rapid development of residential, industry, fisheries, tourism, and resource mining. In Bangka Island, following the coastal zonation plan 2020-2040 [1], additional pressures are expected due to the substantial region allocated for offshore tin mining activities. Previous studies have addressed the potential negative influence of tin mining on the physical environment [2], economic [3], and social aspects [4]. However, more studies are required to provide a baseline understanding of human activities' impacts in coastal areas and to assess the policy needed for maintaining the sustainability of the coastal environments.

The mining of offshore material is reckoned to have an impact on the environment physically and biologically. Some of the impacts are damage to the seafloor, inducing suspended sediment plumes, and also impacting the prevailing current and wave conditions [5]. In addition, it also creates additional pressures, promoting siltation, substances release, and noise. They are leading to multiple impacts, such as disturbance to the ecosystem, changes in the dynamic's morphology equilibrium, and influence on the primary production of organisms [6]. In Bangka Island, prior work has found that suspended sediment induced by tin mining may be deposited at the conservation area [2]. This mechanism is expected to be responsible for resulting in high contaminations of heavy metals in adjacent waters, as reported in a field study [7]. Other studies also underline that based on satellite-derived shorelines, the majority of the east coast of Bangka Island experiences erosion [8].

One of the impacts being concerned by the local people is the coastal erosion, that could be exacerbated by the offshore tin mining activities. In this study, the intra- and inter-annual variability of shoreline dynamics at Pesaren Beach is investigated. This study adopts a scoping approach, employing open-source data and tools to establish a preliminary understanding and formulate hypotheses. By integrating satellite-derived shoreline delineation (CoastSat), ERA5 wave data analysis, and coastal hydrodynamic modelling (Delft3D), this research will present a baseline understanding of coastal processes at Pesaren Beach, which is located at the north-western tip of Bangka Island.

2. Method

2.1 Study area

Pesaren Beach, the shoreline of interest in this study, is located at 105.850 East, 1.521 South, at the North-Eastern tip of Bangka Island as shown in Figure 1. This site is within the Karimata Strait, which connects the Natuna Sea and the Java Sea. It is characterized by sandy beaches, which have relatively fewer shoreline features (i.e. rocks and artificial coastal defenses), which provides an ideal location to investigate nature-driven shoreline dynamics. In the recent zonation plan [1], nearshore and offshore areas to the North and Northwest of Pesaren Beach are permitted for offshore tin mining activities.

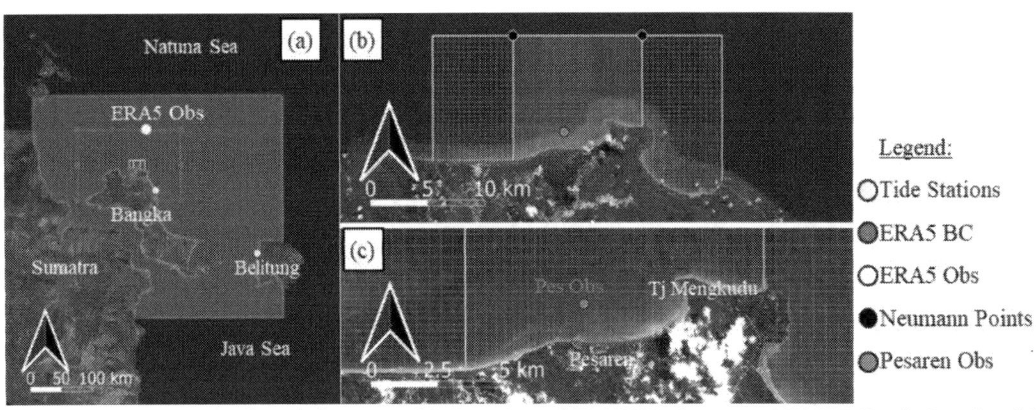

Figure 1. Description of study location. (a) Domains of FLOW Karimata and WAVE North Bangka, (b) Domains of FLOW Pesaren and WAVE Pesaren, and (c) Shoreline at Pesaren Beach.

2.2 CoastSat satellite-derived shoreline

CoastSat [9], a technique that become more widely used, is capable of automatically generating shorelines from publicly available satellite images (Landsat and Sentinel-2). CoastSat handles satellite image retrieval, image sub-setting, image pre-processing, delineation of shorelines based on the MNDWI, pixel classification (into sand, water, and shorelines), tidal correction, and basic analysis. It has shown a promising performance when compared to field measurements at sandy beaches around the world [10]. The tool is Python-based and equipped with user-friendly guidelines for setting the Python and using the script [9]. In this study, the duration of interest is from 1980 to 2022 at 3 km long Pesaren Beach, as shown in Figure 2. Several important settings used are the cloud threshold (0.01), cloud distance (300), pan_off (true), sand colour (default), and beach slope (0.1). For tidal correction, hourly tidal elevation from the NaoTide is used.

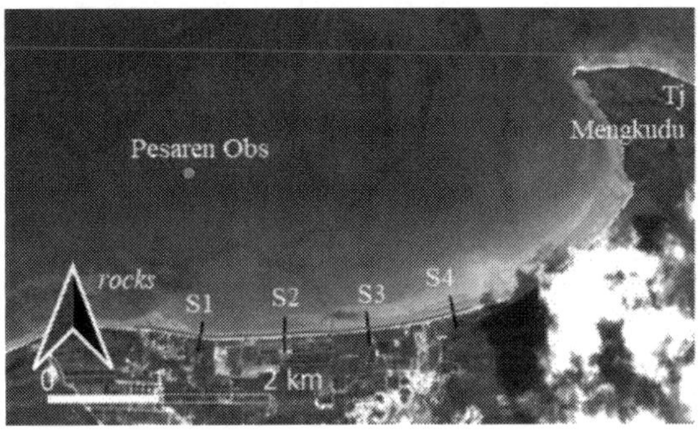

Figure 2. Shoreline of interest (indicated with red line) for CoastSat analysis and four cross-sections of interest, S1 to S4 (indicated with black lines).

2.3 *ERA5 wave data*

Wave climate at Pesaren Beach is characterized by analysing the hourly ERA5 wave data [11] at an offshore site (106 E, 1 S) from 1993 to 2022, a period of 30 years. This site is named ERA5 Obs, denoted as a white dot in Figure 1(a). Since the ERA5 wave dataset is coarse with a distance between points of 0.5 degrees, it is important to use a point at the offshore site in the characterization. The variables used are the significant wave height of combined wind and swell (*Hs*), the mean wave period (*Tm*), and the mean wave direction (*Dir*).

2.4 *Delft3D modelling*

2.4.1 *Model description*

Tidal and wave conditions at Pesaren Beach are described using numerical modelling. The model is built using the (structured-grid) Delft3D-4 developed by Deltares. Delft3D-FLOW [12] and Delft3D-WAVE [13] are used to construct the flow and wave simulations, respectively. Note that the model is not yet coupled in the present work. Since the study is still at a preliminary description, a standalone model is justifiable. For more details on the numerical tools, readers are referred to software manuals [14,15]. These tools have been proven in many cases, including in modelling ocean and coastal hydrodynamics [16,17] and morphodynamics [18]. The advantage of Delft3D-4 is its availability as an open-source, user-friendly interface and many functionalities that have been demonstrated successfully, such as Quickplot and Delft Dashboard [19].

2.4.2 *Grids and bathymetry*

The model is set up using the rectilinear grid, built in the WGS84 coordinate system (in longitude and latitude). The details of the four models (that are FLOW Karimata, FLOW Pesaren, WAVE North Bangka, and WAVE Pesaren) in Figure 1 are presented in Table 1. The FLOW model has two domains, the Karimata and the Pesaren, each of which is the parent and child model. The

IOP Publishing
doi:10.1088/1755-1315/1464/1/012021

WAVE model also has a pair of parent-child models, the North Bangka and the Pesaren. FLOW Karimata is designed to replicate the tide regime around Bangka Island. WAVE North Bangka is sufficiently large to accommodate the nearest ERA5 points to be used as wave boundaries. The description of hydrodynamics and waves at Pesaren Beach will be carried out in FLOW Pesaren and WAVE Pesaren.

In Figure 1, the FLOW and WAVE models are inside the green and blue lines, respectively. WAVE Pesaren has an offshore extent identical to FLOW Pesaren, as shown in Figure 1(b). However, WAVE Pesaren has a wider alongshore extent to anticipate any discrepancies near the lateral boundary. GEBCO_2023 dataset [20] is interpolated to the grid to build the bathymetry for the four domains. The nearshore contour of Pesaren Beach is adapted and idealized from the Navionics Webmaps. The offshore shelf at Pesaren Beach sit at 20-30 m depth.

Table 1. Details on domains of FLOW and WAVE models.

Domains	Type	Resolution	M x N	Boundary
FLOW Karimata	FLOW Parent	1/2800 m	146 x 125	TPXO 8.0
FLOW Pesaren	FLOW Child	1/60 m	128 x 116	FLOW Parent
WAVE North Bangka	WAVE Parent	1/2000 m	82 x 55	ERA5 Wave
WAVE Pesaren	WAVE Child	1/240 m	72 x 40	WAVE Parent

2.4.3 Forcing and other parameters

The tidal boundaries of FLOW Karimata are indicated by green lines in Figure 1. Tidal elevation harmonics generated from the TPXO 8 with 13 constituents [21] are imposed on these boundaries, using Delft Dashboard software [19]. The outputs of FLOW Karimata at two Neumann points, denoted as black dots in Figure 1(b), are fed to the FLOW Pesaren model. Note that the nesting is offline, which means the parent and child models are run separately. With a smaller domain, the FLOW Pesaren use the combination of time-series water levels at offshore boundary and two lateral Neumann boundaries [22]. This configuration avoids unstable jets forming near boundaries, which is often observed when the model uses the time-series water level or current velocity boundaries generated using the NestHD feature. In FLOW Pesaren, the time step is 1 minute, the spin-up period is 4 days, using 0.02 manning coefficient, and other default settings.

ERA5 Wave Dataset [11] at 8 points are prescribed as boundary conditions for WAVE North Bangka. Nesting to WAVE Pesaren is automatic, as online nesting has been implemented. The ERA5 variables used are the significant wave height of combined wind and swell (*Hs*), mean wave period (*Tm*), and mean wave direction (*Dir*). The model uses the SWAN third-generation with the three dissipation parameterisations (depth-induced breaking, bottom friction, and white-capping) activated with default values. Refraction, frequency shifts, and triads and quadruplets are also activated. Since the WAVE domain is relatively small, only local wave propagation is being considered, and the wind and pressure fields are yet to be included. Short-term simulations of one to two months were conducted. Based on available water level observations, both the FLOW and WAVE models were run for June 2022, with an additional WAVE model simulation for December 2022.

2.4.4 Validation

Two tide stations maintained by the Badan Informasi Geospatial – BIG [23] are used to validate the FLOW models. Sungai Liat Station (at Bangka Island) and Belitung Station (at Belitung Island) are shown as yellow dots in Figure 1(a). The result of water level validation is given in Figure 3. Visually, the model produces a good agreement against the measured water level. The water level amplitude and phase in the model and measurement match nicely. The error indexes (which are the correlation coefficient (r) and the root-mean-squared-error (RMSE) [24]) also show a reasonable result.

The stations do not cover the whole tide phase, where the low tides are not fully captured. When the low tide is removed from the RMSE calculation, a more satisfactory validation is achieved, indicated by the bolded RMSE values. At Sungai Liat station, prediction data from the BIG is appended to the comparison to ensure validation quality. The relatively low r-value at Sungai Liat between the model (D3D) and the measured data is due to the location of the monitoring station, which is within semi-enclosed water and is affected either by the dry measurement during low tide or the influence of river processes. Regarding the wave model, no wave monitoring is available; thus, no validation was carried out. A sensitivity analysis of the wave grid resolution shows that the present wave model setup is a stable simulation.

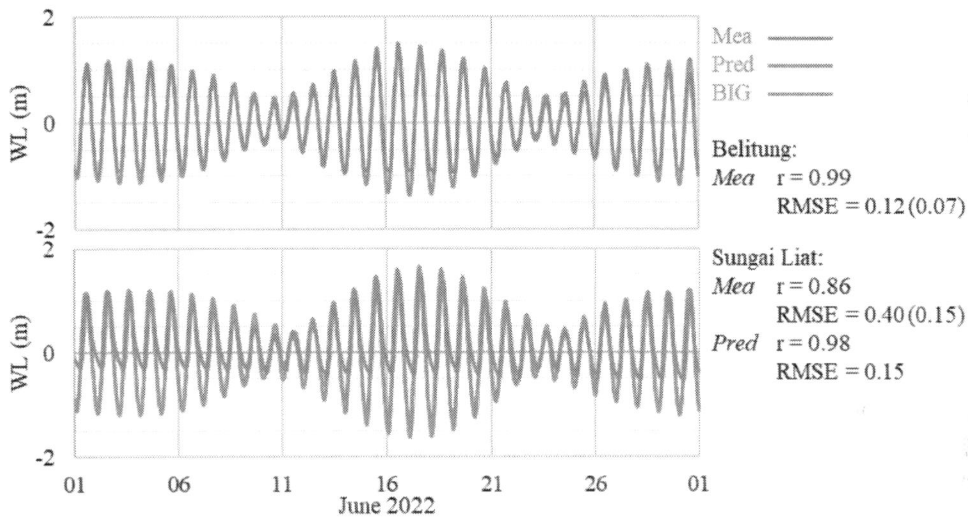

Figure 3. Comparison between the modelled (D3D = Delft3D), predicted (Pred), and measured (Mea) water level at Tanjung Pandan, Belitung and Sungai Liat, Bangka.

3. Results

3.1 Shoreline dynamics

From 1980 to 2022, there are a total of 1099 images from satellite Landsat 5, 7, 8, 9, and Sentinel-2. However, only 104 images (< 10%) show a clear and undisturbed shoreline. The disturbances encountered are sand colour, missing pixels, cloud cover, and cloud shadow, which are common issues when using satellite data. In addition, the imagery time is very early in the morning, which sometimes blurs the limit between sand and water. CoastSat application, in this case, also saw that the majority of the available figures are from December to April. June and July are the periods with the least images. This trend is highly likely similar at other locations and is expected to be influenced by the monsoon atmospheric characterises (i.e., rain and cloud cover among the variables).

The time series of shoreline position (W (m)) at S1 resulting from the CoastSat is presented in Figure 4. The data is scattered, most often available around December to February (DJF). Many years, particularly before 2013, are seen as not having enough data points that are suitable for analysis. From 2013 to 2022, the density of data points increased due to the availability of Sentinel-2 images. The shoreline positions during this period reveal a cyclic fluctuation with evident interannual variability. The observed cycle of shoreline dynamics is the maximum shoreline recession that occurred at DJF while the shoreline advanced from July to September (JAS). The data shown in Figure 4 is based on results in one section, S1 (indicated in Figure 2). Results in other sections produce a minor difference from S1 but have an identical trend. Ideally, this finding is verified against shoreline field measurement [10], however, no such data is available in the study area. Nonetheless, this finding is deemed acceptable, supported by reports from a relevant study [8].

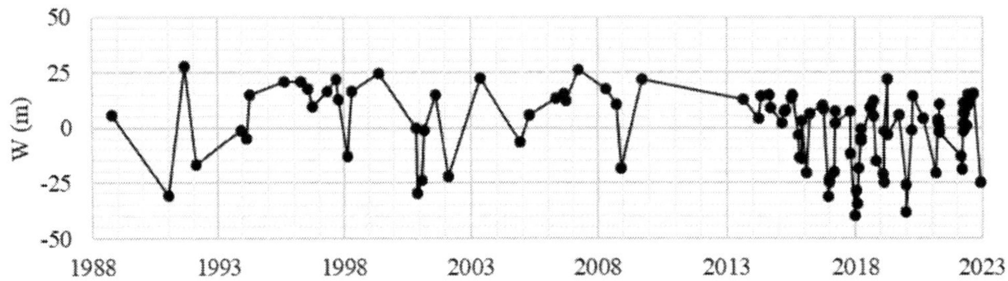

Figure 4. Shoreline position dynamics resulted from CoastSat at S1 from 1988 to 2022.

3.2 *Wave climates – long-term*

The long-term wave climates at Pesaren Beach are characterized by analysing the 30 years of ERA5 data at ERA5 Obs, denoted as a white dot in Figure 1(a). Statistical analysis of the data shows that the maximum and average *Hs* are 2.22 m and 0.51 m, respectively. For the *Tm*, the maximum and average are 3.98 s and 9.22 s. The wave intra-annual seasonality is affected by the Australian-Indonesian monsoons. Based on this ERA5 data, wave seasonality may be described as Northwest Monsoon from December to February, March to June being the 1st transition, East Monsoon from July to September, and October to November being the 2nd transition. During the dry season/east monsoon, in JAS (July, August, September), the east monsoon takes place, thus resulting in a dominant wave from the ESE and SE quadrants. During the peak of the rainy season or northwest monsoon, in DJF (December, January, February), the prevailing waves are from NNW to NNE quadrants, a characteristic of the northwest monsoon. This is summarized in Figure 5, which shows the wave roses for all months, east monsoon, and northwest monsoon periods.

When observing the data as a time series, an inter-annual variability is also seen. Figure 6 displays the time series of the monthly mean *Hs* and the Oceanic Nino Index (ONI) from 2013 to 2022. The ONI is obtained from [25], where ONI > 0.5 means a warm period (indicating *el-nino*) and ONI < -0.5 means a cool period (indicating *la-nina*). From Figure 6, wave-monsoon seasonality is consistent in the background, where the strongest season is DJF (northwest monsoon), and the second strongest is from JAS (east monsoon). Waves from April to June and October to November are in a transition period; thus, the *Hs* is lower. Further, the strength of both monsoons is seen as influenced by the ONI. Most notably, during the strong *el-nino*, the difference between waves in JAS and the consecutive DJF is relatively small. The two highest DJF waves, in 2014 and 2021, occur within a *la-nina* period. Also, if JAS months are exposed to the *la-nina* period, the waves tend to be calmer. While not entirely consistent, the trends may explain the wave climate at Pesaren and the ONI index.

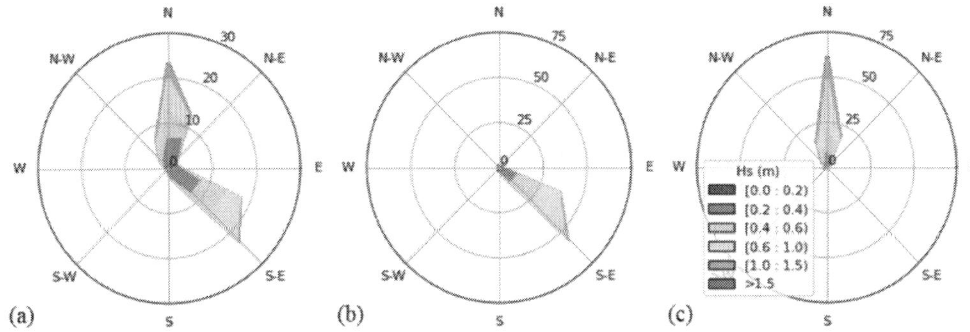

Figure 5. Wave rose diagrams for significant wave height (combined swell and wind) at ERA5 Obs (60 km from P) from 1993 to 2022: (a) Total, (b) during July, August, and September, (c) during December, January, and February.

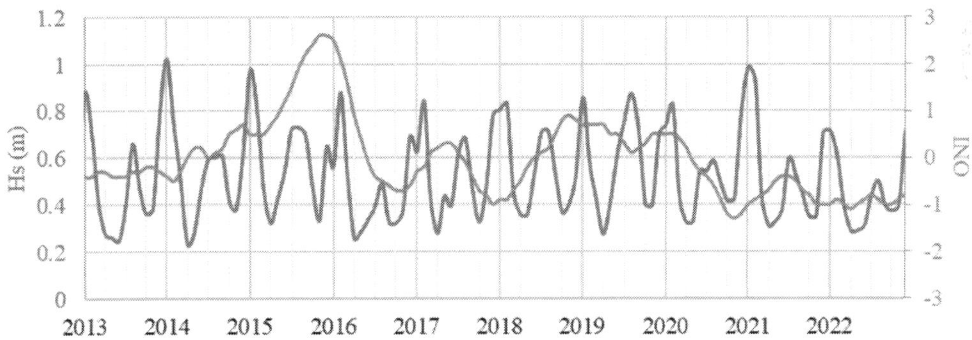

Figure 6. Time series diagram of monthly-mean significant wave height (combined swell and wind) in red and the Oceanic Niño Index (ONI) in blue.

3.3 Tidal characteristics

By conducting harmonic analysis on the water level data at the West Neumann Point (shown in Figure 1), it is established that the tidal regime is diurnal. The most dominant constituent is K1, and later followed with P1, O1, and M2. The interplay between three diurnal harmonics (K1, P1, and O1) results in a large difference in tidal range at the neap and spring phases. The lowest neap tidal range is below 20 cm. While the largest spring tidal range is around 300 cm.

Figure 7 displays the tidal curves at spring tide condition, June 16th, 2022, data are obtained from Pesaren Obs shown as a purple dot in Figure 1. The depth-averaged current speed shown in Figure 7 can be positive or negative, which indicates eastward or westward-directed current, respectively. The eastward ebbing current starts 1 h before High Water, reaches the peak at MSL and switches to the westward flooding current exactly at Low Water.

Flow patterns at each of the peak ebb and flood conditions are shown in Figure 8. As also mentioned in Figure 7, regionally, the peak flow speed in ebb is higher than in flood, as also shown in Figure 8. A strong rocky point and island situated east of Pesaren Beach, named Tanjung Mengkudu and Mengkudu Island, induce a significant difference between both conditions. At both conditions, tidal flow is contracting, thus amplifying around this area. The significant difference is in flood condition, in Figure 8(b), some part of Pesaren Beach at the lee-side of Tanjung

Mengkudu is within a low-velocity zone, which will further be occupied by a headland-induced eddy. The contrasting flow pattern in ebb and flood conditions will result in a more complex residual formation and, thus a sediment transport mechanism within the area.

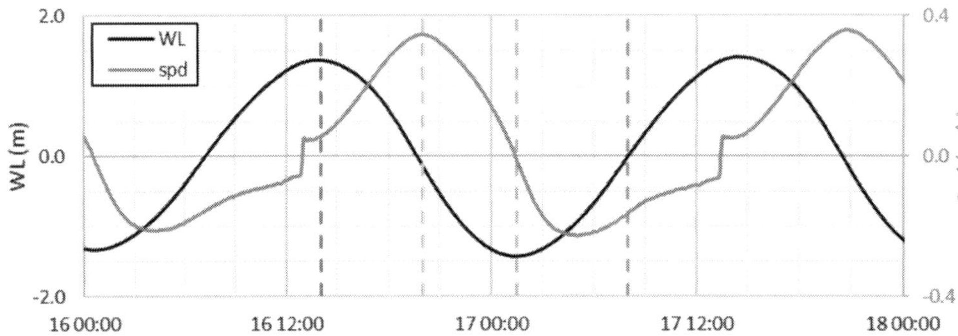

Figure 7. Tidal curves of spring tide's water level (in black) and depth-averaged current speed (in purple). The dotted vertical line indicated the high-water, mid-water at ebb, low-water, and mid-water

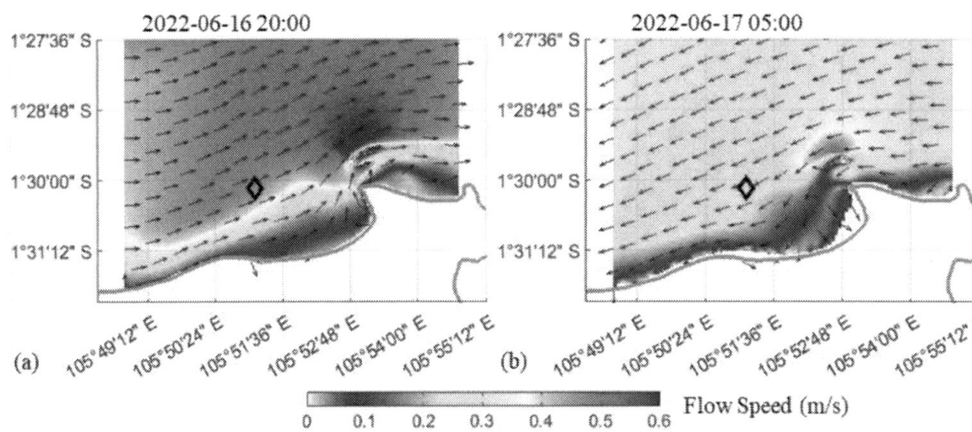

Figure 8. Flow field around Pesaren Beach at (a) ebb and (b) flood phases.

3.4 *Wave characteristics – short term*

The wave seasonality is further described in Figure 9. Wave parameters obtained from ERA5 Dataset at 106 E, 1 S (at ERA5 Obs, white dot in Figure 1(a)) and WAVE Pesaren at 105.852 E, 1.510 S (at Pesaren Obs, purple dot in Figure 1(c)). Each observation point is 60 km and 1.4 km away from Pesaren Beach, respectively. Waves in December 2022 were seen from ERA5 Obs to Pesaren Obs, which shows that the wave height is reduced due to the wave dissipations, while the Dir is in a similar trend. However, in June 2022, the *Hs* and *Dir* changed more significantly when comparing ERA5 Obs to Pesaren Obs. ERA5 Obs indicates the waves approach from 90-180 degrees in June 2022 (also in the majority of east monsoon). Considering the shoreline features around Pesaren Beach, waves arriving at Pesaren Beach will have experienced a strong disturbance, including in *Hs* and *Dir,* as displayed in Figures 9(a) and 9(b).

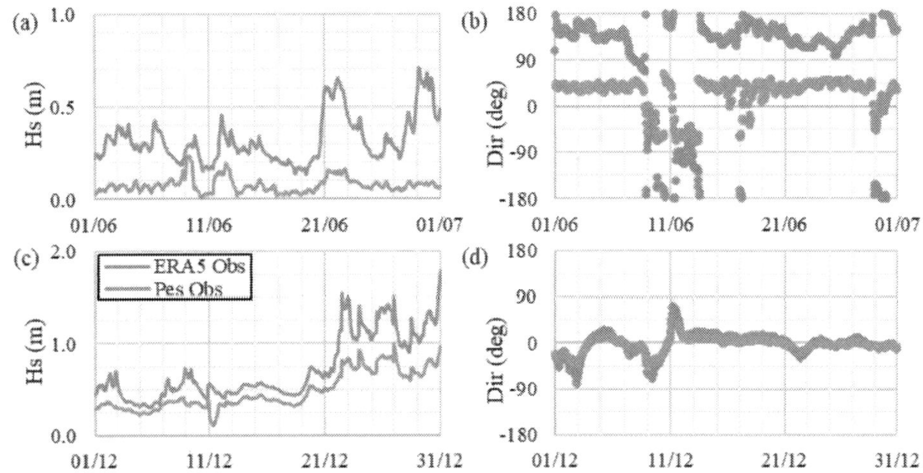

Figure 9. Comparison between wave parameters in ERA5 and Pesaren Observation Points. (a, b) for June 2022 and (c, d) for December 2022.

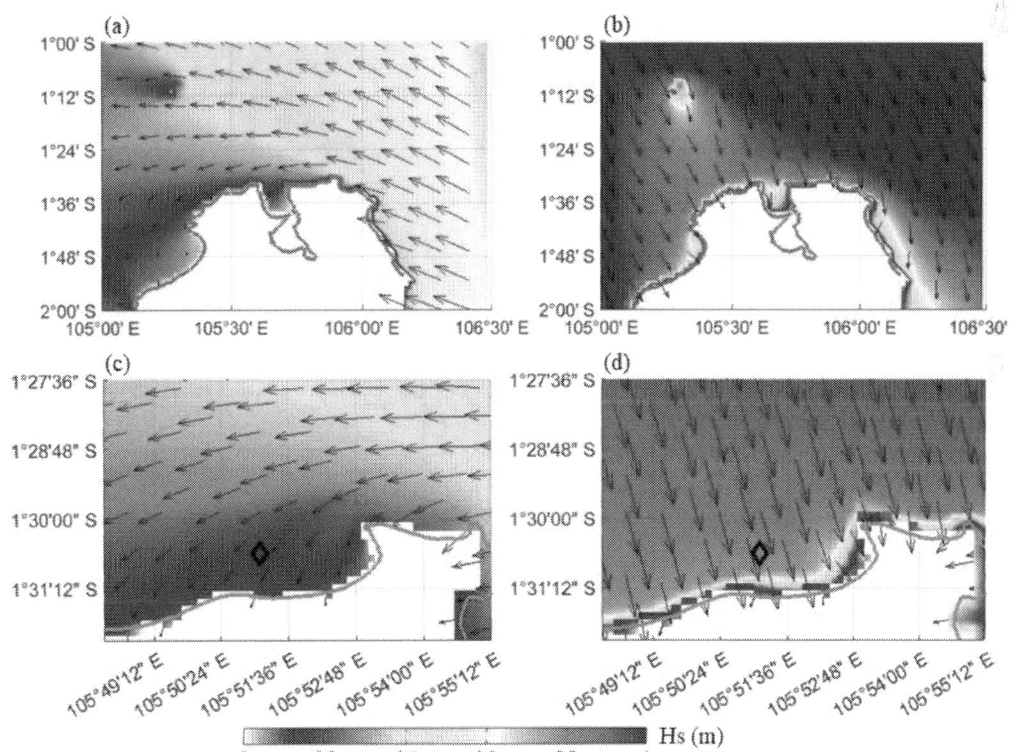

Figure 10. Wave patterns in (a, c) June 22nd, 2022 12:00 and (b, d) December 22nd, 2022 12:00. The visuals are for (a, b) North Bangka and (c, d) Pesaren Beach.

Figures 10(a) and 10(b) visualize the distinction in regional wave propagation patterns in June and December 2022. Further, Figures 10(c) and 10(d) depict the wave field in the vicinity of Pesaren Beach. These figures restate that the influence of the monsoonal wave regime at Pesaren Beach is greater than anticipated, owing to the beach geometry characteristics. Tanjung

IOP Conf. Series: Earth and Environmental Science **1464** (2025) 012021
doi:10.1088/1755-1315/1464/1/012021

Mengkudu and the other western headlands protect the eastern end of Pesaren Beach from waves in the east monsoon. Waves will have a greater alongshore variation when arriving at the beach and subsequently induce a larger potential radiation stress gradient. While the east monsoon waves are the second strongest type, strong *el-nino* conditions may potentially form strong JAS waves, thus inducing a strong sediment movement at the beach, presumably westward.

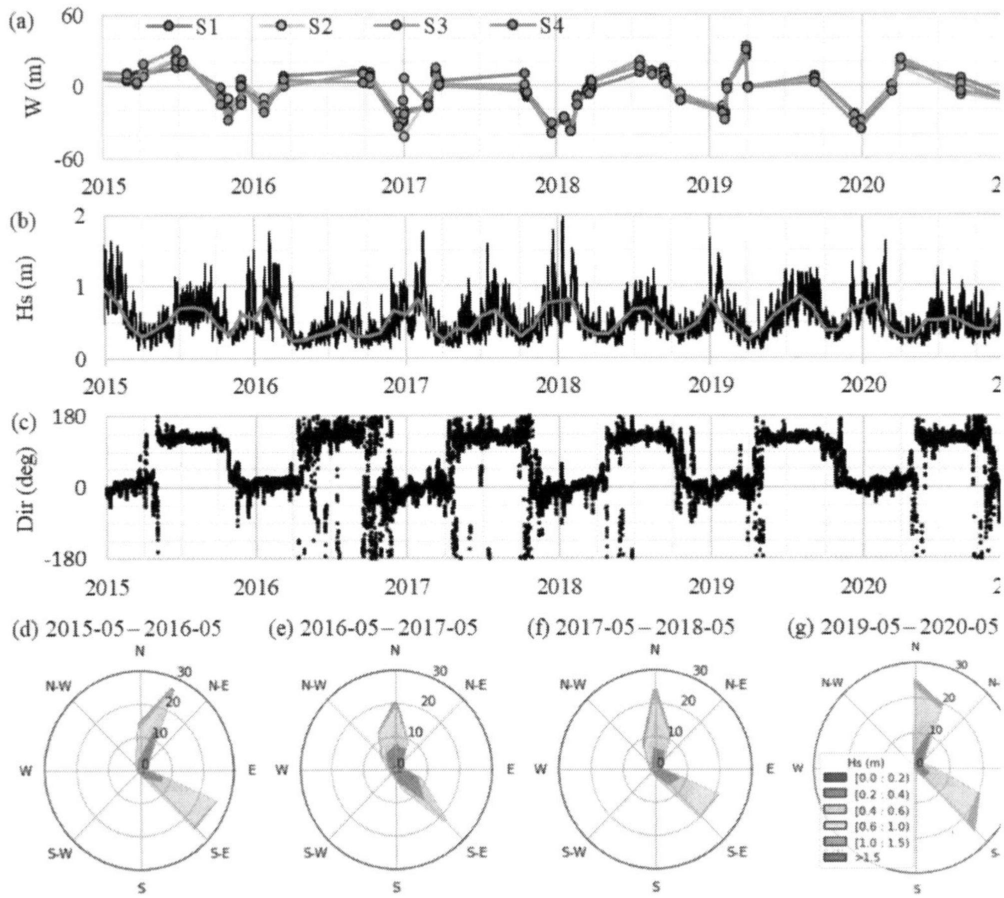

Figure 11. Shoreline dynamics at S1 to S4 and wave climate at ERA5 Obs from 2015 to 2021. (a) Shoreline positions [W (m)] are offset to their average positions between 2013 to 2022. Wave climate presented as (b, c) time series of hourly *Hs* and *Dir* and (d, e, f, g) wave roses for four periods in 2016, 2017, 2018, and 2020.

4. Discussion

Fluctuations of shoreline positions at four sections (S1-S4) resulting from CoastSat are provided in Figure 11(a), where the W (shoreline position) is offset against their average values from 2013 to 2022. Its comparison against *Hs* and *Dir* provided in Figures 11(b) and 11(c) shows that the receding or eroding shoreline is observed in the northeast monsoon period, around DFJ. In this period, the wave intensity is higher than in other months and is approaching perpendicular to the beach, as shown in Figure 11. In JAS (west monsoon), while the shoreline positions are sometimes not captured, it is likely to be the period where sediment is deposited at the beach, advancing the shoreline. This pattern is cyclic during the observation periods, although some evident outliers are seen.

In Figure 11(a), only DJF months consistently contain data points. Therefore, the interannual variability discussion is only limited to the shoreline recession in DJF. The highest shoreline recession may be expected when the monthly *Hs* is the largest. However, this relationship is not applicable here. While the *Hs* monthly around DJF did not change significantly from 2015 to 2020, each year saw a varying rate of shoreline recession. The largest *Hs* also could not correlate consistently to the shoreline recession. The year 2018, which has the larger peaks, revealed the largest erosion. However, the year 2016 had a slightly larger peak *Hs* than the year 2017 but ended up seeing a smaller rate of recession.

An interesting pattern emerged when including the wave direction in the discussion. Figures 11(d) to 11(g) presents the wave roses at DFJs in 2016, 2017, 2018, and 2020. These years are selected since they have multiple data at their DFJ period. The observed trends are when, in DFJ, the waves approach from the N quadrant, the result is a high erosion, as seen in 2017 and most evidently in 2018. Meanwhile, when the dominant incident wave direction slightly shifts to the NNE quadrant, even when the wave intensity is high, the resulting erosion is slightly smaller. This is still a hypothesis. To confirm this, follow-up studies with morphology modelling combined with adequate field measurement are required. The trend at JAS and other transition months is not yet discussed, since this study highly depends on the availability of clear images.

5. Conclusion

This study has combined data acquisition and coastal area modelling techniques to establish a baseline understanding of coastal processes at Pesaren Beach, Bangka. Tidal modelling revealed a shore parallel diurnal tidal current with a spring peak flow speed of 0.4 m/s. Wave climates analysis indicated pronounced seasonal asymmetry, with waves predominantly approaching from the NNW to NNE quadrant during the December-February (DJF) monsoon and from the SE to SSE quadrant during the June-August (JAS) monsoon. Output from the Delft3D model described the wave propagation and tidal flow pattern at the beach. Results suggest that Tanjung Mengkudu, located at the eastern end of Pesaren Beach, significantly influences the beach's wave and tidal dynamics.

The long-term shoreline dynamics exhibit a cyclical pattern of erosion during DJF and sedimentation during JAS. Signals of erosion/sedimentation from CoastSat qualitatively correlate with the wave pattern. However, it is still unclear how sediment transport works, which is a potential direction for further research. The present model has yet to include a sediment transport simulation, as this would necessitate more detailed field data to yield a result with reduced uncertainty. Furthermore, from a comparison between shoreline dynamics, wave parameters, and wave roses, the interannual variability at DJF is evident and can be explained as influenced not primarily by the wave height but rather by the dominant incident wave direction.

Overall, the study is the first to describe the seasonal and interannual shoreline dynamic at a sandy beach for Bangka Island, that subjected to highly asymmetrical monsoonal wave climates and with complex geometry due to multiple strong points (rocks and headlands). The trend may also be analogously applicable to other beaches in Bangka Island. To mitigate the potential adverse impacts of offshore tin mining, further research, particularly those including field studies, is necessary.

References

[1] Gubernur Kepulauan Bangka Belitung 2020 *Lampiran I: Peraturan Daerah Provinsi Kepulauan Bangka Belitung Nomor 3 Tahun 2020 tentang Rencana Zonasi Wilayah Pesisi dan Pulau-Pulau Kecil Provinsi Kepulauan Bangka Belitung Tahun 2020-2040* (Pangkalpinang: Gubernur Kepulauan Bangka Belitung)
[2] Pamungkas A and Husrin S 2020 The Modelling of Suspended Sediment Distribution due to Tin Mining in Bangka Seas *J. Ilmu dan Teknol. Kelaut. Trop.* **12** 353–68
[3] Jihan D, Yuniarti F R, Monalisa J T, Panjaitan P M, Aprila M, Khasanah U, Meliala Y C and Ningrum Y 2022 Analysis the impact of tin mining at sea on marine ecosystems and fishermen's income in Tanjung Binga village *J. Kelaut. dan Perikan. Terap.* **4** 79
[4] Adrian K, Winarno W and Hartanto R V P 2021 Analisis dampak aktivitas proyek tambang timah di perairan laut Pulau Bangka terhadap hak atas pekerjaan nelayan tradisional: Perspektif Inclusive Citizenship *J. Pendidik. Kewarganegaraan* **11** 76–84
[5] Peukert A, Petersen S, Greinert J and Charlot F 2018 Seabed Mining *Springer Geology* ed A Micallef, S Krastel and A Savini (Springer Cham) pp 481–502
[6] Kaikkonen L, Venesjärvi R, Nygård H and Kuikka S 2018 Assessing the impacts of seabed mineral extraction in the deep sea and coastal marine environments: Current methods and recommendations for environmental risk assessment *Mar. Pollut. Bull.* **135** 1183–97
[7] Kurniawan, Supriharyono and Sasongko D P 2014 Effects of tin mining activities to quality of sea water in the coastal area of Bangka Regency, Province of Bangka Belitung Archipelago *Akuatik-Jurnal Sumberd. Perair.* **8** 13–21
[8] Ginanjar Y C, Yonvitner Y and Nurjaya I W 2021 Evaluation of changes in the coastal line of East Bangka using the Digital Shoreline Analysis System Method *J. Perikan. dan Kelaut.* **11** 162–74
[9] Vos K, Splinter K D, Harley M D, Simmons J A and Turner I L 2019 CoastSat: A Google Earth Engine-enabled Python toolkit to extract shorelines from publicly available satellite imagery *Environ. Model. Softw.* **122** 104528
[10] Vos K, Harley M D, Splinter K D, Simmons J A and Turner I L 2019 Sub-annual to multi-decadal shoreline variability from publicly available satellite imagery *Coast. Eng.* **150** 160–74
[11] Hersbach H, Bell B, Berrisford P, Biavati G, Horányi A, Muñoz Sabater J, Nicolas J, Peubey C, Radu R, Rozum I, Schepers D, Simmons A, Soci C, Dee D and Thépaut J-N 2023 ERA5 hourly data on single levels from 1940 to present *Copernicus Clim. Chang. Serv. Clim. Data Store*
[12] Lesser G R, Roelvink J A, van Kester J A T M and Stelling G S 2004 Development and validation of a three-dimensional morphological model *Coast. Eng.* **51** 883–915
[13] Booij N, Ris R C and Holthuijsen L H 1999 A third-generation wave model for coastal regions: 1. Model description and validation *J. Geophys. Res. Ocean.* **104** 7649–66
[14] Deltares 2023 *Delft3D-FLOW: User Manual* (Delft, the Netherlands: Deltares)
[15] Deltares 2024 *Delft3D-WAVE: User Manual* (Delft, the Netherlands: Deltares)
[16] Pratama M B 2019 Tidal flood in Pekalongan: Utilizing and operating open resources for modelling *IOP Conference Series: Materials Science and Engineering* vol 676 (Institute of Physics Publishing) p 012029
[17] Pratama M B, Venugopal V, Ajiwibowo H, Ginting J W and Novico F 2020 Modelling Tidal Flow Hydrodynamics of Sunda Strait, Indonesia *ILMU Kelaut. Indones. J. Mar. Sci.* **25** 165–72
[18] Luijendijk A P, Ranasinghe R, de Schipper M A, Huisman B A, Swinkels C M, Walstra D J R and Stive M J F 2017 The initial morphological response of the Sand Engine: A process-based modelling study *Coast. Eng.* **119** 1–14
[19] van Ormondt M, Nederhoff K and van Dongeren A 2020 Delft Dashboard: a quick set-up tool for hydrodynamic models *J. Hydroinformatics* **22** 510–27
[20] GEBCO Compilation Group 2023 GEBCO 2023 Grid
[21] Egbert G D and Erofeeva S Y 2002 Efficient Inverse Modeling of Barotropic Ocean Tides *J. Atmos. Ocean. Technol.* **19** 183–204
[22] Roelvink J A and Walstra D J R 2004 Keeping it simple by using complex models *Proceeding of the 6th International Conference on Hydro-Science and Engineering* ed M S Altikanar, S S Y Wang, K P Holz and M Kawahara (Brisbane, Australia: National Center for Computational Hydroscience and Engineering) pp 335–46
[23] Badan Informasi Geospasial 2024 Stasiun Pasang Surut
[24] Pratama M and Venugopal V 2024 Modeling Tidal Hydrodynamics of the Montrose Tidal Inlet System: Ebb Jet and Eddy Formation at a Tidal Inlet with a High-Angle Half-Delta *J. Waterw. Port, Coastal, Ocean Eng.* **150** 04024016
[25] NOAA National Weather Service 2024 Cold & Warm Episodes by Season *Clim. Predict. Cent.*

IOP Conf. Series: Earth and Environmental Science **1464** (2025) 012022 doi:10.1088/1755-1315/1464/1/012022

Numerical dynamic analysis of single point wave energy converter

Abid Arham[1*], Philipp Thies[1] and Ajit Pillai[1]

[1] Renewable Energy, Department of Engineering, University of Exeter, Penryn, UK

*E-mail: aa1550@exeter.ac.uk

Abstract. In this work, numerical dynamic analysis of Single Point Wave Energy Converter (SPWEC) is conducted using the OrcaFlex software package. Maxsurf software package is used to create both the hydrostatic and hydrodynamic properties used in the OrcaFlex dynamic analysis. The numerical model replicates the WaveEl buoy WEC prototype project by Waves4Power which was held in Runde, Norway. The input data and model validation sources are elaborated with existing research. This work evaluates the use of a commercial software package OrcaFlex in wave energy converter modelling and analysis. The use of a 6D spar buoy in OrcaFlex can be used to simulate the motion of SPWEC. However, as the 6D spar buoy in OrcaFlex can only consider Morison's equation, wave data adjustment should be made to keep the numerical model accurate and eliminate the possibility of numerical issues occurring. Unlike the past work which uses a heave plate to simulate the Power Take Off (PTO) device, this present work simulates the PTO by the use of winches within OrcaFlex. The numerical model is validated using the decay test which compares the resonant period and damping coefficient with existing research. The decay test undertaken in this project leads on model property with a 1.004 ratio of the damping coefficient and 0.9497 ratio of the resonant period compared with the existing research. Overall, regardless of the limitations, the results in this work can demonstrate that a 6D spar buoy and winches within commercial finite element software OrcaFlex is capable of representing the full-scale WEC and therefore supports this as a suitable methodology for marine renewable energy analysis especially floating wave energy converters such as SPWEC.

1. Introduction

Marine wave energy is currently at the stage where several large-scale installations are being planned [1]. The UK alone has a wave energy resource potential of 40-50 TWh/year which equates to 15% of the annual electricity generation in 2019 [2]. However, wave energy conversion is considered relatively immature and not is commercially ready compared with other renewable energy technologies [3]. Hence, technological development is needed to tackle the challenges in wave energy conversion.

At the 9th European Wave and Tidal Energy Conference, [4] presented an overview of the hydrodynamic performance of different types of WECs. On average, the hydrodynamic efficiency of a modern WEC is between 20 and 30%. It can be said that a WEC device could absorb around a quarter of the wave energy to which it is exposed. Many research has been done to take the challenges of wave energy conversion, with most of it focusing on engineering discipline especially the hydrodynamics and behaviour of the Wave Energy Converter (WEC) devices as the hydrodynamic response and the behaviour of the device are one of the main factors of the effectiveness of the WEC device.

SPWEC is one kind of point absorber WEC that harvests the wave energy by utilizing the heave motion. Typically, a point absorber can be a single-body that generates energy by reacting against a fixed base, or a multiple-body device that generates energy from the relative motion between each bodies [5]. To realistically model the loads experienced by the SPWEC, it is vital to include the Power take-off (PTO) and the dynamic response of the device due to wave and ocean current interaction. Besides these factors, the effect of mooring lines is also an important part of the floating WEC system as it can also have a significant coupling influence on the dynamic response and the power captured by the floating WEC [6]. This work aims to evaluate the use of a 6D buoy and winch within the commercial hydrodynamic software package OrcaFlex in wave energy converter modelling and analysis.

2. Numerical Modelling of SPWEC

The data used in this project mainly referred to the experimental study of a taut-moored WEC by referenced work [7] and an ongoing prototype WEC designed by Waves4Power which is still being tested in Runde, Norway. Table 1 and Figure 1 below present the SPWEC buoy dimensioning and basic properties used in the analysis.

Table 1. SPWEC buoy basic properties.

Parameter	Unit	Value
Mass	kg	451751
Draft*	m	34.9
Centre of Gravity (COG) (x,y,z)	m	(0, 0, -14.1)
Roll inertia relative to COG (I_{xx})	kgm^2	$5.160x10^7$
Pitch Inertia Relative to COG (I_{yy})	kgm^2	$5.160x10^7$
Yaw Inertia relative to COG (I_{zz})	kgm^2	$2.154x10^5$

* The plane of water surface at the WEC centre is acting as a cartesian coordinate reference

2.1 SPWEC Buoy Finite Element Model

The buoy model is made by using the Maxsurf modeler. The panel method is used and implemented by dividing the surface of the buoy into 1771 panels. A buoy model is made by assembling several symmetrical bodies into a whole buoy body model. The buoy model generally can be divided into three parts, bottom part, mid part, and top part with each corresponding mesh and panel size. Triangle mesh is used to define the model shell which later will be in contact with seawater and external load. Detailed model dimensioning and trimesh used are presented in Table 2 and Figure 2 below.

Maxsurf Motion is used to calculate the RAO of the SPWEC buoy. Strip theory is used to calculate the global SPWEC buoy added mass, damping and cross-coupling terms.

IOP Conf. Series: Earth and Environmental Science **1464** (2025) 012022 doi:10.1088/1755-1315/1464/1/012022

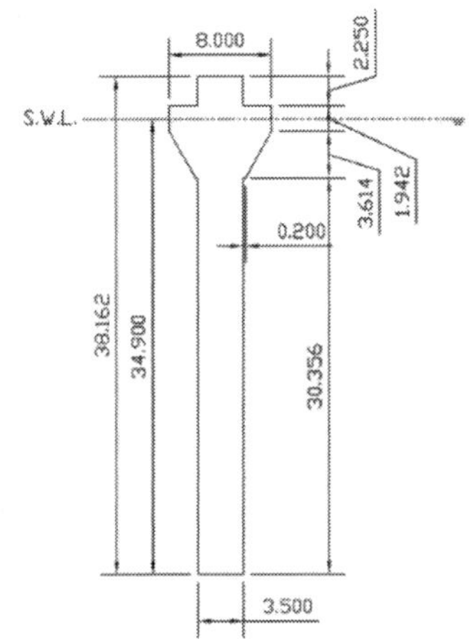

Figure 1. SPWEC buoy dimensioning.

Table 2. Finite element model dimensioning.

Buoy Part	Diameter (m)	Trimesh Dimensioning (m)
Bottom	3.5	0.074 - 1.048
Mid	3.9 - 8	0.111 - 0.967
Top	3.5	0.073 - 1.049

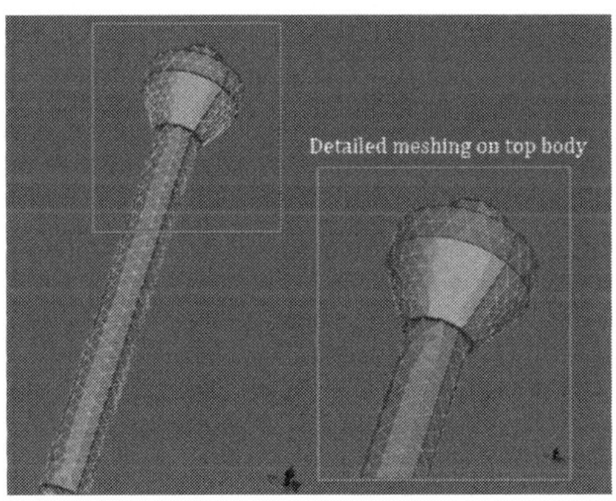

Figure 2. Detailed buoy trimesh.

The coupled heave and pitch equations are solved to obtain the vertical plane RAOs of the buoy. Figure 3 below presents the response amplitude operator for three translational motions Surge, Sway, and three rotational motions Roll, Pitch, and Yaw. It can be seen that the only rotational motion experienced by the SPWEC buoy is Pitch motion. This is the result of the buoy body shape arrangement which is symmetrical and circular so the body will not experience the roll and yaw motion in every load direction faced.

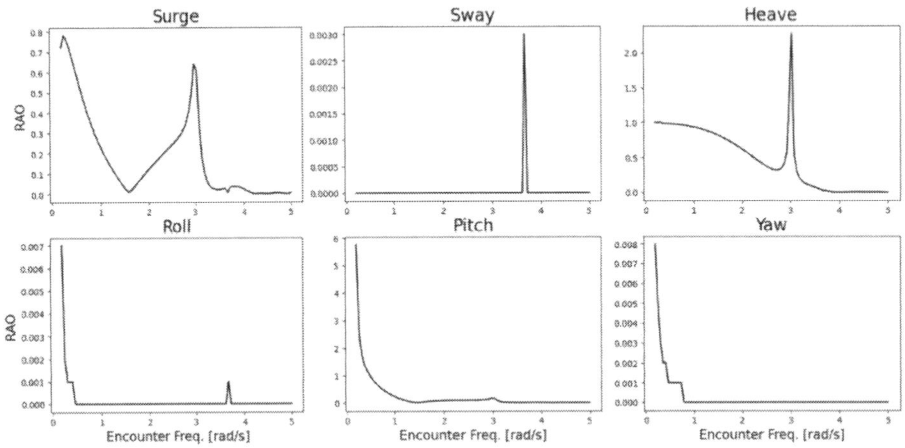

Figure 3. SPWEC buoy RAOs.

2.2 Global analysis model

A 6D Spar buoy is used in OrcaFlex to represent the SPWEC as it provides better flexibility and accuracy on how the SPWEC is modelled compared to the lumped buoy and vessel object. The SPWEC buoy is modelled as a series of axisymmetric cylinders placed end to end along the local z-axis. Hydrodynamic loads and motion response of the SPWEC buoy are calculated using Morison's equation as the SPWEC buoy is considered as a small structure compared with the wavelength.

Table 3 and Figure 4 below present the mooring system properties used in the project with L1, L2, and L3 expressing the mooring lines.

Table 3. Mooring lines properties.

Parameter	Unit	Value
Depth of the anchor	m	80
Radius of the anchor	m	125.109
Nominal Diameter	m	0.08
Dry mass	kg/m	4.9
Submerged wight	N/m	35.868
Axial Stiffness	N	5.754x10^6

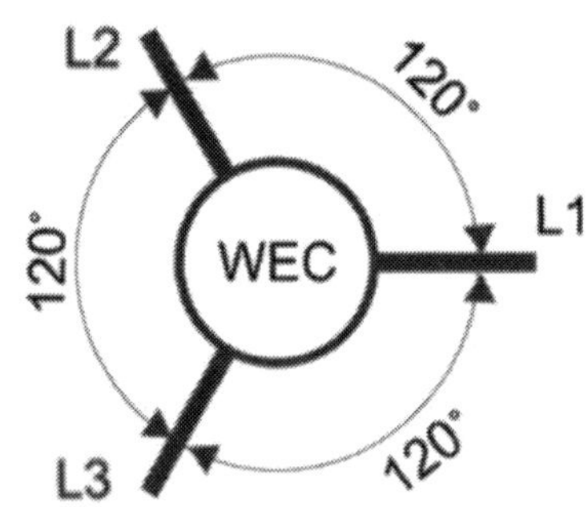

Figure 4. Mooring lines configuration.

Added mass and drag forces are applied only to those parts of the buoy which are instantaneously below the water level. For partly submerged cylinders, added mass and drag are scaled according to the proportion of the individual cylinder volume that is submerged. General 6D Spar buoy configuration is presented in Figure 5 below. Unlike the former research which used a heave plate to simulate the Power Take Off (PTO) device, this work simulates the PTO by the use of winches wire which attached to the bottom part of the SPWEC buoy within OrcaFlex.

IOP Conf. Series: Earth and Environmental Science **1464** (2025) 012022 doi:10.1088/1755-1315/1464/1/012022

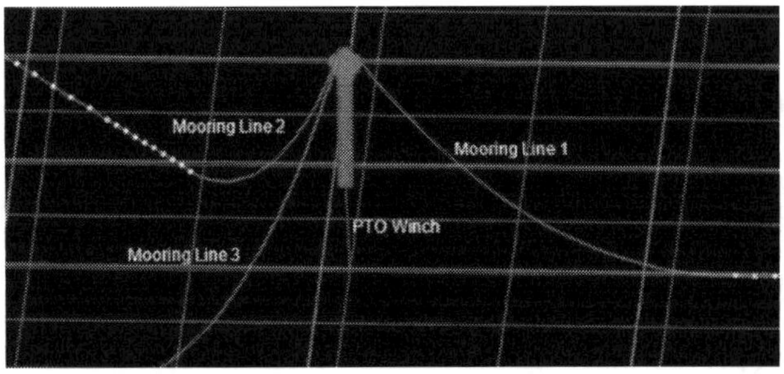

Figure 5. Model arrangement in OrcaFlex.

2.3 Load case matrix

This project used a similar load case matrix as the reference study held by referenced work [7] with additional adjustment on the wave parameter to match the Waves4Power WEC prototype testing in Runde.

Table 4. Load case matrix.

Load Case	Type of Wave	Wave Period (T) or (T_P) (s)	Wave Height (H) or (H_s) (m)	Current Speed (m/s)
1	Regular	3.185	0.238	
2	Regular	5.37	0.675	
3	Regular	6.37	0.95	
4	Regular	6.37	1.9	
5	Regular	7.37	1.272	0.514
6	Regular	6.37	3.801	
7	Regular	12.74	3.679	
8	Irregular	6.5	2.5	
9	Irregular	9.6	4.5	
10	Irregular	13.8	9	

2.4 Wave data adjustment

Two kinds of wave theory are used in this project. Airy wave theory is used for regular wave load cases and JONSWAP spectrum is used to define the wave properties used in the irregular wave load cases. To cover the high-frequency tail of the spectrum, some short-period waves are included in the calculation. In some cases, they can go down to the condition that the wavelength is shorter than the diameter of the spar buoy which is no longer applicable for Morison's equation and could possibly introduce uncertainties in the modelling. Although the issue is only relevant to the near-surface buoy part since short waves did not penetrate far down in the water column, this limitation should raise awareness of the small wave's effect on the irregular wave buoy modelling in OrcaFlex.

A wave whose length is short compared to the buoy diameter will have effects that vary significantly across the diameter of the buoy body and as in OrcaFlex the spar buoy can only be divided in the axial direction and cannot be divided in the longitudinal direction. Hence, the adjustment of the wave data in load case 8, 9 and 10 should be done to eliminate the modelling issue and ensure the accuracy of the results. This adjustment is done by removing a small number of individual waves with wavelengths smaller than three times of the buoy diameter at the waterline (24m) from the spectrum which is used in irregular wave load cases. This leads to the 11.9% removal of wave data from case 8 spectrum, 7.9% for case 9, and 3.3% for case 10 from the total of 218 wave data for case 8, 215 wave data for case 9, and 207 wave data for case 10. This methodology is supported by Orcina and explained in the example document related to the mooring and buoy system [8].

3. Results and Discussion

3.1 Model validation and motion responses

The expected outcome is that the final model, after calibration through an iterative approach by varying input of geometry approach, trimesh dimensioning, drag coefficient, and added mass coefficient, will be as similar to the results as possible. The model built should give promising similarity both in its hydrostatics and hydrodynamic properties.

Table 5 below present the hydrostatic model validation results. Among eight parameters compared in Table 5, the vertical centre of gravity became the parameter with the lowest similarity with the prototype. The numerical model centre of gravity lies 0.54 m higher than the prototype centre of gravity. This is possibly caused by different sea densities used in the numerical model and prototype buoy calculation. SPWEC finite element numerical model calculated using default seawater density of 1,250 kg/m^3 while in the prototype Runde project, the density applied is an actual sea density at the project location, North Sea. Besides that, the actual PTO attached in the prototype SPWEC device can affect its hydrostatic properties by lowering the centre of gravity point, while hydrostatic properties in the finite element model are calculated without considering the PTO attachment below the buoy.

Table 5. Hydrostatic model validation results.

Measurement	Unit	Value	
		Model	Prototype
Displacement	kg	45377	451751
Volume (displaced)	m^3	443.30	443.26
Draft	m	34.9	34.9
Waterplane Area	m^2	50.14	50.27
Waterplane Transverse Moment inertia	m^4	200.02	201.06
Waterplane Longitudinal Moment Inertia	m^4	200.28	201.06
Vertical Centre of Buoyancy	m	-13.518	-13.65
Vertical Centre of Gravity	m	-13.56	-14.1

With the hydrostatic properties verified, the dynamic model validation was held by performing the decay test or free-release simulation. The purpose of the decay test is to investigate the resonant period and damping coefficient of the SPWEC system. To perform a decay test in OrcaFlex, the dynamic calculation was performed with the buoy displaced about 1 meter above its equilibrium floating position as its initial position. The buoy is then released, and the motion of the buoy is plotted. The length of testing time should be enough to include two periods, and that decay tendency must be clearly indicated. Figure 6 below illustrates the data evaluation of a decay test result. The resonant period and damping coefficient then can be calculated from the motion result plotted by using the following equation.

$$T_r = \frac{t_5 - t_1}{2} \tag{1}$$

$$\zeta = \frac{\ln(u_1/u_5)/2}{\sqrt{4\pi^2 + [\ln(u_1/u_5)/2]^2}} \tag{2}$$

Where,

t_i : the time at which the ith peak or trough occurs in the free oscillation.

u_1 : the amplitude of the ith peak or trough occurs in the free oscillation.

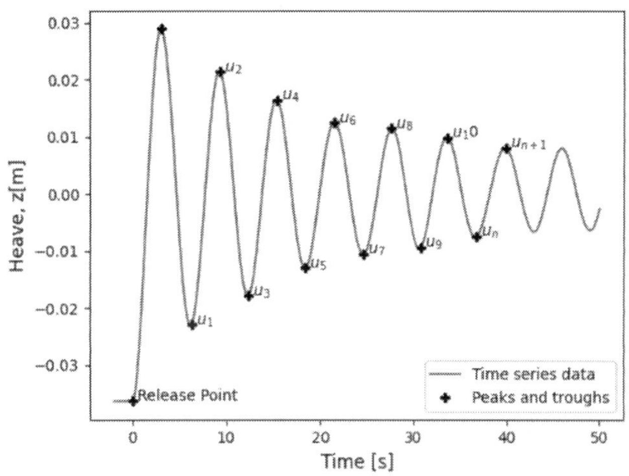

Figure 6. Decay test results.

All degrees of freedom were included in the dynamic calculation. Initial drag and added mass coefficient are stated as. The decay test was performed multiple times by changing the drag and added mass coefficient until the resonance period T_r and damping coefficient ζ is matched with the experimental decay test held by the referenced work [7]. The decay test was held for the full SPWEC system with the mooring lines and the PTO.

Within simulations, the heave, surge, and pitch responses are observed to behave linearly as the incident wave, the higher incident wave will produce a higher heave and surge motions of the buoy as seen in Figure 7 below. However, it is worth noting that linear trends are only applicable to linear waves as random wave results behave differently. The discrepancies between linear and random wave results presented in Figure 6 come from the wave regime applied. On the linear wave cases, time domain analysis with linear waves with non-diffracting 6D buoy is considered. On the other hand, frequency domain analysis is considered within random wave cases considering diffraction theory. The research would benefit from additional random wave cases to find the SPWEC dynamic response under random waves which work in the frequency domain.

IOP Conf. Series: Earth and Environmental Science **1464** (2025) 012022 doi:10.1088/1755-1315/1464/1/012022

Table 6. Decay test validation results.

Parameter	Simulation	Experiment/Prototype	Ratio
Drag force coefficient (Normal)	0.7	-	-
Drag force coefficient (Axial)	1.13	-	-
Added mass coefficient (Normal)	0.5	-	-
Added mass coefficient (Axial)	0.92	-	-
Damping coefficient (ζ)	0.04518	0.045	1.004
Simulation resonant period (T_r)	6.050	6.37	0.949

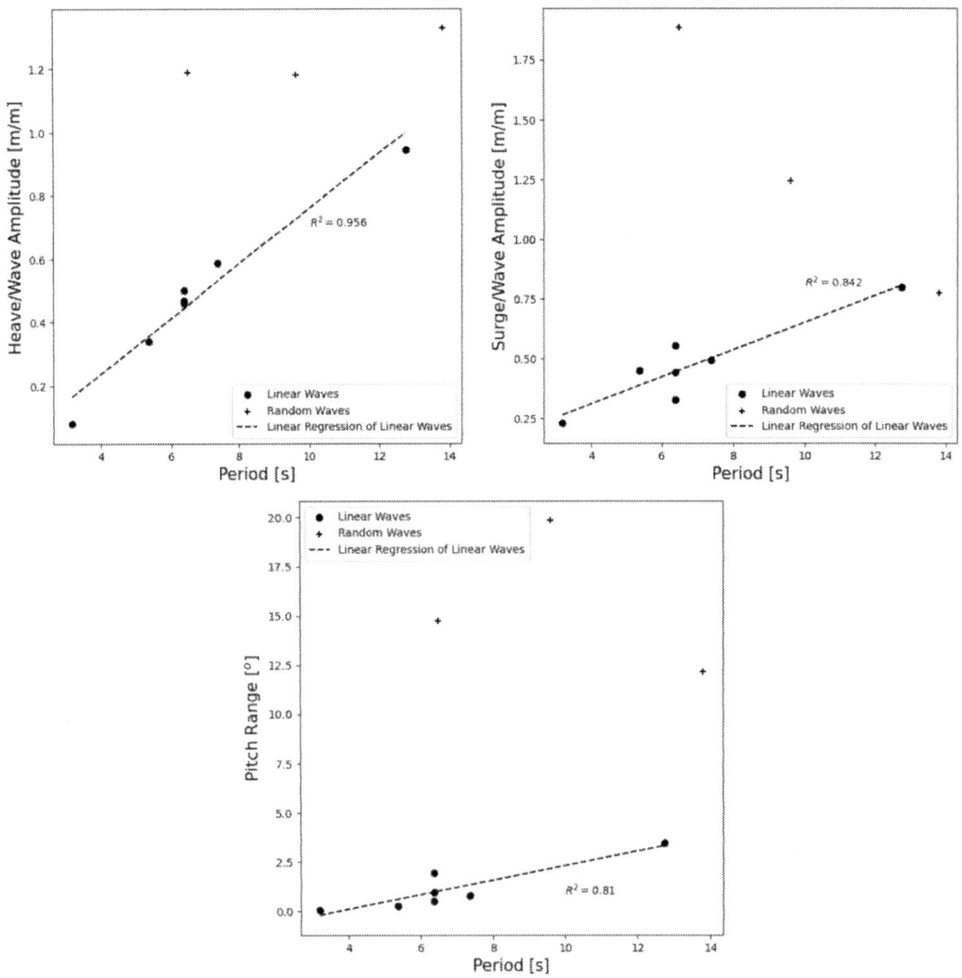

Figure 7. SPWEC motion results.

3.2 Approach Limitation

Approaches used in this project have several limitations due to the way physical prototype device is modelled and the limitation of the software and method used. One limitation of using the OrcaFlex 6D spar buoy to simulate the SPWEC buoy is that the buoy can only calculate the loads applied based on Morison's equation. Hence, wave data adjustment should be done by removing a small amount of wave data which lies in high-frequency tail of the spectrum.

Removing high-frequency tail from the spectrum can affect the accuracy of the model, in this case, the wave data adjustment ends up with 7.73% of wave data removed on average. One of the methods to accommodate the high-frequency tails is by creating a model which can be divided both in axial and longitudinal axes so Morison's equation can be fully applied to the SPWEC model. The other way to accommodate the high-frequency tail is by separating the high-frequency tail modelling using diffraction theory calculation on a vessel object within OrcaFlex in different analysis runs, then combining the result as one at the end of the simulations.

4. Conclusion

The dynamic response of the SPWEC is affected by many variables such as self-weight, water depth, wave height, and wave period, as well as by the coupled effect from mooring lines and the PTO. The use of 6D spar buoy in OrcaFlex can simulate the SPWEC perfectly. However, as the 6D spar buoy can only consider Morison's equation, the wave data adjustment on the random wave analysis should be done to keep the numerical model in good accuracy and eliminate the possibility for the modelling issue to happen. A better approach on the buoy geometry should be done so the full wave spectrum can be included in the analysis. One of the suggestions to improve the approach method is by dividing the model into smaller parts in both axial and longitudinal axis or by separating the simulation for the small waves.

The decay test held in this project gave promising results as it compared with former research held by the referenced work [7]. The iterative method by varying the added mass and drag coefficient can be used to compare the damping coefficient and resonant period in another type of the heaving body wave energy converter. The decay test held in this project ended up with a 1.004 ratio of the damping coefficient and a 0.9497 ratio of the resonant period. The accuracy of the model however can be improved by conducting further free pitch release tests. Both regular and irregular wave simulations give promising results as well. However, no result comparison with the physical model can be done as the physical model held in previous research by referenced work [7] was conducted in different mooring line arrangements.

This research shows that winches wire in OrcaFlex can be used as PTO substitution and can simulate the motion experienced by the PTO itself. However, improvement in the way the PTO damping is approximated is important to simulate the PTO in more detail. Overall, this research proves that commercial finite element software OrcaFlex is capable of being used in marine renewable energy analysis especially floating wave energy converters such as SPWEC.

Acknowledgements

The authors would like to thank Orsted Wind Power A/S and Fulgor Single Member S/A. Hellenic Cable Industry for their financial support of this work through a funded PhD studentship with the University of Exeter.

References

[1] M. Göteman, M. Giassi, J. Engström, and J. Isberg, 'Advances and Challenges in Wave Energy Park Optimization—A Review', Mar. 06, 2020, Frontiers Media S.A. doi: 10.3389/fenrg.2020.00026.

[2] S. Jin and D. Greaves, 'Wave energy in the UK: Status review and future perspectives', Jun. 01, 2021, Elsevier Ltd. doi: 10.1016/j.rser.2021.110932.

[3] B. Drew, A. R. Plummer, and M. N. Sahinkaya, 'A review of wave energy converter technology', Dec. 01, 2009. doi: 10.1243/09576509JPE782.

[4] A. Babarit and J. H. Todalshaug, 'On the maximum and actual capture width ratio of wave energy converters', Southampton: 9th EWTEC, Sep. 2011.

[5] Y. Li and Y. H. Yu, 'A synthesis of numerical methods for modeling wave energy converter-point absorbers', Aug. 2012. doi: 10.1016/j.rser.2011.11.008.

[6] M. Karimirad, S. Weller, Y. Debruyne, M. A. Bhinder, and M. Guérinel, 'Modelling mooring line non-linearities (material and geometric effects) for a wave energy converter using AQWA, SIMA and Orcaflex Modelling mooring line non-linearities (material and geometric effects) for a wave energy converter using AQWA, SIMA and Orcaflex Madjid Karimirad', Nantes: 11th European Wave and Tidal Energy Conference, Sep. 2015.

[7] S. H. Yang, J. W. Ringsberg, E. Johnson, Z. Hu, L. Bergdahl, and F. Duan, 'Experimental and numerical investigation of a taut-moored wave energy converter: A validation of simulated buoy motions', Proceedings of the Institution of Mechanical Engineers Part M: Journal of Engineering for the Maritime Environment, vol. 232, no. 1, pp. 97–115, Feb. 2018, doi: 10.1177/1475090217735954.

[8] 'C06 CALM buoy'. [Online]. Available: www.orcina.com

IOP Conf. Series: Earth and Environmental Science **1464** (2025) 012023 doi:10.1088/1755-1315/1464/1/012023

Effect of fish nets hydrodynamics on aero-hydro-servo-elastic aspect of new concept floating wind turbine combine for an offshore aqua culture unit

Murdjito[1*], Gumilang Alhafiz[2], Salsabila Tsaniatur Rahma[2], Eko Budi Djatmiko[1], and Fajar Rachmadiarto[2]

[1] Offshore Engineering, Institut Teknologi Sepuluh Nopember, Surabaya, Indonesia
[2] Ocean Engineering, Institut Teknologi Sepuluh Nopember, Surabaya, Indonesia

*E-mail: murdjito@oe.its.ac.id

Abstract. Over the last few decades, studies found that relying on conventional fossil fuel sources has been giving so many negative impacts to the environment. Depletion of coal along with natural gas and crude oil are also the reasons why we need some new energy alternatives that less harmful and sustainable. Offshore wind energy can be an effective solution due to abundant wind resources and space availability. Unfortunately, the cost of offshore wind energy is expensive. Therefore, an innovative concept of FOWT integrated with an aquaculture is proposed here. This concept is expected to evolve continuously because of the huge profit gained from offshore aquaculture trade. To be qualified, its dynamic behaviour should be evaluated and investigated while considering various environmental conditions. Arafura Sea, which has both fish and wind potential, is chosen as the suitable location for this innovative concept. The environmental conditions, representing different sea state are adopted for the analyses. One incident angle of combined wind, current, and wave loads are being considered in this paper for dominant wind direction. Similarly, a numerical model with dynamic simulation is carried out. Then, the dynamic responses of the FOWT integrated with fish nets, including the periods, motions, tower loads, and mooring tension are numerically examined. While the maximum mooring line tension in the FOWT with fish nets is observed, the expected maximum tension should occur at the mooring line aligned with the direction of the environmental load. Additionally, the highest mooring line tension should occur during operational conditions. This is due to the wind turbine system actively generating electricity during operation, with the rotor facing the incoming wind, resulting in relatively substantial aerodynamic load along with hydrodynamic load. The FOWT with nets motions result shows decreased amplitudes compared with those without nets. This reduction of all the motion amplitudes is the result of the damping effect caused by the presence of nets.

1. Introduction

In recent decades, many studies have found that dependence on conventional fossil fuel sources has caused many negative impacts on the environment. These reports show that the environmental damage caused by dependence on these sources will impact human health and create difficulties in the future. The depletion of coal along with natural gas and crude oil is also the reason why we need new energy solutions [1], [2], [3]. Wind is a renewable and clean energy solution generated by wind turbines. It has been developed and applied in several onshore and offshore areas.

Along with the congestion of onshore areas, wind energy is starting to develop in offshore areas. In addition, more abundant wind resources in offshore areas will provide higher power generation efficiency [2], [3], [4]. Starting with traditional fixed foundations, the foundations of wind turbine structures are increasingly developing by using the concept of floating foundations. It is shown that wind turbines with floating foundations provide a number of advantages at deeper water depths (above 50 m depth) compared to wind turbines using fixed foundations [5]. However, the cost of floating foundation wind turbines is very expensive [2], [5]. Therefore, many studies have been conducted to reduce the cost of offshore wind projects, especially for FOWTs, by developing innovative concepts, such as integrating FOWTs with fish cages [2], [3], [4]. This concept is proposed because marine fish farming offers a number of advantages.

The advantages of FOWT integrated with fish cages have been demonstrated in a study conducted by Lei et al. [4]. The study explained that the steel mesh helps to rise the damping level for the whole system and reduce the RAO of the motion when the wave period is around the natural period of the motion. In 2022, Zhang et al. [2], [3] proposed two designs of offshore wind turbines integrated with an aquaculture cages (FOWT-AC) to accelerate the payback time of offshore wind farms. The designs were analyzed using a fully coupled aero-hydro-servo-elastic numerical model. The analysis results show that the design has a good ability to maintain overturning conditions caused by wind and wave loads. However, the study is proposed to direct the conceptual design, modeling, and simulation analysis for an integrated FOWT-AC system. Hence, it is necessary to conduct a more detailed strength check to ensure the efficiency and safety of the design under actual conditions. Further research on this has been conducted recently by Ivanov et al. [6] in 2023. This study conducted an analysis of 13 types of floaters with different hull shape claims, designed as a support for the same turbine based on 3 different design load parameters. The strength analysis results show that tri-column semi-submersible floats for offshore wind turbines provide the best performance.

Practically speaking, offshore wind energy is increasingly generated in areas with high wind resources, while areas with low resources are still neglected. Indonesia is categorized as a country with low wind resources, with an average 100 m wind speed of only 4 m/s [7]. However, several previous studies [7], [8], [9] show that Indonesia has offshore wind potential that can reach 14,722 TWh. According to research conducted by Hesty et al. [10], the Arafura Sea and Indian Ocean have high average wind speeds (at an altitude of 50 m above sea level) of 8 m/s or more. Wind Power Density (WDP) in the region ranges from 500-600 m/s^2. This value of wind speed and WDP is categorized as a promising area to be used as a wind power plant. Therefore, a semi-submersible floating offshore wind turbine (FOWT) supporting a 5 MW of electricity, which integrated with a floating nets cage, is proposed to be constructed in the Arafura Sea.

1.1 Environmental Data

The structure is proposed to be installed in the Arafura Sea with location coordinates of 9°13'07.8"S 132°08'49.6"E and a water depth of 234 m. Since the Arafura Sea offshore area has the highest wind potential in Indonesia, based on the Indonesian wind speed map published by the Ministry of Energy and Mineral Resources (MEMR), this location has been chosen as the FOWT integrated with an aquaculture operation site.

A 20-year hourly wind speed data has been collected to conduct a wind hindcast analysis, which generates a windrose diagram. Ultimately, the range of wind speeds and dominant wind directions can be seen in the windrose diagram as shown in Figure 1.

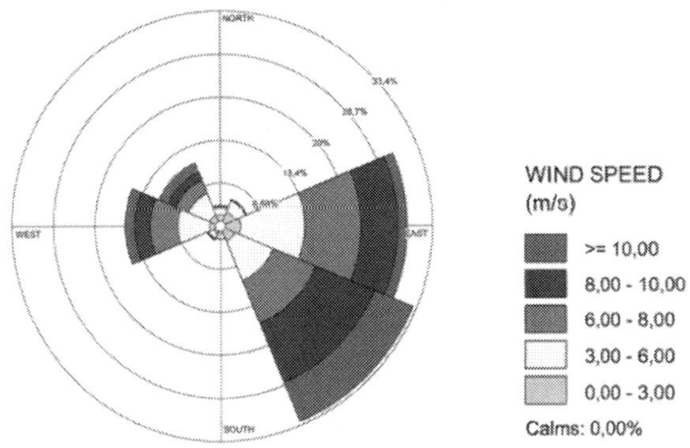

Figure 1. Wind rose diagram.

This paper considered two load cases which are the operation case and the parking case, in accordance with the previous research [3]. Each load case reflects the environmental data based on the dominant wind direction. Based on the windrose in the figure above, which is obtained from the wind hindcasting process, the environmental data of 135 degrees was used and calculated for each of the operating and parking conditions as listed in Table 1. Case 1 corresponds to the operating condition while Case 2 corresponds to the parking condition.

Table 1. Environmental conditions for simulations.

Case No.	Wind Speed (m/s)	Wave		Current Speed (m/s)
		Significant Height (m)	Peak Period (s)	
Case 1	17.25	3.55	9.11	0.22
Case 2	25	3.98	9.44	0.29

Specifically, the operating situation is the condition when the wind turbine is in production, where the blades are rotating, and the controller is activated. In addition, to keep away from potential structural damage when facing extreme conditions, all blades are set to the feathering position as well as the turbine system is shut down. To obtain environmental data during operating conditions, a 10-year return period was used, while to obtain environmental data during parking conditions, a 50-year return period was used.

1.2 Structural Model

1.2.1 Floating Platform and Wind Turbine

Currently, there are many possible floating platform concepts for offshore wind turbines. However, a floating platform concept called "OC-DeepCwind", developed by Jason Jonkman of NREL, has been selected in this paper. This concept was taken because of its simplicity in design, its compatibility with modelling, and its tendency to be commercialized. Jason Jonkman also slightly adapted aspects of the original data, making the platform design suitable for supporting NREL's 5-MW baseline turbine and farmed fish cages. Figure 2 is illustrated the details of the FOWT-AC sub-structure and Table 2 is described the properties of the FOWT-AC sub-structure.

IOP Conf. Series: Earth and Environmental Science **1464** (2025) 012023 doi:10.1088/1755-1315/1464/1/012023

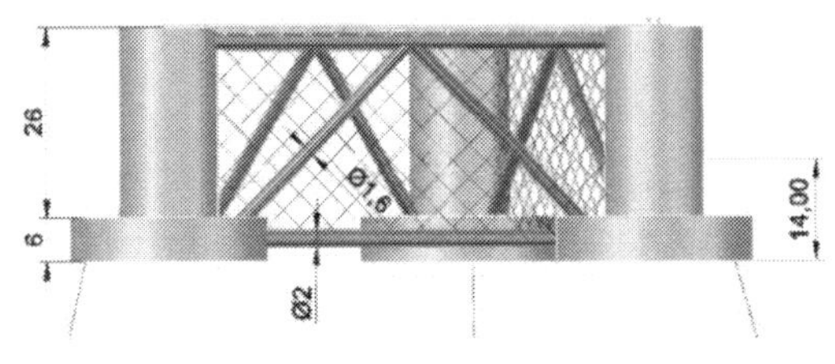

(a)

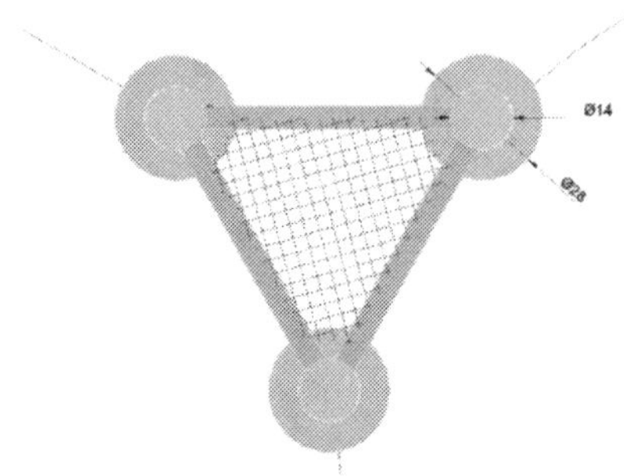

(b)

Figure 2. Semi-submersible details from (a) front view and (b) top view.

Table 2. Semi-submersible details parameter.

Paramaeter	Value [Unit]
Platform Type	Semi-submersible
Draft (with turbine)	14 [m]
Displacement	10,438 [tonne]
Upper Column Diameter	14 [m]
Upper Column Length	26 [m]
Heave Plate Diameter	28 [m]
Heave Plate Length	6 [m]
Horizontal Brace Diameter	2 [m]
Horizontal Brace Length	70 [m]
Diagonal Brace Diameter	1.6 [m]
Diagonal Brace Length	37.5 [m]

The power and thrust curves used in this paper is referred to the NREL's 5 MW offshore wind turbine. In concise, this wind turbine is a conventional wind turbine with three blades which adaptable for variable wind speed and variable pitch. This wind turbine has a cut-in-wind-speed of 3 m/s and cut-out wind speed of 10.77 m/s. It means that the wind turbine will start to operate and generates power when there is a minimum 3 m/s wind speed exerted to the turbine blade and will automatically shutting the operation down when the exerted wind speed reaches or exceed the value of 25 m/s. Meanwhile, wind turbine will reach its maximum of 2100 kN at the rated wind speed of 10.77 m/s. Table 3 shows the main properties of the NREL offshore 5 MW baseline wind turbine.

Table 3. Undistributed tower properties.

Paramaeter	Value [Unit]
Overall tower mass	347.46 [Tonne]
Tower base diameter	6.5 [m]
Tower top diameter	3.87 [m]
Structural damping ratio	1 [%]

1.2.2 Mooring Line System

An evenly distributed three-row mooring system is applied in an integrated system. The fairlead is placed on the outer heave plate, at a depth of 35 m below sea level and 62 m away from the center of the cage. The anchor is at a depth of 200 m and at a radius of 891.6 m from the center of

IOP Conf. Series: Earth and Environmental Science **1464** (2025) 012023 doi:10.1088/1755-1315/1464/1/012023

the cage. The mooring ropes are made of grade 4 studdles chain, with a nominal chain diameter of 0.153 m. The detailed configuration of the mooring line system is shown in Table 4.

Table 4. Configuration of mooring line system.

Paramaeter	Value [Unit]
Number of mooring lines	3 x 1 [ea]
Angle between adjacent lines	120 [deg]
Depth from fairlead to MSL	14 [m]
Pre-tension angle	55 [deg]

2. Methodology

Figure 3 shows a flowchart, which present the analysis stage of this research in sequence.

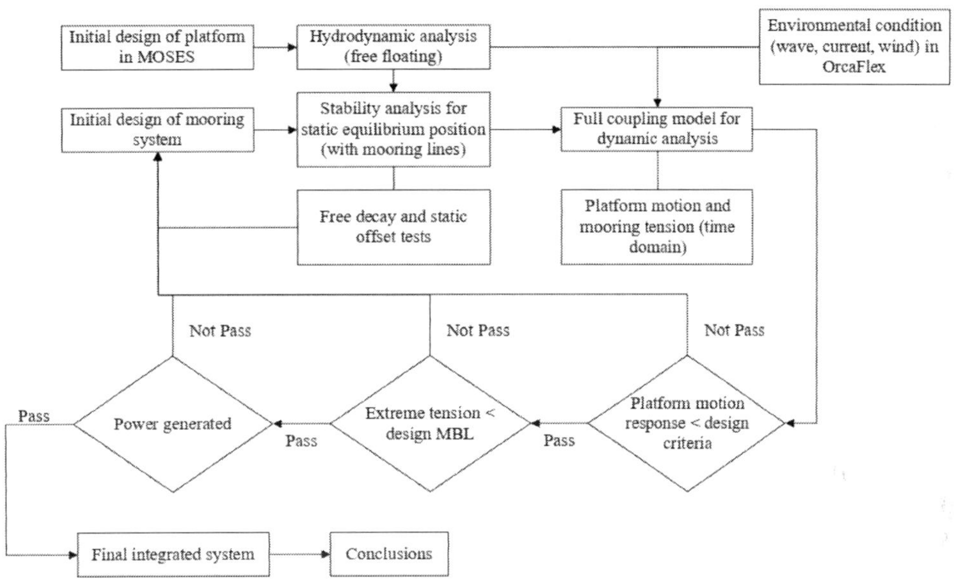

Figure 3. Analysis flow chart.

2.1 Mooring Analysis

A spread mooring system with three mooring lines is used to integrate the system at a fixed location between a floating platform, a wind turbine and a fish cage. Calculations are based on three-dimensional diffraction and radiation in hydrodynamics. The calculation of the performance of the floating structure in the time domain is referred to the following equation of motion:

$$M(p, a) + C(p, v) + K(p) = F(p, v, t) \qquad (1)$$

Where $M(p, a)$ is the inertia force of the floater, $C(p, v)$ is damping force, $K(p)$ is stiffness load, $F(p, v, t)$ is the external force acting on the floater, p, v, a are the vectors of displacement, velocity, and acceleration, respectively, and t is simulation time. The external forces acting on the FOWT consist of hydrodynamic and aerodynamic loads, as illustrated in Figure 4. The hydrodynamic loads are primarily due to wave and current acting on the submerged components, while the aerodynamic loads result from wind forces acting on the turbine blades and nacelle.

(a) (b)

Figure 4. Load definition on OrcaFlex (a) aerodynamic and (b) hydrodynamic loadings.

Chung and Hulbert [11] applied a time domain integration method by using the generalized-α integration scheme to solve the motion's equation. Those equations used the iterative method to solve the p, v, a, displacement, velocity, and acceleration vectors. A finite element method, namely lumped mass, is used to model the mooring lines. In this method, a single mooring line is divided into segments. Subsequently, the points where the segments intersect are referred to as nodes, which account solely for the axial and torsional characteristics of the line. Other attributes, such as mass, weight, and buoyancy, are presumed to be incorporated within the nodes. The length of each segment can be adjusted according to the needs of.

2.2 Net Hydrodynamics

The cage net must also be modelled correctly. The net is flexible and undergoes varying degrees of deformation according to wave and current conditions. The hydrodynamic forces acting on the net and the coupling effect between the net and the cage support structure have strong nonlinearities.

In compared to the traditional flexible collar fish cages, this semi-submersible rigid fish cage has greater rigidity and support structure. Consequently, this preliminary study models the equivalent net with Morison elements, which are considered rigid. This concept demonstrates the rigid connection between nets and cage as a whole system. Accordingly, the relative movement between the nets and the cage is omitted. This simplification may cause the hydrodynamic forces and the viscosity effects of the net to be overestimated. Nonetheless, this approach is still a good initiation to evaluate the global dynamic response of an integrated structure using a net modelled by a simplified rigid model.

2.3 Morison Theory

Morison equation is suitable when the structure's diameter is small in comparison to the wavelength. Consequently, the Morison equation is used here to simulate the hydrodynamic forces acting on the netting and cage columns, as their diameters are relatively small compared to the wavelength. The Morison equation is expressed as follows:

$$F(t) = \frac{1}{2}\rho C_d D u(t)|u(t)| + (1 + C_m)\rho\frac{\pi D^2}{4}\dot{u}(t) \tag{2}$$

In this context, $u(t)$ represents the horizontal velocity of wave-induced water particles, while $\dot{u}(t)$ is their horizontal acceleration. The parameters C_m and C_d denote the added mass coefficient and drag coefficient, respectively. Since the netting is relatively light, inertial forces are minimal, so only drag forces are considered. Figure 5 presents a force diagram of the Morison model for two perpendicular equivalent net lines, where and U_{rel} shows the horizontal relative velocity between the water particles and the structure and A illustrates the area of a net line. Subsequently, the overall drag force exerted to the fish cage nets is determined by summing all the drag force vectors across the entire net lines.

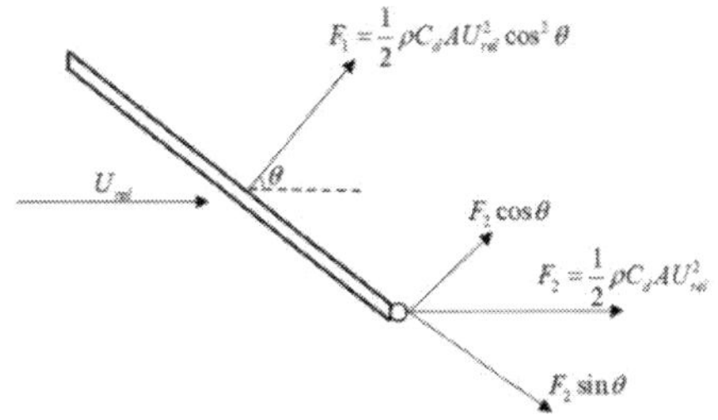

Figure 5. Force diagram for the Morrison model of orthogonal lines.

3. Results and Discussions

3.1 Motion Responses

The surge, heave, and pitch motion characteristics of the FOWT without or with nets under operating conditions are shown in Figure 6 below. It can be seen that the maximum energy occurs at a frequency around 0.08 rad/s corresponding to the surge, heave, and pitch responses and can be detected in both sea states. For the heave motion, the peak frequency of 0.11 rad/s is also captured with the coupling frequency of 0.11 rad/s of the surge and pitch motions, although it is not very significant. In addition, the presence of fish net, reduce the wave energy concentrated in frequency around 0.08 rad/s. In other words, Hydrodynamic load has given more significant effect to the FOWT's motions compared to the aerodynamic load. This can be proven by graphics shown in Figure 7, which shows a matching frequency of the FOWT and the wave at 0.08 Hz.

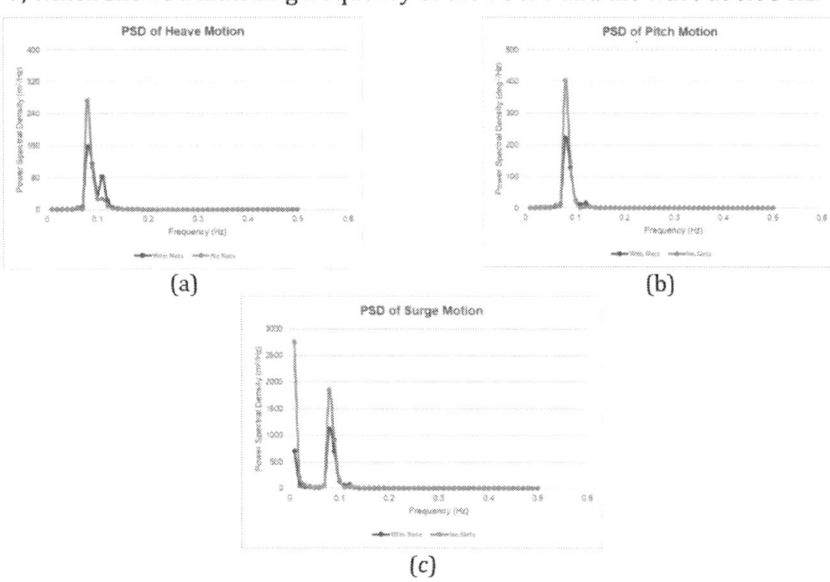

Figure 6. Power spectra response of the FOWT with and without fish nets for (a) heave motion (b) pitch motion, and (c) surge motion.

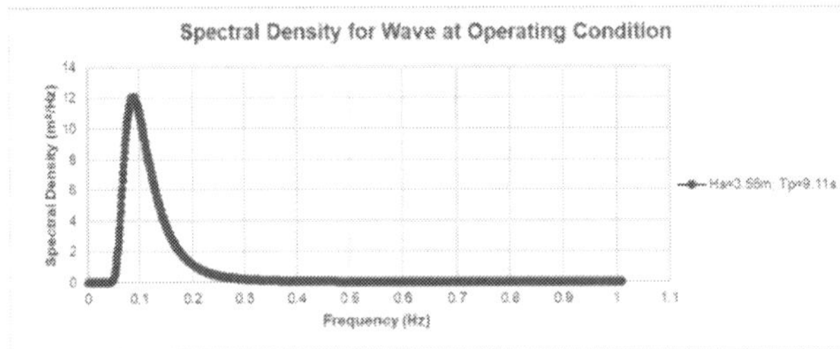

Figure 7. Power spectral density for wave at operating conditions.

3.2 Blade Deflection

Based on Figure 8, the existence of fish nets does not significantly impact the wind turbine blade deflections, both in-plane and out-of-plane. This is because the FOWT's behaviours is predominantly influenced by aerodynamic loads.

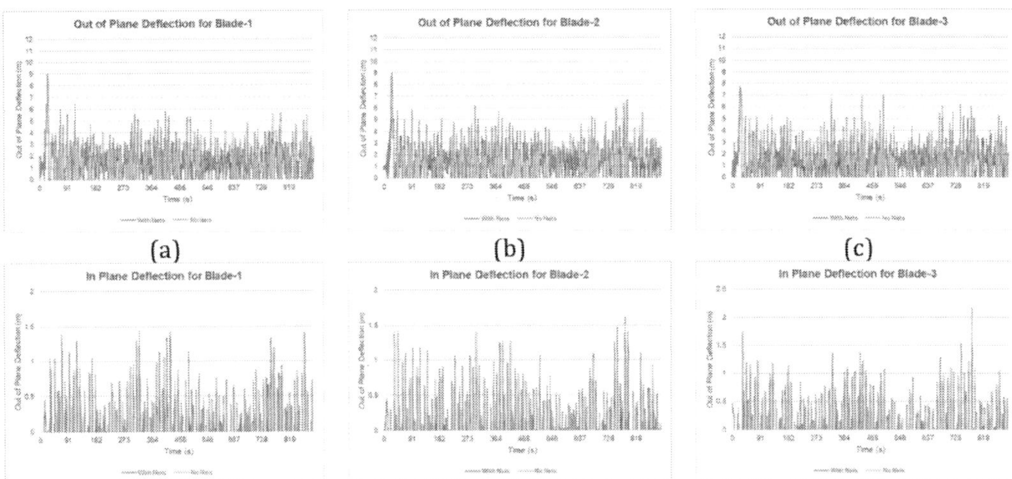

(a) (b) (c)

(d) (e) (f)

Figure 8. Blades deflection.

3.3 Power Generated

The power generated by a wind turbine is derived from the rotation of its blades, driven by the force of the wind. A stronger wind load increases power output, while a weaker wind load reduces it. To optimize power generation, a blade pitch control mechanism is used to adjust the blade angle, ensuring maximum efficiency under varying wind conditions.

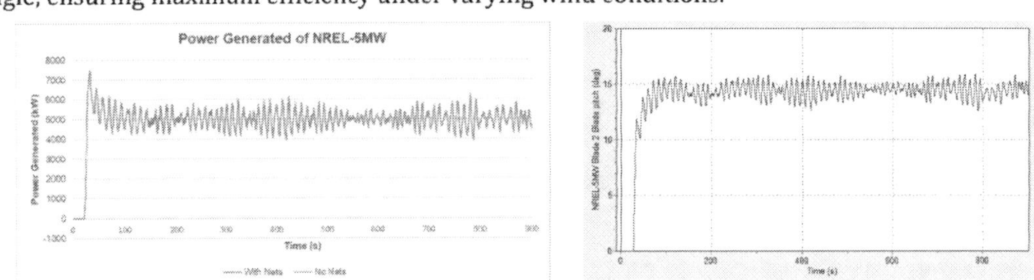

(a) (b)

Figure 9. Diagram of (a) power generated by NREL-5MW (b) time history NREL-5MW blade pitch.

Based on Figure 9 (a), the presence of fish nets does not give any significant effect to the average power generated by FOWT. This condition occurred because of the FOWT structure is mostly affected by the aerodynamic load. A small difference of the total energy generated by FOWT with and without nets in 900-seconds simulation is also validate the previous statement. The total power generated at 135 degrees with the nets and without the nets are 4849.468 kW and 4849.561 kW respectively with a very small difference, with a disparity of 0.093 kW. However, Figure 9 (b) shows that Pitch control angle give a significant effect to the average power generated by FOWT because it controls the aerodynamic load.

3.4 Mooring Line Tension

The maximum tension on the mooring line occurs on Mooring Line 1 in the FOWT without fish nets, which is about 2317.48 kN. This value still meets the requirements applied on API RP 2SK. Therefore, all mooring lines can be considered safe for the operation of floating wind turbines integrated with fish cages in the Arafura Sea (depth of about 234 m).

From Figure 10 below, there is a significant difference in the mooring line tension between the FOWT with fish nets and without fish nets. The difference in Mooring Line 1 tension is 10% under operating conditions and 38% under parking conditions. It can be concluded that the presence of fish nets reduces mooring tension in both operating and parked conditions (shown in blue graph).

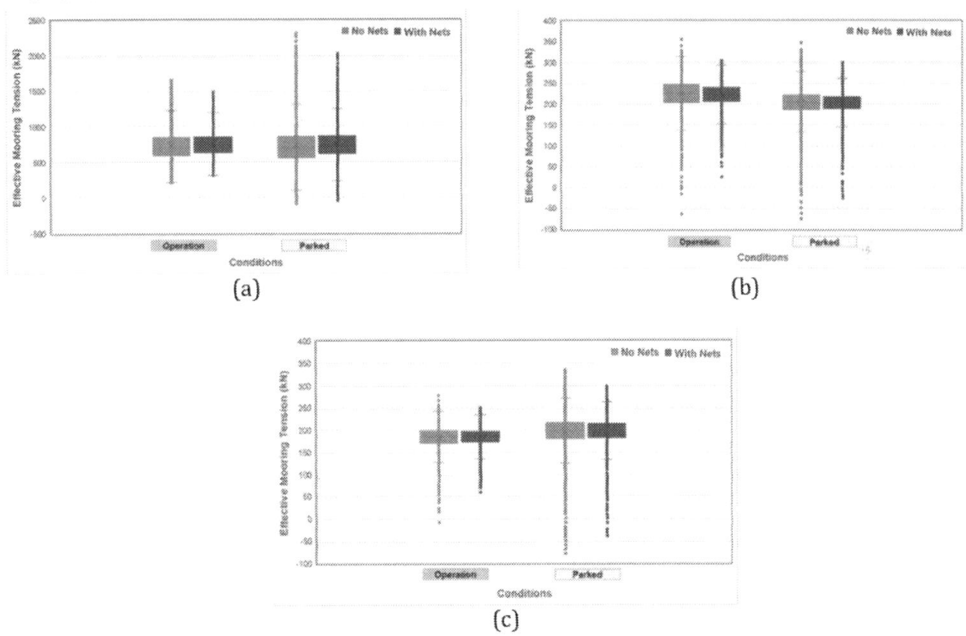

Figure 10. Mooring chain tension of the FOWT and the FOWT-AC at (a) mooring chain-1 (b) mooring chain-2 and (c) mooring chain-3.

4. Conclusions

In this study, the dynamic behaviours of an innovative semisubmersible floating wind turbine with aquaculture cages were explored and investigated under operating and parking conditions (50-year extreme sea state). Correspondingly, the motions, generated power, and mooring tension of the FOWT both without and with fish nets are analysed and discussed under irregular waves and wind and wave combinations. In addition, the influence of the nets on the dynamic response of the new FOWT is also investigated under different sea conditions. Finally, the analysis results can be summarized as follows.

1. The presence of fish nets has a significant effect to the FOWT-AC's surge, heave, and pitch motions compared to the system not integrated with the nets. It reduces the wave energy concentrated in the wave frequency at 0.08 Hz and introduce additional drag.
2. The nets integrated in the FOWT do not give a significant impact on the power generated compared to the system not integrated with the nets, with a disparity of 0.093 kW.
3. Fish nets reduce mooring tension in both operating and parked conditions, they act as an additional damping mechanism, provides a significant difference in the mooring line tension of the FOWT-AC, with tensions decrease of 10% and 38% for operating and parking conditions, respectively.

In the combination of wind and waves, the presence of the nets can serve as a dampening factor or excitation to the motion, acceleration and peak load on the tower. Although the research findings can be utilized to enhancing the innovative design of the FOWT, there are still numerous challenges remains before the new FOWT can be applied, such as the combined effect of wind and wave directions, current considerations, and potential wave slamming effects on the FOWT, etc. Therefore, the dynamic behaviours of the new FOWT under more complex conditions needs to be investigated in the future.

References

[1] K. K. Jaiswal *et al.*, "Renewable and sustainable clean energy development and impact on social, economic, and environmental health," *Energy Nexus*, vol. 7, no. April, 2022, doi: 10.1016/j.nexus.2022.100118.

[2] C. Zhang *et al.*, "Modeling and dynamic response analysis of a submersible floating offshore wind turbine integrated with an aquaculture cage," *Ocean Eng.*, vol. 263, no. August, p. 112338, 2022, doi: 10.1016/j.oceaneng.2022.112338.

[3] C. Zhang *et al.*, "Preliminary Study on an Integrated System Composed of a Floating Offshore Wind Turbine and an Octagonal Fishing Cage," *J. Mar. Sci. Eng.*, vol. 10, no. 10, 2022, doi: 10.3390/jmse10101526.

[4] Y. Lei, S. X. Zhao, X. Y. Zheng, and W. Li, "Effects of Fish Nets on the Nonlinear Dynamic Performance of a Floating Offshore Wind Turbine Integrated with a Steel Fish Farming Cage," *Int. J. Struct. Stab. Dyn.*, vol. 20, no. 3, pp. 1–31, 2020, doi: 10.1142/S021945542050042X.

[5] X. Y. Zheng and Y. Lei, "Stochastic response analysis for a floating offshore wind turbine integrated with a steel fish farming cage," *Appl. Sci.*, vol. 8, no. 8, 2018, doi: 10.3390/app8081229.

[6] G. Ivanov, I. J. Hsu, and K. T. Ma, "Design Considerations on Semi-Submersible Columns and Pontoons for Floating Wind," *Proc. Int. Conf. Offshore Mech. Arct. Eng. - OMAE*, vol. 8, 2023, doi: 10.1115/OMAE2023-101503.

[7] J. Langer *et al.*, "How offshore wind could become economically attractive in low-resource regions like Indonesia," *iScience*, vol. 25, no. 9, 2022, doi: 10.1016/j.isci.2022.104945.

[8] J. Simanjuntak, "Techo-Economic and Institutional Assessment of Wind Energy in Indonesia," Delf University of Technology, 2021.

[9] A. Vidinopoulos, J. Whale, and U. Fuentes Hutfilter, "Assessing the technical potential of ASEAN countries to achieve 100% renewable energy supply," *Sustain. Energy Technol. Assessments*, vol. 42, no. August, 2020, doi: 10.1016/j.seta.2020.100878.

[10] N. W. Hesty, D. G. Cendrawati, Aminuddin, B. Pranoto, S. R. Fithri, and A. Fahim, "Estimasi potensi energi angin indonesia Menggunakan Model Weather Research and Forecast-Four Dimension Data Assimililation (WRF-FDDA)," *Jurnal Sains Dirgantara*, vol. 19, no. 2. pp. 11–20, 2022.

[11] J. Chung and G. M. Hulbert, "A time integration algorithm for structural dynamics with improved numerical dissipation: The generalized-α method," J. Appl. Mech. Trans. ASME, vol. 60, no. 2, pp. 371–375, 1993, doi: 10.1115/1.2900803.

IOP Conf. Series: Earth and Environmental Science **1464** (2025) 012024 doi:10.1088/1755-1315/1464/1/012024

Scaling and modeling strategy for two-dimensional physical model of near-shore floating photovoltaic structure

Maria Angelin Naiborhu[1*], Ahmad Mukhlis Firdaus[2], Farid Putra Bakti[2], Eko Charnius Ilman[2], Ricky Lukman Tawekal[2]

[1]Civil Engineering Program, Faculty of Civil and Environmental Engineering, Institut Teknologi Bandung (ITB), Bandung, Indonesia,

[2]Ocean Engineering Program, Faculty of Civil and Environmental Engineering, Institut Teknologi Bandung (ITB), Bandung, Indonesia,

*E-mail: 35022003@mahasiswa.itb.ac.id

Abstract. Floating Photovoltaics (FPV) is a prospective renewable energy-based power plant for the future. Several large-scale FPVs have been installed in inland water bodies, such as reservoirs, lakes, and hydropower dams. There is an opportunity to install the PV in the nearshore area. The FPV structure consists of numerous individual floaters connected to one another by various types of connections. The PV panels are installed in the modular structure of the floater called an island. Each island typically has hundreds of meters in length and width, while the height of the floaters is less than a meter, resulting in a highly flexible and deformable structure against the external loads from the environment. Shifting the existing FPV design to marine nearshore areas has its own challenges as the environmental loads differ from those of inland water bodies. For the last decade, there has been a growing interest globally in studying the dynamic behaviour of FPV structures under working loads. Several numerical tools were developed to analyze the modularity effect on floating structure responses. However, there is a lack of operational FPV data in marine environments to validate the numerical results. Appropriate physical models, therefore, play an essential role in validation. The complex system of nearshore FPV has raised new challenges for the physical model as the system is considered a fully elastic floating structure, and local deformation is dominant. Elasticity is related to the bending stiffness of the structure, and achieving model elasticity to the equivalent of the prototype while satisfying geometry and mass scaling requirements is challenging. This paper provides a literature review of scaling and modelling strategies for two-dimensional (2D) small-scale physical modelling of floating bodies, as well as the application of a multi-body FPV structure physical model in nearshore areas to investigate structural movement against wave loads.

1. Introduction

Large-scale floating PV (FPV) development in Indonesia is growing in interest as it may conserve land while enhancing PV efficiency. The FPV combines photovoltaics with light floating structure technologies, making it economically viable for deployment in inland water bodies. While PV technology has advanced and proven to convert solar energy into electricity, it has not for its structure. The recent FPV structure, which is generally made up of multiple individual floaters, is indeed suitable for most inland water bodies. However, long-term structural performance remains questionable, as several failures due to wind force have been reported [1], [2]. The potential root causes of failures in FPV systems are further discussed in Ref. [3].

Though the FPVs have been commercially applied on a large scale in inland water bodies, surface water utilisation for FPVs remains limited in order to prioritise the essential public functions. Given Indonesia's features as an archipelagic country with large cities located along its coastline, then the marine environment offers an appealing option for the development of large-scale Floating PV (FPV) plants. Marine FPV can be classified as Offshore FPV (OFPV) and Near-shore FPV (NFPV). Both differ in terms of location, depth, accessibility and environmental conditions. The OFPVs are deployed in deep water, far from the shoreline, where environmental conditions are severe. It also requires a long power evacuation distance as the location is remote from the onshore power grid. Furthermore, due to their remote location, accessing, monitoring, and maintaining offshore installations can be difficult and costly. All of these conditions cause the OFPV to be less economically than the NFPV. The NFPVs, on the contrary, are deployed in relatively shallow and less harsh environmental conditions. Near-shore installations are generally more accessible for maintenance and operation, and shorter power evacuation makes them more economically feasible. Though the NFPV is preferred, shifting FPV development to the marine environment raises additional challenges as the structure is more exposed to varying dynamic environmental conditions than FPV in inland water bodies deployment. Furthermore, near-shore areas tend to have higher levels of maritime traffic and activity, such as shipping lanes, fishing vessels and fish farms. Ensuring the safety of NFPV can be challenging as so many stakeholders are engaged.

Despite the challenges, there has been a growing interest globally in understanding the dynamic behavior of large FPV structures under working loads. Large-scale FPV is considered a Very Large Floating Structure (VLFS) because it has a very large length-to-thickness ratio, hundreds of meters for length, in contrast to less than a meter for thickness [4]. Several numerical tools with different approaches were developed to analyze the structural responses of large scale FPV. A common approach is to represent the structure as a continuous thin plate, and the response structures are analyzed by coupling the Finite Element and Boundary Element Methods [5-9]. Nonetheless, the Computational Fluid Dynamic (CFD) models would present a realistic analysis [10]. Several studies using CFD model for FPV are discussed in Ref. [11], [12]. However, there is a lack of operational FPV data in marine environment to validate the numerical results. Appropriate physical models therefore play an essential role in validation. Several studies on the physical model of floating PV can be found in Ref. [9], [13], however it is difficult to find physical model literature that incorporates the hydro-elasticity effect of the modular floating PV system.

This paper aims to provide a literature review of scaling and modeling strategies for a small-scale two-dimensional (2D) physical model of floating bodies, as well as its application for a physical 2D model for a multi-body NFPV structure to investigate structural motion against the wave load. Particular attention is given to capture and model the hydro-elasticity effect to the motion response of the modular NFPV structure.

2. Overview of Large-Scale NFPV System

The NFPV system consists of four basic components; (i) the floating structure to support the PV array and other necessary equipment, (ii) a PV array, (iii) electrical cables, and (iv) a mooring system to limit the movement and maintain the position of the FPV system. The floating structure is typically made of light and buoyant materials such as high-density polyethylene (HDPE) [10], whereas mooring lines are typically steel chain or mooring rope [14].

The DNVGL defines FPV arrays into three main categories: pure floats, modular rafts, and membranes [15]. The pure float array is a type of FPV in which solar panels are directly installed on the floaters or floating platforms without additional support structures. The modular raft refers to a system where solar panels are installed on pre-assembled floating modules or rafts, which are then linked together to form a larger floating array. The membrane is characterised by attaching the PV-generating technology to a reinforced membrane which is supported by additional structures, such as tubular rings, to provide sufficient buoyancy. This paper will further discuss the NFPV as the modular raft type, as shown in Figure 1(a).

Large FPV plants may consist of multiple large block PV arrays (also known as islands) as shown in Figure 1(b). Each island comprises thousands of floating units, most commonly made of HDPE, that are connected by various types of connections to form a large floating island with dimensions of more than a hundred meters in length and width but less than a meter in height. Figure 1(b) shows an example of an inland large FPV plant in Cirata's hydropower dam, West Java-Indonesia. It consists of 13 islands, where island no. 1-12 has 420 m length and 220 m width, as measured by Google Earth [16], while the depth of the floater is 0.25 m. These islands are equipped with a mooring system. A similar configuration of large FPV plants is intended to develop in the marine environment as well.

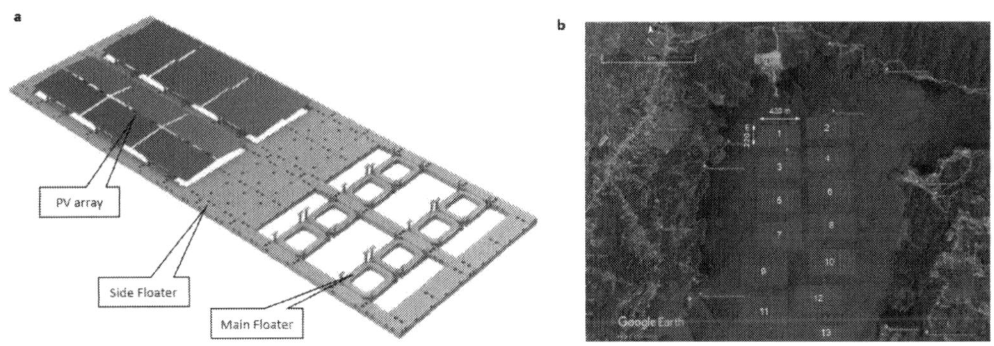

Figure 1. (a) Illustration of the FPV with modular raft type (image property of Flot Indonesia); (b) Number of islands and its arrangement for FPV in Cirata's hydropower dam in West Java-Indonesia

3. Wave Characteristics in the Near-shore Area

Wind waves can be the dominating force received by NFPV structures in open sea areas since the fetch length is significantly longer than in inland water bodies, resulting in higher waves. To define the wave load on NFPV structures, it is necessary to understand the water wave theory that needs to be applied.

The selection of regular water wave theory depends on the wave height (H), wave period (T), wavelength (L), and water depth (h) of the model. Figure 2(a) depicts Le Méhauté guidelines for determining the application range of linear and non-linear regular wave theories by using two dimensionless parameters H/gT^2 and h/gT^2. The graph in Figure 2(a) is updated based on a newly developed 5th-order solution for Stokes wave theory, as shown in Figure 2(b), by introducing H/L and h/L as two dimensionless parameters to determine the range limit of each wave theory [17]. The symbol of λ in Figure 2(a) and L in Figure 2(b) are referring to wavelength. The updated graph provides clear boundary between Stokes and Cnoidal wave theory. For the case of irregular waves, the JONSWAP spectrum is commonly applied for both offshore and nearshore area [18].

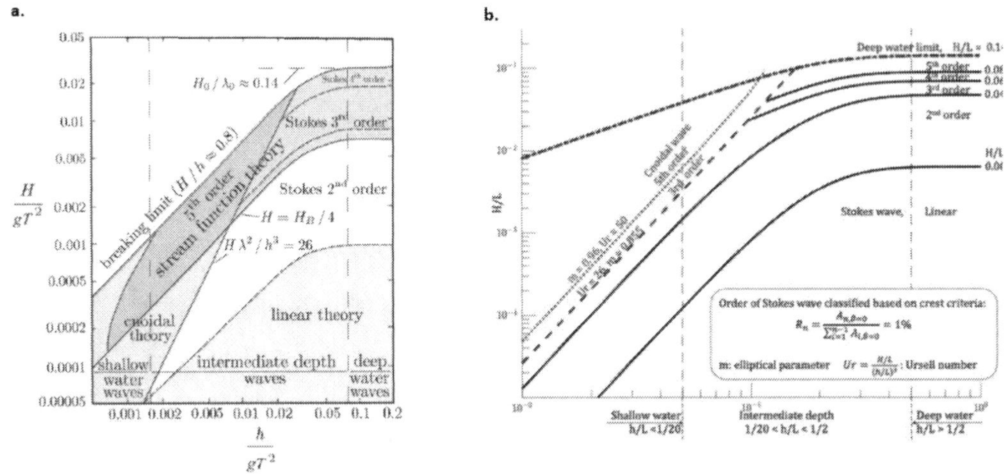

Figure 2. Ranges of applicability of water wave theories: (a) based on Le Méhauté [10]; (b) Updated graph for the applicable range of different wave theories [17].

4. Two-Dimensional Physical Model

A 2D physical model represents the marine structure and its surroundings in a simplified, planar form (two axes) to reduce the complexity of the model while preserving essential geometric and dynamic characteristics relevant to its behaviour. It is generally conducted in a wave flume, where a small-scale model representing the prototype is given certain wave loads (other loads such as wind and current are also applicable, depending on the wave flume's capacity), and its response is measured through various types of sensors [19].

The physical model is critical for validating and calibrating the numerical model to assess the dynamic behaviour of VLFS. The scaling of NFPV as VLFS will be challenging due to several issues: (i) the dimension of the floating structure, (ii) the scaling of bending stiffness [19], and (iii) the structure modularity. Because of its enormous ratio between structural length and ratio, then the NFPV exhibits elastic deformation when subjected to wave loads. By considering its small rigidity, the structure is mostly modelled as an elastic and deformable thin plate. The draft of the plate can be either negligible or non-negligible, depending on the modelling purpose [20].

Scaling the bending stiffness of the modular NFPV prototype is challenging as it has two bending stiffness values, each for the individual floater unit and the connection between floaters, with the individual floater unit having a higher bending stiffness than the connection between floaters, as presents in Ref. [6]. However, such conditions can be simplified as further discussed in the following sub-sections.

4.1. Similitude and scaling law

The physical model must be similar to its prototype in terms of geometry, kinematics, and dynamics. Geometric similitude requires the dimensions of the model to be scaled proportionally to the prototype. It is important that the prototype be geometrically scaled as an undistorted model with the correct weight and mass distribution so the model has correct buoyancy and mass moments of inertia. Kinematic similitude concerns similarity in the relative motion patterns and velocities between corresponding points of the model and the prototype. Finally, dynamic similitude requires that the forces, accelerations, frequencies, and other dynamic characteristics in the model are scaled proportionally to those in the prototype [19].

The NFPV as floating structure is subject to gravitational, inertial, elastic surface tension, and viscous shear force. No scale model will be able to attain dynamic similarity of all these forces. Fortunately, complete similitude is not necessary because only a few of the forces are dominant in the dynamic response of the floating structure [20].

To achieve the required similitudes for the NFPV model, it is important to investigate the required scaling laws for the system. Scaling laws provide a systematic approach to understanding and predicting how properties and behaviours of the prototype change as it scales. A physical model for rigid marine structures with free surfaces followed Froude's scaling law. The Froude number (*Fr*) indicates the importance of the inertia forces of the object to gravitational forces, i.e:

$$F_r = \frac{v}{\sqrt{gL}} \tag{1}$$

The notations v and L represent the reference velocity and length, respectively, while g denotes the gravitation acceleration. However, the NFPV is not a rigid structure due to its size and hydro-elasticity. Hydro-elasticity by means is the interaction between fluid flow (hydrodynamics) and the structural response (elasticity) of a flexible or deformable floating body in or moving through a fluid. For large NFPVs, it is important to understand the floating structure response by taking hydro-elasticity into account, as this might affect the structural integrity and reliability, as well as the performance of the PV panel mounted on it. Thus, the elasticity of the prototype should be maintained in the model. Therefore, in addition to the Froude', the Cauchy's scaling law is desired [22]. The Cauchy number (*Ca*) is a dimensionless parameter to define the ratio of inertia to elastic force and is generally applied when structural flexibility requires, i.e:

$$C_a = \frac{\rho v^2}{E} \tag{2}$$

where ρ and E represent fluid density and Young Modulus of the material.

As g cannot be scaled and fluid density is assumed to remain constant at all times, by applying Froude-Cauchy similitude, the scale factors for the main variables involved in an NFPV physical model are obtained and summarized in Table 1.

The example of scaling result for a multi-body floating PV prototype with scale factor 10 is shown in Table 2. By applying the scale factor, then the mass and bending stiffness of the model will be 10^3 and 10^5 smaller than the prototype. Sometimes, it is not possible finding the material to meet the mass criteria and bending stiffness at once. Such a condition together with wave flume capacity is major consideration while selecting the scale factor for the model.

4.2. Dimensional analysis

Dimensional analysis plays important role in physical model as it offers a method for reducing complexity of physical problems and enable the relevant factors influencing a certain problem to be combined in a manner to ensure the model are properly scaled [23]. For example, the 2D physical model to examine the heave motion of the NFPV structure will be a function of the wave characteristics, water depth, mass and geometry of the floating body, and material properties such as stiffness and density.

On the basis of the above, and following the dimensional analysis principles, then the potential relevant dimensionless parameters for the heave amplitude (*Zh*) are:

$$Z_h = f\left\{\frac{H}{L}, \frac{H}{gT^2}, \frac{h}{gT^2}, \frac{h}{L}, \frac{HL^2}{h^3}, \frac{\rho v^2}{E}, \frac{v}{\sqrt{gL_t}}, \frac{L}{L_t}, \frac{mg}{\rho_w v^2 Bd}, \frac{mgL_tB}{EI}\right\} \tag{3}$$

where f = some functions, E_f = the modulus elasticity of the structure, L_t = total length of the structure, m = mass of the structure, B = width of the structure, d = the height of the structure, and EI = structural bending stiffness. Eq. (3) implies that the heave amplitude will be the same at both model and prototype when all the non-dimensional parameters in Eq. (3) have the same numerical values at model and prototype.

Table 1. The scale factor for the main variables in an NFPV physical model based on Froude-Cauchy's similitude.

Variables	Unit	Scale Factor	Variables	Unit	Scale Factor
Length	m	$\frac{L_p}{L_m} = \Upsilon$	Fluid Acceleration	m/s^2	1
Fluid Velocity	m/s	$\frac{v_p}{v_m} = \left[\frac{L_p}{L_m}\right]^{0.5}$	Force	N	$\frac{F_p}{F_m} = \left[\frac{L_p}{L_m}\right]^3$
Mass	kg	$\frac{m_p}{m_m} = \left[\frac{L_p}{L_m}\right]^3$	Stress/Pressure	N/m^2	$\frac{s_p}{s_m} = \frac{L_p}{L_m}$
Time	s	$\frac{\boldsymbol{T_p}}{\boldsymbol{T_m}} = \left[\frac{\boldsymbol{L_p}}{\boldsymbol{L_m}}\right]^{0.5}$	Bending Stiffness	$N.m^2$	$\frac{EI_p}{EI_m} = \left[\frac{L_p}{L_m}\right]^5$
Angle	rad	1	Frequency	s^{-1}	$\frac{f_p}{f_m} = \left[\frac{L_p}{L_m}\right]^{-0.5}$

Table 2. Principal parameters for a multi-body 2D NFPV model with scale factor 10.

Parameter	Unit	Prototype	Model	Remark
Hydrodynamic				
Regular Wave Height	m	1.0; 1.5; 2.0	0.1; 0.15; 0.2	
Regular Wave Period	s	4.7; 6.3; 8.0; 9.5	1.5; 2.0; 2.5; 3.0	
Water depth	m	6.0; 8.0	0.6; 0.8	
Unit Floater Properties				
Height	m	0.2	0.02	These are properties for a main floater as shown in Figure 1(a).
Width	m	1.1	0.11	
Length	m	1.4	0.14	
Mass (Floater and PV Panel)	kg	42.32	0.042	
EI	$N.m^2$	519.75	0.0052	

4.3. Modeling strategy

In 2D physical model, the NFPV is commonly represented as a single flexible thin plate structure. The plate simplifies the structure as either a solid or perforated continuous body with uniform material properties throughout [11], [24], [25] or with composite material [26], [27]. Such arrangement is suitable for analyzing structure dynamic responses, such as wave-induced motions and mooring system dynamics. However, this approach neglects the interactions between floaters.

For the case where the interaction between floaters is important, for example, to investigate internal force and stress in connection points between floaters, a multi-body structure model using composite materials is applied. In this model, a continuos thin plate is divided into multiple interconnected bodies and the connecting joints are represented using materials that differ from the main bodies [28]. Such an arrangement is adopted based on a common approach in the physical model of large vessels as in Ref. [29].

For multi-body or modular structure model, the NFPV is represented by utilizing two different materials. The material for the floaters model is chosen to meet the mass scale requirement. In order to capture structural deformation, connector beam made of material that meets the bending stiffness scale is chosen and attached on the floaters surface to replace the connection in the prototype as illustrated in Figure 3. This figure is a simplified geometry of the FPV modular raft type as shown in Figure 1(a) for the 2D physical model.

In real conditions, the wave energy tends to disperse along the side of the structure [30]. However, in a 2D model, such a condition cannot be accurately represented as the wave flume has a finite width. This limited width restricts the extent to which waves can propagate and interact with the model in a manner that simulates actual conditions accurately. Waves may reflect off boundaries (the side walls of the wave flume) instead of dispersing naturally as they would in the open sea. To avoid the boundary effect, there should be no gap between the model and the side walls of the flume.

4.4. Instruments

The modelling purpose determines the required type and number of instruments. For example, typical required sensors to investigate the structural response and stresses in floater connections due to external forces (in this case, wave force) for 2D NFPV model includes wave gauges, strain gauges to measure strain and stress, accelerometers to measure floating body movement and high-resolution camera to verified the wave gauge and accelerometer measurement result.

Since different types of sensors are required, sensor installation will be challenging for a small-scale model as their weight may add unintended mass and disrupt the origin model's centre of mass. For example, accelerometers and strain gauges are tiny sensors, but they come along with amplifiers, microcontrollers and accompanying electrical components. Figure 3(a) shows an example of the layout arrangement of strain gauges, accelerometers and mooring load cells that are directly connected to the multi-body 2D NFPV model. The layout should consider mass distribution while achieving the measurement purposes at the critical points of interest. In many cases, it is advantageous to utilise an optical-based motion sensor instead of the accelerometer to assess model motion in order to avoid adding mass to the model, but this sort of sensor is often more expensive.

4.5. Modeling Schemes and Model Setup

Modeling schemes are determined by the purpose of the modeling and developed to obtain correlation between key variables. For example, for a 2D NFPV model aims to in investigate relationship between wave loads to the structure responses, then the modeling schemes should incorporate varying wave periods and heights for regular and/or irregular waves. For modeling in the nearshore area condition, it is also important to varying the water depth to evaluate the effect of bottom changes on waves. Table 2 shows examples of varying wave parameters and water depth for the 2D NFPV model, aims to capture relationship between waves and heave motion of the NFPV structure.

IOP Conf. Series: Earth and Environmental Science **1464** (2025) 012024
doi:10.1088/1755-1315/1464/1/012024

Model setup defines the configuration and model arrangement, as well as how data is collected accurately by using various types of sensors. Sensors arrangement as shown in Figure 3(a) and model positioning inside the flume are part of model setup.

4.6. Modeling workflow

Based on the description mentioned above, a simplified modelling workflow for the 2D NFPV physical model under wave load condition is introduced here. As shown in Figure 4, model simplification is the first and crucial stage as it reduces the complexity of physical phenomenas while still capturing critical factors that affected the floating structure response; nonetheless, it is determined by the purpose of the modeling and the prototype condition. Once the simplification has been defined, the proper scaling law is applied to the selected key variables of the model. The selection of scale factor mostly dictates by the laboratory facilities and material availability, which also affected the accuracy and precision of the modeling result. Model setup and modelling schemes are determined by modelling purpose, which is defined early on.

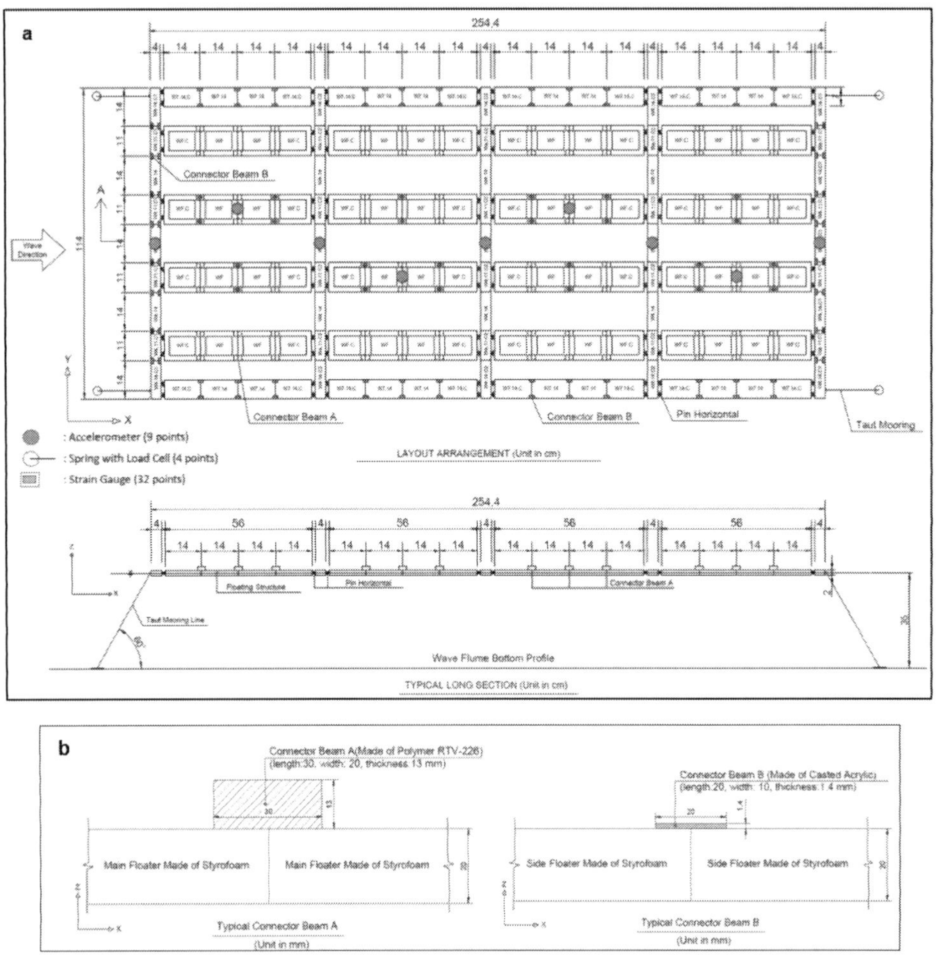

Figure 3. (a) Sensors layout arrangement on a modular NFPV model (units in cm); (b) Typical cross section of modular NFPV model with connector beam.

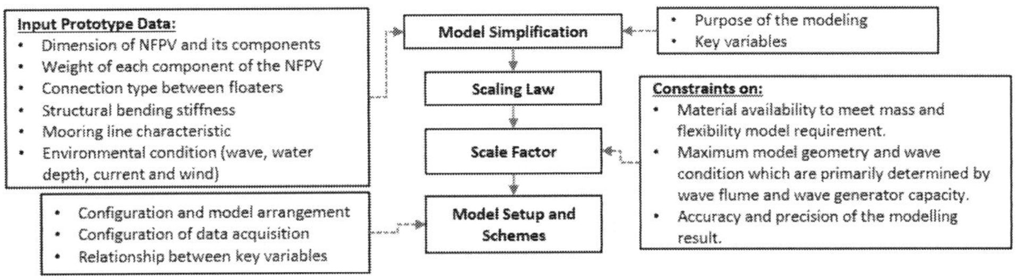

Figure 4. A simplified modelling workflow for the 2D NFPV physical model.

5. Conclusions

This paper mainly reviews scaling and modelling strategies for the 2D modular NFPV physical model, incorporates the hydro-elasticity effect. The Froude-Cauchy's scaling law is applied when structural elasticity is taken into account in motion responses analysis of the NFPV structure model. Constraints in selecting the scale factor and modelling strategy have also been identified and discussed. An example of an ongoing research that incorporates hydro-elasticity is also included to provide a better understanding. The performance of the model will be discussed in separate paper.

It is hoped that the results presented in this paper will help to determine and provide guidance for a better modelling technique for the 2D modular NFPV model. A proper modelling technique provides valuable insight into the performance and verification of the numerical tool that is used in the design of such a system in a marine environment.

Acknowledgement

Acknowledgement for the financial support provided by the Center for Higher Education Fund (*Balai Pembiayaan Pendidikan Tinggi*)-Ministry of Education, Culture, Research and Technology of the Republic of Indonesia, Flot Indonesia for the floating PV prototype arrangement, and OC Enviro for sensors assembly and arrangement.

CRediT authorship contribution statement

Maria Angelin Naiborhu: Conceptualization (equal), Data Curation (lead), Formal Analysis (lead), Visualization (lead), Writing-Original Draft Preparation (lead), Funding acquisition (lead). **Ahmad Mukhlis Firdaus**: Conceptualization (lead), Supervision (equal), Writing-Review & Editing (equal), Funding acquisition (support). **Farid Putra Bakti**: Conceptualization (co-lead), Supervision (equal), Writing-Review & Editing (equal). **Eko Charnius Ilman**: Conceptualization (equal), Supervision(equal), Writing-Review & Editing (equal). **Ricky Lukman Tawekal**: Supervision (lead), Conceptualization (equal), Writing-Review & Editing (equal).

References

[1] Bellini, E. (2021). Japan's largest floating PV plant being reconstructed after Typhoon impact. PV Magazine. https://www.pv-magazine.com/2021/02/22/japans-largest-floating-pv-plant-reconstructed-after-typhoon-impact/

[2] Times of India (2024). Storm damages world's biggest floating solar plant in Madhya Pradesh. https://timesofindia.indiatimes.com/city/indore/storm-damages-worlds-biggest-floating-solar-plant/articleshow/109231119.cms
[3] Zhao, S., Low, Y. M., David Rodríguez-Gallegos, C., & Reindl, T. (2023). Potential root causes for failures in floating pv systems. Proceedings of the ASME 2023 42nd International Conference on Ocean, Offshore and Arctic Engineering OMAE2023
[4] Beinema, T. A. (2022). Floating solar a hydroelastic method. [Online]. Available: http://repository.tudelft.nl/
[5] Ni, X., Cheng, X., Wu, B., & Wang, X. (2018). Coupled analysis between mooring system and VLFS with an effect of elastic deflection of floater. Ocean Engineering, 165, 319–327. https://doi.org/10.1016/j.oceaneng.2018.07.044
[6] Dai, J., Zhang, C., Lim, H. V., Ang, K. K., Qian, X., Wong, J. L. H., Tan, S. T., & Wang, C. L. (2020). Design and construction of floating modular photovoltaic system for water reservoirs. Energy, 191. https://doi.org/10.1016/j.energy.2019.116549
[7] Ni, X., Cheng, X., Wu, B., & Wang, X. (2018). Coupled analysis between mooring system and VLFS with an effect of elastic deflection of floater. Ocean Engineering, 165, 319–327. https://doi.org/10.1016/j.oceaneng.2018.07.044
[8] Shirkol, A. I, Nasar, T (2017). Coupled boundary element method and finite element method for hydroelastic analysis of floating plate, Journal of Ocean Engineering and Science, vol. 3, 19-37, https://doi.org/10.1016/j.joes.2017.11.003
[9] Yang, R.-Y.; Yu, S.-H. A. (2021). Study on a Floating Solar Energy System Applied in an Intertidal Zone. Energies, 14, 7789. https://doi.org/10.3390/en14227789
[10] R. Claus and M. López (2022). Key issues in the design of floating photovoltaic structures for the marine environment, Renew. Sustain. Energy Rev., vol. 164, doi: 10.1016/j.rser.2022.112502.
[11] Sree, D. K. K., Law, A. W. K., Pang, D. S. C., Tan, S. T., Wang, C. L., Kew, J. H., Seow, W. K., & Lim, V. H. (2022). Fluid-structural analysis of modular floating solar farms under wave motion. Solar Energy, 233, 161–181. https://doi.org/10.1016/j.solener.2022.01.017
[12] Trapani, Kim & Millar, Dean. (2016). Hydrodynamic Overview of Flexible Floating Thin Film PV Arrays.
[13] Delacroix, S., Bourdier, S., Soulard, T., Elzaabalawy, H., Vasilenko, P. (2023). Experimental Modelling of a Floating Solar Power Plant Array under Wave Forcing. Energies, 16, 5198. https://doi.org/10.3390/en16135198
[14] Villa-Caro, R., Carral, J. C., Fraguela, J. Á., López, M., & Carral, L. (2018). A review of ship mooring systems. In Brodogradnja, Vol. 8, Issue 1, pp. 123–149, Brodarski Institute. https://doi.org/10.21278/brod69108
[15] DNVGL (2021). Design, Development and Operation of Floating Solar Photovoltaic Systems, pp. 1–152
[16] Google earth (March 19, 2024). Cirata Solar Power Plant
[17] Kuifeng Zhao, Yufei Wang, Philip L.-F. Liu. (2024). A guide for selecting periodic water wave theories - Le Méhauté (1976)'s graph revisited. Coastal Engineering, 188. https://doi.org/10.1016/j.coastaleng.2023.104432
[18] Shi, Y., Wei, Y., Tay, Z. Y., & Chen, Z. (2023). Hydroelastic analysis of offshore floating photovoltaic based on frequency-domain model. Ocean Engineering, 289. https://doi.org/10.1016/j.oceaneng.2023.116213
[19] Ruzzo, C., Muggiasca, S., Malara, G., Taruffi, F., Belloli, M., Collu, M., Li, L., Brizzi, G., & Arena, F. (2021). Scaling strategies for multi-purpose floating structures physical modeling: state of art and new perspectives. In Applied Ocean Research, Vol. 108, Elsevier Ltd. https://doi.org/10.1016/j.apor.2020.102487
[20] Hughes, Steven A. (1993). Physical Models and Laboratory Techniques in Coastal Engineering. WORLD SCIENTIFIC. https://doi.org/10.1142/2154
[21] Karperaki, A. E., Belibassakis, K. A., & Papathanasiou, T. K. (2016). Time-domain, shallow-water hydroelastic analysis of VLFS elastically connected to the seabed. Marine Structures, 48, 33–51. https://doi.org/10.1016/j.marstruc.2016.04.002
[22] Chakrabarti, S. (n.d.). (1998). Physical Model Testing of Floating Offshore Structures. Offshore Structure Analysis, Inc.
[23] Vassalos, D. (1999). Physical modelling and similitude of marine structures. In Ocean Engineering (Vol. 26).
[24] Feng, M. wei, Sun, Z. chen, Liang, S. xiu, Li, Z., Lv, X., Jia, S. lin, & Hu, X. yue. (2022). Experimental Study of Mooring Type Effect on the Hydrodynamic Characteristics of VLFS. China Ocean Engineering, 36(1), 155–166. https://doi.org/10.1007/s13344-022-0014-y
[25] Sree, D. K. K., Law, A. W. K., Pang, D. S. C., Tan, S. T., Wang, C. L., Kew, J. H., Seow, W. K., & Lim, V. H. (2022). Fluid-structural analysis of modular floating solar farms under wave motion. Solar Energy, 233, 161–181. https://doi.org/10.1016/j.solener.2022.01.017
[26] Yago, K., & Endo, H. (n.d.). On the Hydoroelastic Response of Box-Shaped Floating Structure with Shallow Draft (Tank Test with Large Scale Model).
[27] Jiang, C., el Moctar, O., & Zhang, G. (2024). Nonlinear modeling of wave-structure interaction for a flexible floating structure. Ocean Engineering, 300. https://doi.org/10.1016/j.oceaneng.2024.117489
[28] Li, S., Fu, S., Zhang, S., & Moan, T. (2022). Second-order hydroelastic analysis of a flexible floating structure under spatially inhomogeneous waves. Marine Structures, 86. https://doi.org/10.1016/j.marstruc.2022.103306
[29] Rajendran, S., & Guedes Soares, C. (2016). Numerical investigation of the vertical response of a containership in large amplitude waves. Ocean Engineering, 123, 440–451. https://doi.org/10.1016/j.oceaneng.2016.06.039

[30] Min, Z., & Sebastian, S. (2022). Review of wave interaction with continuous flexible floating structures. Ocean Engineering, 264. https://doi.org/10.1016/j.oceaneng.2022.112404.

IOP Conf. Series: Earth and Environmental Science **1464** (2025) 012025 doi:10.1088/1755-1315/1464/1/012025

Resource assessment and performance analysis of the tidal bridge in Larantuka Strait, Indonesia

Michael Suryaprawira[1]*, Ahmad Mukhlis Firdaus[1] and Alamsyah Kurniawan[1]
[1] Ocean Engineering Program, Institut Teknologi Bandung, Bandung, Indonesia

* E-mail: msuryaprawira@gmail.com

Abstract. Located in East Nusa Tenggara Province, Larantuka Strait possesses one of the highest current velocities among Indonesian straits. With a maximum current speed of 4.1 m/s, it possesses a huge potential for tidal current energy. In this simulation, a series of Tocardo tidal turbine power generators will be installed in a single-line formation under an 800-meter bridge stretching perpendicularly over the strait connecting Flores Island and Adonara Island. This research calculates the potential energy produced when the tidal farm operates in the strait. A 2-dimensional, finite element method, a depth-averaged numerical hydrodynamic model, is created using TELEMAC-2D software to represent the existing flow of the channel. The turbine farm is shaped into a single line of 26 turbines with a diameter of 8 meters each. The distance between the tips of the turbines is 8 meters (1D), and they are installed under a bridge. The turbine is deployed constantly 2 meters below sea level since the section of the bridge with tidal turbines floats. The simplified undisturbed kinetic flux calculation (undisturbed current flow) is often used to calculate the potential of the tidal turbine in a basin (P=0.5ρAv3). However, in this research, the hydrodynamic impact of the tidal turbines is considered to produce a more realistic calculation. The increased drag force introduced in the model represents the physical presence of the turbines blocking the tidal current flow in the channel, reducing the energy produced by the tidal power generators. This method shows a drop of energy potential by 24% per turbine, presenting a more representative model, method, and a more realistic channel energy potential.

1. Introduction

1.1 Tidal stream energy as the emerging source of sustainability problems

Harvested from the ocean using the tidal turbine, tidal stream energy has much potential due to the predictable pattern of tidal waves' speed and direction, which offers efficiency in extracting power. The tidal turbine must be installed in a set of arrays to achieve the maximum results. As the most mature type of ORE, the development of tidal turbines in many countries has advanced [1]. [2] shows that the most promising areas for extracting tidal current energy are in the Mid-Eastern part of Indonesia, as shown in Figure 1.

Previous research on tidal energy conversion has been conducted in several straits in Indonesia using a standard idealized undisturbed kinetic flux method. However, the hydrodynamic impact is not included in the model. Thus, the extracted energy calculation needs to be more accurate, and, as this research has proven, it is far from the actual result.

The research conducted by [3] shows that the presence of the turbine needs to be included in the model to calculate the extracted energy more accurately. Two acceptable approaches are possible to model the hydrodynamic impacts of a tidal turbine over a strait. It is possible to model the drag force induced by a TEC opposing the flow (this can be thought of as being similar to increasing the ground friction), or one can either use the actuator disk method, where the influence of a TEC is modeled through a head drop [4]. The first method was successfully attempted in TELEMAC-2D, a Discontinuous Galerkin Finite Element hydrodynamic modeling software developed in the UK. This research aims to model the hydrodynamic impact of the tidal bridge's presence and the device's extracted energy.

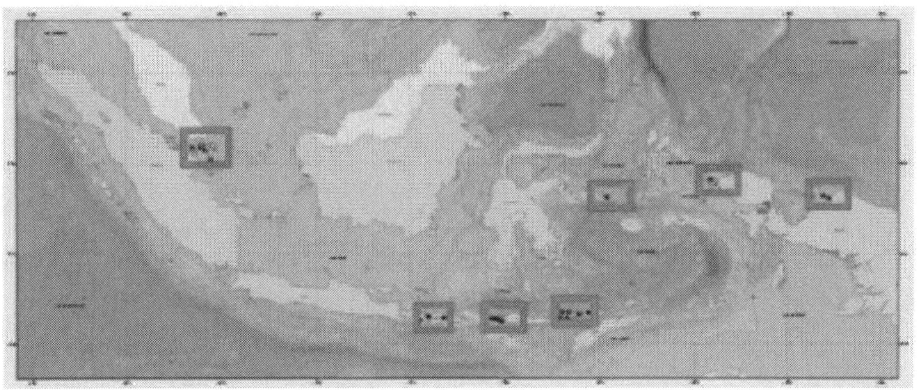

Figure 1. Tidal Current potential location for extraction in Indonesia by The Ministry of Energy and Mineral Resources in 2014.

1.2 Research overview

The research is structured as follows: In the first chapter, a brief introduction about the research is given to tell the purpose of this research along with the area of study and datasets used. Next, the numerical model is constructed, and the result is compared with the field-measured data to prove the model is valid. The next step, perhaps the most important and distinctive feature of this research, is to simulate the presence of the turbine in the model to see the impact they give on the flow of the strait and how it affects the energy generated from the turbine farm. Lastly, we compare the amount of power generated from the current model with the one using the simplified undisturbed kinetic flux equation.

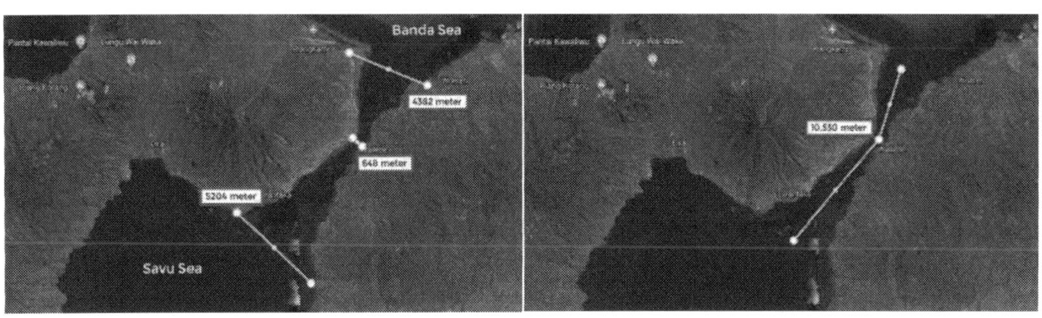

Figure 2. Detailed measurement of the Larantuka Strait, the focus area of the study.

IOP Conf. Series: Earth and Environmental Science **1464** (2025) 012025 doi:10.1088/1755-1315/1464/1/012025

1.3 Area of study

The Larantuka strait stretches between Flores and Adonara islands, connecting the Banda and Savu Seas. The details of this measurement is shown in **Error! Reference source not found.**.

Previous work by [5] proved that this strait is full of energy potential to be extracted using a tidal turbine. However, a different approach is taken in this research as there will only be a single file of tidal turbine farms instead of multiple staggered farms.

More detailed research on the Larantuka Strait by [6] shows the location in the strait with the highest current velocity and energy potential, which will be used in this research to serve as the location of the tidal current energy converter farm shown in Figure 3.

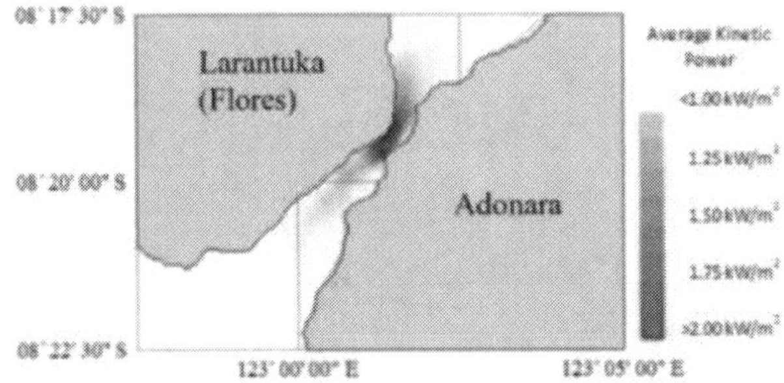

Figure 3. Potential location for tidal energy converter based on undisturbed kinetic flux in Larantuka Strait.

1.4 Research data

The data used in this research is a mix of first-hand field measurement primary data and secondary ones available online. Four main datasets are used in this research: Bathymetry, Water Elevation, Current Speed, and Tide Harmonics Constituents. Bathymetry data is the core of the modelling. The Larantuka Strait bathymetry profile, the finer area of the mesh, is gathered using ADCP by P3GL (Marine Geology Research and Development Center). On the other hand, Savu and Banda Sea were acquired from BATNAS (National Bathymetry), developed by BIG (Geospatial Information Agency). Field water elevation and current speed measurement are required to validate the model. The model has to represent the actual condition of the strait closely. This research also uses water elevation (by P3GL and BIG) and current speed data (by P3GL). The Tidal Harmonic Constituents is extracted from the TPXO Indian Ocean Atlas 2010 developed by [7].

2. The Larantuka Strait Model

2.1 TELEMAC-2D

TELEMAC is Python-based open-source hydrodynamic modeling software that solves shallow water equations, continuity, and momentum for every node in the domain. The TELEMAC-2D solves the second-order partial differential equations derived from the three-dimensional Navier-Stokes equation for the depth-averaged free surface flow equations, also known as the Barré de Saint-Venant equation (1871). It uses the Finite Element Method to solve the shallow water equation. The main result at each node of the computational mesh is the water depth and horizontal components of the vertically averaged velocity (u,v) [8].

TELEMAC-2D allows users to add modules to the calculation, called the steering file, written in Fortran. The presence of the turbine was added to the calculation using the steering file based on one of many modules offered by TELEMAC-2D.

2.2 BlueKenue

BlueKenue is advanced open-source software for performing hydrodynamic pre-modeling and post-modeling activities such as data preparation, mesh generation, analysis, and visualization tools. It produces interface, model input, and data results as the post-modeling products are directly imported from TELEMAC model results. The domain of the research can be seen in Figure 4.

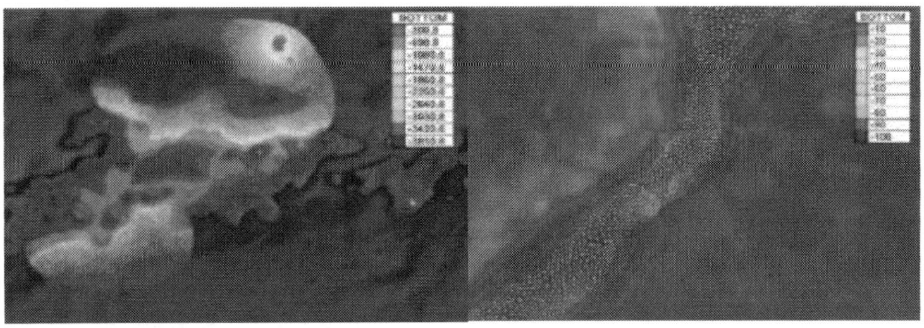

Figure 4. *(left)* Mesh domain of the research. *(right)* A detailed and smaller mesh size in the focus area of Larantuka Strait. All generated using BlueKenue software.

2.3 Model calibration and validation

The strait's free-surface elevation and tidal current in the strait area were calibrated against the P3GL's ADCP results. Since there was no official location information for the P3GL datasets, it was compared with datasets extracted from 9 points in the model around the tidal bridge area. The locations are shown in Figure 5.

The model will be calculated using 3 Manning's Law of Bottom Friction coefficients of 0.03, 0.035, 0.04, and 3 Chezy's Law of Bottom Friction coefficients of 60, 55, and 50. The chosen bed friction law is Manning's Law with a coefficient value of 0.05, which shows the best agreement with the field measurement data. The datasets obtained from each location will be analyzed using the four statistical representations: Standard Error, Mean Bias Error, Pearson Correlation, and Root Mean Square Error. The summary of the important values is shown in Table 1.

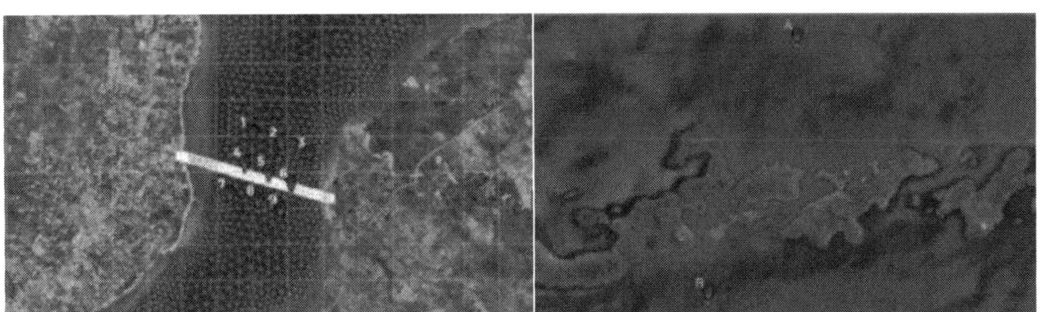

Figure 5. Nine points of observation in the Larantuka Strait (left) and the open water measurement points (left).

Table 1. Statistical summary with Manning's coefficient value of 0.035.

Date	25/6/2009 (10 a.m. GMT +8) – 10/7/2009 (09 a.m. GMT +8)	
Location	Larantuka Strait	
Coordinate	-8.316 °N, 123.025 °E	
Depth (m)	30.28	
	Free Surface Validation	Velocity Validation
Error	8.84%	7.75%
MBE	-0.058	0.095
R	0.875	0.913
RMSE	0.302	0.312

The result shows a good correlation between both data sets. The time-series comparison and correlation plot between ADCP data and TELEMAC results is visualized in Figure 6. Based on this validation process, it is concluded that the modeling result represents the observation data very well, with all values of chosen statistical parameters being under the desired threshold.

Tidal current velocity examines the relationship between the velocity model and observed data. The velocity model validation procedure is similar to the water level validation with qualitative (similarities between both sets of time series) and quantitative (statistical representation) approaches. Nine points with the same coordinate as the ones used in the free surface. The time-series comparison between ADCP data and TELEMAC results at Data Extraction Point 2 for the validation period and full period are shown in Figure 7.

As observed in the validation process, the validation of both the free surface elevation and current velocity shows good agreement with all the datasets satisfy the following conditions:

- The rate of error is below 10%,
- Mean Bias Error of under 0.2 m for free surface elevation and under 0.1 m/s for current velocity,
- Pearson Correlation value is above 0.8, and
- RMSE value between 0.2 - 0.5.

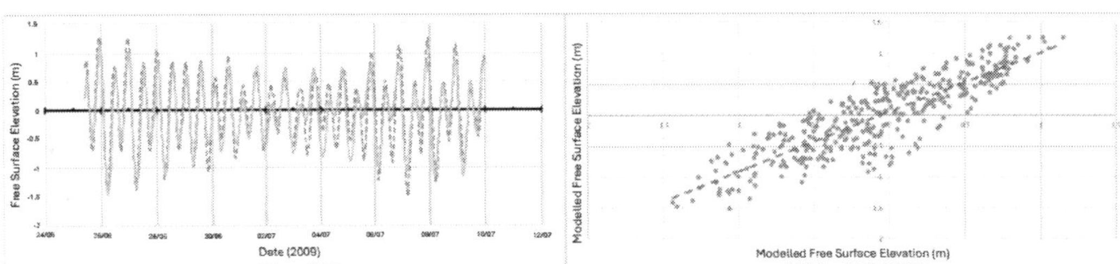

Figure 6. Comparison between ADCP Field Measurement and TELEMAC Result (left). Correlation graph between TELEMAC model result and ADCP Data (right) at Data Extraction Point 9.

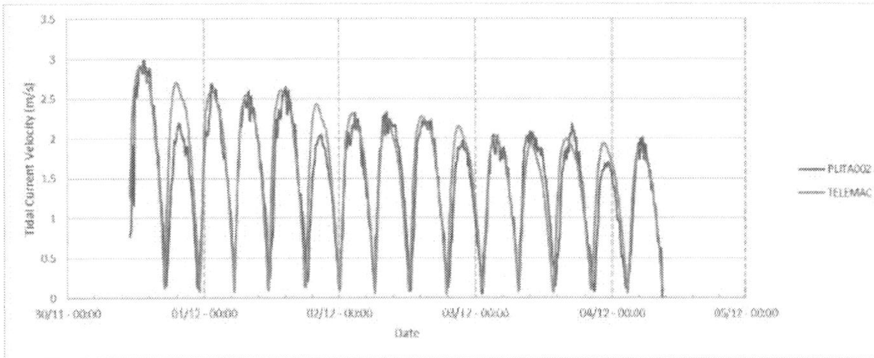

Figure 7. Comparison between observed data and Manning model result at data extraction point 2 for current velocity within the validation period; 2007.

3. Representation of the Turbines in the strait

The maximum average tidal current energy extraction rate is often assumed to be the average kinetic energy flux in the idealized undisturbed state through the most constricted cross-section of a channel where the currents are the strongest. However, there is no simple relationship between the maximum average power and the average kinetic energy flux in the undisturbed state [3]. The calculation shows that the presence of the tidal turbine in the channel will disrupt the flow. Thus, there is a maximum limit to the amount of power that turbines can generate in tidal channels, as too many turbines will block the flow and reduce the current velocity, resulting in a smaller amount of energy extracted. The strait is represented in Figure 8.

[3] describes a flow through a channel with various cross-section profiles, as pictured in Figure 8. The flow u is a function of t and x along the channel.

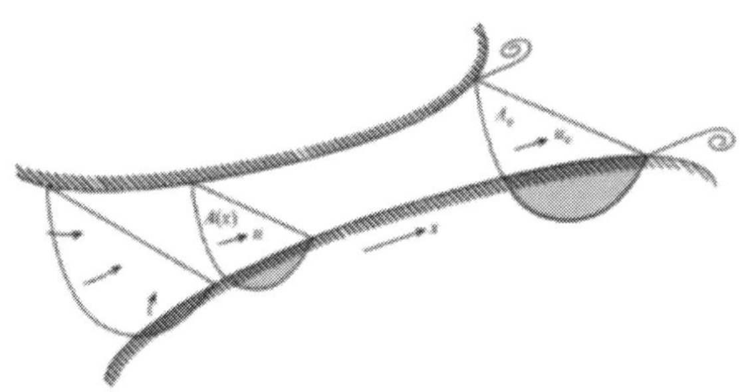

Figure 8. Illustration of a channel by [3].

The power generated at the turbines is the integral along the channel $\int_0^L F\,dx$ of the product of water density, current, cross-section, and the local frictional force that represents the turbine. Assuming that the turbine drag is linearly proportional to the current $u = Q/A$ at any cross-section, it is possible to write $\int_0^L F\,dx = \lambda Q$ and $P = \lambda\rho\overline{Q^2}$, where λ is related to the number of turbines and their location along the channel. The maximum power, in which $\lambda = c\omega$ is expressed as

$$P_0 = 0.25\rho g^2 a^2 c\omega^{-1} = 0.25\ \rho c\omega Q_0{}^2 = 0.25\ \rho g a Q_0 \quad (1)$$

The peak flow is reduced by 71% of its original value. The head loss associated with the turbines is also 71% of the original head difference along the channel.

4. Turbine Efficiency in the Strait

Turbine configuration efficiency is tested using the actuator disc theory which had been attempted previously by [9]. [4] develops the calculation steps and equations further to solve the LMADT equation. The model is divided into 5 stations according to Figure 9 defined as:

1. Far upstream of the turbine where the flow has not interacted with the turbine
2. Directly upstream of the turbine
3. Directly downstream of the turbine
4. Downstream of the turbine where the pressure may be treated as uniform although the flow is not
5. Far downstream of the turbine where the flow may be treated as uniform with and regained velocity equal to the velocity far upstream of the turbine.

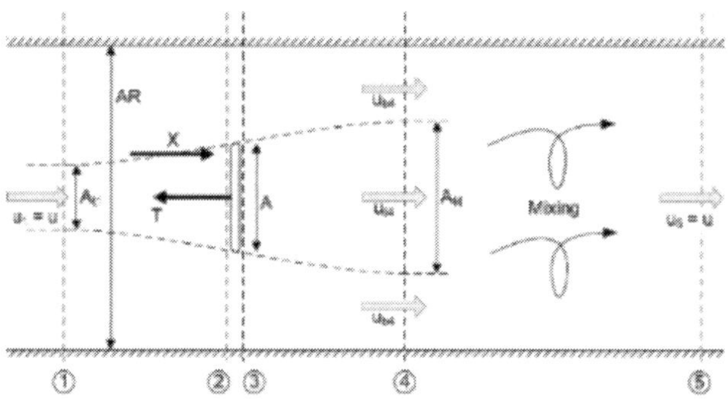

Figure 9. Illustration of turbine presence in the strait by [4].

The turbine constricts the channels creating a blockage. The ratio of the channel blocked by the turbine is defined as Blockage Ratio, B. This condition divides the flow into two parts, the *through flow* and the *bypass flow*, which are identified by a subscript "t" and "b" respectively.

α_4 is a dimensionless velocity factor that has a range of value of $0 \leq \alpha_4 \leq 1$. This value is a percentage ratio between u_4 and u, representing how much velocity is reduced after passing the turbine.

The velocity that passes on either side of the turbine changes due to the constriction of the channel caused by the turbine. This change in velocity is also represented by the constant β_4 which is expressed in the equation:

$$\beta_4 = \frac{1 + B\alpha_2}{1 + B\alpha_2/\alpha_4} \tag{2}$$

Where α_2 is a constant that reduces u to u_2 defined as:

$$\alpha_2 = \frac{1 + \alpha_4}{(1 + B) + \sqrt{(1 - B)^2 + B(1 - \frac{1}{\alpha_4})^2}} \tag{3}$$

Where,

$$T = 0.5\rho A u^2 C_T \quad (4)$$
$$C_T = \beta_4 - \alpha_4^2 \quad (5)$$
$$P = 0.5\rho A u^3 C_P \quad (6)$$
$$C_P = \alpha_2 C_T \quad (7)$$

[4] stated that the maximum power is extracted when the power coefficient C_P is maximized as a function of α_4. Simple differentiation reveals that this occurs when $\alpha_4 = \frac{1}{3}$.

5. Turbine Configuration

The original idea of the bridge was developed by [10]. Unlike the normal deck on the pile, the bridge's turbine section will float. Thus, no pile between the turbines will further disturb the flow. The parameters of the turbines in this model are adapted from a commercially available turbine, the Tocardo T2si Turbine, which cut-in and cut-out water speed are 0.5 m/s and 3.5 m/s respectively, suitable for the research.

The bridge will be installed with a single line of 26 turbines underneath. This configuration is chosen to maximize the potential of high energy production in the strait. The top view and the cross-section of the tidal bridge can be seen in Figure 10. There will be 16 meters (2D) spacing between each tip. The turbines are installed 6 meters above the water level at MSL as shown in Figure 10.

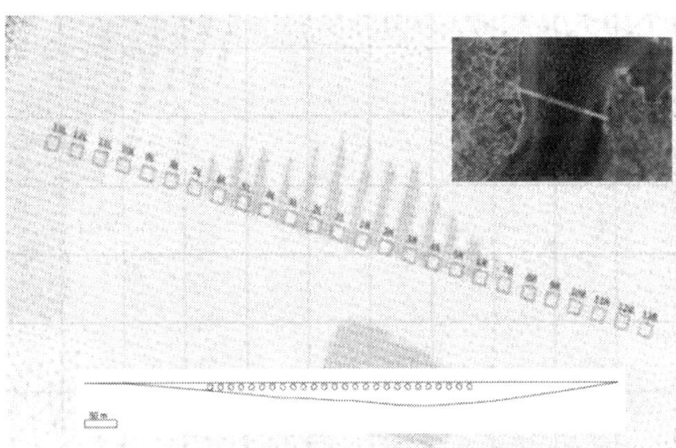

Figure 10. Top view and cross-section view of the tidal turbine bridge.

The turbines are numbered to help identify them easily. The set of turbines is split into two sections: right and left. The numbering starts from the two turbines in the middle, identified as 1R and 1L. The number progressively increases as it moves towards the far-most left and right of the bridge.

Using this configuration, the blockage ratio achieved is 10%, meaning that the turbines blocked 10 percent of the strait. The correlation between the parameters produces a condition where the channel efficiency is at its maximum, where the values of the power and drag coefficient are 0.731 and 1.21, respectively. The correlation can be seen in Figure 11.

IOP Conf. Series: Earth and Environmental Science **1464** (2025) 012025 doi:10.1088/1755-1315/1464/1/012025

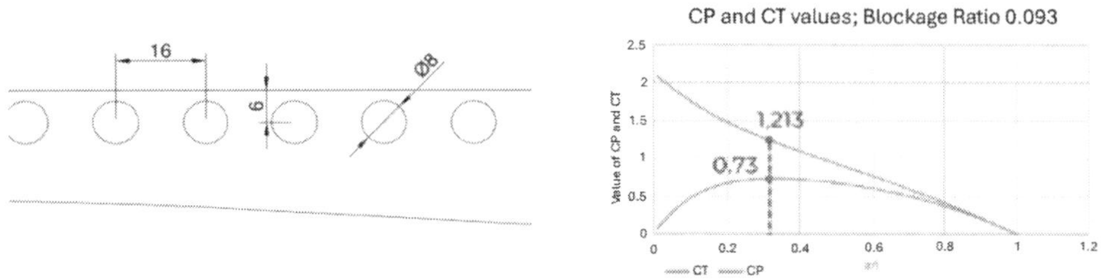

Figure 11. Measurement of the tidal turbines used in the research (left) and CP and CT values of the tidal bridge to optimize channel efficiency (right).

The turbines are then implemented in the model. Using the method developed by [11], a module called DRAGFO is integrated within the TELEMAC environment to model the presence of the turbines in the strait.

6. Hydrodynamic Impact Model

Once the model setup is proven to be valid, a test case will be conducted in which the model will be run for 30 days, starting from 1 January 2022 to 31 January 2022. The result from this model will act as a control model to visualize the strait condition before the turbines are introduced into the calculation and to find the potential location for the turbine bridge to be installed (The highest calculated velocity throughout the month). Figure 12 shows the impact of the turbines on the flow in comparison to the flow without the turbines when the velocity is the highest (05/01/2022 - 9:00:00 GMT).

Velocity data were taken at each turbine to see the difference in current velocity. On average, the velocity is reduced by 12.76% at each observation point. The comparison of average velocity for 6 days is shown in Figure 13.

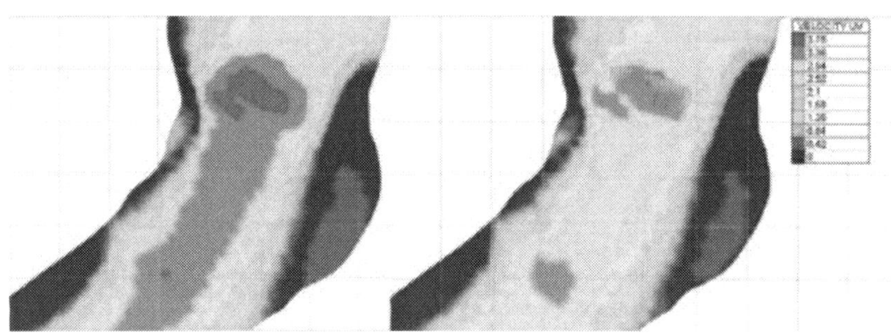

Figure 12. Current velocity at Larantuka Strait without the presence of the turbines (left) and the hydrodynamic impacts from the presence of the turbines (right).

In this research, other parameters affect each turbine's power production, producing a more realistic analysis. One of the most influential factors is the local blockage ratio. The average power produced by each turbine for five days and the comparison of average power production is shown in Figure 14.

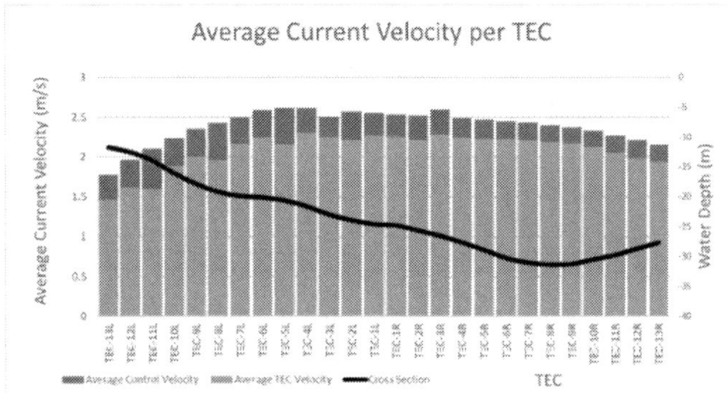

Figure 13. The comparison of current velocity without the turbines (blue) and with the turbines (grey) on every turbine points.

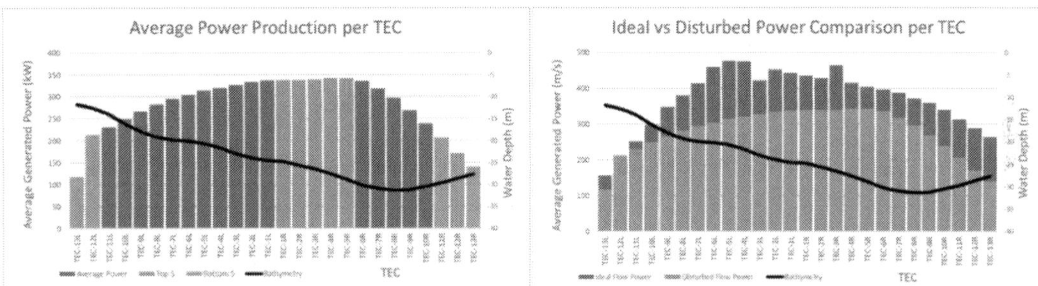

Figure 14. *(left)* Average power production for five days per Tidal Energy Converter. *(right)* The comparison of 5-days average power production per turbine using the undisturbed kinetic flux (blue) and the disturbed ones (orange).

On average, the disturbed kinetic flux method produces 24.3% less power than the undisturbed kinetic flux. This is due to the velocity used to calculate the power having already been reduced by the blockage of the turbine. The channel efficiency due to the blockage ratio is also included in the calculation, producing a more realistic power potential analysis.

The local blockage ratio plays an important part in the turbine power generation rate. As seen in Figure 15, it shows that the highest average velocity in the undisturbed kinetic flux model passes turbine 5L. However, it shows that using the disturbed kinetic flux model, the turbine that produces the highest power average is turbine number 4R. This shows that turbines that have higher current velocity pass through it do not necessarily produce more power.

As shown in Figure 16, the highest velocity passes through turbine 4L. However, turbine 4R produces the highest average power. This shows that local blockage affects the amount of power a single turbine produces besides current velocity.

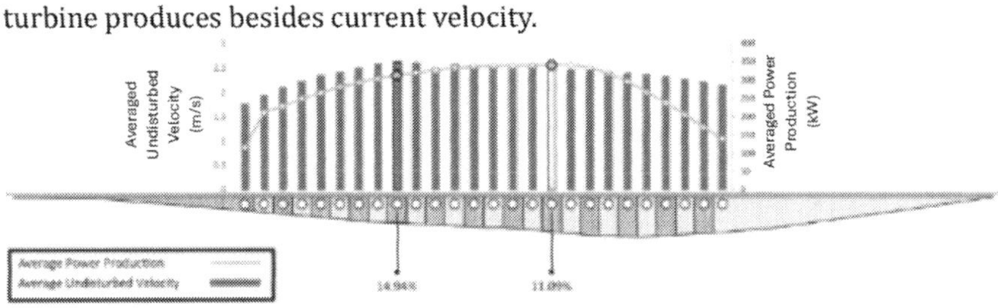

Figure 15. Average undisturbed velocity (bars) and power generated using disturbed kinetic flux method (orange line). Percentage number shows the blockage ratio.

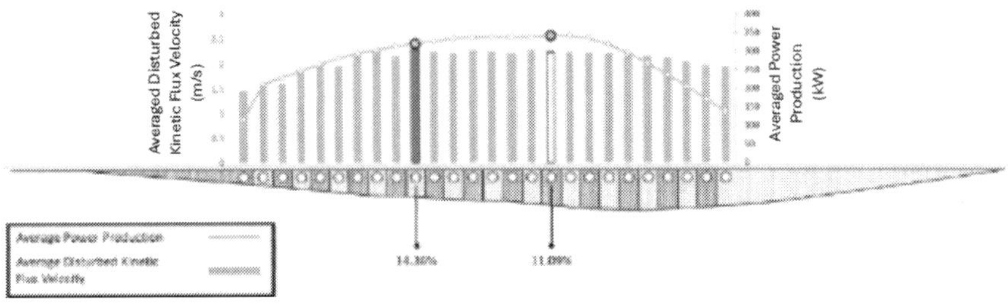

Figure 16. Average disturbed velocity (bars) and power generated using disturbed kinetic flux method (orange line). Percentage number shows the blockage ratio.

7. Conclusions

This research demonstrates a numerical presentation of hydrodynamic impact due to the presence of the turbine and its efficiency in the channel. This shows that the standard undistrubed kinetic flux equation described as P=0.5 ρAv^3, is not representative enough. As shown in the results, the velocity changes both upstream and downstream of the turbines where the velocity decreases more in the latter case. The model result shows a strong correlation between the local blockage ratio of each turbine and power conversion efficiency. The configuration of the array is important to maximize channel efficiency and the distance between turbines plays an important role in increasing efficiency. More experiment, preferably in a 3-Dimensional approach, needs to be performed as to whether a single-line turbine farm is the best configuration for Larantuka Strait.

Furthermore, there are some assumptions made in this research that might differ the results from the real condition, such as:

- Some turbines are too close to the seafloor. This needs to be addressed, as there should be an ideal setting for the turbine's location in the vertical domain.
- The velocity used in this research is depth average.
- The floating bridge creates a situation where the turbine is always two meters below sea level
- The calculation of the turbine does not include the displacement of the water level due to the pressure head loss related to the presence of the turbine.
- The minimum reference velocity included in the research to calculate the drag force of the turbine is 0.5 m/s due to the limitation of the TELEMAC subroutine code. Thus, any drag force that occurs when the current velocity is under 0.5 m/s is considered zero.

References

[1] IRENA, "Ocean Energy: Technology Readiness, Patents, Deployment Status, and Outlook," IRENA, Abu Dhabi, 2014.

[2] A. M. Firdaus, G. T. Houlsby and T. A. A. Adcock, "Opportunities for Tidal Stream Energy in Indonesian Waters," Cork, Ireland, 2017.

[3] C. Garrett and P. Cummins, "The Power Potential of Tidal Currents in Channels," London, 2005.

[4] G. T. Houlsby, S. Draper and M. L. G. Oldfield, "Application of Linear Momentum Actuator Disc Theory to Open Channel Flow," University of Oxford, Oxford, 2008.

[5] K. Orhan and R. Mayerle, "Potential Hydrodynamic Impacts and Performances of Commercial-Scale Turbine Arrays in the Strait of Larantuka, Indonesia," *Journal of Marine Science and Engineering,* 2020.

[6] A. M. Firdaus, "Tidal Energy in Indonesia: Opportunities and Challenges," Melbourne, 2023.

[7] G. D. Egbert and S. Y. Erofeeva, "Efficient Inverse Modeling of Barotropic Ocean Tides," *Ocean Technology,* vol. 19, no. 2, pp. 183-204, 2002.

[8] A. Pérez-Ortiz, J. Pescatore and I. Bryden, "A Systematic Approach to Undertake Tidal Energy," in *European Wave and Tidal Energy Conference*, Aalborg, Denmark, 2013.

[9] C. Garrett and P. Cummins, "The Efficiency of a Turbine in a Tidal Channel," *Fluid Mechanic*, vol. 588, pp. 243-251, 2007.

[10] T. B. BV, "Tidal Bridge Projects," 2021. [Online]. Available: https://www.tidalbridge.com/projects/tidal-power-plant-larantuka-indonesia-3/. [Accessed July 2021].

[11] C.-T. Pham, M. Andreewsky, S. Saviot and Lauriane, "Using the DRAGFO Subroutine to Model Tidal Energy Converters in TELEMAC-2D," Warrington, 2015.

IOP Conf. Series: Earth and Environmental Science **1464** (2025) 012026 doi:10.1088/1755-1315/1464/1/012026

Tidal energy storage simulation in Larantuka strait using non-orthogonal 3D hydrodynamic model

Arung Bahari Muslim[1*], Muslim Muin[2] and Fatwa Islami Azfa[3]

[1] MuTeknologi Komputasi Hidraulika, Bandung, Indonesia
[2] Ocean Engineering Research Group, Faculty of Civil and Environmental Engineering, Bandung, Indonesia
[3] Department of Ocean Engineering, Faculty of Civil and Environmental Engineering, Bandung, Indonesia

*E-mail: arungbahari15@gmail.com

Abstract. Indonesia is a maritime country, and because it is a maritime country, Indonesia has the potential for renewable energy from Sea or tidal currents. One of the straits in Indonesia is Larantuka Strait. So, in this study, Tidal Energy in Larantuka Strait, Nusa Tenggara Timur, Indonesia, is investigated using a three-dimensional Non-orthogonal Boundary-Fitted (NOBF) hydrodynamic model called MuHydro3D. MuHydro3D is suitable for this study because it has a flexible grid system that accommodates the complexity of Larantuka Strait and a robust computational system. Since tidal energy is not constant and depends on the tide, energy storage must be built to stabilize the energy supply. This study proposes to use a water reservoir as storage to stabilize and meet the energy supply. An optimization for the water reservoir is also conducted to choose the best elevation and size of the reservoir where it can obtain the minimum energy losses and cost. The hydrodynamic model simulation results show that Larantuka Strait has annual velocity quadratic ranges from 0.2 to 1.9 m^2/s^2 and the annual maximum ranges from 0.8 to 2.8 m/s, which also shows a good comparison with the observation data. The analysis shows that higher-diameter pipes give higher energy gain in the same reservoir head. With the same diameter, a higher reservoir head gives higher energy gain. The optimization reservoir shows that to accommodate the 2048 Larantuka electricity consumption projection, the reservoir head must be at 100 m with a diameter pipe of 5.05 m, giving a constant energy of 6.03 MW, which also accommodates the 2050 Larantuka electricity consumption projection. The optimum reservoir size is 269.26 million m^3 with an area reservoir of 6731 Ha (Hectare) and a depth reservoir of 4 m.

1. Introduction

Indonesia is a vast country surrounded by sea, estuaries, beaches, and straits. One of the potentials is to harvest tidal energy from straits in Larantuka Strait, NTT. The project to harvest this tidal energy is called The Palmerah Tidal Bridge Project. The project is to create a bridge in the Larantuka Strait that connects the Flores-Adonara Islands. It consists of 4 tidal bridge elements with 32 fish flow tides that power turbines attach to it. The installed capacity of this bridge is 30 MW and has power generation at 80 GWh/year [1].

Since tidal energy is not constant, there are times when the energy supply is zero. Maintaining an energy supply is a top priority for a power plant based on tidal energy. This problem can be solved by creating energy storage. One of the energy storages is a reservoir that is built near the sea or strait area. The reservoir will be filled with seawater using the pump, which operates when the tide or ebb occurs. This concept has been applied in Okinawa, Japan. The storage uses a pure pumped storage facility where the seawater comes from the Philippine Sea. The specification of this storage has a maximum flow of 26 m^3/s, and it has 25 meters depth where the effective storage capacity is 564000 m^3. At the same time, the head of the reservoir is approximately 150 meters above sea level [2].

Figure 1. Sea pump storage in Okinawa [2].

The principle of pumped hydro storage is that the surplus electricity from coal, fossil fuel, or other energy source plants is used for pumping the seawater when periods of low energy consumption occur. If high energy consumption occurs, the reservoirs are opened, allowing water to pass through the hydroelectric turbines to generate the electricity to supply the energy consumption. The illustration of the principle of this pump is shown in Figure 2.

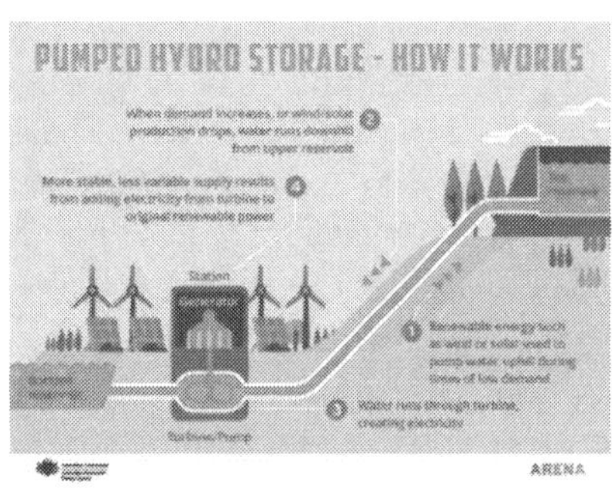

Figure 2. Illustration pump sea storage [3].

The difference between Larantuka sea water storage and Okinawa Sea water is that sea turbines power the pump from ocean currents, not excess electricity from coal, fossil fuel, and other energy resources. The pump is operated when the current ocean speed meets the cut-in speed from the turbine, which is 1 m/s [4]. When the current ocean speed is exceeded, the speed cut-in speed of the excess electricity is supplied to Larantuka. So, the energy is purely from ocean currents. The criteria for when the reservoir must be released is when the turbine does not meet the cut-in speed. The reservoir must be released not based on flood or ebb but when the ocean current at Larantuka Strait is 1 m/s or more. The cut-in speed graph is shown in Figure 3.

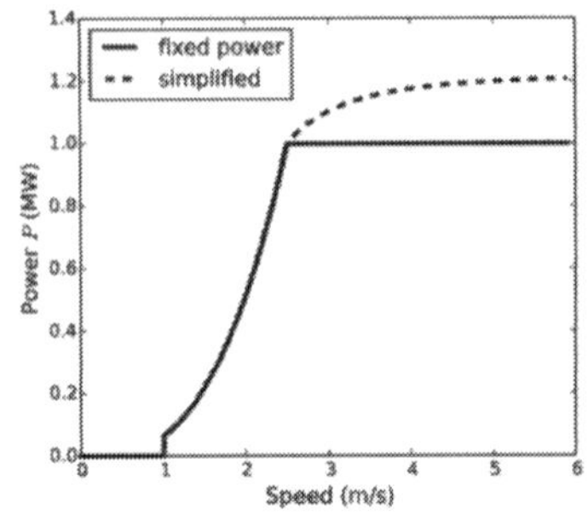

Figure 3. The cut in-speed graph on turbine sea [4].

Finding the best place to put the turbine in the Larantuka strait using a Doppler or current meter is a complex challenge since it takes much operational cost and time. The best way to determine the best location is by using the hydrodynamic model. Previous studies were conducted on potential tidal energy in Larantuka Strait using a hydrodynamic model by Arijibowo [5], Orhan [6], and Firdaus [7]. The disadvantage of these studies is that the hydrodynamic model is not developed in spherical coordinates, and one of them only uses a two-dimensional hydrodynamic model that does not represent the field. Muin [8] developed a 3D hydrodynamic model using a boundary-fitted technique called MuHydro3D. The model was developed in spherical coordinates and transformed to non-curvilinear coordinates suitable for accommodating complex geometry: shorelines, estuaries, and strait. The model results are water level and velocity in three dimensions. The model has been validated using observation data and analytical solutions, which is shown in great comparison [9],[10]. The same model has been used under a different name by American Consultants, where it has been applied in San Fransisco Bay [11], Buzzard Bay [12], and other areas.

Before optimizing seawater storage occurred, there were limitations, such as reservoir location, energy availability from the current tides, energy consumption on Larantuka, and flow speed in the pipe. After that, process optimization occurred based on several factors, such as parameter optimization on the reservoir (charge and discharge of the reservoir, pipe specification, volume of the reservoir, and reservoir head), optimization of the volume reservoir (finding the best reservoir width and length and reservoir head), and optimization of cost reservoir construction.

2. Methodology

2.1 Hydrodynamic model

MuHydro3D is a Three-Dimensional Non-Orthogonal Boundary Fitted Hydrodynamic Model that solves continuity and momentum equations in spherical coordinates on the non-orthogonal boundary-fitted grid to simulate a time-varying field of surface elevations and three-dimensional ocean currents. The forcing includes tides, winds, river flows, and density differences. A detailed description of the model with associated test cases can be found in [8], [9], and [13], where the test cases have undergone extensive testing against the analytical solution. Muin also included the water exchange between the wet/dry area (mangrove) and the main channel in the computation.

IOP Conf. Series: Earth and Environmental Science **1464** (2025) 012026 doi:10.1088/1755-1315/1464/1/012026

This ocean circulation model has been applied widely worldwide, including the Providence River [10], Hudson River [14], Nerang River [15], and many more.

The dependent and independent variables in the governing equations are transformed from spherical to curvilinear (boundary-fitted) coordinates. Semi-implicit and split mode techniques are employed to solve the equations. The governing equations are divided into exterior (vertically averaged) and interior modes (vertical structure). The exterior mode is solved implicitly using a sparse matrix solution technique to predict the surface elevation for each grid. The exterior mode velocity is then determined explicitly using the momentum equation. The vertical structure of velocity is calculated using the tridiagonal matrix technique. This methodology allows time step is not limited by wave speed, therefore results in computationally efficient and inexpensive. The momentum and continuity equation can be seen in [9], [10].

2.2 Equation Parameter Optimization for Reservoir

Determining reservoir requires parameter physics and is based on equation. The equation used in this research includes tide energy, discharge pump, Pipe Speed, Head Loss, Energy Loss, Tidal energy Gain, Energy Turbine from Reservoir, and Reservoir Volume. Tide energy determines how much energy can be produced from the turbine sea. The equation for tide energy ($P_{current}$) is shown in Equation 1. This equation depends on 3 parameters: which amount of turbine($N_{Turbine}$), the density of the sea (ρ_{sea}) the area of the turbine ($A_{turbine}$) , and the sea current (uThe tide energy equation assumes that extractable energy is theoretical tidal energy. The area turbine is simplified as the area of a rectangle with a width and height of 20 m. The number of turbines used is 32, and this number comes from the specification of the bridge [1].

$$P_{current} = \frac{N_{Turbine} * \rho_{sea} * A_{turbine} * u^3}{2} \tag{1}$$

The next equation is to determine the discharge pump (Q_{pump}) for the reservoir, as shown in Equation 2. The equation parameter is depends on the tide energy, gravity (g), reservoir level ($Head$), and the density of the sea. After that, pipe speed ($Pipe_s$) is determined using Equation 3, where the discharge pump is divided by the area of the pipe (A_{pipe}).

$$Q_{pump} = \frac{P_{current}}{g * Head * \rho_{sea}} \tag{2}$$

$$Pipe_s = \frac{Q_{pump}}{A_{pipe}} \tag{3}$$

The following equation, which is Equation 4, determines the head loss because pumping the seawater to the reservoir (H_L). The equation depends on the Darcy friction factor, the length of the pipe to the head reservoir ($Lpipe$), the Pipe Speed, and the Pipe Diameter (D_{pipe}). In this study, the Darcy friction factor value is 0.2. The equation to calculate the energy loss (E_L) is shown in Equation 5. Energy loss depends on the discharge pump, the density of the sea, gravity, and head loss.

$$H_L = \frac{0.2 * Lpipe}{2 * g * D_{pipe}} / \tag{4}$$

$$E_L = Q_{pump} * \rho_{sea} * g * H_L / \tag{5}$$

The next equation for optimization is to determine the tidal energy gain (TEG) from the Larantuka Strait. The equation is shown in Equation 6. Equation 7 shows that the Energy Turbine from the Reservoir (ETR) depends on reservoir discharge (RD), reservoir level, seawater density, and gravity. The last equation is to determine the reservoir volume (RV). Equation 8 shows that this equation depends on the discharge pump and reservoir discharge for one hour. The optimization process determines the best physical parameters, such as reservoir level, pipe diameter, and reservoir discharge, which affect the reservoir volume.

$$TEG = P_{current} - E_L \tag{6}$$

$$ETR = RD * head * \rho_{sea} * g \tag{7}$$

$$RV_t = RV_{t-1} + \left(\left(Q_{pump} * 3600\right) - \left(RD * 3600\right)\right) \tag{8}$$

3. Result and Discussion

3.1 Results and validation model

Figure 4 shows the grid system used in this study. Then, the validation model is conducted to determine if it can represent the field data. The validation compares water elevation from the model and observation data. The water elevation from the model and observation data is shown in Figure 5, which shows a good comparison since the model and observation data overlap with each other.

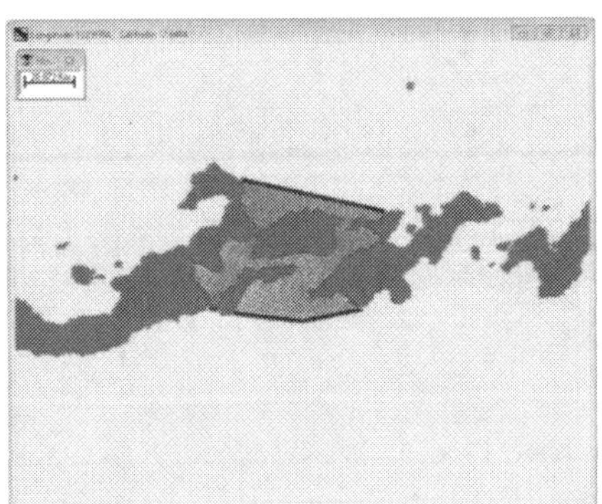

Figure 4. Grid system in the study area.

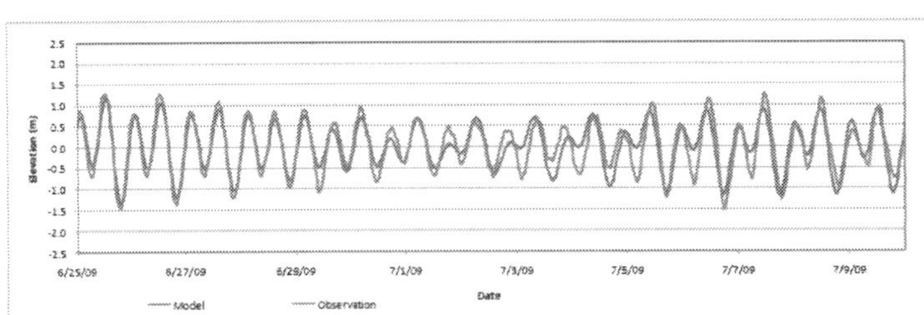

Figure 5. Comparison water level between model and observation.

The results of the hydrodynamic simulation are presented in a contour plot. The plots illustrate the spatial extent of the Yearly Average Velocity Quadratic. Figure 6 shows the contour of the yearly average and maximum in Larantuka Strait (between Flores and Solor Island). The model indicates that the annual velocity quadratic in Larantuka Strait at the area between Larantuka and Adonara ranges from 0.2 m^2/s^2 to 1.9 m^2/s^2, and the yearly maximum speed ranges from 0.8 m/s to 2.8 m/s. With maximum water, velocity is about 3.5 m/s. A comparison of the water velocity from the model with observation data is shown in Figure 7. The comparison shows good results, so MuHydro3D is suitable for predicting reservoir optimization in Larantuka Strait.

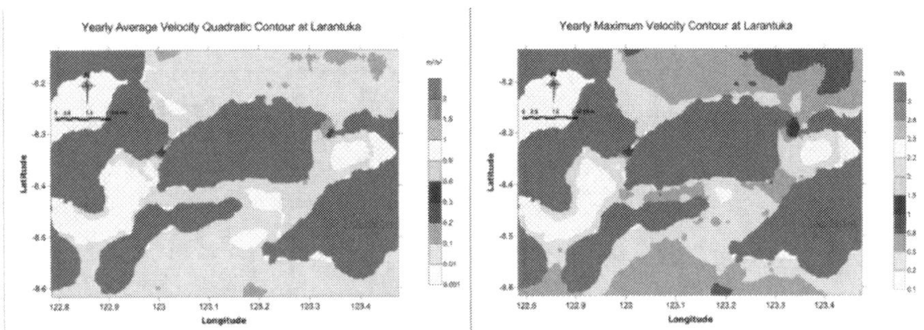

Figure 6. Yearly average and maximum velocity contour at Larantuka Strait.

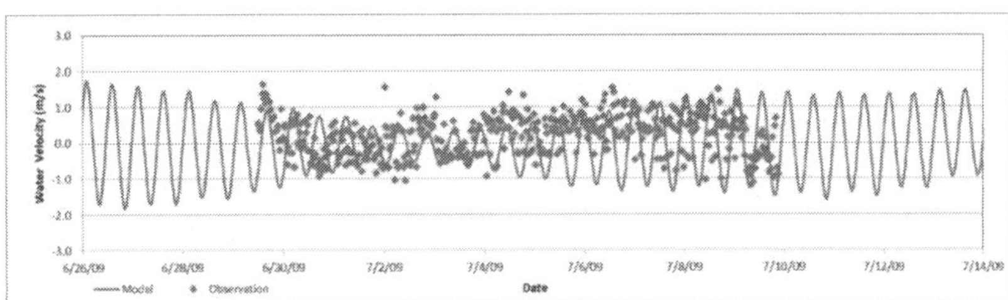

Figure 7. Comparison water velocity between model and observation.

3.2 Reservoir Optimization

In this research, several scenarios are conducted to determine the best parameters for the reservoir. Table 1 shows the parameter reservoir with the variation of pipe diameter from the head reservoir of 60 m. Table 1 also shows that at the same elevation, the higher the pipe diameter, the higher the energy gain. The analysis also shows that the minimum pipe diameter is 7 m for the Larantuka bridge to have a constant output of tidal energy with a reservoir head of 60 m. This is because the energy gain is negative when the pipe diameter is smaller than 6 m. Figure 8 also shows that the tidal energy gain is not constant but is time series or depends on time. There is a time when the minimum energy gain is 0, just like in Table 1. This means the reservoir must be released to operate the turbine and have a constant energy supply.

The next scenario is to vary the head reservoir with the same diameter pipe. Table 2 shows the variation of the head reservoir with a diameter pipe of 7 m. It also shows that with the same diameter, the higher the reservoir head, the higher the energy gained from the tidal energy. Table 1 and 2 show that the same energy gain can be accommodated with a higher reservoir head with a smaller diameter pipe. The analysis can also be interpreted that a lower reservoir head with a bigger diameter pipe can accommodate the same energy gain.

Table 1. Variation of Pipe Diameter for reservoir head of 60 m.

No	Pipe Diameter (m)	Average Percentage Gain (%)	Maximum Energy Gain (MW)	Minimum Energy Gain (MW)	Average Energy Gain (MW)
1a	5	23.55	20.06	-223.61	1.87
2a	6	36.69	31.65	-32.84	8.96
3a	7	41.43	46.53	0	11.52
4a	8	43.44	64.92	0	12.60
5a	9	44.37	78.46	0	13.11

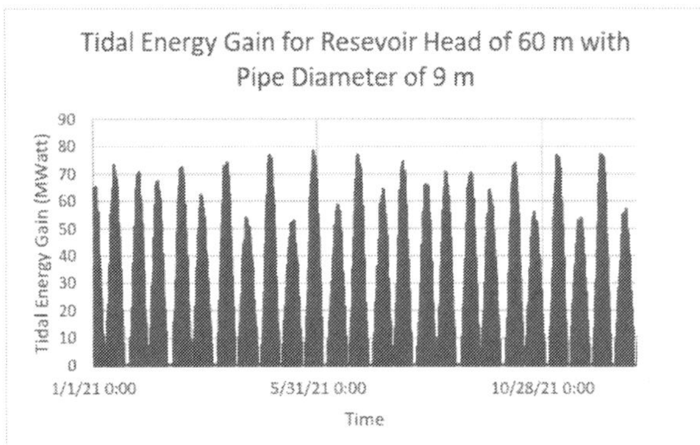

Figure 8. Tidal energy gain for 1 year using reservoir head of 60 m with pipe diameter of 9 m.

Table 2. Variations of reservoir head for diameter pipe of 7 m.

No	Reservoir Head (m)	Length Pipe (m)	Average Percentage Gain (%)	Average Energy Gain (MW)
1b	60	1520	41.43	11.52
2b	70	1710	42.64	12.17
3b	80	1860	43.42	12.59
4b	90	1990	43.95	12.88
5b	100	2100	44.31	13.07

According to Azfa [16], the projection of electricity consumption in Larantuka for 2048 is estimated at 4.74 MW; there is an increase in consumption of 6.7% for each year where the projection graph is shown in Figure 9. From Table 2, the average energy gain for a reservoir head of 60 m is 11.52 MW. So, the reservoir head can be lower than 60 m to accommodate the 2048 Larantuka electricity consumption. The field data shows that if the reservoir head is lower, the village in Larantuka must be relocated. Table 3 shows the variation of the head reservoir and pipe diameter with the minimum energy gain from Tidal that can accommodate 2048 Larantuka electricity consumption. The optimization shows that a reservoir head of 100 m with a pipe diameter of 5.05 m is suitable for accommodating the 2048 Larantuka electricity consumption.

This is because the average energy gain from the other reservoir head is the highest. Also, the minimum energy gain from the tidal using a 100 m reservoir head is 4.78 MW, which is the highest and meets the 2048 Larantuka Electricity consumption prediction.

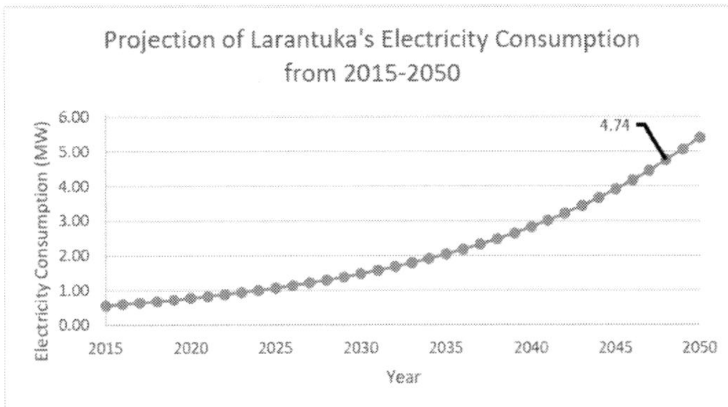

Figure 9. Projection of Larantuka's Electricity Consumption from 2015-2050

Table 3. Variation of reservoir head and pipe diameter to accommodate 2048 Larantuka electricity consumption

No	Reservoir Head (m)	Pipe Diameter (m)	Average Percentage Gain (%)	Minimum Percentage Gain from Tidal (%)	Average Energy Gain (MW)	Minimum Energy Gain from Tidal (MW)
1c	60	6.43	39.28	4.88	10.36	4.66
2c	70	6	39.28	4.75	10.36	4.53
3c	80	5.63	39.27	4.59	10.35	4.37
4c	90	5.32	39.28	4.84	10.36	4.61
5c	100	5.05	39.30	5.01	10.37	4.78

The last scenario is an optimization of discharge from the reservoir to the turbine, which can affect the reservoir volume. The optimization discharge is conducted to create a constant output energy. Table 4 shows optimized reservoir volume variations with a depth reservoir of 4 m. The unoptimized condition is when the reservoir volume is negative, which means the reservoir discharges cannot generate electricity from the turbine reservoir. The analysis shows that the higher the reservoir head, the lower the reservoir volume needed to achieve the 2048 Larantuka electricity consumption. In this scenario, an assumption is made that the reservoir shape is square, so the required length and width have the same number. The next analysis should be optimizing the cost of constructing the reservoir. The optimization of construction costs can be seen in Azfa [16].

Table 4. Minimum required volume reservoir for variation head from 60 until 100 m with depth reservoir of 4 m

No	Head (m)	Energy Turbine from Reservoir (MW)	Minimum Required Volume Reservoir (Million m^3)	Required Area of Reservoir (10^4 m^2 or Ha)	Required Length and Width for Reservoir (10^3 m or km)
1d	60	6.64	374.46	9361	9.68
2d	70	6.05	346.63	8665	9.31
3d	80	6.03	304.10	7602	8.72
4d	90	6.06	269.26	6731	8.20
5d	100	6.03	243.36	6081	7.80

4. Conclusion

The potential tidal energy for Larantuka Strait fluctuates because it depends on the sea current, and the maximum energy gain is not at the maximum speed from the tidal. By constructing a reservoir, the energy supply can be constant. The analysis of the reservoir concludes that with the same head, a higher pipe diameter gives higher energy gain. With the same diameter, a higher reservoir head gives higher energy gain, although the length of the pipe needed for the reservoir is longer than when used with the lower head reservoir. To accommodate the Larantuka 2048 electricity consumption projection (4.71 MW), the tidal energy storage must be placed with a head of 100 m with a pipe diameter of 5.05 m, creating a constant energy supply of 6.03 MW. So, the Larantuka Tidal Bridge cannot provide constant tidal energy of 30 MW nor maximum tidal energy of around 80 MW instead of 6.03 MW with the help of Tidal Energy Storage. This constant energy also accommodates the 2050 Larantuka Electricity Consumption (5.4 MW).

The Required volume to achieve a constant energy supply from the Larantuka Strait is 243.36 million m^3 with a reservoir depth of 4 m. The required area of the reservoir is 6081 Hectares (Ha) with a specification width and length of 7.8 Kilometers (Km). This study also concludes that a higher head gives a lower reservoir volume to achieve the same energy. The deeper the reservoir, the more it can reduce the reservoir area.

5. Future Study

For future studies, the environmental impact analysis will be included because it can affect the natural habitats and displacement of local communities around the Larantuka Strait. A detailed cost-benefit analysis should also be addressed, especially during the storage system's operation, and a comparison between alternative energy solutions should also be addressed.

References

[1] Tidal Bridge. Tidal Power Plant Larantuka, Indonesia - Tidal Bridge [Internet]. 2019 [cited 2024 Sep 11]. Available from: https://www.tidalbridge.com/projects/tidal-power-plant-larantuka-indonesia-3/

[2] Oshima K, Kawai J, Otsuka S, Wada T, Imano H. Development of Pump-Turbine for Seawater Pumped Storage Power Plant. In: Waterpower '99. Reston, VA: American Society of Civil Engineers; 1999. p. 1–6.

[3] The Australia Renewable Energy Agency. What is pumped hydro and how does it work? - Australian Renewable Energy Agency [Internet]. 2017 [cited 2024 Aug 31]. Available from: https://arena.gov.au/blog/what-is-pumped-hydro-and-how-does-it-work/#

[4] Martin-Short R, Hill J, Kramer SC, Avdis A, Allison PA, Piggott MD. Tidal resource extraction in the Pentland Firth, UK: Potential impacts on flow regime and sediment transport in the Inner Sound of Stroma. Renew Energy. 2015;76(April 2015):596–607.

[5] Ajiwibowo H, Lodiwa KS, Pratama MB, Wurjanto A. Field measurement and numerical modeling of tidal current in larantuka strait for renewable energy utilization. Int J GEOMATE. 2017;13(39):124–31.

[6] Orhan K, Mayerle R. Assessment of the tidal stream power potential and impacts of tidal current turbines in the Strait of Larantuka, Indonesia. Energy Procedia. 2017;125:230–9.

[7] Firdaus AM, Houlsby GT, Adcock TAA. Tidal energy resource in Larantuka Strait, Indonesia. Proc Inst Civ Eng Energy. 2020;173(2):81–92.

[8] Muin M. A three-dimensional boundary fitted circulation model in spherical coordinates. University of Rhode Island; 1993.

[9] Muin M, Spaulding M. Three-Dimensional Boundary-Fitted Circulation Model. J Hydraul Eng. 1997 Jan;123(1):2–12.

[10] Muin M, Spaulding M. Application of Three-Dimensional Boundary-Fitted Circulation Model to Providence River. J Hydraul Eng. 1997 Jan;123(1):13–20.

[11] Sankaranarayanan S, McCay DF. Application of a two-dimensional depth-averaged hydrodynamic tidal model. Ocean Eng. 2003;30(14):1807–32.

[12] Sankaranarayanan S. Modeling the tide and wind-induced circulation in Buzzards Bay. Estuar Coast Shelf Sci. 2007 Jul 1;73(3–4):467–80.

[13] Muin M, Spaulding M. Two-Dimensional Boundary-Fitted Circulation Model in Spherical Coordinates. J Hydraul Eng. 1996 Sep;122(9):512–21.

[14] Swanson C, Mendelsohn D, Kim Y, Crowley D, Applied Science Associates I. Hydrothermal Modeling of the Cooling Water Discharge from the Indian Point Energy Center to the Hudson River. 2011.

[15] Zigic S. Modelling The 3-Dimensional Flow Between The Nerang River Estuary And Burleigh Lakes System , Gold Coast. 2004;(July):2–9.

[16] Azfa FI. Optimasi Waduk Penyimpan Energi Untuk Pembangkit Listrik Tenaga Arus Laut Di Kawasan Larantuka. Institut Teknologi Bandung; 2017.

IOP Conf. Series: Earth and Environmental Science **1464** (2025) 012027 doi:10.1088/1755-1315/1464/1/012027

Identification of tidal current renewable energy potency: study case of Moa Strait, Southwest Maluku, Indonesia

Fatchur Rahman[1*], Anasya Arsita Laksmi[1], Munawir Bintang Pratama[2], Muhamad Weldy Hermawan[3], Muhammad Hamzah Fansuri[1], Pungky Dharma Saputra[1]

[1] Civil Engineering, Republic of Indonesia Defense University, Bogor Regency, Indonesia
[2] School of Engineering, The University of Edinburgh, Edinburgh, United of Kingdom
[3] Ocean Engineering, Institut Teknologi Bandung, Bandung, Indonesia

*E-mail: adikfatur1@gmail.com

Abstract. A study based on a numerical modelling approach was carried out to evaluate the potency of harnessing tidal current renewable energy in Moa Strait (8°11' S, 127°46.5' E), at Southwest Maluku Regency of Maluku Province, Indonesia. The research objectives are to build a preliminary tidal hydrodynamic model using Delft3D and analyse of the potential power production from tidal currents at the strait. Observation data verify the constructed model. The model output shows that the sites with higher flow speeds are localized at the bottleneck of the strait on the East side of Leti Island. The averaged and peak flow speeds are around 0.5 m/s and 1.5 m/s, respectively. However, these values are still relatively lower compared to other straits in Indonesian water that have been studied before. Nonetheless, the implementation of tidal energy may have a higher potential due to the nearest two islands being strongly dependent on diesel energy. A hypothetical tidal farm is conducted to investigate the implementation of a Darrius Turbine, a device suitable for regions with low speed. The study revealed that twenty-one Darrieus turbines are estimated to produce up to 18.94 MWh annually. This number is sufficient to meet the demand at Moa and Leti Islands. However further studies are required to take on this hypothesis.

1. Introduction

On the global mission toward energy transition, the Indonesian Government aims to achieve a 31% renewable energy contribution in the 430 GW national energy mix by 2050 [1]. Indonesia has an abundant renewable energy potential, including hydropower, geothermal, solar, wind, and ocean energy. Tidal is one of the promising forms of ocean energy, being predictable and abundant in nature. Tidal energy is a promising option, particularly in remote areas like Moa and Leti Islands (8°11' South, 127°46.5' East), at Southwest Maluku Regency of Maluku Province, which often relies on diesel power plants. Several studies have explored the tidal energy potential in Indonesian waters, such as Larantuka [2,3], Toyapakeh [4], Sunda [5], and Patinti Straits [6]. These straits exhibit strong tidal currents, with peak flow speeds reaching up to 3.5 m/s in some areas. While tidal energy costs are currently higher than conventional sources, it offers an alternative and cleaner energy generation for remote regions with limited grid access. Here, one case is presented, which is the Moa Strait, located between Moa and Leti Islands.

Moa Strait is expected to have tidal energy potential and also heavily relies on diesel energy. Both islands face challenges related to power supply, including reliance on diesel power plants and frequent outages. According to the Saumlaki Customer Service Implementation Unit (UP3), five PLTDs efficiently supply a total of 1,720 kW to 5,524 customers in Moa. Further, according to a sailing direction [7], the strait between Leti Island and Moa Island has strong tidal currents and often causes large waves in the sea.

This research aims to assess the available tidal energy resources at Moa Strait and their potential to fulfil the energy demand at Moa and Leti Islands. A numerical modelling approach is used, employing Delft3D-FLOW tidal hydrodynamics modelling. The modelling output is further analysed to determine the best tidal turbine location, suitable turbines, and produced power. This exercise presents insights into the potential of tidal current energy in providing energy needs in the Moa and Leti Islands.

2. Materials and methods

2.1 Model description

This paper uses the Delft3D modelling tool [8] to build the hydrodynamic model. Features such as RGFGRID, QUICKIN, and Delft3D-Flow are used to build the grids and bathymetry and prescribe the physical parameters, initial conditions, boundary conditions, and observation points. The set-up hydrodynamics model is made in Delft3D-Flow, and detailed explanations are provided in [9] and [8]. The variables used in assembling the model may be obtained from field measurement. However, this case is only sometimes applicable, especially in this case, since Moa Strait is a remote location and is currently still understudied. To facilitate a quick model set-up in the Delft3D framework, users can use Delft Dashboard (DDB), an open-source, user-friendly, enabling access to local and global datasets [10]. Prior studies at sites within Indonesian Water have shown a successful model set-up using DDB [5].

2.2 Modelling domain

The area of interest is located at Moa Strait, as indicated by the red lines in Figure 1, which lies between Moa and Leti Islands in the Southwest Maluku Province. However, the modelling domain coverage must be sufficiently large to accurately replicate the flow hydrodynamics at Moa Strait. The first modelling stage aims to test the sensitivity of domain coverage with a relatively coarser resolution. Three global domains are evaluated, each with a different area: Domain-A (19,593 km^2), Domain-B (11,213 km^2), and Domain-C (1,904 km^2). The Global Domain indicated within the yellow lines in Figure 1(a) represents the smallest domain, Domain-C.

One of the three alternatives of Global Domains is selected with the selection process provided in Section 3.1. From the selected Global Domain, a Local Domain focusing on Moa Strait is built by applying a nesting method. The domain decomposition (DD) online nesting method has been chosen for this purpose. In this DD approach, the Global and Local Models are run parallel, exchanging parameters at each time step. This method has been successfully implemented in numerous previous studies, including in [5,11].

2.3 Model input and parameters

The domain coverages of the selected Global Domain and Local Domain are shown in Figure 1(a) as yellow and red lines, respectively. The grids exhibit a rectilinear structure, generated using DDB and constructed in geographic coordinate systems. The Global Domain measures 43.4 km in both the easting and northing directions, with a cell resolution of 220 m and 197 cells in each spatial dimension. In contrast, the Local Domain at Moa Strait, as shown in Figure 1(b), extends 10.8 km in the easting direction and 10 km in the northing direction. Additionally, the resolution is 74 m with 147 and 135 cells in the easting and northing directions, respectively.

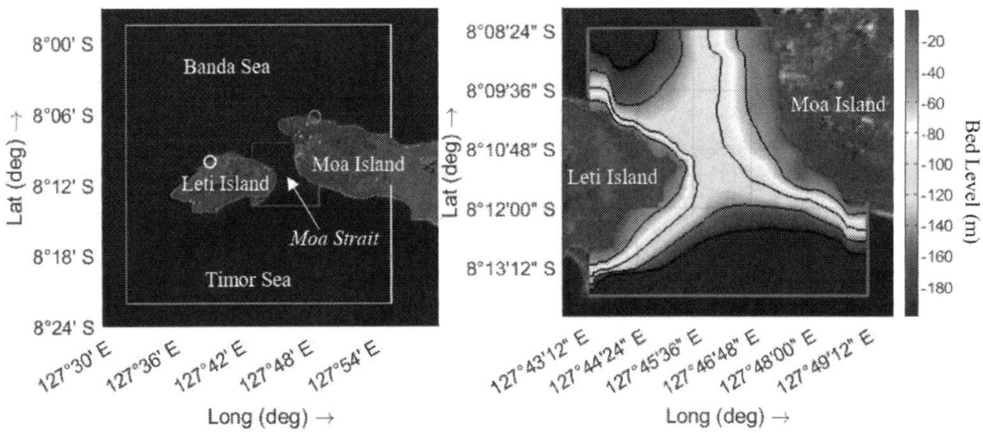

Figure 1. Overview of modelling domain: (a) Global (within yellow line) and Local Domain (within red lines), (b) Bathymetry of Moa Strait. Yellow and Red dots indicate the location of Serwaru and Kaiwatu Harbours, where water level observations are available.

The bathymetry for the Global and Local Domains is generated using the gridded GEBCO_23 dataset available in DDB. Figure 1(b) presents the bathymetry at Moa Strait, constructed using GEBCO 2023 [12]. The open boundaries of the Global Domain are prescribed with thirteen tidal harmonics from TPXO 7.2 [13]. Meanwhile, the Local Domain automatically receives the hydrodynamics parameters at the interface from the Global Domain, indicated by red lines in Figure 1(a). In this study, the transition of bathymetry from global and local domains is crucial to establishing a stable model. Here, the interface is priorly determined carefully where the depth change is not dramatic. Then, a manual depth assimilation, as in point-by-point, is conducted to ensure a smooth transition.

The present model only considers the tidal hydrodynamics, and other forcings have yet to be accounted for. At the initial attempt, the Global Domain also saw a strong jet near the open boundaries, which was solved by applying a buffer zone, a very rough seabed for several grid cells near the boundaries, and smoothly transitioned into a uniform roughness coefficient. This study employs the Chezy coefficient, which value is selected by trial and error, 55, 60, 65, and 75 $m^{1/2}/s$, as discussed in Section 3.1. The viscosity is set in default values as well as the numerical parameters and the initial condition uses a still water condition.

2.4 Data for model validation

There are two water level observation data used to validate the model output, as shown in Figure 1(a), indicated with yellow and red dots. A water-level data from the hydro-oceanographic measurement campaign at Kaiwatu Port is available from August 3rd to 17th, 2018, at hourly intervals. In addition, tidal water level observation at Serwaru Port is also obtained from the Agency of Geospatial Information (*Badan Informasi Geospasial*) from June 1st to 30th, 2021, in hourly intervals. Field current velocity data is also available, which is measured for two days using a current meter at Tiakur Harbour and Kaiwatu Harbour on Moa Island. The model will be simulated during this period. However, the simulation starts five days earlier to facilitate a warming period from an initial still water condition.

3. Results and discussion

3.1 Model validation

Figure 2 presents the validation of the numerical model, where the modelled water is compared against the measured water level at Kaiwatu Harbour and Serwaru Harbour. The result is satisfactory. The model nicely replicates the amplitude, phase, and mixed diurnal-semidiurnal type of the measured data. This shows the suitability of the numerical model for further analysis. Since the site is remote and rarely studied, the flow speed validation is still unavailable here. Therefore, a comprehensive sensitivity analysis was conducted before the validation to ensure the model is at optimum setting. This involved systematically testing the influence of various parameters, including bathymetry, domain size, boundary conditions, and roughness coefficient [5]. To reduce computational costs, the sensitivity analysis was initially performed using a simplified Global Domain, excluding DD to the Local Domain.

GEBCO 2023 provides the most suitable dataset of bathymetry as the two other datasets, GEBCO 2019 and BATNAS, are less accurate in depicting the geometry of Moa Strait. Furthermore, combinations in [5] are tested for the tidal boundaries, and the optimum setting is water level at the North and South sides and current velocity at the East and West sides. Regarding domain size, the Global Domain shown in Figure 1(a) was deemed sufficient, as its water level differed by less than 1 cm compared to the alternative domains, which are 2 and 3 times more comprehensive and longer. Three Chezy roughness values were compared, with 55 $m^{1/2}/s$ producing the closest similarity against the observed water level. Finally, a very rough bed is applied at a few cells along the open boundaries to dampen any unrealistic jet that may be formed.

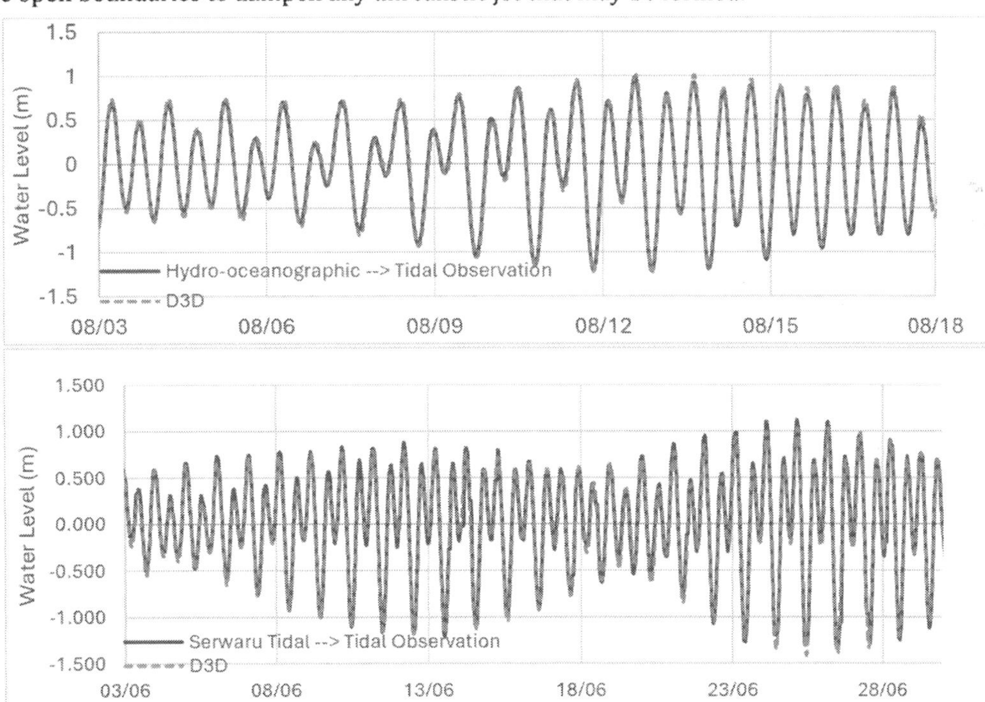

Figure 2. A comparison between water levels was achieved using the Delft3D numerical model (dashed orange line) and measurement data (solid blue line) at (a) Kaiwatu Harbour in 2018 and (b) Serwaru Harbour in 2021.

3.2 Flow patterns at Moa Strait

Figure 3 shows the water level and depth-averaged flow speed curves at spring tide conditions from June 26th – 28th, 2021, observed at one site near Leti Island. The flow speed represents Northward or Southward magnitudes for each positive and negative value. Based on the water level curves, there are two high- and two low-tides in one day. The flow speed curves also have two high- and low-speed cycles within one day. The figure shows that the peak flow to the South or North occurs at or just after the Low- and High-Tides, respectively.

Furthermore, four-time points on June 27th, 2021, are observed to explain the tidal flow patterns in more detail. They are displayed in Figure 4. At 00:00 and 07:00, as shown in Figures 4(a) and 4(b), the southward and northward flows are the strongest, respectively. Sites with the stronger flow speed are seen along the shoreline of Leti Island, which located at the west area of the strait. The strong southward and northward flows occur under a relatively smaller fluctuation from high to low and back to high tides.

Meanwhile, the flow field is weaker when fluctuation is higher, as shown at 14:00 and 18:30 in Figures 4(c) and 4(d), respectively. The flow field also saw a high intensity of eddy along the shoreline of Leti Island. Tidal flow experiences an intense contraction along this shoreline and further produces a shedding mechanism, thus resulting in eddies at both southward or northward flow conditions.

3.3 Potential sites

The location of potential sites for tidal energy harvesting has been hinted at in Figure 4(b), which shows the strongest flow field during a spring tide and are located near Leti Island. The mean velocity during a month is calculated to select a more detailed site, and the result is displayed in Figure 5. The highest mean flow speed is about 0.5 m/s near Leti Island. Further, three siting areas: site-A, -B, and -C, are selected as the potential sites for placing tidal turbines. The sites are indicated in Figure 5(a) inside the white polygons, each has 10, 2, and 6 grid cells, respectively.

In term of the mean flow speed, these sites may be categorized within the lower potential for tidal energy sites, having a similar range of mean flow speed as Lepar Strait [11]. Meanwhile, potential sites at Larantuka [2,3], Toyapakeh [4], and Capalulu Straits [14] possess mean flow speeds of 1-2 m/s. Nonetheless, while power production is cubic of flow speed, the implementation of tidal energy also depends on the availability of other energy sources. In the case of Moa Strait, people at Leti and Moa Islands heavily rely on diesel energy, thus elevating the competitiveness of tidal energy, similar to the case of Patinti Strait [6].

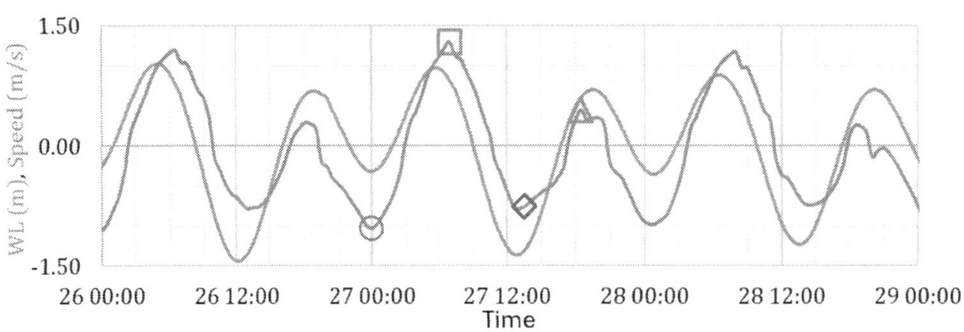

Figure 3. Time-series of water level and current velocity at 55,70 with 30-minute data interval. The four markers indicate the time frame of interest for describing flow patterns at Moa Strait.

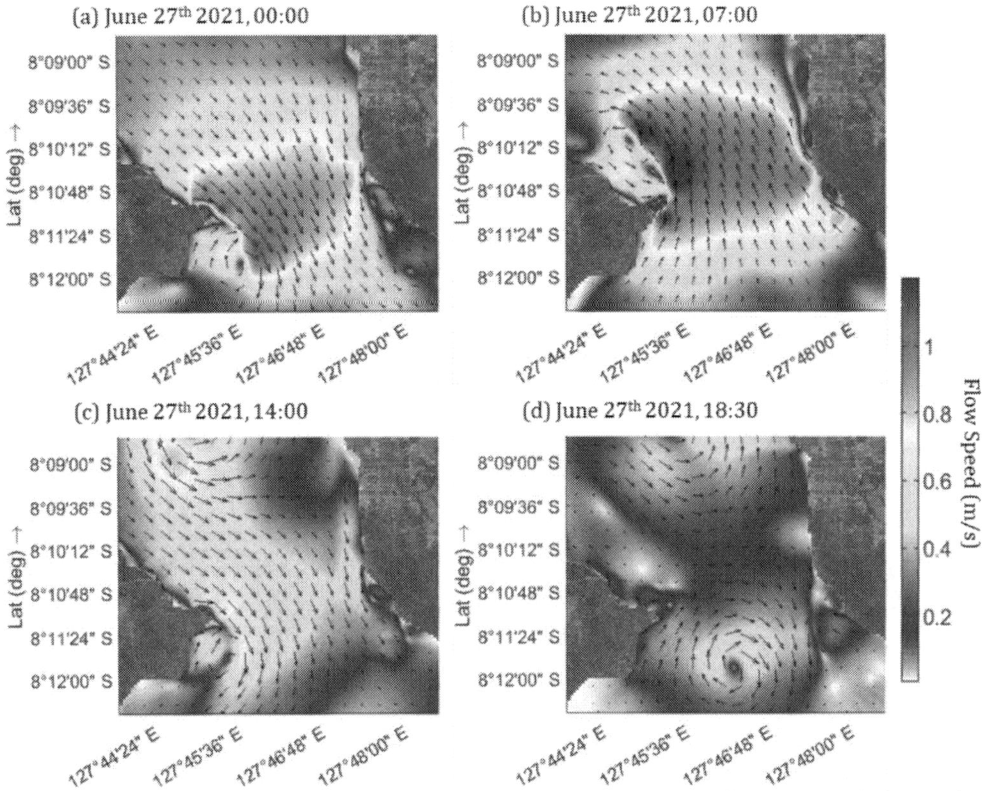

Figure 4. Flow patterns (in-depth-averaged) at Moa Strait on June 27th, 2021, at (a) the 1st low tide [00:00], (b) the 1st high tide [07:00], (c) the 2nd low tide [14:00], and (d) the 2nd high tide conditions.

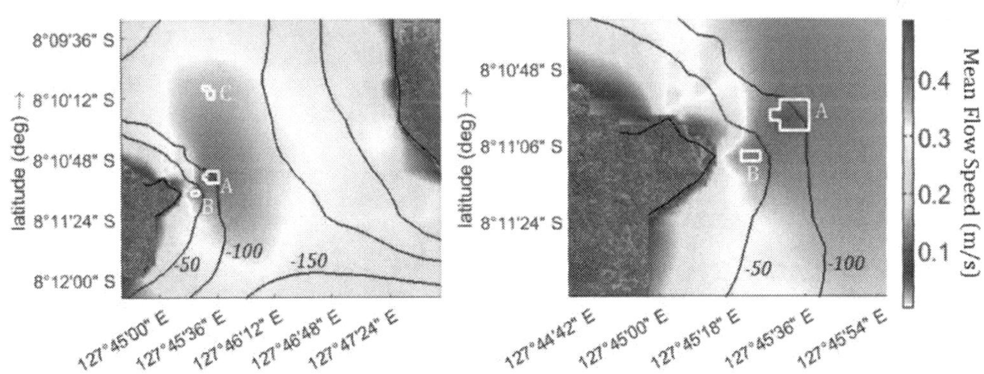

Figure 5. Spatial distribution of mean flow speed during June 2021.

Figure 6 displays the statistics of flow speed at the potential sites at sites A, -B, and -C. As reflected in the mean velocity, it is expected that 55%-80% of the speed falls in 0.01 – 0.5 m/s and between 17%-25% in 0.50 – 0.75 m/s. The significant difference between Site-A, -B, and -C can be seen when observing the percentages of speed in higher class (> 0.75 m/s). Sites A and B have a stronger flow speed.

3.4 Power production

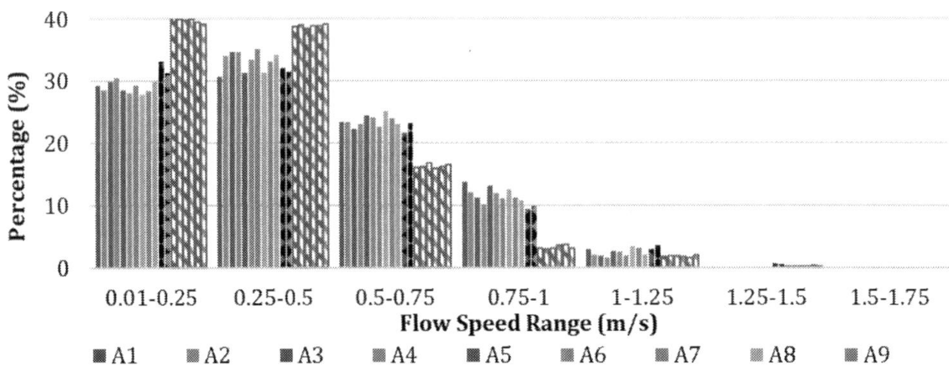

Figure 6. Statistics distribution of flow speed at potential sites.

For each site, the annual power production is calculated. The model is run for three months, and the power at each site is calculated and multiplied by a factor of 4 to estimate the annual power production. Two turbines with a relatively low cut-in speed are used in this analysis, the Darrieus turbine adapted from a study at Suramadu Bridge [15] and the Twin-Hull turbine from a study at Lirung Strait [16]. The specifications of each turbine are presented in Table 1.

Table 1. Specification of Darrieus and Twin Hull Turbines.

Specification	Darrieus	Twin Hull
Diameter	2 m	1.85 m
Area	3.141 m^2	2.68 m^2
Efficiency	35%	35.5%
Cut in Speed	0.5 m/s	0.6 m/s

The power is then calculated using the formula in [11], considering the cut-in speed, turbine swept area, and efficiency. The annual power production in kWh per unit for each site is presented in Figure 7. Comparing the two turbines, the Darrieus Turbine shows a larger power production owing to its larger swept area and lower cut-in speed, although the efficiency is slightly smaller. Also, reflected in the flow speed statistics in Figure 6, the power production displayed in Figure 7 shows that site-C has a lower productivity. Sites with the highest production are A1, A5, A8, and B2, each with annual power production per unit of 891.56 kWh, 883.43 kWh, 902.13 kWh, and 862.88 kWh, respectively. Rough estimation reveals that taking A8 for tidal turbine implementation, twenty-one Darrieus Turbines are sufficient to fulfill the annual electricity demand at Moa and Leti Island, which is about 18.94 MWh.

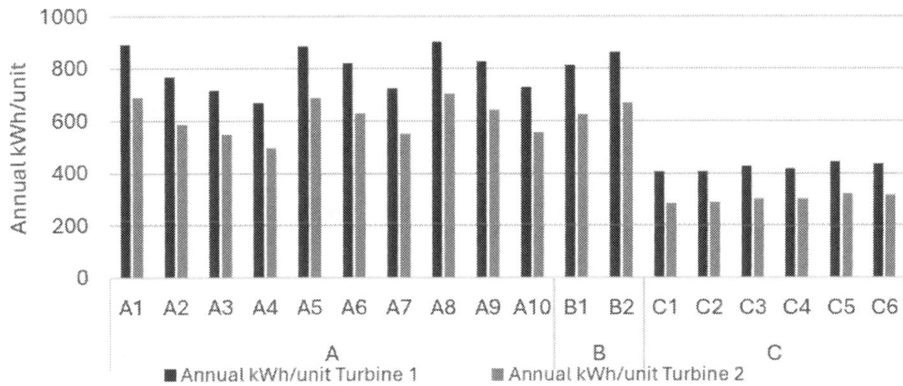

Figure 7. Power production at each potential site for Turbine 1 (Darrieus) and Turbine 2 (Twin Hull).

4. Conclusion

In this study, tidal hydrodynamics numerical modelling using Delft3D-Flow has been conducted at Moa Strait. The model is validated against two water level measurement datasets and shows a good agreement. Although no current velocity validation was performed, the model is still deemed appropriate for the current purpose, providing a preliminary assessment of tidal energy resources at Moa Strait. From the tidal-standalone simulation, it noted that only tidal forcing is being considered, the best siting coordinates for tidal turbines are found around the East coastline of Leti Island. These regions have the highest flow speed, average, and peak at 0.50 m/s and 1.5 m/s, respectively. Further, hypothetical tidal farms are evaluated, and the investigation reveals that 21 units of Darrieus Turbine at 8.184° S, 127.759° E can meet the electricity demands of the islands, producing about 18.94 MWh annually. Again, this is still an initial attempt to utilize tidal energy devices at Moa Strait. The implementation is certainly subjected to a detailed field campaign and modelling exercise and followed by an in-depth feasibility study.

References

[1] Indonesia P R 2014 *Peraturan Pemerintah Republik Indonesia Nomor 79 Tahun 2014 tentang Kebijakan Energi Nasional* (Jakarta: Presiden Republik Indonesia)
[2] Ajiwibowo H, Lodiwa K S, Pratama M B and Wurjanto A 2017 Field measurement and numerical modeling of tidal current in Larantuka Strait for renewable energy utilization *Int. J. GEOMATE* **13** 124–31
[3] Firdaus A M, Houlsby G T and Adcock T A A 2020 Tidal energy resource in Larantuka Strait, Indonesia *Proc. Inst. Civ. Eng. - Energy* **173** 81–92
[4] Kurniawan A, Barli P P A, Pratama M B and Fitriadhy A 2021 Potential Study of Tidal Stream Turbine Farm at Toyapakeh Strait, Bali *Ilmu Kelaut. Indones. J. Mar. Sci.* **26** 155–62
[5] Pratama M B, Venugopal V, Ajiwibowo H, Ginting J W and Novico F 2020 Modelling Tidal Flow Hydrodynamics of Sunda Strait, Indonesia *ILMU Kelaut. Indones. J. Mar. Sci.* **25** 165–72
[6] Novico F, Sudjono E H, Egon A, Menier D, Methew M and Pratama M B 2021 Tidal Current Energy Resources Assessment in the Patinti Strait, Indonesia *Int. J. Renew. Energy Dev.* **10** 517–25
[7] National Geospatial-Intelligence Agency 2023 Sector 4 - The Banda Sea and Off-lying Islands *Pub. 164 Sailing Directions (Enroute) New Guinea* (Virginia, United States: National Geospatial-Intelligence Agency) pp 71–102
[8] Deltares 2023 *Delft3D-FLOW: User Manual* (Delft, the Netherlands: Deltares)
[9] Lesser G R, Roelvink J A, van Kester J A T M and Stelling G S 2004 Development and validation of a three-dimensional morphological model *Coast. Eng.* **51** 883–915
[10] Nederhoff K, Dongeren A van and Ormondt M van 2016 *Delft Dashboard: a MATLAB- based rapid tool for setting*

up coastal and estuarine models (Delft: Deltares)
[11] Ajiwibowo H and Pratama M B 2022 Hydrodynamic Model and Tidal Current Energy Potential in Lepar Strait, Indonesia *Int. J. Renew. Energy Dev.* **11** 15–25
[12] GEBCO Compilation Group 2023 GEBCO 2023 Grid
[13] Egbert G D and Erofeeva S Y 2002 Efficient Inverse Modeling of Barotropic Ocean Tides *J. Atmos. Ocean. Technol.* **19** 183–204
[14] Kurniawan A, Azmiwinata M, Pratama M B and Kusuma C 2024 Tidal current power in Capalulu strait, North Maluku: A feasibility study *Int. J. Renew. Energy Dev.* **13** 375–85
[15] Kasharjanto A, Rahuna D and Aditya R B 2017 Uji kinerja Turbin arus laut-Twin Hull 10 Kilowatt di Jembatan Suramadu *Kapal J. Ilmu Pengetah. dan Teknol. Kelaut.* **14** 79
[16] Rachmat B and Ilahude D 2016 Penentuan Lokasi Turbin Pembangkit Listrik Tenaga Arus Laut Skala Kecil di Perairan Selat Lirung, Talaud, Sulawesi Utara *J. Geol. Kelaut.* **13** 127–42

Empowering local sustainability: assessing small-scale wind energy potential offshore Enggano Island for independent renewable communities

Mochammad Fathurridho Hermanto[1*], Elsa Rizkiya Kencana[1], Nafisa Nandalianadhira[1], Revie Aditia[2], Fadlillah Ihsan[2]

[1] Ocean Engineering Program, Institut Teknologi Sumatera, South Lampung, Indonesia
[2] Ocean Engineering Program Student, Institut Teknologi Sumatera, South Lampung, Indonesia

*E-mail: mochammad.hermanto@kl.itera.ac.id

Abstract. Enggano is an island in North Bengkulu Regency. Last year, the island only had electricity for 14 hours a day, supplied by generators using fuel oil (BBM). The use of fuel oil is considered inefficient because it is difficult to transport to Enggano. In addition, fuel oil is a non-renewable energy source. The use of new and renewable energy can be a solution to these problems because harvesting energy from around the Enggano site itself does not require a lot of money for shipping. One of the new and renewable energy potentials to be analyzed in Enggano is wind energy. Enggano Island has wind speeds of more than 4.0 m/s or medium wind speeds. Therefore, the wind around Enggano Island deserves further energy analysis. The research started with the selection of data collection points. Secondary wind speed and direction data were then obtained from ECMWF over 20 years. The data will then be compared to NASA Power data. The wind data will be used as a reference for the calculation of the wind energy potential offshore Enggano Island. In addition to the wind potential data, the energy needs of the island will be analyzed in the coming years. In this way, the planning of the wind turbines will be in line with the needs of the Enggano Island community to become an energy-independent community. Based on all the analyses carried out, 2,422 MWh of energy will be required every year to meet the needs of around 1300 houses on Enggano Island. To meet this demand, 40 wind turbines with a capacity of 20 kW and a diameter of 13.1 m can be used.

1. Background

Indonesia, according to the Minister of Energy and Mineral Resources in 2023, revealed that oil reserves in Indonesia are estimated to only last around 9-10 years [1]. Fossil energy materials are limited energy to meet future energy needs and fossil fuels can damage the environment [2]. Therefore, to overcome future energy challenges, it is time to develop renewable energy sources. Renewable energy is energy that can be produced quickly through natural processes [3]. Indonesia's electricity production originating from renewable energy has increased until 2020, reaching 13.21%, however, Indonesia's Renewable Energy is still lagging behind neighboring countries, where Cambodia's electricity capacity in 2020 has reached 49.87% and Vietnam's 36.72% in 2021 [4]. One of the renewable energies that have developed rapidly globally in recent years and the fastest growing renewable energy is renewable energy that comes from wind [5], [6], [7]. According to data from the Ministry of Energy and Natural Resources (ESDM), in 2008 Indonesia had quite a large Renewable Energy Potential, one of which is wind, the speed of wind energy in Indonesia according to the Ministry of Energy and Mineral Resources is 3-6 m/s [8]. This wind energy will be converted into electricity to meet the electricity needs of the people around Enggano District. According to Das et al coastal areas have sufficient wind speed to drive large turbines [9]. In this research, the area analyzed is the offshore area of the Enggano sub-district. In recent years global offshore installations have also increased [10].

Enggano District in North Bengkulu Regency is an island that previously only had electricity for 14 hours until mid-2023 and has been electrified for 24 hours with the addition of a generator and fuel oil (BBM) tank [11]. Factors causing the lack of electricity distribution in rural areas include villages that have sparse populations and difficult terrain to reach areas [12].

2. Methodology

This research aims to determine the potential for wind energy and the conversion of wind energy into electrical power in the waters of Enggano Island, so to determine the potential for wind speed, several stages are required as listed in the flow diagram in Figure 1.

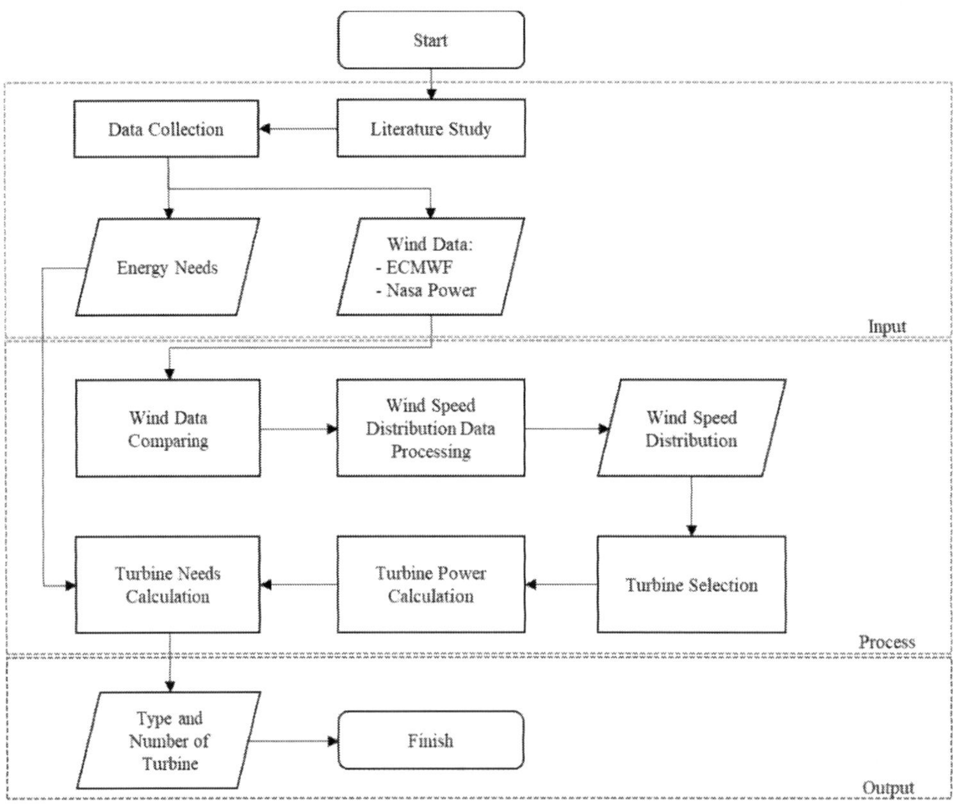

Figure 1. Research Flowchart.

Based on Figure 1, the workflow is divided into three main sections: Input, Process, and Output. The Input phase begins with the data collection and literature study. This phase involves gathering information, such as energy needs and wind speed data from open sources like ECMWF and NASA Power. The collected data serves as the foundation for subsequent calculations and decisions. Wind speed data undergoes a detailed analysis by comparing, processing, and generating the distribution. Following this, turbine selection identifies the suitable turbine type based on the wind speed distribution. The analysis concludes with turbine needs and turbine power to ensure the selected turbine type meets the energy demand. Finally, the results define the type and number of turbines required.

The secondary data collection sources are shown in Table 1. The wind speed data was collected from ERA-5 ECMWF for 2013-2023 and NASA Power for January 2021. These provide a basis for assessing wind potential. The energy sales data for different customer types, sourced from PLN statistics in 2022, offers insights into energy demand. Additionally, the number of households and public facilities, sourced from BPS Enggano in 2023, helps estimate the community's energy needs. This compilation of secondary data enables a comprehensive evaluation of energy requirements and resources for the area.

Table 1. Secondary data.

No	Data	Time	Source
1	Wind speed	2013-2023	ERA-5 ECMWF
		January 2021	NASA Power
2	Energy sold per customer type	2022	PLN statistics 2022
3	Number of households	2023	BPS Enggano
4	Number of public facilities		

After that, ECMWF and NASA Power wind data in January 2021 were compared using the RMSE method and the equation (1) as follows [13]. To find out the level of RMS error, it can be seen in Table 2 [14].

$$RMSE = \sqrt{\frac{1}{n}\sum_{i=1}^{n}(y_i - x_i)^2} \tag{1}$$

Explanation:

- n = Number of data
- y_i = NASA Power wind speed data (m/s)
- x_i = ECMWF wind speed data (m/s)
- i = Sequence of data

Table 2. Interpretation of RMSE.

RMSE	Error Rate
0%-29.9%	Low
30% - 59.9%	Moderate
60% -89.9%	High
>90%	Very High

The data should be compared to assess its suitability for proceeding to the next stage of data processing, which involves determining the Probability Density Function (PDF) of 10-year wind data using the Weibull distribution method. Wind speeds are grouped into intervals, such as 0–1 m/s, 1–2 m/s, and so forth. [15], [16], [17].

The Weibull distribution has 2 important parameters, namely the shape parameter (k) and the scale parameter (c) [18], to calculate the Probability Distribution Function (PDF) Weibull uses an equation (2) [19].

$$f(U) = \left(\frac{k}{c}\right)\left(\frac{vm}{c}\right)^{k-1} exp\left[-\left(\frac{vm}{c}\right)^{k}\right] \tag{2}$$

Explanation:

- Vm = Average wind speed for each hour (m/s)
- k = Shape factor
- c = Scale factor

After that, to find out how suitable the distribution used is by validating the probability results of the Weibull distribution against the measured distribution. Measured distribution is a direct measurement of probability using an Excel pivot which is used as validation of the Weibull distribution using the chi-square test method. The chi-square test equation is as follows [20].

$$\chi^2_{count} = \sum_{i=1}^{n} \frac{(y_i - x_i)^2}{x_i} \tag{3}$$

Explanation:

- x^2 = Chi-square
- yi = Probability measured distribution
- xi = Weibull distribution probability
- i = Sequence of data to

To evaluate the chi-square results, calculate the degrees of freedom using equation (4). That result will be compared to the chi-square distribution table. This step is essential for confirming the results.

$$db = (r-1)(c-1) \tag{4}$$

Explanation :

- db = degrees of freedom
- r = number of rows
- c = number of columns

After knowing the value χ^2_{table}, if $\chi^2_{count} \leq \chi^2_{table}$, then Chi-square is acceptable. If $\chi^2_{count} \geq \chi^2_{table}$, then Chi-square can be rejected. Once the distribution validation results are obtained, a suitable wind turbine can be selected for analysis. Wind turbine technical specifications can be collected through output data from the wind turbine manufacturer itself, which includes technical specifications for wind turbines including rotor power output, rotor diameter, hub height, cut-in wind speed, and so on [21]. Two wind turbines, named Bergey Excel 15 and C&F CF20, that will be analyzed are shown in Table 3 and pictures of them are shown in Figure 2.

Table 3. Interpretation of RMSE.

No	Specification	Bergey Excel 15	C&F CF 20	Unit
1	Hub height	18	20	m
2	Output power	15.6	20	kW
3	Rated speed	11	11	m/s
4	Cut-in speed	2.5	3.5	m/s
5	Rotor diameter	9.6	13.1	m

Figure 2. Wind Turbine Bergey Excel 15 (left), and C&F CF 20 (right) [22], [23].

Then, calculate the probability using the Weibull distribution based on the specifications of the relevant wind turbine. After that, it can convert wind energy into electrical energy. The first stage is to find out the annual probability based on wind groups. Here is the probability formula for wind speed occurring within one year, measured in hours per year, as shown in equation (5).

$$f_{Annual} = f_{wind\ groups} \times 24\ hour \times 365\ day \quad (5)$$

Explanation :

- f_{Annual} = annual probability (hours/year)
- $f_{wind\ groups}$ = probability of each wind group

To determine the annual power output of each wind group, use the equation (6), as follows:

$$f_{Annual\ Yield} = f_{Annual} \times Power\ Output\ of\ each\ wind\ group \quad (6)$$

The power output of each wind group is obtained from the power curve of each turbine, measured in watts or kilowatts.

Then, add up the annual power of all groups of wind using the equation (7), as follows:

$$Power_{total} = \sum_{i=1}^{n} Power_{group} \quad (7)$$

IOP Conf. Series: Earth and Environmental Science **1464** (2025) 012028 doi:10.1088/1755-1315/1464/1/012028

Next, the energy needs of the Enggano District community can be determined. Once the community's energy requirements are established, the necessary number of wind turbines can be calculated using the appropriate equation.

$$N_{turbine} = \frac{Power_{need}}{Power_{total}} \quad (8)$$

The number of wind turbines to meet the electricity needs of the Enggano sub-district is based on the wind turbines that have been selected.

3. Result

3.1. Wind data comparison

The results of the comparison of wind speed data between ECMWF and NASA Power secondary data for 1 month in January 2021 at a height of 10 m are in Figure 3.

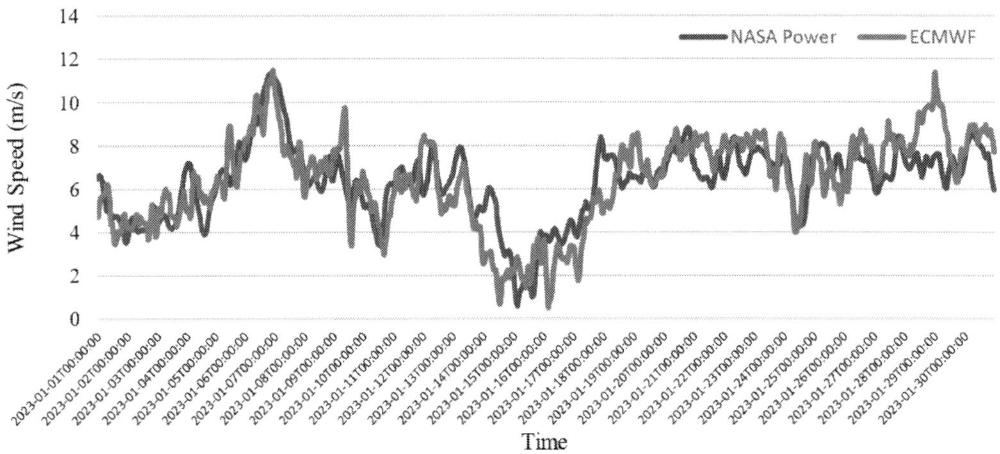

Figure 3. Comparison of ECMWF and NASA power.

The results of the comparison of wind speed data between ECMWF and NASA Power using the RMSE method in the equation (1) are 6.70%, based on Table 2. The results of the RMSE fall into the category of small errors. For the ECMWF average wind speed of 6.3 m/s, the average wind speed at this location has a lower value compared to research in Egypt, which has an average wind speed of 12 m/s [7].

This graph optimistically highlights the strong correlation between wind speed data from NASA Power and ECMWF, showcasing consistent patterns and trends over time. Despite minor discrepancies at specific moments, both sources largely align, reinforcing the reliability of the data. Notably, ECMWF captures slightly higher wind speed peaks, which could indicate enhanced sensitivity in detecting wind variations, adding depth to our understanding of wind behavior across the period analyzed.

3.2. Wind data distribution

The wind speed histogram and theoretical probability density function, using the Weibull distribution in the equation (2) at various heights for 10 years, from 2013-2023 are shown in Figure 4.

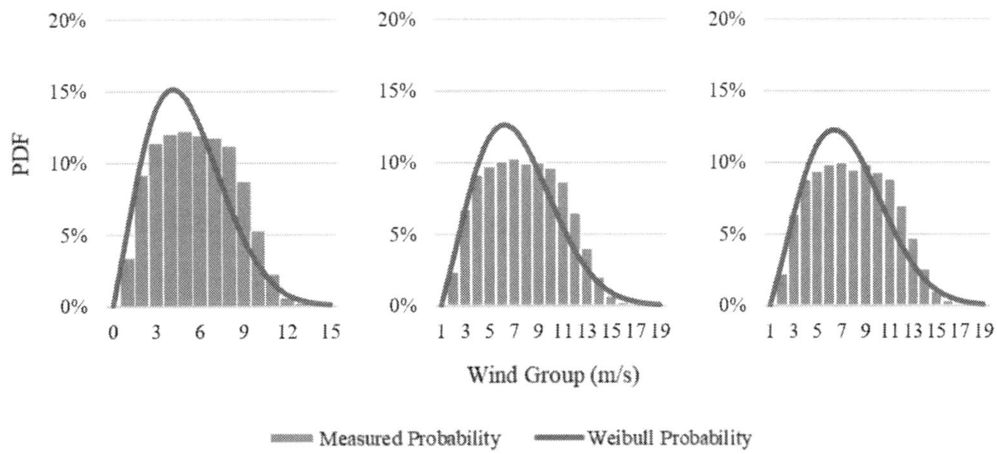

Figure 4. Wind speed histogram and theoretical Weibull Distribution at 10, 18, and 20 m heights.

Figure 4 shows that there is a strong potential for electricity generation, with a data probability indicating 96% and a Weibull probability indicating that 94% of all wind speeds fall within the 2-15 m/s range. This result suggests that these wind speeds are highly reliable for energy production. Notably, the Weibull distribution is dominated by wind speeds of 4-5 m/s. This range domination is higher than had been studied in Chad, where the dominant speed range is only 2-3 m/s [5]. This favorable wind profile highlights a promising opportunity for wind energy generation in the area.

3.3. Validation of probability data between Weibull and measured distributions

To validate the probability data between the Weibull and measured distributions, a Chi-square test (x^2) was performed. The calculated Chi-square value (x_{count}^2) helps determine if the Weibull distribution is a good fit. The chi-square calculation is shown in equation (3). This allows it to be used in power conversion calculations. First, the Chi-square table value (x_{table}^2) is determined based on the degrees of freedom, calculated using equation (4). With degrees of freedom value of 15, x_{table}^2 is found to be 14.3, which serves as a threshold for decision-making.

For wind speed probabilities at a height of 10 meters, the Chi-square validation test produced a x_{count}^2 of 0.078, which is well below the x_{table}^2 value of 14.3. This result indicates that the Weibull distribution is an acceptable model for this data. Previous research by Rasham et al [18] is aligned with this finding. Their study also found that measured wind speeds closely matched the Weibull estimates, suggesting the high accuracy of this method.

3.4. Wind turbine power

After getting the probability value (PDF), calculate the annual probability using equation (5), after that, the annual yield output is based on the wind group, namely using equation (6), thus producing power output based on the wind turbine which is analyzed using equation (7) for the Weibull distribution. All the calculation is summarized in Table 4.

In the Table 4, the wind group column represents the wind speed categories. For example, wind group 1 m/s refers to wind speeds below 1 m/s, while wind group 4 m/s represents wind speeds between above 3 m/s and up to 4 m/s. The probability column defines the likelihood of each wind speed category occurring within a year. The probability function of the Bergey Excel 15 and C&F CF20 turbines differs due to variations in hub height; the Bergey Excel has a height of 18 meters, while the C&F CF20 has a height of 20 meters. The power of each turbine is also adjusted based on its power curve, resulting in the annual power output shown in the annual yield column.

Table 4. Conversion Bergey Excel 15 and CF 20.

Wind Group (m/s)	Bergey Excel 15			CF20		
	Probability (%)	Power (kW)	Annual Yield (kWh)	Probability (%)	Power (kW)	Annual Yield (kWh)
1	3.29%	0	0	3.10%	0	0
2	6.78%	0	0	6.41%	0	0
3	9.74%	0.11	92.11	9.24%	0	0
4	11.73%	0.68	697.69	11.23%	0.70	688.52
5	12.58%	2.07	2,285.03	12.17%	2.07	2,206.72
6	12.31%	3.82	4,123.51	12.07%	3.92	4,145.48
7	11.14%	6.09	5,943.68	11.11%	6.72	6,538.36
8	9.40%	8.50	6,999.66	9.55%	10.59	8,859.81
9	7.43%	11.27	7,327.61	7.71%	14.89	10,058.05
10	5.51%	13.66	6,593.62	5.86%	18.94	9,729.86
11	3.85%	15.61	5,259.51	4.21%	20.08	7,404.50
12	2.53%	16.88	3,740.47	2.86%	20.08	5,023.82
13	1.57%	18.21	2,505.12	1.83%	20.08	3,224.85
14	0.92%	19.10	1,538.74	1.11%	20.08	1,960.01
15	0.51%	20.25	902.72	0.64%	20.08	1,128.50
16	0.27%	20.25	471.75	0.35%	20.08	615.65
17	0.13%	20.25	232.93	0.18%	20.08	318.27
18	0.06%	20.25	108.68	0.09%	20.08	156.02
Total			48,822.82			62,058.40

For the Bergey Excel 15 turbine, the probability of wind speed occurrence is highest in the 5 m/s wind group, at 12.58%. This wind speed group generates power amounting to 2285.03 kWh/year. The Bergey Excel 15 operates at maximum capacity at a wind speed in the 15 m/s group, with a power curve value of 20.25 kW. However, because the occurrence of wind speeds in the 15 m/s group is only 0.51%, the annual yield produced is only 902.72 kWh/year. The total power output generated by the Bergey Excel 15 in one year is 48,822.82 kWh/year.

For the C&F CF20 turbine, the probability of wind speed occurrence is also highest in the 5 m/s wind group, at 12.17%. This wind speed group generates power totaling 4145.48 kWh/year. The C&F CF20 operates at maximum capacity at a wind speed in the 11 m/s wind group, with a power curve value of 20.08 kW. Although the occurrence of wind speeds in the 11 m/s group is only 4.21%, the power generated is higher, reaching 7404.50 kWh/year. The highest annual yield is in the 9 m/s wind group, with a result of 10,058.05 kWh/year. Thus, the total annual yield from the C&F CF20 turbine is 62,058.40 kWh/year.

For each turbine, the load factor can be calculated. The load factor is determined by dividing the annual yield by the maximum power output that should be achieved. For example, the annual yield for the Bergey Excel 15 turbine is 48,822.48 kWh/year, while the maximum power output that should be achieved is 20.25 kW multiplied by 8,760 hours in a year, resulting in a load factor of 0.275. This is also related to the turbine's efficiency, which is 27.5%. The load factor for the C&F CF20 turbine is 0.353, with a turbine efficiency of 35.3%. This aligns with the Lanchester-Betz limit theory, which states that a wind turbine describes that only about 59% of the flow can ideally be captured and converted by the turbine into electricity [22]. The results show that the efficiency of both turbines is below 59%.

Table 5. BPS number of households.

Village	The number of House using electricity		House Without Electricity	Total
	PLN	Non-PLN		
Banjarsari	236	8	50	294
Meok	233	0	6	239
Apoho	96	5	0	101
Malakoni	235	0	8	243
Kaana	205	30	17	252
Kahyau	132	34	1	167
Total	1137	77	82	1296

Table 6. Energy needs calculation.

No	Building	Amount	Energy needs (kWh)
1	Mosque/Church	29	44,461.25
2	School	9	17,100.48
3	Hospital	2	3,420.09
4	Bank	3	9,607.17
5	Hotel	7	43,673.41
6	Restaurant	65	473,128.61
7	House	1296	1,679,106.22
8	Other	1	8,844.03
	Total		2,421,770.53

3.5. Energy needs

The parameters needed to calculate the electricity needs of Enggano District include data sourced from BPS, the number of households as well as the number of buildings and lighting. Data on the number of households is in Table 5. The total energy needs of Enggano Island are shown on Table 6. Total energy needs are 2,422 MWh every year. This value will be needed in further wind turbine amount calculation.

3.6. Wind turbine needs

To find out the number of wind turbines that will be used, the total electricity needs of the Enggano sub-district is divided by the annual amount of wind turbine power output that has been selected. The number of turbines that will be used to meet the needs of electricity users in the Enggano sub-district is 40 wind turbines, more than the research in the city of N'Djamena, namely 10 turbines producing 50,420 GWh of power per year at a height of 100 m [5].

4. Conclusion

The conclusion from the results of the research analysis is that the wind speed around Enggano Island has an average wind speed of 5.24 m/s and a maximum speed of 14.85 m/s. Enggano Island requires a total amount of electricity of around 2,422 MWh every year. The CF20 wind turbine produces a total annual power of 62,058.36 kWh and the number of wind turbines needed is 40.

References

[1] V. Setiawan Nano, "So Sad.. Cadangan Minyak RI Hanya Bisa Bertahan 10 Tahun Lagi," CNBC Indonesia. [Online]. Available: https://www.cnbcindonesia.com/news/20230210141333-4-412802/so-sad-cadangan-minyak-ri-hanya-bisa-bertahan-10-tahun-lagi#:~:text=Pasalnya%2C cadangan minyak di RI,9 sampai 10 tahun saja.

[2] B. Kilic, "Evaluating of renewable energy potential in Turkey," *International Journal of Renewable Energy Research*, vol. 1, no. 4, pp. 259–264, 2011.

[3] I. Daut, M. Irwanto, Suwarno, Y. M. Irwan, N. Gomesh, and N. S. Ahmad, "Potential of wind speed for wind power generation in Perlis, Northern Malaysia," *Telkomnika*, vol. 9, no. 3, pp. 575–582, 2011, doi: 10.12928/telkomnika.v9i3.750.
[4] D. Fathur Rahman, "13% Listrik RI Berasal dari Energi Terbarukan pada 2020," Katadata Green. [Online]. Available: https://databoks.katadata.co.id/datapublish/2022/04/12/13-listrik-ri-berasal-dari-energi-terbarukan-pada-2020
[5] M. A. Abdraman, A. M. Tahir, D. Lissouc, M. Y. Kazet, and R. M. Mouangue, "Wind resource assessment in the city of N'Djamena in chad," *International Journal of Renewable Energy Research*, vol. 6, no. 3, pp. 1022–1036, 2016, doi: 10.20508/ijrer.v6i3.4066.g6885.
[6] M. S. Aziz, S. Ahmed, U. Saleem, and G. M. Mufti, "Wind-hybrid power generation systems using renewable energy sources-A review," *International Journal of Renewable Energy Research*, vol. 7, no. 1, pp. 111–127, 2017, doi: 10.20508/ijrer.v7i1.5102.g6971.
[7] A. Ismaiel, "WindPACT 1.5 MW Wind Turbine Rotor Dynamic Loads Under the Effect of Atmospheric Turbulence," *International Journal of Renewable Energy Research*, vol. 13, no. 1, pp. 14–21, 2023, doi: 10.20508/ijrer.v13i1.13551.g8656.
[8] ESDM, "Potensi Energi Baru Terbarukan (EBT) Indonesia," ESDM.
[9] H. S. Das, A. Dey, T. C. Wei, and A. H. M. Yatim, "Feasibility analysis of standalone PV/wind/battery hybrid energy system for rural Bangladesh," *International Journal of Renewable Energy Research*, vol. 6, no. 2, pp. 402–412, 2016, doi: 10.20508/ijrer.v6i2.3411.g6800.
[10] U. S. Satpute, D. R. Joshi, and S. M. Inamadar, "Optimal design of electric system for off-shore wind power plant (OWPP)," *International Journal of Renewable Energy Research*, vol. 9, no. 1, pp. 493–503, 2019, doi: 10.20508/ijrer.v9i1.9018.g7614.
[11] Bisri, "Pulau Enggano Segera Teraliri Listrik 24 Jam," RRI.co.id. [Online]. Available: https://www.rri.co.id/daerah/159244/pulau-enggano-segera-teraliri-listrik-24-jam
[12] N. Bashir and B. Modu, "Techno-economic analysis of off-grid renewable energy systems for rural electrification in Northeastern Nigeria," *International Journal of Renewable Energy Research*, vol. 8, no. 3, pp. 1217–1228, 2018, doi: 10.20508/ijrer.v8i3.7126.g7422.
[13] E. DOKUR and M. KURBAN, "Wind Speed Potential Analysis Based on Weibull Distribution," *Balkan Journal of Electrical and Computer Engineering*, vol. 3, no. 4, pp. 231–235, 2015, doi: 10.17694/bajece.72748.
[14] K. F. Astari, A. Hendri, and M. Fauzi, "Analisis pasang surut perairan Dumai menggunakan metode admiralty," *Jom FTEKNIK*, vol. 8, no. 9, pp. 52–57, 2018.
[15] M. F. Hermanto *et al.*, "Pesisir Barat Lampung Offshore Wind Characteristic in Local Renewable Energy Development Studies," *IOP Conf Ser Earth Environ Sci*, vol. 1298, no. 1, 2024, doi: 10.1088/1755-1315/1298/1/012024.
[16] T. Hardianto, B. Supeno, A. Saleh, D. K. Setiawan, Gunawan, and S. Indra, "Potential of Wind Energy and Design Configuration of Wind Farm on Puger Beach at Jember Indonesia," *Energy Procedia*, vol. 143, pp. 579–584, 2017, doi: 10.1016/j.egypro.2017.12.730.
[17] D. Derome, H. Razali, A. Fazlizan, and A. Jedi, "Application of a New Method 'AGM' to Estimate Weibull Parameters for Low Wind Speed," *International Journal of Renewable Energy Research*, vol. 12, no. 3, pp. 1362–1371, 2022, doi: 10.20508/ijrer.v12i3.13164.g8520.
[18] A. M. Rasham and J. M. Mahdi, "Performance of wind turbines at three sites in Iraq," *International Journal of Renewable Energy Research*, vol. 8, no. 3, pp. 1327–1338, 2018, doi: 10.20508/ijrer.v8i3.7715.g7432.
[19] M. Daoudi, A. S. M. Abdelaziz, E. Mohammed, and E. Elmostapha, "Wind speed data and wind energy potential using weibull distribution in zagora, morocco," *International Journal of Renewable Energy Development*, vol. 8, no. 3, pp. 267–273, 2019, doi: 10.14710/ijred.8.3.267-273.
[20] A. K. Khamees, A. Y. Abdelaziz, M. R. Eskaros, M. A. Attia, and A. O. Badr, "The Mixture of Probability Distribution Functions for Wind and Photovoltaic Power Systems Using a Metaheuristic Method," *Processes*, vol. 10, no. 11, pp. 1–16, 2022, doi: 10.3390/pr10112446.
[21] S. Pookpunt, "Optimal design for hybrid renewable energy system using particle swarm optimization," *International Journal of Renewable Energy Research*, vol. 9, no. 4, pp. 1616–1625, 2019, doi: 10.20508/ijrer.v9i4.9896.g7764.
[22] C. Bergey Windpower, "Bergey Windpower Co." [Online]. Available: https://www.bergey.com/products/grid-tied-turbines/excel-15/
[23] M. Wind Turbine, "C&F CF 20 - 20,00 kW - Wind turbine," 2012.
[24] G. Kuik, "The Lanchester-Betz-Joukowsky Limit," *Wind Energy*, vol. 10, pp. 289–291, Nov. 2007, doi: 10.1002/we.218.

IOP Conf. Series: Earth and Environmental Science **1464** (2025) 012029 doi:10.1088/1755-1315/1464/1/012029

Effectiveness of jet grouting for liquefaction remediation – a numerical case study in Cilegon, North Coast of Java Island

Sugeng Krisnanto[1*], Muhammad Faizal[2], I Wayan Sengara[1], Yuamar Imarrazan Basarah[1], and Dayu Apoji[1]

[1] Geotechnical Engineering Research Group, Civil Engineering Program, Faculty of Civil and Environmental Engineering, Bandung Institute of Technology (ITB)
[2] Graduate Student, Civil Engineering Program, Faculty of Civil and Environmental Engineering, Bandung Institute of Technology (ITB)

*E-mail: sugeng.krisnanto@itb.ac.id

Abstract. Saturated loose sand in coastal areas is prone to a danger of liquefaction during earthquakes. On the other hand, the infrastructure development in such region is unavoidable at the present time. This is due to the increase in the need for land for infrastructure development, particularly on the north coast of Java Island. This condition highlights the necessity of liquefaction remediation especially in highly seismicity areas. A numerical study was performed using 2-D dynamic analysis to appraise the effectiveness of jet grouting in Cilegon, an area on the north coast of Java Island. The soil parameters from a site in Cilegon area were used. A time history of earthquake acceleration was used in the analysis. The pore-water pressure ratio, r_u was used as the liquefaction criterion. Different jet grouting configurations were considered to appraise the effects of depth and diameter of jet grouting on reducing pore-water pressure ratio. The diameters of 1.2 and 1.5 meters were used in an arrangement with spacing of 1.5 and 3 meters. The analysis showed that the presence of jet grouting reduced the values of r_u ranges from the values of 0.9 (arrangement 1) to the value of 0,5 (arrangement 4) The variation of the diameters and distances of jet grouting was obtained. The effectiveness of the jet grouting design was appraised in terms of the replacement ratio that can be used for design purposes.

1. Introduction

Saturated loose sand in coastal areas is prone to a danger of liquefaction. This is because loose sand tends to compress during shaking. This compression causes a pore-water pressure increase, Δu. This increase results in a reduction in the effective stress. Since the effective stress governs the shear strength [1], this increase in pore water pressure results in a reduction in the shear strength and causes liquefaction to occur.

Infrastructure development in such regions is unavoidable at present (e.g., [2]-[5]). This is due to the increased need for land for infrastructure development, particularly on the north coast of Java Island. This condition highlights the necessity of liquefaction remediation.

Several earthquakes occurred in Indonesia (e.g., [6],[7]). This shows that considering earthquakes in design is essential, particularly in mitigating liquefaction. Jet grouting is one of the soil improvement methods for liquefaction mitigation. It works by mixing soil with cement thus increasing soil cohesion. This method of jet grouting has been used in several projects in coastal areas (e.g., [8],[9]).

In this paper, an analysis to appraise the effectiveness of jet grouting for liquefaction remediation is presented. A case study in Cilegon, north coast of Java Island was considered. The study was performed using 2-D numerical analysis. While there are several other soil improvement methods, this paper focuses on the analyses of jet grouting.

2. Methodology

This study uses a comprehensive 2-D dynamic numerical analysis to appraise jet grouting effectiveness in Cilegon, Java Island. Incorporating site-specific conditions, advanced soil models, and specifid ground motion parameters. The methodology outlines procedures for modeling soil behavior, configuring boundary conditions, and processing ground motion data to assess jet grouting performance in mitigating soil liquefaction risks.

2.1 Site condition

The area of study in Cilegon, on the north coast of Java Island, is depicted in Figure 1. The stratigraphy is dominated by sandy and silty soils at various depths, varying from loose to dense across different depths. Jet grouting columns are placed to mitigate soil liquefaction risks.

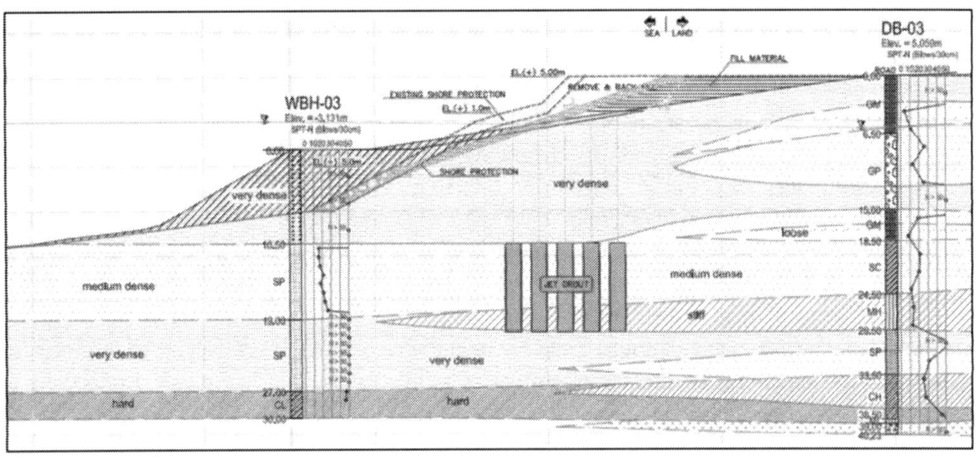

Figure 1. Site and soil condition.

2.2 Constitutive model and soil parameters

Soil constitutive model UBCSAND [10] was used in the numerical analysis to model the liquefiable sand. The dynamic calculation was performed after the initial stress was obtained. In the dynamic calculation, Rayleigh damping of 0.2% was used in the frequency domain ranging from 0.57 to 5.25 Hz. In addition, hysteretic damping (i.e., the decrease of shear modulus and damping ratio) was added to the soil material. The hysteretic damping and damping ratio were calibrated using Seed and Idriss [11] for sandy soils, EPRI [12] for gravelly soils, and Ishibashi and Zhang [13] for cohesive soils. The soil parameters used for the analysis are shown in Table 1. The soil parameters for the analysis were obtained from the relationships shown in Table 2. Jet grouting material was modeled using solid material of concrete for concrete with f_c' equal of 21 MPa which complies with SNI 8460: 2017 (BSN [14]) with elastic linear stress-strain relationship.

Table 1. Soil parameters used in the analysis.

Depth (m)	Soil Type	Consistency or Density	N-SPT	Unit Weight, γ (kN/m³)	Undrained Cohesion, c_u (kN/m²)	Friction Angle, ϕ (deg)	Permeability, K (m/s)	Shear Velocity, v_s (m/s)	Shear Modulus, G_{max} (kN/m²)	Bulk Modulus, K (kN/m²)
0 – 2	Poorly graded SAND	Loose	8	16	0	29.4	1×10^{-4}	236	90833	196805
2 – 10	Poorly graded SAND	Medium	18	17	0	32.4	1×10^{-4}	289	145000	208793
10 – 15	Poorly graded SAND	Loose	9	16	0	29.7	1×10^{-4}	243	96366	208793
15 – 20	Fine SAND	Loose	9	16	0	29.7	1×10^{-5}	243	96366	208793
20 – 23	Clayey SILT	Medium	7	17	35	0	1×10^{-7}	225	87814	190264
23 – 25	Fine SAND	Medium	16	19	0	31.8	1×10^{-5}	281	152755	330969
25 – 30	SILT with sand	Hard	43	22	215	0	1×10^{-7}	362	294201	637436
30 – 40	SILT with sand	Hard	50	22	250	0	1×10^{-7}	377	318395	689858

Table 2. Correlations of soil parameters used in the analysis.

Parameters and Correlations	Unit
Friction angle of sand (Shioi and Fukui, [15]): $\phi = 0.36N + 27$	Degree
Cohesion of clay and silt (Stroud, [16]): $c = 5N$	(kN/m²)
Shear velocity in sand (this study): $v_s = 140.03N^{0.251}$	(m/s)
Shear velocity in clay (this study): $v_s = 135.2N^{0.262}$	(m/s)
Shear modulus: $G_{max} = \rho v_s^2$	(kN/m²)
Bulk modulus: $K = \dfrac{2G(1+v)}{3(1-2v)}$	(kN/m²)

2.3 Numerical modelling and boundary conditions

The analysis was performed using 2-D numerical analysis. In the numerical analysis, the pore-water pressure was calculated using the software FLAC [17]. Using the calculated pore-water pressure, the excess pore pressure generation during the shaking can be estimated using the UBCSAND constitutive model, and hence the excess pore pressure ratio, r_u can be obtained. The ratio r_u is defined as the ratio between the excess pore pressure and the initial effective stress. Liquefied soils have a ratio r_u equal to 1.0 whereas non-liquefied soils have a ratio r_u equal to or less than 0.9.

The mesh for the numerical model and the location of the observational point are shown in Figure 2. The lateral boundary condition is free field. This free field boundary can absorb the waves toward the model boundaries. The boundary condition was also applied at the base of the model to avoid the reflected wave propagation into the model. The value of r_u is calculated in an observational point (Figure 2).

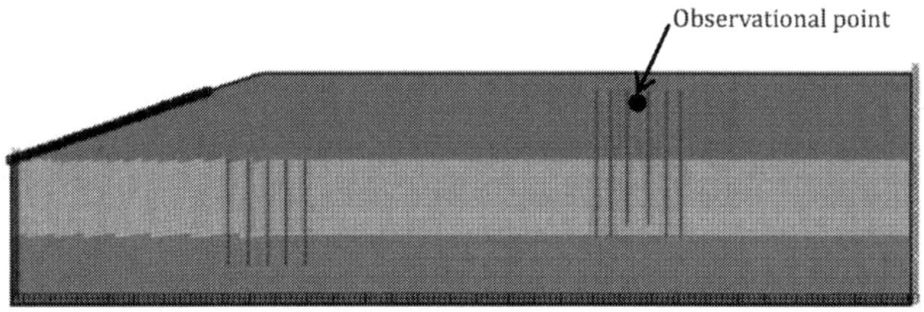

Figure 2. Mesh for numerical model and location of the observational point.

2.4 Ground motion

The ground motion parameters used in the analysis are presented in Table 3. This ground motion was obtained from probabilistic seismic hazard analysis (PSHA) and site-specific response analysis (SSRA) performed in the project area that represents a 2500-year return period earthquake. To reduce the computation time without reducing the accuracy, the duration used in the analysis was only taken as 40 seconds by cutting the insignificant amplitude. In numerical modeling, fitting was performed: (i) Conversion of the unit of acceleration, velocity, and displacement; (ii) Baseline correction, and (iii) Filtering. In this study, the lowest shear wave velocity was 230 m/s. Taken the maximum largest zone as 1 m, the maximum frequency of the input motion that can be propagated accurately throughout the model is defined as [17]:

$$f = \frac{C_s}{10\Delta l} = 23\text{ Hz} \tag{1}$$

where C_s is the speed of the shear wave through a medium and Δl is the size of the zone.

Table 3. Ground motion parameters.

PGA (g):	0.338
Total duration (s):	150
Significant duration D5-95 (s):	50

Time histories of acceleration are presented in Figure 3. Filtering was then performed for these time histories, (Figure 4). From Figure 4, it was obtained that each input motion does not has significant values for frequency greater than 23 Hz, and hence the mesh size of 1 m was sufficient for wave propagating in the soil layers. Therefore, filtering as in Equation (1) is not required.

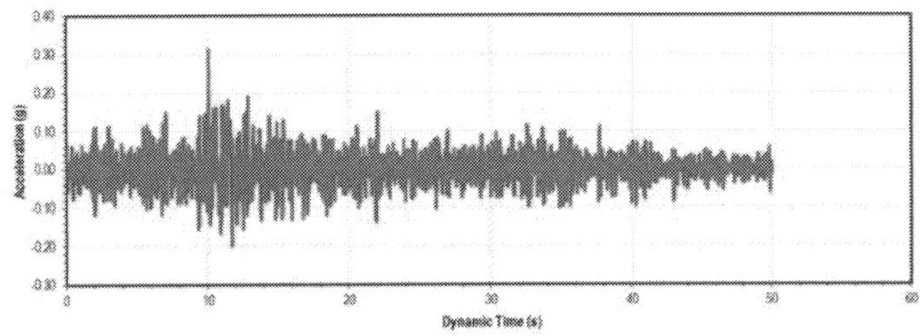

Figure 3. Acceleration time histories.

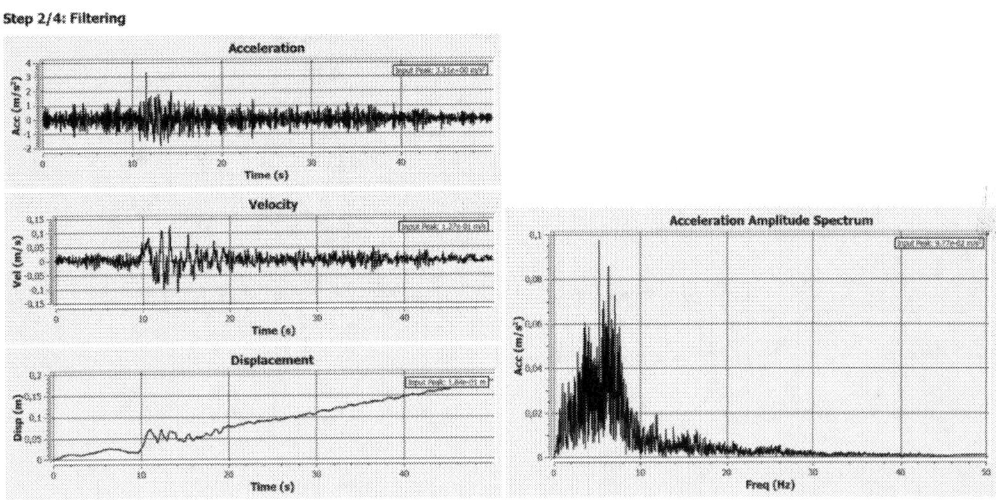

Figure 4. Input records, amplitude spectrum, and filter for acceleration time histories.

3. Results and Discussions

The findings from the liquefaction analysis is presented in this section, by first examining the soil without improvement, showing liquefaction occurrence. It then appraises the effectiveness of four jet grouting arrangements in reducing pore-pressure ratios during seismic events, highlighting the optimal design for mitigating liquefaction.

3.1 Liquefaction analysis without soil improvement

The results of the analysis in terms of effective stress are presented in Figure 5. This figure indicates that the effective stress at shallow depth near the ground surface is around zero Pa, which indicates liquefaction. The values of the ratio r_u versus time are shown in Figure 6. The values or r_u is obtained at the observational point (Figure 2). From these figures, it can be observed that the ratio r_u increases as the time increases until it reaches 1.0 at about the time of 15 seconds. The ratio r_u equal to 1.0 indicates the occurrence of liquefaction.

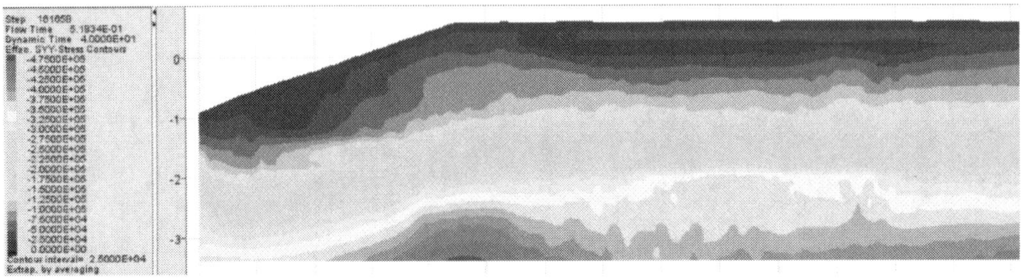

Figure 5. Results of analysis of site without jet grouting in effective vertical stress.

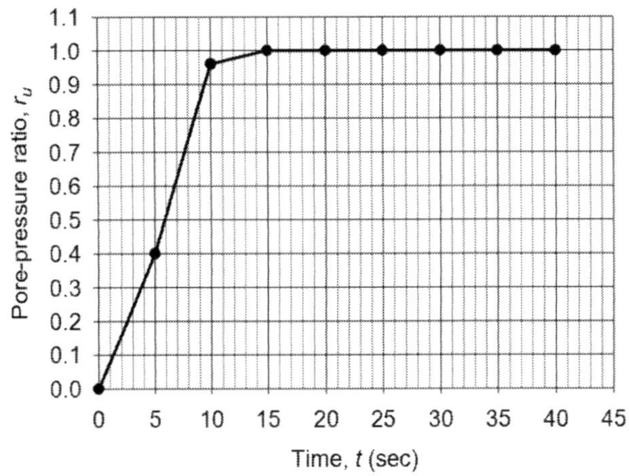

Figure 6. Pore-water pressure ratio versus time for acceleration time histories.

3.2 Liquefaction analysis with jet grouting improvement

There are four configurations of jet grouting design, as shown in Table 4. The jet grouting reached a depth of 26 m. The results of the numerical analyses for the different jet grouting configurations are presented in Figure 7: (a) Arrangement 1 with a diameter and a spacing of 1.2 m and 3 m, respectively; (b) Arrangement 2 with a diameter and a spacing of 1.2 m and 1.5 m, respectively; (c) Arrangement 3 with a diameter and a spacing of 1.8 m and 3 m, respectively; (d) Arrangement 4 with a diameter and a spacing of 1.8 m and 1.5 m, respecitively. The results are presented as the generated pore-water pressure ratio. These results show that the ratio r_u within the study area (dashed black box) was reduced due to the presence of jet grouting. The reduction value varies depending on the jet grouting arrangement.

Table 4. Arrangement of jet grouting design.

Arrangement	Diameter of Jet Grouting (m)	Depth of Jet Grouting (m)	Spacing s_x of Jet Grouting (m)	Spacing s_z of Jet Grouting, (m)
1	1.2	26	3	3
2	1.2	26	3	1.5
3	1.8	26	3	3
4	1.8	26	3	1.5

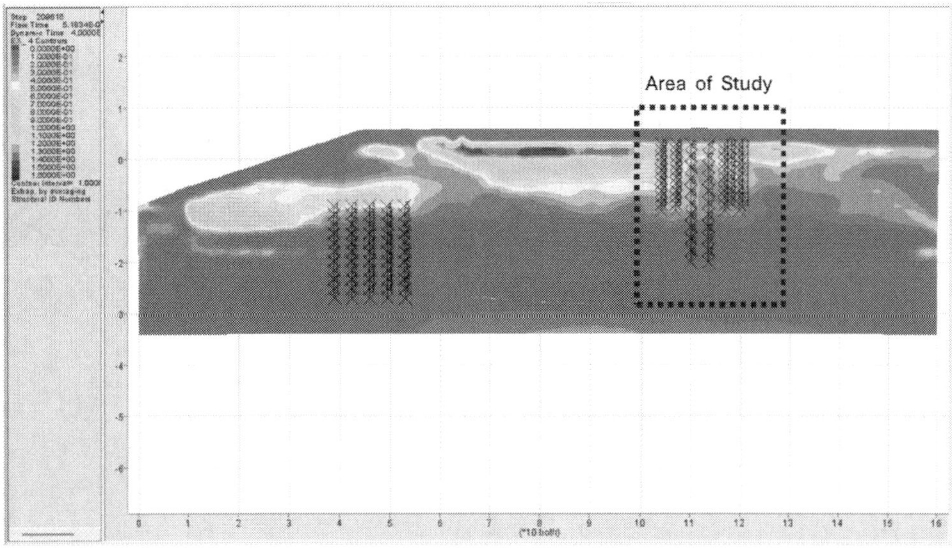

(a)

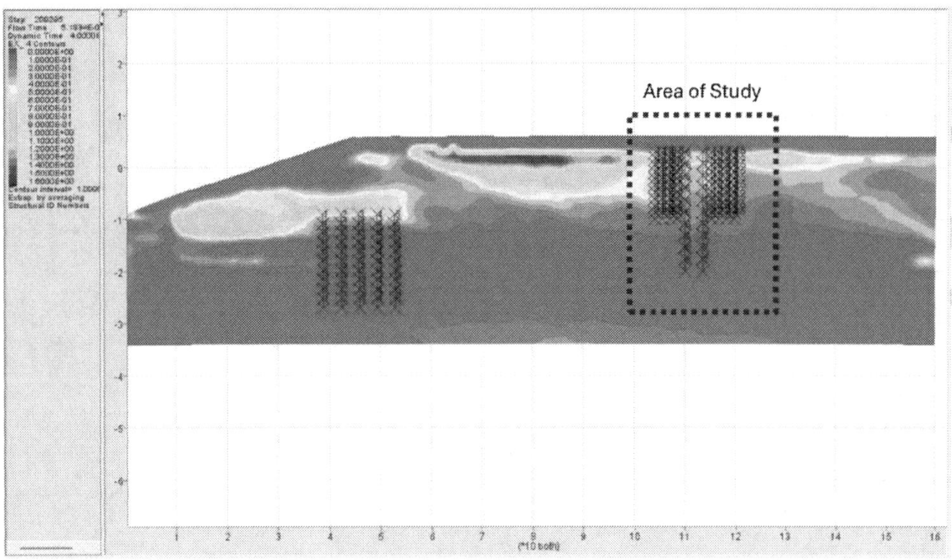

(b)

IOP Conf. Series: Earth and Environmental Science **1464** (2025) 012029 doi:10.1088/1755-1315/1464/1/012029

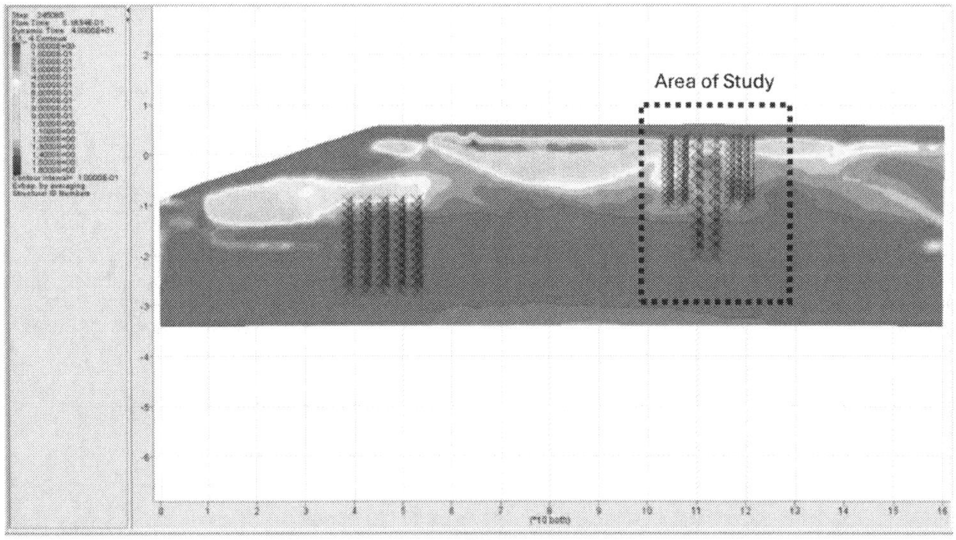

(c)

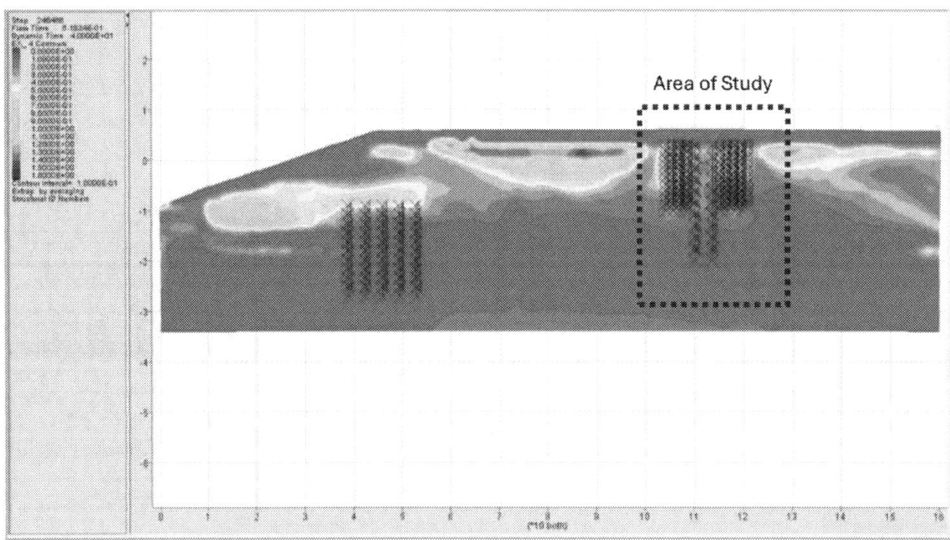

(d)

Figure 7. Pore-water pressure ratio, r_u of soil with jet grouting: (a) Arrangement 1 with a diameter and a distance of 1.2 m and 3 m, respectively; (b) Arrangement 2 with a diameter and a distance of 1.2 m and 1.5 m, respectively; (c) Arrangement 3 with a diameter and a distance of 1.8 m and 3 m, respectively; (d) Arrangement 4 with a diameter and a distance of 1.8 m and 1.5 m, respectively.

The ratio r_u of soil with jet grouting during the simulated earthquake is presented in Figure 8. The figure indicates that with four arrangements of design considered in this study, the ratio r_u is below 0.9. This shows the effect of jet grouting in mitigating the potential of liquefaction. The figure also shows that for the same diameter of jet grouting, the decrease in spacing of jet grouting reduces the ratio r_u during seismic loading. In addition, for the same spacing of jet grouting, the increase in diameter of jet grouting reduces the ratio r_u during seismic loading.

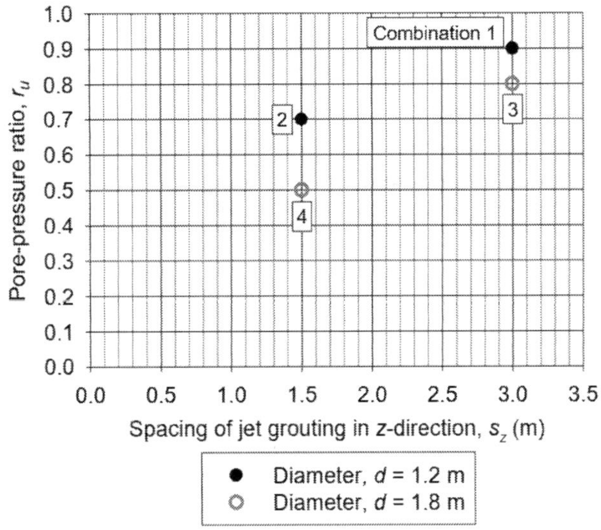

Figure 8. Pore-water pressure ratio, r_u vs. jet grouting diameter and distance at the observation point.

The ratio r_u at the observation point is presented in Figure 9, further demonstrating the effectiveness of the jet grouting improvement. Jet grouting cross-section area/tributary area represents the cost of construction, whereas the ratio r_u represents the target of the soil improvement. It is shown that among the arrangements considered in this study, arrangement 1 (jet grouting with a diameter and spacing of 1.2 m and 3 m, respectively) corresponding to the ratio r_u of 0.9 was the least effective arrangement whereas arrangement 4 (jet grouting with a diameter and spacing of 1.8 m and 1.5 m, respectively) correspond to the ratioe r_u of 0.5 was the most effective arrangement. This comparison can be used to assess other design arrangements to achieve the target soil improvement to remedy liquefaction.

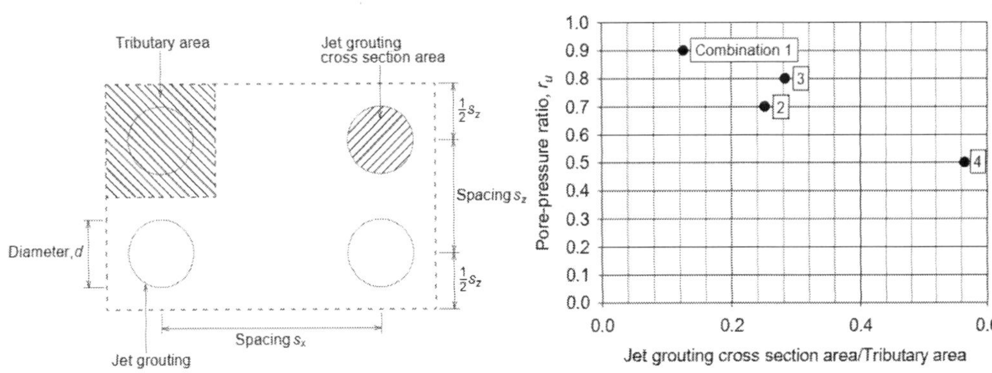

(a) (b)

Figure 9. Consideration of the effectiveness of jet grouting design: (a) Plan view of jet grouting; (b) Plot of jet grouting cross-section area/tributary area vs. pore-pressure ratio, r_u.

4. Conclusions

The conclusions of this study highlight the effectiveness of jet grouting in mitigating soil liquefaction, as listed in the following:

1. For ground soil without soil improvement, it was found that the ratio r_u reaches 1. The ratio r_u equal to 1 indicates the occurrence of liquefaction.
2. For all four design arrangements considered in this study, the ratio r_u is below 0.9. This shows the effect of jet grouting in mitigating the potential of liquefaction.
3. It is shown that among the arrangements considered in this study, arrangement 1 (jet grouting with a diameter and spacing of 1.2 m and 3 m, respectively) corresponds to the ratio r_u of 0.9 was the least effective arrangement whereas arrangement 4 (jet grouting with a diameter and spacing of 1.8 m and 1.5, respectively) correspond to the ratio r_u of 0.5 was the most effective arrangement.

References

[1] Terzaghi K 1936 The Shear Resistance of Saturated Soils and the Angle between the Planes of Shear, Proc. 1st Int. Conf. Soil Mechs.Found. Eng., Cambridge, Massachusetts, 1, 54-56.

[2] Wurjanto A, Mukhti J A, Ayuningtyas S 2019 Alternative Designs for Semarang-Demak Coastal Dike and Toll Road, J. Eng. Tech. Sci., 51(3), 337-354.

[3] Deros M S N, Din Md N, Norzeli S M, Omar R C, Usman F, Hamim S A 2022 Land Subsidence Susceptibility Projection for Palembang Slum Area by Complex MCDM-AHP Technique, J. Eng. Tech. Sci., 54(1), 42-56.

[4] Pham Q D, Ngo V L, Tran T V, Trinh M T 2022 Settlement Prediction of Vacuum Preloading Combined with Prefabricated Vertical Drains in Soft Ground Treatment, J. Eng. Tech. Sci., 54(5), 859-872.

[5] Krisnanto S, Kabila M P, Amiruddin Moch D, Wijayadi S, Adryan F, Ikrimah 2024 Analysis of the Variability of Axial Bearing Capacity of Piles in Cohesionless Soil in the Development of Meyerhof Formula, J. Teknik Sipil, 31(1), 77-84. https://doi.org/10.5614/jts.2024.31.1.9

[6] Krisnanto S, Irsyam M, Asrurifak M, Alatas I M, Adi A D, Daryanto H, Prakoso W, Pramono S 2018 Chapter 14 Geotechnical Aspect and Seismic Hazard, Quick Report Lombok Earthquake, West Nusa Tenggara Province, Research and Development Center for Housing, Ministry of Public Work and Housing (PUPR), September, 127-136, Bandung, Indonesia, ISBN: 978-602-5489-13-6.

[7] Center for Indonesian Earthquake Study/PUSGEN 2017 Indonesian Earthquake Source Map 2017, Research Center for Housing and Settlement, Jakarta (in Indonesian).

[8] Tsai K W, Chou C K, Chang J C, Wang W H 1993 Jet Grouting to Reduce Liquefaction Potential, Proc. 3rd Int. Conf. Case Histories in Geotech. Eng., St. Louis, Missouri, 609-611.

[9] Olgun C G, Martin J R 2008 Effectiveness of Jet-Grout Columns for Mitigation of Liquefaction during Earthquakes, Geotech. Eng. for Disaster Mitigation and Rehabilitation, Liu, H., Deng, A., Chu, J. (eds), Springer, Berlin, Heidelberg.

[10] Beaty M H, Byrne P M 2011 UBCSAND Constitutive Model, Report UBCSAND Constitutive Model on Itasca UDM Web Site.

[11] Seed H B, Idriss I M 1970 Soil Moduli and Damping Factors for Dynamic Response Analysis, Report No. UCB/EERC-70/10, Earthquake Engineering Research Center, University of California, Berkeley, 48 p.

[12] Electric Power Research Institute (EPRI) 1993 Guidelines for Determining Design Basis Ground Motions, Electric Power Research Institute/EPRI, Palo Alto.

[13] Ishibashi I, Zhang X 1993 Unified Dynamic Shear Moduli and Damping Ratios of Sand and Clay, Soils and Found., 33(1), 182-191.

[14] Indonesian Standardization Board (BSN) 2017 SNI 8460: 2017 Requirement for Geotechnical Design, BSN, Jakarta (in Indonesian).

[15] Shioi Y, Fukui J 1982 Application of N-Value to Design of Foundations in Japan, Proc. 2nd European Symp. on Penetration Testing (ESOPT), 1, 159-164.

[16] Stroud M A 1974 The Standard Penetration Test in Insesitive Clays and Soft Rocks, Proc. European Symp. on Penetration Testing, Schockholm, 2(2), 367-375.

[17] Itasca 2018 FLAC: Geomechanical Software for Analyzing and Simulating Dynamics.

Effect of cracking in fly ash concrete on chloride diffusion rate

Sri Murti Adiyastuti[1,2*], Briyandi Iqbal[1], Eko Charnius Ilman[1,2]

[1]Ocean Engineering Program, Faculty of Civil and Environmental Engineering, Institut Teknologi Bandung, Bandung, Indonesia
[2]Offshore Engineering Research Group, Faculty of Civil and Environmental Engineering, Institut Teknologi Bandung, Bandung, Indonesia

*E-mail: adiyastuti@itb.ac.id

Abstract. Reinforced concrete structures in coastal and marine environments are particularly vulnerable to chloride ion damage from seawater, leading to reinforcement corrosion. Chloride ion penetration is a key factor in this process, with the problem becoming more pronounced when cracks develop in the concrete. Cracks significantly influence the rate of chloride penetration, with wider cracks allowing easier access for chloride ions, thereby accelerating the corrosion process. This experimental study investigates how cracks impact chloride diffusion in reinforced concrete, elates the chloride concentration and crack width. The research involves testing concrete samples from both actual structures and laboratory-cast cylinders. Nine concrete cylinders with different compressive strengths were prepared, including K-250 and K-400 grades, both with and without 30% fly ash. Fly ash is known to enhance concrete's resistance to chloride penetration. Concrete cylinders are cut into a 50 mm thick discs and subjected to a steady state migration test in line with NT BUILD 355 standards. Chloride diffusion rates were calculated using the Nernst-Planck formula. Cracks were introduced into the discs with splitting test, with crack widths of 200, 400, 600, and 800 μm. The results demonstrate a clear correlation between crack width and content of chloride in the concrete. As crack width and area increase, chloride concentration within the concrete rises. This trend was particularly evident in K-400 concrete, where the presence of fly ash significantly reduced chloride concentration compared to K-400 without fly ash, especially when crack widths exceeded 0.6 mm. The study underscores the importance of crack management and the utilisation of fly ash in mixtures of concrete to mitigate chloride-induced corrosion in marine environments.

1. Introduction

Reinforced concrete is one of fundamental material in modern structures, particularly for structures exposed to harsh environments such as marine settings. However, these structures are highly susceptible to degradation affected by aggressive nature of chloride present in marine environment. Corrosion induced by chloride in steel bar inside concrete is significantly affect durability issues and the lifespan of concrete in seawater environments [1],[2]. The chloried ion ingress into concrete initiating and accelerating the corrosion process of the steel bar, ultimately compromising structural integrity and safety [3], [4].

The chloride penetration mechanism into concrete is primarily governed by diffusion, particularly in the absence of any external forces or gradients [5]. However, the situation becomes more complex when cracks are present in the concrete. Cracks serve as preferential pathways for chloride ions, allowing them to penetrate the concrete more rapidly and in greater concentrations [6], [7]. The width of cracks are significant in defining the chloride penetration rate. Wider cracks leading to higher rates of diffusion and, consequently, faster corrosion of the reinforcement [8],[7].

To mitigate this issue, various strategies have been employed, including the utilization of addition of cement-based materials such as fly ash. Fly ash has been recognized due to its ability to enhance the concrete durability by refining its pore structure, decreasing permeability, and enhancing resistance to chloride ion penetration [9], [10]. The incorporation of fly ash inside concrete mixtures shows significantly decrease the diffusion coefficient of chlorides, thereby offering a protective measure against chloride-induced corrosion [10], [11].

Despite the recognized benefits of fly ash, interaction between crack width, chloride diffusion and fly ash content in concrete remains inadequately explored. While several studies have examined the chloride resistance of fly ash-based concrete, the specific effects of cracking on the chloride diffusion rate in fly ash concrete require further investigation [12], [13]. Understanding these interactions is crucial for the development of more durable concrete structures in marine environments [11], [14], [15].

This study aims to address this gap by assessing the effect of cracks on the chloride diffusion rate in reinforced concrete with and without the addition of fly ash. The research focuses on determining the chloride diffusion rate in both actual concrete structures and laboratory-cast concrete cylinders. This study is not intended as a comparative analysis but rather focuses on examining and analyzing the results of the experimental procedures. The findings will provide insights into the relationship between crack width, fly ash content, and chloride concentration, ultimately contributing to improved design and maintenance practices for reinforced concrete structures in oceanic environments.

2. Methodology

In this experiment, the concrete mix design is according to SNI 03-6468 [16] and ACI 211.1-91 [17] standards, targeting a strength range of 33 to 35 MPa (K-400). This strength level is typically specified by various codes for applications in harsh environments, such as concrete used in marine conditions. The samples were prepared to reflect the strength commonly utilized in actual marine structures. A total of nine concrete cylinders were used: 2 cylinders with 0% fly ash (K-250), 3 cylinders with 0% fly ash (K-400), and 4 cylinders with 30% fly ash (K-400). A splitting test was employed to create cracked concrete samples with varying crack widths. Chloride penetration was accelerated using a steady-state migration test, and the diffusion rate was calculated using the Nernst-Planck equation, which can be found in most textbook in electrochemistry.

$$J(x) = -D_e\frac{\partial c}{\partial x} + D_e\frac{zFE}{RTL}c + cv(x) \tag{1}$$

where D_e represents concrete diffusion rate (m^2/s), c represents chloride concentration (mol/m^3), $J(x)$ refers to chloride ions fluxes (mol/m^2s), z is the valency of chloride ion (z = 1), F stands for the Faraday constant (F= 96480 J/(V mol)), E is the actual potential drop (V), R

represents the gas constant (R = 8.3144 J/(mol K)), T is temperature (K), L is the concrete specimen thickness (m), and $v(x)$ indicates velocity of the solution (m/s).

When the concrete is fully saturated, the solute velocity can be disregarded, allowing the equation to be simplified to Equation 2.

$$D_e = \frac{LRT}{czFE}J \tag{2}$$

The concentration gradient needs to be maintained during the test in order to reach a steady state, necessitating the continuous replenishment of the solute. The flux (J) is determined by utilizing a regression line's constant slope, as shown in Eq. 3.

$$J = \frac{V}{A}\frac{\Delta C_{Cl^-}}{t} \tag{3}$$

where V is the volume of the anode compartment's solution volume (m^3), A represents the cross-sectional area of the concrete (m^2), ΔC_{Cl^-} refers to the rise in chloride concentration at the anode (mol/m^3), and t refers to the time that has passed (s). The term $\frac{\Delta C_{Cl^-}}{t}$ represents the slope (see Figure 1). The time it takes for the chloride to penetrate the specimen, also known as the lag time (T_{lag}).

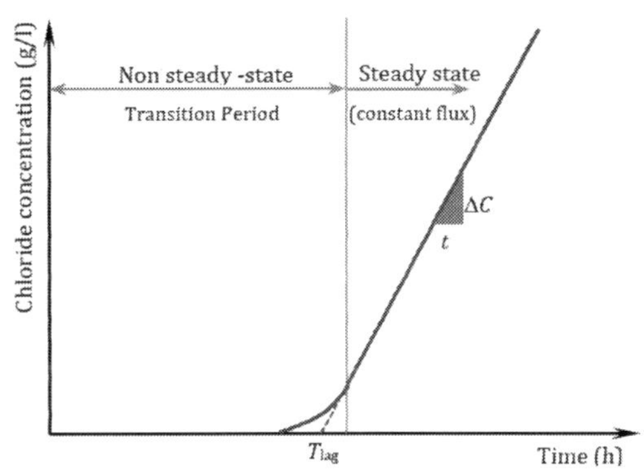

Figure 1. Sketch of chloride evolution with time chloride migration in non-steady state and steady state conditions schemes.

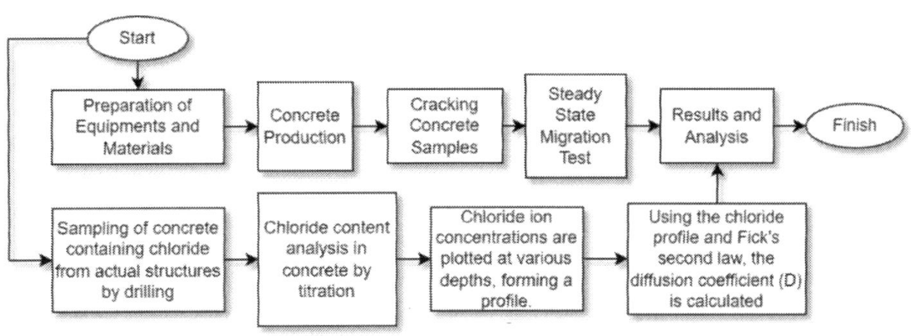

Figure 2. Experiment methodology flowchart.

The experiment methodology is shown in Figure 2. The study began by collecting concrete powder samples from PT. Pelabuhan Indonesia II Cirebon Branch and PT. Cirebon Power Jetty, as shown in Figure 3. After determining the chloride diffusion rates at both locations, concrete test specimens were prepared in the laboratory, followed by the execution of Steady State Migration experiments.

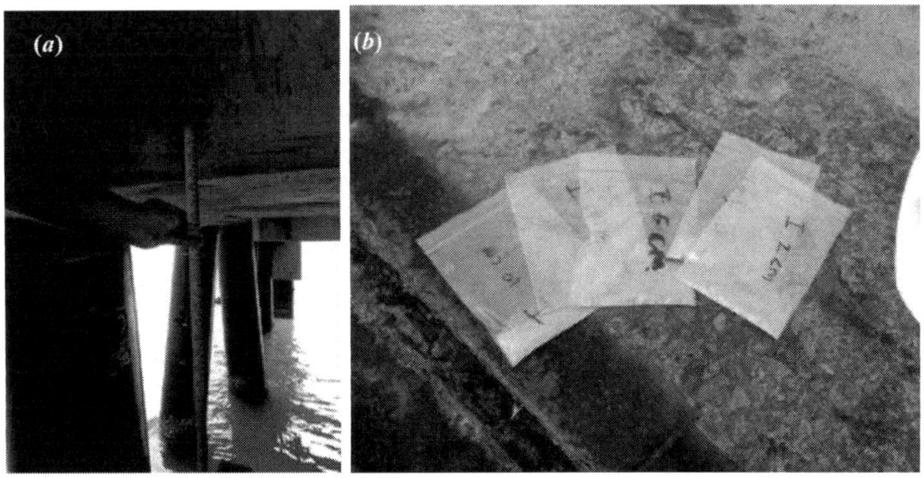

Figure 3. *In-situ* sample concrete powder collection: (*a*) location of sample collection I Jetty with elevation of 3 m from mean sea level; (*b*) collected concrete powder samples.

The concrete mixture was prepared in three batches because the desired strength was not initially achieved. The poor quality of the aggregates used in the first mix significantly affected the results. The initial mix produced a concrete compressive strength of 20 MPa (K-250), which was lower than the target. However, this mix was included in the study for comparison with higher-strength concrete (K-400). A total of nine concrete cylinders were prepared: 2 cylinders with 0% fly ash (K-250), 3 cylinders with 0% fly ash (K-400), and 4 cylinders with 30% fly ash (K-400). Table 1 below summarizes the concrete mixes and their corresponding compressive strengths used in this study.

Table 1. Concrete mix design.

Concrete mixes	Concrete Compressive strength	Total
First batch	K-250	2 Cylinder
Second batch	K-400	4 Cylinder
Third batch	K-400 + 30% Fly Ash	3 Cylinder

The concrete cylinders were sliced into discs, each 50 mm thick, with for the splitting tensile test, four discs were removed from every cylinder. To guarantee that chloride flowed through the discs in a single direction, the sides of the discs were sealed with two coats of epoxy resin after cutting. The cracking of the concrete discs was carried out using splitting test equipment and a digital dial gauge.

The cracking process was conducted at the Concrete Structure Laboratory of Institut Teknologi Bandung. Two digital dial gauges with a measurement range of up to 800 μm were used to measure lateral displacement. The epoxy-coated concrete disc was placed in a splitting test device equipped with a digital dial sensor. The Crack Opening Displacement (COD) was calculated as the average crack width on both sides of the concrete disc. The splitting test was performed by applying load until the cracks reached widths of 200, 400, 600, and 800 μm. Figure 4 provides an illustration of the concrete cracking process.

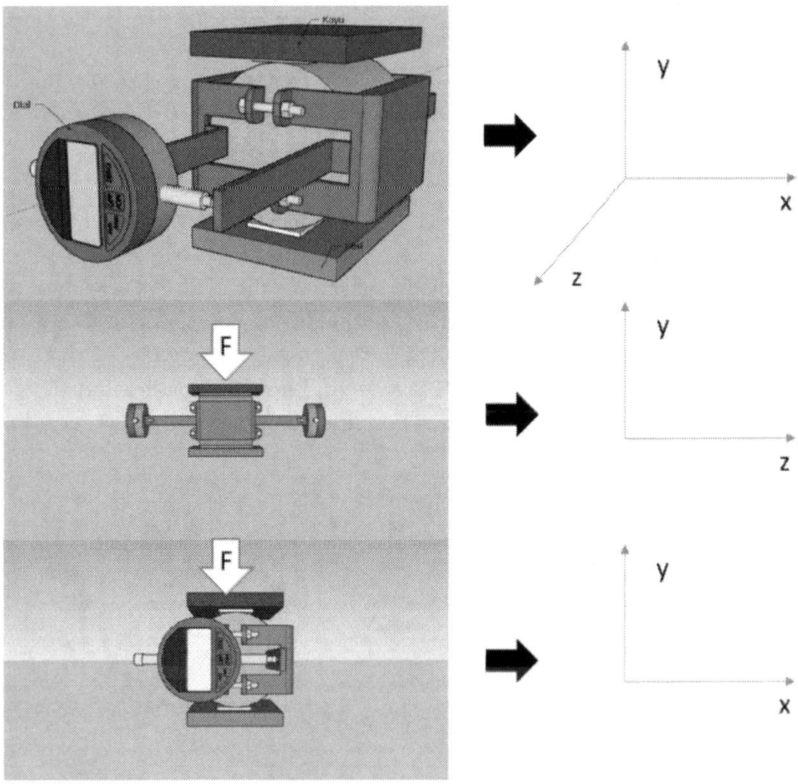

Figure 4. Illustration of concrete cracking process with digital dial equipment.

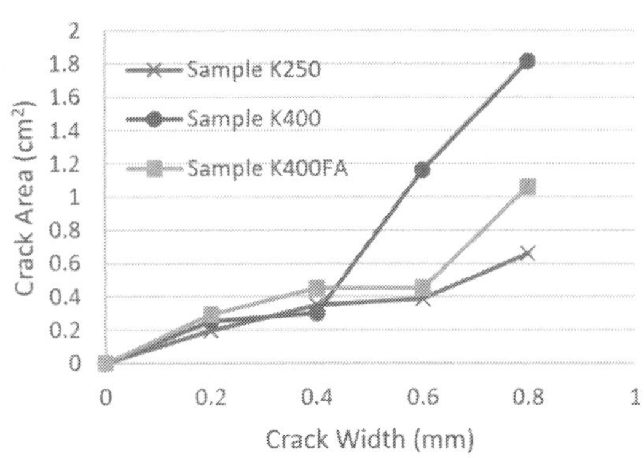

Concrete Mixing	Concrete Specimen	Crack Width (mm)	Crack Area (cm^2)
K250	A.1.2	0.2	0.1949
	A.1.3	0.4	0.3487
	A.1.4	0.6	0.3871
	A.1.5	0.8	0.6598
K400	B.1.2	0.2	0.2509
	B.1.3	0.4	0.2987
	B.1.4	0.6	1.1603
	B.1.5	0.8	1.816
K400 + 30% FA	C.1.2	0.2	0.2915
	C.1.3	0.4	0.4505
	C.1.4	0.6	0.4523
	C.1.5	0.8	1.059

Figure 5. Relationship between crack width and crack area for each sample.

Figure 5 illustrates that as crack width increases, the crack area in the concrete disc also expands. For all samples, the crack area remains stable up to 0.4 mm crack width. In the K-250 (concrete discs A.1.2 to A.1.5) and K-400FA (concrete discs C.1.2 to C.1.5) samples, changes in crack width do not significantly affect the crack area. However, in the K-400 samples (concrete discs B1.2 to B.1.5), the crack area shows a notable increase of 0.6 mm and 0.8 mm crack widths. These observations suggest that the correlation between width of crack and crack area cannot be accurately determined using only the splitting test and digital dial.

3. Results and discussion

3.1 Determination of chloride diffusion coefficient from actual structures

Concrete powder samples were collected from four different locations within the actual structure, specifically from PT. Pelabuhan Indonesia II Cirebon Branch and PT. Cirebon Power, with two samples taken from each site. The samples are obtained by boring perpendicularly into the concrete member at various depths. The chloride content in the concrete was determined using the titration method, which enabled the plotting of a chloride profile, with depth at x-axis and chloride content at y-axis. The diffusion coefficient was then calculated by fitting the best curve to the chloride profile in accordance with Fick's second law about diffusion. The diffusion of chloride coefficients calculated from the PT. Pelabuhan Indonesia II Cirebon and PT. Cirebon Power samples are shown in Table 2.

Table 2. Chloride diffusion coefficient from the actual structure.

No	Location	Chloride diffusion (m^2/s)
1	Beam 1 PT. Cirebon Power	2.134×10^{-12}
2	Beam 2 PT.Cirebon Power	8.507×10^{-12}
3	Muara Jati III PT. Pelindo II Cirebon	6.462×10^{-12}
4	Pelita II PT. Pelindo II Cirebon	20.670×10^{-12}

The lowest chloride diffusion rate was observed in Beam 1 at PT. Cirebon Power, where the concrete powder was collected 3 meters above sea level. At this height, chloride penetration was largely restricted by evaporation processes, unlike in samples collected from other locations. In contrast, the highest chloride diffusion rate, measured at 20.6×10^{-12} m^2/s , was found at Pelita II, likely due to damage to the concrete cover on the pier.

3.2 Determination of chloride diffusion rate from Steady State Migration test

The experiments conducted on K-250 concrete using the steady-state migration method were unsuccessful due to several errors and inconsistent data. The expected trend of increased chloride penetration with wider cracks was not observed in the K-250 samples, rendering the data unsuitable for analysing the chloride diffusion rate.

IOP Conf. Series: Earth and Environmental Science **1464** (2025) 012030 doi:10.1088/1755-1315/1464/1/012030

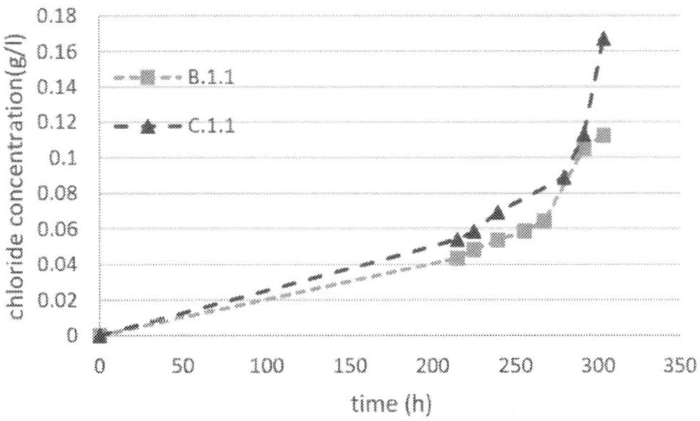

Figure 6. Comparison of chloride concentration in K-400 vs. K-400FA Concrete, without cracks.

For samples A.1.5, B.1.5, and C.1.5, which had 0.8 mm cracks, special handling was required, necessitating sampling within 12 hours. This urgency arose because the large cracks allowed the NaCl and NaOH solutions to mix rapidly, leading to saturated data that failed to capture the chloride penetration process. As a result, the data from these 0.8 mm cracks are not usable.

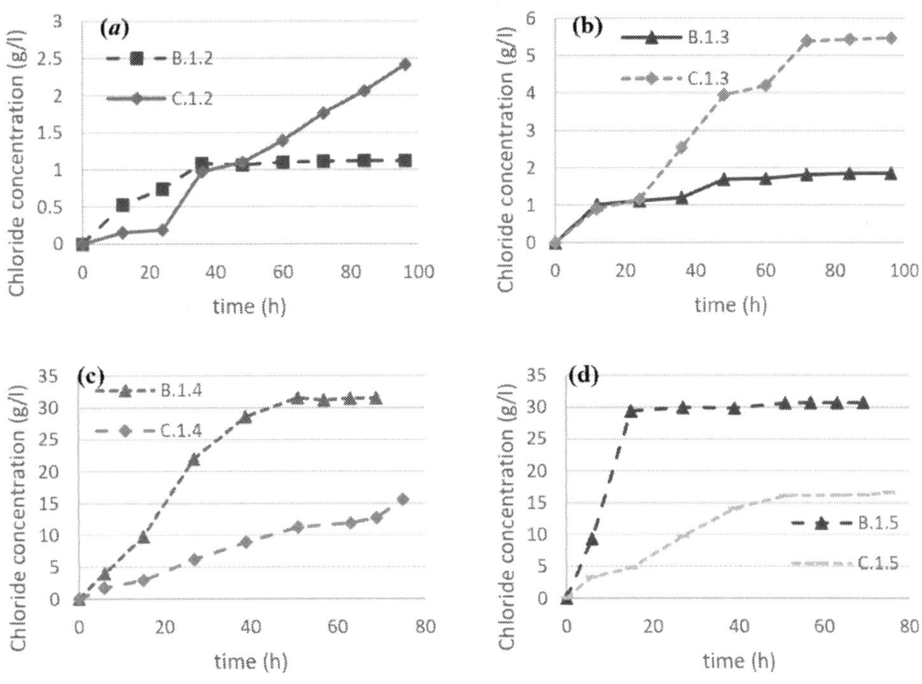

Figure 7. Comparison of Chloride Concentration in K-400 vs. K-400FA Concrete: (*a*) 0.2 mm crack; (*b*) 0.4 mm crack; (*c*) 0.6 mm crack; (*d*) 0.8 mm crack.

Figure 6 above compares the chloride concentrations in uncracked K-400 and K-400FA samples. As shown, the chloride concentrations in both concrete samples are quite similar, suggesting that the mix's resistance to chloride penetration is not considerably impacted by the

30% fly ash addition. This trend is also evident in samples with 0.2 mm and 0.4 mm crack widths, as shown in Figures 7(*a*) and 7(*b*). However, for samples with 0.6 mm and 0.8 mm crack widths, the chloride concentration in the K-400FA sample is considerably lower than in the K-400 sample (without fly ash), as shown in Figures 7(*c*) and 7(*d*). The results show that, particularly in concrete with larger cracks, the addition of fly ash significantly increases the concrete mix's resistance to chloride penetration.

The diffusion rate for concrete for K400 and for K400 FA without crack and for various crack width is presented in Table 3 and Figure 8 below. *D* represents the diffusion coefficient rate of a concrete sample. It is evident that concrete K-400 is significantly affected by crack width, whereas concrete K-400 FA is less affected by crack width.

Table 3. Chloride Diffusion Coefficients from laboratory concrete specimens.

Concrete	Without Crack	0.2 mm	0.4 mm	0.6 mm	0.8 mm
	m^2/s	m^2/s	m^2/s	m^2/s	m^2/s
K400	2.44745E-15	9.77E-14	1.84E-13	6.78E-12	2.27E-11
K400FA	4.2952E-15	2.66E-13	7.43E-13	1.99E-12	3.21E-12

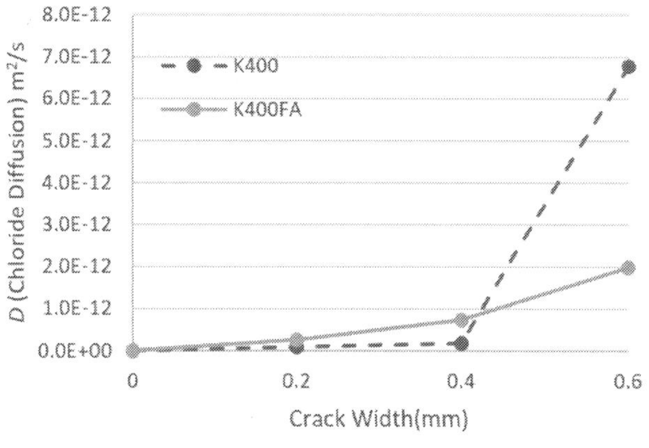

Figure 8. Crack width vs. chloride diffusion coefficient comparison.

4. Conclusions

The study findings indicate that for 0.2 mm and 0.4 mm crack widths, chloride diffusion rate in K-400 concrete is slightly lower than in K-400FA, suggesting better resistance to chloride penetration; however, this resistance diminishes as crack widths increase to 0.6 mm or more. When cracks widen to 0.6 mm and 0.8 mm, the K-400FA sample exhibits significantly lower chloride concentrations compared to the K-400 sample without addition of fly ash, indicating that the addition of fly ash to the concrete mix substantially enhances chloride penetration resistance, particularly in concrete with larger cracks. Consequently, the chloride diffusion coefficient is significantly lower for K-400FA than for K-400. Additionally, the chloride diffusion coefficients derived from the K-400 and K-400FA samples are lower than those obtained from the actual structure. However, due to the unavailability of specific details regarding the concrete mix used in the actual structure, such as its strength and whether fly ash was included, the comparison is somewhat limited. Overall, these findings underscore the importance of incorporating fly ash in concrete mixtures, particularly for structures in chloride-rich environments and those susceptible to cracking.

Acknowledgement

The authors acknowledge the funding provided by the P3MI program of Institut Teknologi Bandung.

CRediT authorship contribution statement

Sri Murti Adiyastuti: Conceptualization (lead), Supervision (lead), Funding Acquisition (lead), Writing – Review & Editing (equal), Resources (lead). **Briyandi Iqbal**: Data Curation (lead), Formal Analysis (lead), Investigation (lead), Writing – Original Draft Preparation (lead), Visualization (equal). **Eko Charnius Ilman**: Supervision (equal), Writing – Review & Editing (lead), Visualization (lead).

References

[1] F. Qu, W. Li, W. Dong, V. W. Tam, and T. Yu, 'Durability deterioration of concrete under marine environment from material to structure: A critical review', *Journal of Building Engineering*, vol. 35, p. 102074, 2021.
[2] G. Xiang, K. Rao, X. Xiang, and X. Yu, 'Overview and analysis on recent research and challenges of dropped objects in offshore engineering', *Ocean Engineering*, vol. 281, p. 114616, Aug. 2023, doi: 10.1016/j.oceaneng.2023.114616.
[3] A. Hameed *et al.*, 'Behavior and Performance of Reinforced Concrete Columns Subjected to Accelerated Corrosion', *Metals*, vol. 13, no. 5, p. 930, 2023.
[4] W. Green, 'Steel reinforcement corrosion in concrete–an overview of some fundamentals', *Corrosion Engineering, Science and Technology*, vol. 55, no. 4, pp. 289–302, 2020.
[5] A. Ababneh and Y. Xi, 'An experimental study on the effect of chloride penetration on moisture diffusion in concrete', *Materials and Structures*, vol. 35, pp. 659–663, 2002.
[6] A. Poursaee and B. Ross, 'The role of cracks in chloride-induced corrosion of carbon steel in concrete', *Corrosion and Materials Degradation*, vol. 3, no. 2, pp. 258–269, 2022.
[7] L. Yu, R. François, V. H. Dang, V. L'Hostis, and R. Gagné, 'Development of chloride-induced corrosion in pre-cracked RC beams under sustained loading: Effect of load-induced cracks, concrete cover, and exposure conditions', *Cement and Concrete Research*, vol. 67, pp. 246–258, 2015.
[8] S. M. Adiyastuti, 'Influence of cracks on chloride induced corrosion in reinforced concrete flexural members.', 2005.
[9] J. Liu, Q. Qiu, F. Xing, and D. Pan, 'Permeation properties and pore structure of surface layer of fly ash concrete', *Materials*, vol. 7, no. 6, pp. 4282–4296, 2014.
[10] R. Dhir and M. Jones, 'Development of chloride-resisting concrete using fly ash', *fuel*, vol. 78, no. 2, pp. 137–142, 1999.
[11] B. H. Ortiz-Salcedo, J. M. Paris, C. C. Ferraro, R. Minkara, and K. A. Riding, 'Evaluation of chlorides in fly ash for use in concrete', *Cleaner Materials*, vol. 5, p. 100098, 2022.
[12] C. Lu, Z. Cui, R. Liu, and Q. Liu, 'Chloride diffusivity in flexural cracked Portland cement concrete and fly ash concrete beams', *Journal of Central South University*, vol. 21, no. 9, pp. 3682–3691, 2014.
[13] A. Djerbi, S. Bonnet, A. Khelidj, and V. Baroghel-Bouny, 'Influence of traversing crack on chloride diffusion into concrete', *Cement and concrete research*, vol. 38, no. 6, pp. 877–883, 2008.
[14] G. Heirman *et al.*, 'Chloride penetration and carbonation in self-compacting concrete', presented at the ConcreteLife'06-International RILEM-JCI Seminar on Concrete Durability and Service Life Planning: Curing, Crack Control, Performance in Harsh Environments, RILEM Publications SARL, 2006, pp. 13–23.
[15] T. Sugiyama, T. Bremner, and Y. Tsuji, 'Determination of chloride diffusion coefficient and gas permeability of concrete and their relationship', *Cement and Concrete Research*, vol. 26, no. 5, pp. 781–790, 1996.
[16] SNI, *SNI 03-6468 Tata Cara Perencanaan Campuran Tinggi dengan Semen Portland dengan Abu Terbang*, 2000.
[17] ACI, *ACI 211.1-91 Standard Practice for Selecting Proportions for Normal, Heavyweight, and Mass Concrete. ACI Commitee 211.*, 2002.

IOP Conf. Series: Earth and Environmental Science **1464** (2025) 012031 doi:10.1088/1755-1315/1464/1/012031

Lateral pressure on piles in soft cohesive soil due to adjacent embankment filling: A comparative study of 2D and 3D finite element analyses

Azka Syarifa Amani[1], Endra Susila[2*]

[1] Department of Civil Engineering, Bandung Institute of Technology, Bandung, Indonesia
[2] Department of Civil Engineering, Bandung Institute of Technology, Bandung, Indonesia

*E-mail: esusila@itb.ac.id

Abstract. The construction of embankments on soft clays near existing structures is generally avoided due to the risk of undrained lateral pressure exerted on existing adjacent piles. However, when such construction is really unavoidable, accurately predicting the influence of the undrained lateral soil movement due to embankment filling becomes crucial for understanding the behavior of the nearby foundation structures. Unfortunately, the impacts of the soil lateral displacement to existing adjacent piles are highly complex due to the interaction complexities of pile and soil, compounded by the three-dimensional nature of the problem. This study investigates the pressures and behaviors associated with soil movements resulting from embankment filling adjacent to pile foundations. The main objectives are predicting the displacements and developed bending moments on piles. A comparison was conducted between the observed behavior from centrifuge model tests performed by other researchers and the results obtained using the finite element method. The study utilized professional finite element softwares: PLAXIS 2D (with embedded beam elements) and MIDAS 3D GTS NX. The findings indicate that embankments significantly influence the internal forces and deformations of pile foundations. The results show that 3D modelling produces smaller internal forces compared to 2D modelling. Additionally, the internal force analysis and deformation curves obtained from this study exhibit similar shapes, allowing for comparative analysis of results from centrifuge models, PLAXIS 2D, and MIDAS 3D GTS NX.

1. Introduction

With the vast constructions in Indonesia specially toll roads, soil embankments on soft and compressible soils become increasingly common, often necessitating work in areas with existing structures. However, the construction of embankments and/or excavations on soft soils near existing structures on piles requires careful consideration due to potential for lateral soil movements induced by active lateral pressure from nearby not only embankments, but also excavations, tunnels, and shifting slopes. These lateral movements inducing lateral pressure to the existing nearby piles, differ significantly from the vertical and horizontal loads directly imposed on piles, presenting challenges in predicting additional horizontal deformation, rotations, bending moments and other related effects.

IOP Conf. Series: Earth and Environmental Science **1464** (2025) 012031 doi:10.1088/1755-1315/1464/1/012031

The complexities involved in predicting and mitigating the effects of lateral soil movements near embankments to existing structures, necessitate thorough analysis. Several researches have been performed by several researchers, such as Bauer et. al. [1], Dao. [2], Goh et. al. [3], Jennings and Naughton [4], Jie et. al. [5], Yi [6], Broms et. al. [7]. This study focuses on analyzing the forces that affect pile foundations adjacent to embankments, particularly examining soil displacements and additional bending moments on piles through numerical simulations using PLAXIS 2D with embedment beam elements [8,9] and MIDAS 3D GTS NX [10] and laboratory test results by Dao [2]. This research quantifies the impact of embankments on adjacent pile behavior. At the end, this study aims to enhance the design and construction practices for infrastructure projects in challenging soil conditions, ensuring the reliability and longevity of pile-supported structures amid embankment-induced soil movements.

2. Research Significance

This study investigates the structural behavior of the piles, focusing on pile displacement, bending moments, and surface settlement due to the staged construction of the embankments. By comparing centrifuge test results with those obtained using finite element methods in PLAXIS 2D with embedment beam elements and MIDAS 3D GTS NX in addition to PLAXIS 3D by Dao [2], the research highlights how embankments influence internal forces and deformations in piles. The results of analysis show that the deformation and internal force curves from different models are relatively similar, enabling a comparative evaluation of results across centrifuge tests and finite element simulations.

3. Methodology

3.1 Centrifuge model

A centrifuge test, known as the "Centrifugeproef GeoDelft," was selected by Dao [2] to validate the PLAXIS 3D model using embedded piles with actual conditions. This test involved a model pile situated near an embankment. The centrifuge test utilized a "model box" with a dimension of 850mm in length, 450mm in width, and 200mm in height. At the base of the box, a 50mm thick sand layer, representing Pleistocene sand, was placed. On top of this, a 100mm thick layer of Speswhite clay was added. Given that the centrifuge test was conducted at a 1:100 scale, this means that 1mm in the test model equates to 100mm in real life. Figure 1 illustrates the cross-section of the centrifuge model at this scale.

The water level was at the ground surface. A pile was installed at a position 400mm in the X-direction. The pile had a total length of 151.5mm, with 10.5mm extending above the ground surface. The model pile had a diameter of 5mm, which corresponds to a diameter of 500mm in real life. The pile head was connected to a framework. In the centrifuge test, soil displacement and bending moments in the pile were measured at several locations using measuring instruments. These measurements are next compared with the outputs from PLAXIS 2D, and MIDAS 3D GTS NX, and PLAXIS 3D by Dao [2].

IOP Conf. Series: Earth and Environmental Science **1464** (2025) 012031 doi:10.1088/1755-1315/1464/1/012031

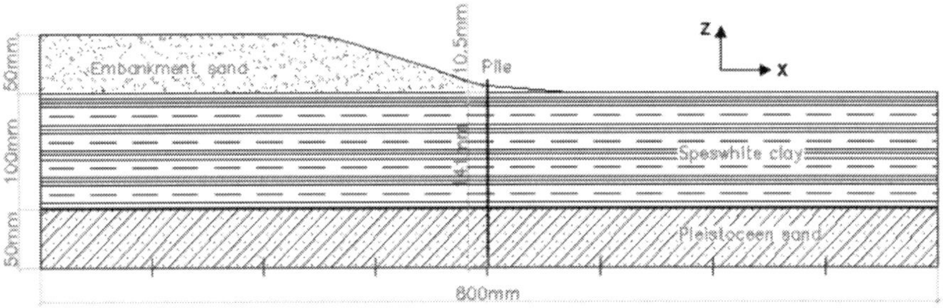

Figure 1. Cross section of centrifuge model with a scale of 1:100 [2]

The soil and pile parameters (Tables 1 to 4) used in this study are based on data directly from the actual centrifuge test report, "Centrifugeproef GeoDelft." [2]. The clay layer is modeled using the soft soil creep model to account for consolidation parameters, while the sand layer is modeled with the Mohr-Coulomb model, assuming no significant consolidation and settlement occurs in this layer. The sand layer is also modeled using the Mohr-Coulomb model.

Table 1. Properties of the embedded pile [2]

Embedded Pile			
Parameter	**Name**	**Value**	**Unit**
Predefined pile type	-	Massive circular pile	-
Young's modulus	E	2.66×10^7	kN/m^2
Unit weight	γ	0	kN/m^3
Pile diameer	d	0.5	m/day
Pile area	A	0.1963	m^2
Inertia moment	I	0.003068	m4
Skin resistance type	Type	Linear	-
Maximum traction allowed at the pile top	$T_{top,max}$	200	kN/m
Maximum traction allowed at the pile bottom	$T_{bot,max}$	500	kN/m
Base resistance	F_{max}	10,000	kN

Table 2. Properties of the embankment sand [2]

Embankment Sand			
Parameter	**Name**	**Value**	**Unit**
Material model	Model	MC	
Drainage type	Type	Drained	
Unit weight	γ_{unsat}	17	kN/m^3
	γ_{sat}	19	kN/m^3
Young's modulus	E'	10,000	kN/m^2
Poisson's ratio	ν'	0.3	-
Cohesion	c'	0.3	kN/m^2
Friction angle	ϕ'	42	°
Dilatancy angle	ψ	3	°
Horizontal permeability	k_h	1	m/day
Vertical permeability	k_v	1	m/day

Table 3. Properties of the clay layer [2]

Clay Layer			
Parameter	**Name**	**Value**	**Unit**
Material model	Model	SSC	-
Drainage type	Type	Undrained	
Modified compression index	λ^*	0.085	-
Modified swelling index	k^*	0.022	-
Modified creep index	μ^*	0.00013	-
Poisson ratio for unloading/reloading	ν_{ur}	0.2	-
Cohesion	c'	0.5	kN/m²
Friction angle	ϕ'	22	°
Dilatancy angle	ψ	0	°
Horizontal/Vertical stress ratio	K^0_{nc}	0.64	-
Horizontal permeability	k_h	0.0004	m/day
Vertical permeability	k_v	0.0002	m/day
Over-consolidation ratio	OCR	1.3	-
Pre-overburden pressure	POP	20	-
Unit weight	γ_{unsat}	16.2	kN/m³
	γ_{sat}	16.2	kN/m³

Table 4. Properties of the sand layer [2]

Sand Layer			
Parameter	**Name**	**Value**	**Unit**
Material model	Model	MC	
Drainage type	Type	Undrained	
Unit weight	γ_{unsat}	18	kN/m³
	γ_{sat}	20	kN/m³
Young's modulus	E'	7.5×10^4	kN/m²
Poisson's ratio	ν'	0.3	-
Cohesion	c'	0.1	kN/m²
Friction angle	ϕ'	35	°
Dilatancy angle	ψ	0	°
Horizontal permeability	k_h	1	m/day
Vertical permeability	k_v	1	m/day

3.2 PLAXIS model

In this study, an analysis was conducted using PLAXIS 2D utilizing the beam elements to anticipate the flow of soils in between piles. The results were compared with those from Dao's [2] study, which utilized PLAXIS 3D. Figures 2 and 3 provide illustrations of the 3D and 2D analysis models. The connection at the pile head in the actual centrifuge test is not clearly detailed, making it difficult to replicate the same connection conditions in PLAXIS. To address this, various models of pile head connections were tested to determine the best method for approximating the actual pile behavior, particularly in terms of bending moment distribution along the pile.

The most accurate approximation was achieved by assuming the pile head is fixed in the X and Y directions but free to move in the Z direction. To simulate this fixed condition in the X and Y directions, a very stiff plate with a Young's modulus of E = 3,107 kN/m^2 was added to the PLAXIS 2D model. The plate is connected on one side to the pile head and on the other side to the model boundary to ensure the pile head is fully constrained in these directions.

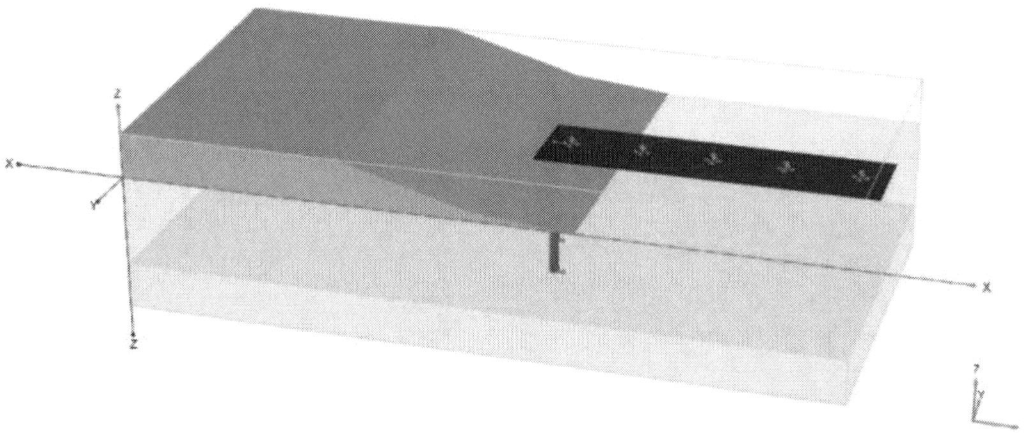

Figure 2. Visualization of the model in PLAXIS 3D [2]

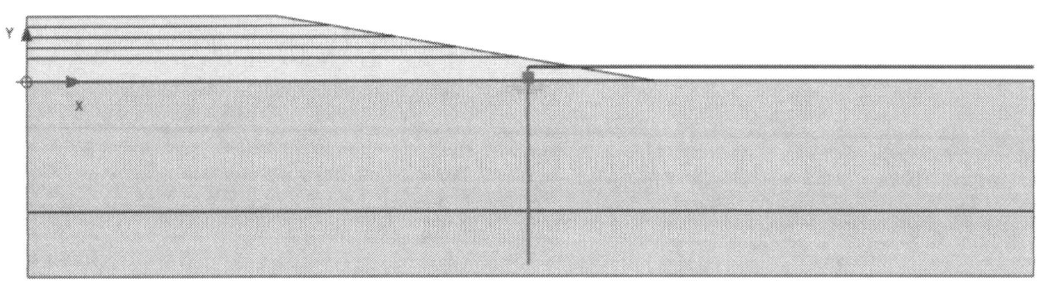

Figure 3. Visualization of the model in PLAXIS 2D

3.3 MIDAS 3D GTS NX model

Modeling using MIDAS 3D GTS NX employs the same geometry (Figure 4) and material properties as the centrifuge model described in subchapter 3.1. What sets MIDAS 3D GTS NX apart from PLAXIS is that MIDAS 3D GTS NX does not require modeling plates as reinforcing elements. Instead, the connections at the pile head are modeled as fixed constraints to replicate the behavior of the centrifuge model.

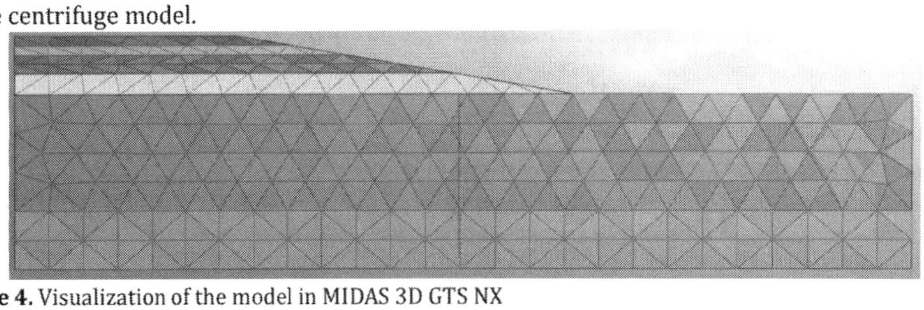

Figure 4. Visualization of the model in MIDAS 3D GTS NX

4. Calculation Phases

The calculation phases in PLAXIS 2D and MIDAS 3D GTS NX are determined based on the construction phases from the actual centrifuge tests that were conducted. Details of the calculation phases in PLAXIS 3D are summarized in Table 5.

Table 5. Calculation phases of the model (after [2])

No	Description	Total Embankment Thickness (m)	Δt (day)	T (day)
0	Initial phase	0.0	0	0
1	Install the pile	0.0	0.5	0.5
2	1st layer of sand: added	1.8	0.5	1
3	1st layer of sand: consolidated	1.8	6.5	7.5
4	2nd layer of sand: added	2.6	0.5	8
5	2nd layer of sand: consolidated	2.6	34.5	42.5
6	3rd layer of sand: added	3.4	0.5	43
7	3rd layer of sand: consolidated	3.4	83.5	126.5
8	4th layer of sand: added	4.2	0.5	127
9	4th layer of sand: consolidated	4.2	244.5	371.5
10	5th layer of sand: added	5.0	0.5	372
11	5th layer of sand: consolidated	5.0	209.5	581.5

5. Results

5.1 Bending moment

In general, the similarity between the curve of the finite element method (FEM) modeling results and the measurement curve suggests that FEM models with embedded piles can provide bending moments comparable to the measurement results (Figure 5). However, there are significant deviations from the actual test curve.

The maximum bending moment value is observed at the top of the pile. It is evident that as the embankment height increases, the bending moment value also rises. This is because a higher embankment results in greater forces acting on the pile, leading to an increase in the bending moment.

Based on the modeling results, the trend of the bending moment-depth curve from the PLAXIS 2D and 3D and MIDAS 3D GTS NX models shows similar results to those from Dao's [2] tests, with the largest deviation in bending moment occurring at top of pile. The modeling results indicate that the bending moment at the top of pile in the software models is higher compared to the centrifuge test results. This discrepancy may be attributed to factors such as the parameters and constitutive models used, which may cause the soil conditions in PLAXIS 2D and MIDAS 3D GTS NX to be stiffer.

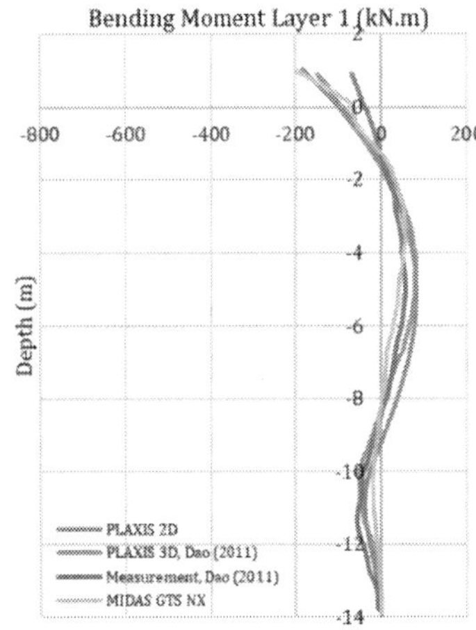

(a) Bending moment-depth (layer 1)

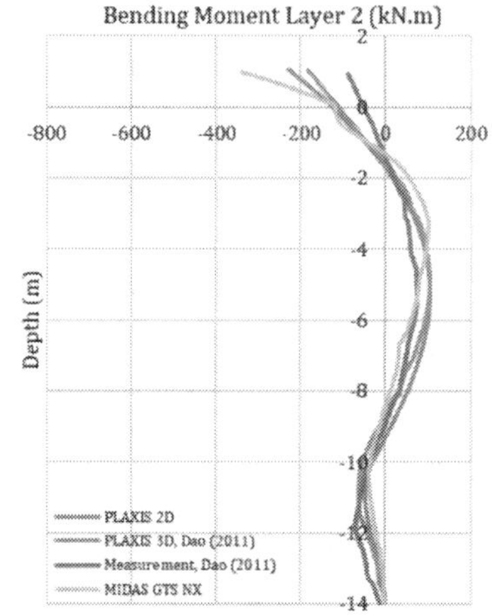

(b) Bending moment-depth (layer 2)

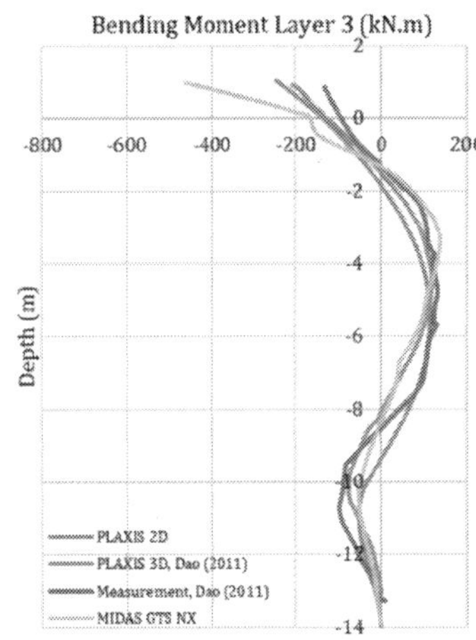

(c) Bending moment-depth (layer 3)

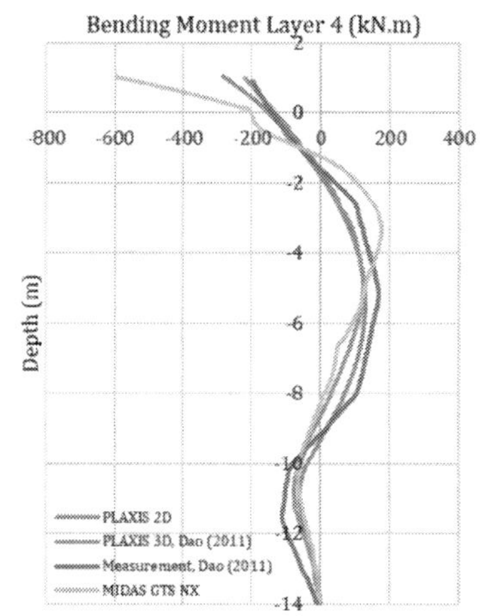

(d) Bending moment-depth (layer 4)

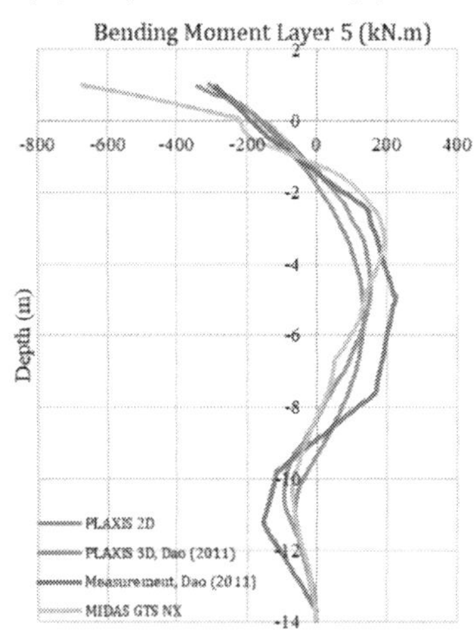

(e) Bending moment-depth (layer 1)

Figure 5. Distribution of bending moments along the pile

A review of the bending moment values at the pile base for the first embankment condition (Figure 5) shows close alignment in both trends and predictions of the bending moment curves for most of the pile foundations. However, notable deviations from the test results occur at the pile top. The calculated bending moments along most of the pile body differ from the test results by only 5.9%, less than 1%, and 2.38% for PLAXIS 3D, PLAXIS 2D, and MIDAS GTS NX, respectively. The significant difference at the foundation top is likely due to the discrepancy between the fixed boundary model and the actual test boundary conditions. Compared to PLAXIS 2D and 3D, the bending moment at the pile base in MIDAS GTS NX is larger, likely because it directly models the pile head as fixed with a constraint. In contrast, PLAXIS 2D and 3D model it as a rigid connection, which is only applicable when the beam connection point coincides with the plate element. Generally, modeling results tend to be higher than actual measurements to ensure a conservative design. The modeling results from tests by Dao [2] in PLAXIS 2D and MIDAS GTS NX indicate relatively good agreement, suggesting that modeling foundations with embedded beam elements can effectively simulate real conditions.

5.2 Settlement at ground surface

The comparison of settlement results at the end of the consolidation period (fifth embankment) reveals notable differences among the modeling methods (Figure 6). For PLAXIS 3D, the settlement shows a 3% deviation from the measurement results, while PLAXIS 2D has a 24% deviation, and MIDAS GTS 3D NX also shows a 3% deviation.

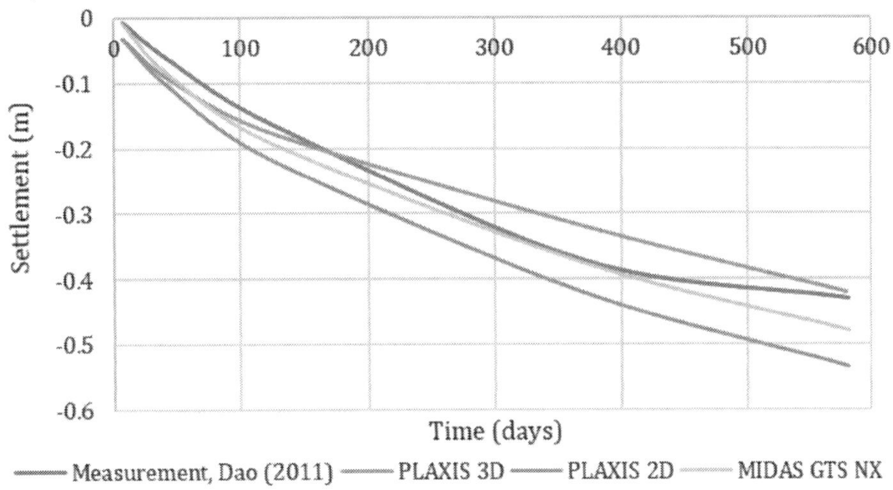

Figure 6. Settlement at ground surface (X=25 m)

Overall, the settlement values on the ground surface from the measurements are smaller compared to those obtained using finite element methods. Notably, 3D modeling with MIDAS 3D GTS NX yields a smaller settlement value compared to PLAXIS 2D utilizing the embedded elements. This is because 3D modeling accounts for forces acting along the Z-axis and distributes the load over a larger area, which helps reduce the influence of these forces. Despite these differences, the general agreement among the modeling methods suggests that the FEM approach provides a reasonable approximation of settlement behavior.

5.3 Horizontal soil displacement

Development of horizontal soil displacements over time (logarithmic scale) at a depth of 1.5 m and 25 m away from the embankment centerline is illustrated in Figure 7. It compares field-measured data obtained from centrifuge testing by Dao [2] with simulated results from three numerical modeling tools: PLAXIS 2D (with embedment beam elements), PLAXIS 3D, and MIDAS GTS NX. The analysis aims to evaluate the predictive accuracy of the finite element models in capturing soil displacement behavior during consolidation phases.

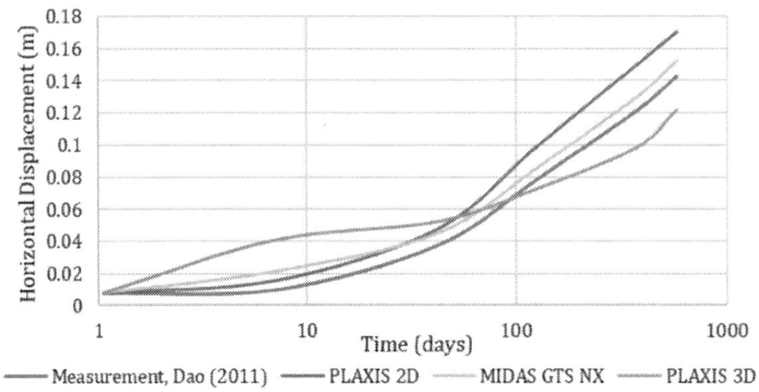

Figure 7. Horizontal soil displacements at X=25m and Z=1.5m

A critical distinction in the data presented is that the centrifuge test provides horizontal displacement measurements only at the conclusion of each consolidation phase, reflecting discrete data points. In contrast, the finite element models, such as PLAXIS and MIDAS GTS NX, generate continuous displacement profiles over time. This fundamental difference makes a direct comparison challenging, as the measured data represents specific moments, whereas the simulation outputs include intermediate values. To address this, the "PLAXIS 3D" curve was modified to reflect only the displacements at the end of the consolidation phases, enabling a more meaningful comparison with the measured data.

Examining the results, the measured and simulated curves exhibit similar trends, but discrepancies arise in both magnitude and timing of displacements. During the initial and early consolidation phases, the numerical models tend to overpredict horizontal displacements compared to the measured values. This trend continues until the end of the third consolidation phase. In the later consolidation phases, the models underestimate the displacements, particularly in PLAXIS 2D and PLAXIS 3D. Such deviations could be attributed to the limitations of the constitutive soil models employed in the simulations, such as the Soft Soil Creep (SSC) model, which might inadequately represent the nonlinear creep and consolidation behavior of the soil under long-term loading.

Quantitatively, the deviations at the end of the final consolidation phase are noteworthy. PLAXIS 2D overestimates horizontal displacements by 19% compared to the measured data, while PLAXIS 3D shows a smaller discrepancy of 14%. MIDAS GTS NX demonstrates the closest agreement, with a difference of only 7%. These variations highlight the importance of model selection, parameter calibration, and consideration of three-dimensional effects when simulating soil-structure interaction problems.

Figure 7 underscores the intricate challenges of accurately predicting horizontal soil displacements during consolidation. Although finite element models offer valuable insights into soil behavior, their precision relies heavily on accurate representation of soil properties, consolidation mechanics, and boundary conditions. Ongoing advancements in constitutive modeling and rigorous validation of numerical predictions against experimental or field data are essential to enhance the reliability of geotechnical analyses.

6. Conclusion

This paper presents an analysis of embankment impact on adjacent pile behavior using numerical simulations with 2D and 3D models, complemented by centrifuge test results. The simulation outcomes from both 2D (utilizing the embedded beam elements) and 3D models show calculated bending moments and calculated settlement in good agreement with the experimental findings by Dao [2], suggesting that modeling foundations with embedded beam elements can effectively simulate real-world conditions. The maximum bending moment, observed at the pile top, increases with embankment height due to the additional forces acting on the pile. Settlement analysis at the end of the consolidation period reveals that PLAXIS 3D and MIDAS 3D GTS NX yield similar results, with only minor deviations from measured values especially at the top of piles, while PLAXIS 2D shows a larger discrepancy. The slightly lower settlement values in MIDAS 3D GTS NX are attributed to its capability to account for forces along the Z-axis and distribute loads over a broader area. Despite some variations among the modeling methods, the overall trends are consistent, indicating that FEM modeling provides a reliable approximation of settlement behavior.

Acknowledgements

The authors would like to thank Dr. Hasbullah Nawir, Dr. Erza Rismantojo, and Dr. Andhika Sahadewa for their valuable inputs to this research.

References

[1] Bauer J, Kempfert H, Reul O 2016 Lateral pressure on piles due to horizontal soil movement *Int. J. Phys. Model. Geotech.* **16** 173-184.
[2] Dao 2011 Validation of PLAXIS embedded piles for lateral loading Delft University of Technology: Geo-Engineering.
[3] Goh ATC, Teh CI, Wong KS 1997 Analysis of piles subjected to embankment induced lateral soil movement *J. Geotech. Geoenviron. Eng.* **123.**
[4] Jennings K, Naughton PJ 2010 Lateral deformation under the side slopes of piled embankments.
[5] Jie YX, Yuan HN, Zhou HD, Yu YZ 2013 Bending moment calculations for piles based on the finite element method *J. Appl. Math.*
[6] Yi SX 2022 Parametric study of passive piles subjected to adjacent surcharge load in extensively deep soft soil *Front. Mater.*
[7] Broms B, Pandey G, Anthony GUH 1987 The lateral displacement of piles from embankment loads *J. Soil Mech. Found. Div. ASCE* No. 388.
[8] CONNECT 2020 *PLAXIS 2D - Reference Manual* Providence: CONNECT.
[9] CONNECT 2023 *PLAXIS Material Models* Providence: CONNECT.
[10] MIDAS 2019 *GTS NX User Manual* Providence: MIDAS.

IOP Conf. Series: Earth and Environmental Science **1464** (2025) 012032 doi:10.1088/1755-1315/1464/1/012032

Engineering critical assessment of subsea pipeline

Santi Frestiqauli[1*], Ricky Lukman Tawekal[2] and Ahmad Taufik[2]

[1] Institute Technology Sepuluh Nopember, Surabaya, East Java, Indonesia
[2] Institute Technology Bandung, West Java, Indonesia

*E-mail: santi@its.ac.id

Abstract. Engineering critical assessment (ECA) or fitness-for-service (FFS) is a fracture mechanics-based approach that is increasingly utilised for assessing the structural integrity of welded steel structures considering various degradation mechanisms which may occur during service. Engineering Critical Assessment (ECA) acceptance criteria the maximum allowable flaw sizes are determined specifically for the service. This study provides ECA level II to know maximum allowable flaw sizes for semi-elliptical weld defect with various size flaws in subsea pipeline based on two conditions, which are hydrotest and installation. Methods of this study is according to API standard 579-1 by define load ratio and toughness ratio then state to ECA diagram. The conclusion state that defect depth has more influence on crack growth. In installation conditions, weld joint number 20 with a defect depth of 4 mm, defect width of 200 mm is unacceptable. While in hydrotest conditions, weld joints number 13 to 20, with a defect depth (a) ranging from 3 mm to 4 mm, are unacceptable. So based on both condition, weld defect more than 3 mm is unacceptable.

1. Introduction

Pipes are used to transport fluid (liquid or gas) from upstream to downstream, or production facilities to the mainland [1]. Pipe serves to drain fluid (liquid, gas) from one or several points to other points. Fluid flow needs a pressure difference from starting point to end point. The friction that occurs between pipe and fluid must be overcome due to when fluid flows, the pipe must not be damaged or leaking [2]. Pipes are used in various industries in a plant, petrochemical plants, fertilizers, nuclear, geothermal, gas and others either on land, onshore or offshore. All of these industries require pipe installations either pipeline or piping [3]. Subsea pipelines have become the most trusted distribution of oil and gas as part of offshore structure and development of marine technology. However, pipelines are susceptible to damage which can result in enormous risk. Pipeline damage could occur both from internal and external influences [4]. Damage and failure in pipes often occur especially for pipes that have experienced aging due to operational processes. If the failure occurs in the main pipeline, it will certainly hamper the fluid transportation process which will cause huge economic losses to the company. Therefore, it is important to minimize the occurrence of disturbances in the pipeline. To ensure pipeline reliability can be done with a series of inspection programs and good inspection data processing starting with determining pipe characteristics from material selection, determining thickness, determining pipe diameter in accordance with fluid load. Then conduct a series of trials conducted by the manufacturing company. Furthermore, conduct several engineering analyses such as pressure analysis, strain analysis, fracture analysis, and so on carried out by the consultant or designated department [5]. Moreover, to obtain acceptance criteria for pipe weld defects to maintain pipe integrity during pipeline laying or installation, an engineering critical assessment (ECA) was conducted [6].

2. Literature Review

Defect generally occurs in the girth weld area is a form of a semi-elliptic undercut defect, so that the geometry of the undercut defect can be modelled as a semi-elliptic defect. These defects are dangerous in high pressure pipes. During initial hydrostatic testing, this defect often causes leaks. However, it is often the case that when the defect has not yet grown to its critical size, it is not recognized during the initial hydrostatic test. However, it is likely that defect growth will occur if the pipe continues to be supplied with high-pressure fluid. Therefore, if a pipe weld has a defect, it must be analysed as soon as possible in order to make a good decision about the incident. Engineering Critical Assessment (ECA) is known as fracture analysis which uses fracture mechanics principles to determine the defect tolerance of critical points [7]. ECA is used to generate the allowable crack size in a welded joint and decide whether the defect is safe from fracture, brittle, possible crack extension due to fatigue or expansion conditions due to high temperature. The result of the ECA can help the company to determine the appropriate repair steps to be implemented so that the company can make budget efficiency for the repair.

Engineering Critical Assessment (ECA) and work criteria are the two acceptance levels when weld defects are found [8]. Workmanship acceptance levels for welding defects in pipe welds can be found in several standards such as BS 4515-1, API 1104 and DNV-OS-F101. These acceptance levels do not correspond to weld defect limits but describe how to perform good welding work. Furthermore, ECA applies fracture mechanics to rationally ensure weld integrity [9]. Most of the ECA procedures used to be not applied in inland pipelines and coastal pipelines, but recently ECA has been widely performed since the latest pipeline designs were introduced with higher complexity such as high temperature and pressure, plastic stressing of pipes during the installation process, installation in deep water and aggressive internal conditions. Another reason is the use of transition technology applications from radiography to automated ultrasonic testing (AUT). ECA is adopted when AUT is needed to control the weld thickness on the pipe. AUT is preferred over X-ray because it is more precise and reliable. It is used as a primary inspection method during construction by generating defect size and location information in 2 dimensions [10].

In current industry practice, ECA is performed throughout the subsea pipeline design work to analyse the acceptable defect size in girth weld joints. ECA is performed through all phases of the pipeline life cycle from installation to the end of the design life. In addition, fracture mechanics-based ECA is also used to evaluate the acceptable defect size in the structure. Usually, defects are present in the weld joints during the pipe fabrication process. The main objective of applying ECA in rigid pipe trusses is to determine the largest envelope size limit of the initial defect size (for defect depth and length) that is acceptable for the design history of the pipe. The basic procedure is to assume the existence of a certain defect size at the weld joint and carry out an ECA to ensure this defect is acceptable without resulting in fracture during the loading process [10]. According to DNV-JIP line and clad pipeline (2013), ECA analysis for line pipes can be categorized into:

1) ECA Category 1: Based on analytical solutions and does not require mechanics finite element-based factors. Category 1 ECA adopts a conservative approach to construct the failure assessment diagram. The conservative approach is based on stress-strain curves.

2) ECA Category 2: Based on the conventional ECA procedure presented in DNV-OS-F101. However, in this category fracture mechanics based on finite element analysis is performed to compare the crack driving force of the developed stress-strain curve. The developed stress-strain curves are handled as equivalent stress-strain curves. In general, it is acceptable to adjust the crack driving force by other processes such as assigning a stress-based solution suitable for modification of the FAD curve shape or assigning a safety factor to determine a conservative crack driving force calculation. If the above procedures are applied, it should be verified that there is no potential for non-conservative results. The traditional ECA approach in accordance with DNV-OS-F101 is acceptable to perform when it is demonstrated that the defect size given the crack driving force for a model presenting the worst-case stress strain curves for all materials is smaller than the model for which only the parent pipe tensile data are specified.
3) ECA Category 3: 3D finite element fracture mechanics analysis was used to represent the maximum size defect allowed in ECA category 3. Due to the specificity and selection of the worst case, geometry and misalignment had to be illustrated in the FE model.

3. Methodology

The data used in this study is data from an underwater pipeline project located on Madura Island with a water depth of about 60 meters. The following is general pipe material data:

Table 1. Pipeline material data.

Parameter	Unit	Description
Material	N/A	API 5L L450MO X65MO
Wall thickness	Mm	21
Young's modulus	GPa	207
SMYS	MPa	490
SMTS	Mpa	507
Diameter	Inch	10
Thickness	Inch	0.5
Length	m/ft	11.9 - 12.5
Hydrotest Pressure	psig	2350

3.1 Define weld defect

During the welding process, it is likely that the pipe will have defects such as cracks, localized corrosion, undercuts, etc. These defects will potentially develop or propagate due to the load and dynamics of the pipe, so there is a possibility that the pipe will fail due to fatigue. When a pipe is deformed, there is stress intensification around the crack tip compared to the surrounding environment. This stress intensity at the crack tip can reach the material strength of either yield strength or tensile strength which will cause failure of the pipe. In this study, the size variation of defects in the weld joint was determined randomly.

3.2 Determining stresses

The main stress in the welded joint to the deformed surface is the internal pressure (hoop stress) caused by the pressure of the fluid flowing inside the pipe. This pressure acts in an area tangential to the pipe wall. The axial stress caused by the pressure load at both ends of the axial force is called longitudinal stress. According to Annex C, Table C.1 on defect geometry, component geometry and loading conditions refer to KCSCLE1 and RCSCLE1. The pressure intensity factor for KCSCLE1 is presented in Annex C, paragraph C.5.10 while the reference pressure for RCSCLE1 is presented in Annex C, paragraph D.5.10. The equation for obtaining hoop stress (σ_h) specified in ASME B31.8 standard:

$$\sigma_h = \frac{Pd}{2t} \tag{3.1}$$

The equation for obtaining the longitudinal stress (σ_l) specified in the ASME B31.8 standard:

$$\sigma_l = \frac{Pd}{4t} \tag{3.2}$$

In hydrotest condition, the pipe is subjected to a pressure of 125% of the operational pressure of the pipe but is not affected by bending stresses. The main pressure on the pipe is the internal pressure (hoop stress) caused by the pressure of the fluid flowing inside the pipe. This pressure acts in an area tangential to the pipe wall. The axial stress caused by the pressure load at both ends of the axial force is called longitudinal stress. Furthermore, the modulus value for uniaxial stress will be assumed to be half of the yield strength (σ_{ys}) stress value. The equation to obtain the uniaxial stress (σ_a) in the pipe:

$$\sigma_a = \frac{Ed}{2Rh} \tag{3.3}$$

Table 2. Pipe dimension data.

Description	Abbreviation	Nominal	Unit
Outer diameter	Do	0.254	Meter
Inner diameter	Di	0.241	Meter
Inner radius	Ri	120.65	Millimetre
Outer radius	Ro	127	Millimetre
Thickness	T	12,7	Millimetre
Pressure	P	1800	Psi
Hydrotest Pressure	Ph	2350	Psi

3.3 Define stress intensity factor (KI)

In fracture mechanics, there is one important parameter that aims to determine the stress field that occurs around the crack tip if a structure with a crack is subjected to a load. The stress intensity factor for semi-elliptic surface defects for both conditions can be shown by solving the equation below. Under installation conditions, where there is no internal pressure but bending stress, according to API standard 579-1 Annex C.5.2.1 on cylindrical wall cracks, circumferential direction, membrane through the wall and bending stress (KCTCC1) equation C.161 or equation C.162:

$$KI = [\{\sigma_m + p_c\}\, G_0 + \sigma_b\, (G_0 - 2G_1)]\, \sqrt{\pi c} \text{ or} \tag{3.4}$$

$$KI = (\{\sigma_0 + p_c\}\, G_0 + \sigma_1\, G_1)\sqrt{\pi c} \tag{3.5}$$

As for hydrotest conditions, where there is internal pressure but no bending stress, then according to API 579 Annex C.5.13 standard regarding cylindrical surface cracking, semi-elliptical circular direction, internal pressure and axial force (KCSCCE1) for cracks in the surface, equation C.193:

$$KI = G_0 \left(\frac{pRo^2}{Ro^2 - Ri^2} + \frac{F}{\pi(Ro^2 - Ri^2)}\right)\sqrt{\frac{\pi c}{Q}} \quad (3.6)$$

$$Q = 1.0 + 1.464\,(a/c)^{1.65};\ a/c \le 1.0 \quad (3.7)$$

$$Q = 1.0 + 1.464\,(c/a)^{1.65};\ a/c \ge 1.0 \quad (3.8)$$

The comparison between the stress intensity factor and the toughness of the material indicates the extent to which the presence of defects of a certain dimension in the material can affect the toughness of the material.

3.4 Define fracture toughness (KIC)
Material toughness is the ability of the material to withstand the amount of stress intensity factor so that there is no increase in crack length. To obtain a valid value of material toughness, namely by conducting laboratory tests. Because KIC data from test results for certain materials are not widely available, many researchers use an empirical formula approach. One of the empirical approaches to estimating material toughness (fracture toughness) is using equation 3.9 which is used to estimate the KIC value for each ferritic steel with the ASME section XI lower bound equation where T is temperature analysis and $Tref$ is reference temperature.

$$KIC = 33.2 + 2.806 \exp\left[0.02\,(T - Tref + 100)\right] \quad (3.9)$$

3.5 Define stress intensity ratio
The comparison between the stress intensity factor (KI) and the material toughness (KIC) indicates the extent to which the presence of defects of a certain dimension in the material can affect the toughness of the material. In the failure analysis diagram, Kr is used to evaluate the material. A brittle fracture condition will occur if the Kr value is greater than 1 (one). This shows that the material cannot withstand the load that hits it, it also shows that the presence of defects will cause material failure. Using equation 3.10 the toughness of the material at each temperature can be determined conservatively. Brittle fracture will occur if these conditions are met:

$$KI \ge KIC \text{ or } Kr = \frac{K_I}{K_{IC}} \ge 1 \quad (3.10)$$

A brittle fracture condition will occur when the Kr value is greater than 1. This is indicated by the material no longer being able to withstand the load applied to it, which means that the presence of the defect causes failure of the material. According to the API 579-1/ASME Fitness for Service 2007 standard on Failure Analysis Diagrams, Kr can be defined using the following equation:

$$Kr = (1 - 0.14(L_r^P)^2)\,(0.3 + 0.7 \exp[-0.65(L_r^P)^6] \quad (3.11)$$

3.6 Define reference stress (σref)
The reference voltage for RCSCLE1 is presented in Annex D, paragraph D.5.10.1 using the following equation:

$$\sigma ref = \frac{g\,Pb + [(gPb) + 9(MsPm(1-\alpha)^2)^2]^{0.5}}{3(1-\alpha)^2} \quad (3.12)$$

$$g = 1 - 20\left(\frac{a}{2c}\right)^{0.75} \alpha^3] \quad (3.13)$$

$$\alpha = \frac{\frac{a}{t}}{1 + \frac{t}{c}} \quad (3.14)$$

IOP Conf. Series: Earth and Environmental Science **1464** (2025) 012032 doi:10.1088/1755-1315/1464/1/012032

3.7 Define Load Ratio

The ratio between stress and material strength which serves to determine how the opportunity for pipe material can experience plastic collapse. From the above stress calculation, the load ration can be determined by:

$$Lr = \frac{\sigma ref}{\sigma ys} \tag{3.15}$$

By knowing the Kr and Lr values that form the coordinates (Lr, Kr) in the failure analysis diagram. The two criteria will be used to predict the reliability and integrity of the pipe and the failure mode of the pipeline, including: brittle fracture, plastic deformation (plastic collapse), or a combination of the two (mixed mode). Determining whether the pipe structure is safe or not against the three types of failure models is seen from a combination of stress, geometry and flaw dimensions, yield strength and material toughness. Each section of pipe at a certain flaw size can be seen whether its position is safe or failure.

4. Analysis and Result

The main pressure on the pipe, namely internal pressure (*hoop stress)*, occurs due to the pressure of the fluid flowing inside the pipe. This pressure acts in an area tangential to the pipe wall. While the axial stress caused by pressure loads at both ends of the axial force is called *longitudinal stress*. There are two conditions of pipeline pressure in this study, which are:

a. Bending condition: In this condition, there is no internal pressure on the pipe, but there is bending stress influenced by the stringer of laying vessel.
b. Hydrotest condition: In this condition, the pipe is subjected to a pressure of 125% of the operational pressure of the pipe but is not affected by bending stresses.

Based on the material data, the stresses acting on the pipe in the presence of defects can be calculated. After dividing by the yield strength of the material, each defect configuration can be determined by the ratio load Lr. Then calculate each stress intensity factor divided by the material toughness factor which results in a stress intensity ratio (Kr). The results of the calculation of load ratio (Lr) and stress intensity factor (Kr) could be seen in table below.

Table 3. Load ratio calculation data (Lr).

Joint no.	a	c	2c	a/c	t	a/t	t/c	α	Σref (3.10)	stres	Lr (3.13)	Kr (3.9)
unit	mm	mm	mm	mm	mm	mm	mm	mm	psi	psi	psi	ksi.in1/2
1	0.5	12.5	25	0.04	12.7	0.0	1.02	0.02	41,893	60,410	0.59	0.93
2	0.5	25	50	0.02	12.7	0.0	0.51	0.03	42,461	60,410	0.60	0.93
3	0.5	50	100	0.01	12.7	0.0	0.25	0.03	42,926	60,410	0.60	0.93
4	0.5	100	200	0.01	12.7	0.0	0.13	0.03	43,241	60,410	0.61	0.93
5	1	12.5	25	0.08	12.7	0.1	1.02	0.04	43,613	60,410	0.61	0.92
6	1	25	50	0.04	12.7	0.1	0.51	0.05	44,833	60,410	0.63	0.92
7	1	50	100	0.02	12.7	0.1	0.25	0.06	45,850	60,410	0.65	0.91
8	1	100	200	0.01	12.7	0.1	0.13	0.07	46,550	60,410	0.65	0.91
9	2	12.5	25	0.16	12.7	0.2	1.02	0.08	47,387	60,410	0.67	0.90
10	2	25	50	0.08	12.7	0.2	0.51	0.10	50,213	60,410	0.71	0.88
11	2	50	100	0.04	12.7	0.2	0.25	0.13	52,672	60,410	0.74	0.86

IOP Conf. Series: Earth and Environmental Science **1464** (2025) 012032 doi:10.1088/1755-1315/1464/1/012032

Joint no.	a	c	2c	a/c	t	a/t	t/c	α	Σref (3.10)	stres	Lr (3.13)	Kr (3.9)
unit	mm	mm	mm	mm	mm	mm	mm	mm	psi	psi	psi	ksi.in1/2
12	2	100	200	0.02	12.7	0.2	0.13	0.14	54,419	60,410	0.77	0.84
13	3	12.5	25	0.24	12.7	0.2	1.02	0.12	51,673	60,410	0.73	0.87
14	3	25	50	0.12	12.7	0.2	0.51	0.16	56,623	60,410	0.80	0.81
15	3	50	100	0.06	12.7	0.2	0.25	0.19	61,137	60,410	0.86	0.75
16	3	100	200	0.03	12.7	0.2	0.13	0.21	64,465	60,410	0.91	0.70
17	4	12.5	25	0.32	12.7	0.3	1.02	0.16	56,567	60,410	0.80	0.81
18	4	25	50	0.16	12.7	0.3	0.51	0.21	64,344	60,410	0.91	0.70
19	4	50	100	0.08	12.7	0.3	0.25	0.25	71,819	60,410	1.01	0.56
20	4	100	200	0.04	12.7	0.3	0.13	0.28	77,572	60,410	1.09	0.44

Table 4. Calculation data of stress intensity ratio (Kr).

Weld Joint No	KI bending	Kr bending	FAD categories	KI hydrotest	Kr hydrotest	FAD categories
1	21,220	0,14	Acceptable	22.548	0.15	Acceptable
2	30,010	0,20	Acceptable	22.658	0.15	Acceptable
3	42,450	0,28	Acceptable	22.694	0.15	Acceptable
4	60,030	0,39	Acceptable	22.705	0.15	Acceptable
5	20,670	0,13	Acceptable	43.996	0.29	Acceptable
6	30,010	0,20	Acceptable	45.095	0.29	Acceptable
7	42,450	0,28	Acceptable	45.317	0.29	Acceptable
8	60,030	0,39	Acceptable	45.388	0.30	Acceptable
9	25,130	0,16	Acceptable	87.552	0.57	Acceptable
10	38,170	0,25	Acceptable	94.960	0.62	Acceptable
11	57,240	0,37	Acceptable	98.457	0.64	Acceptable
12	80,950	0,53	Acceptable	98.940	0.64	Acceptable
13	34,280	0,22	Acceptable	118.327	0.77	Unacceptable
14	35,530	0,23	Acceptable	134.711	0.88	Unacceptable
15	53,980	0,35	Acceptable	143.644	0.93	Unacceptable
16	80,950	0,53	Acceptable	148.087	0.96	Unacceptable
17	34,280	0,22	Acceptable	146.883	0.96	Unacceptable
18	35,530	0,23	Acceptable	175.105	1.14	Unacceptable
19	53,980	0,35	Acceptable	189.920	1.24	Unacceptable
20	80,950	0,53	Unacceptable	196.914	1.28	Unacceptable

The failure analysis diagram according to API 579-1/ASME Fitness for Service 2007 standard could be seen in Figure 1. Regarding the Failure Analysis Diagram is defined using Equation 3.11. Failure analysis diagram for bending and hydrotest condition could be seen in Figure 2 and Figure 3.

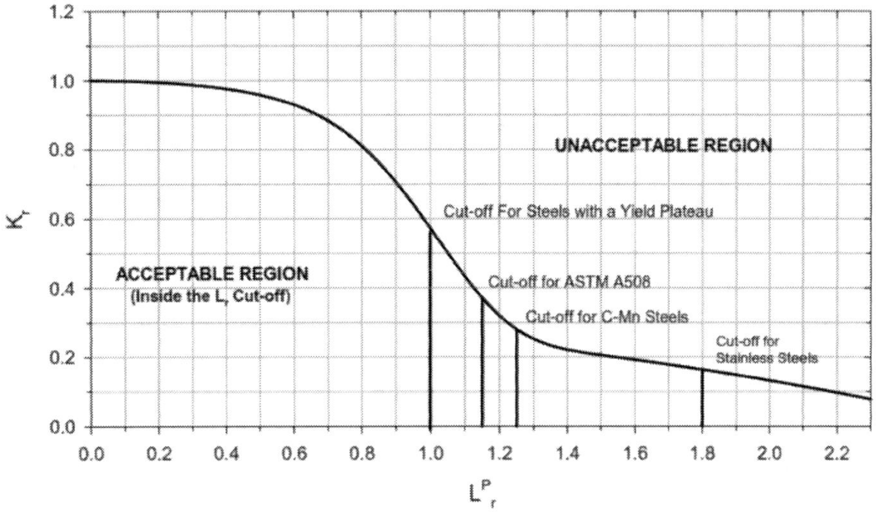

Figure 1. Failure analysis diagram. [11]

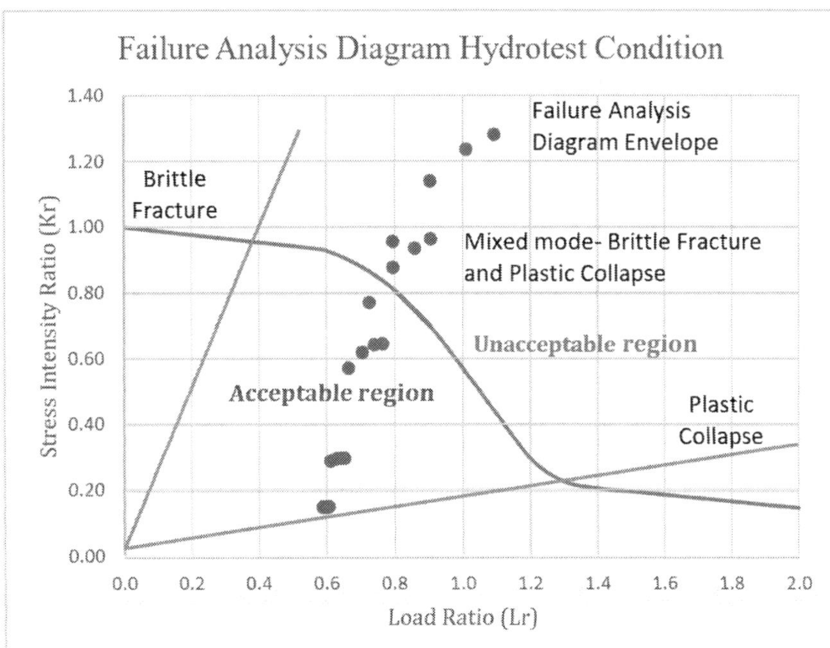

Figure 2. Failure analysis diagram at hydrotest condition.

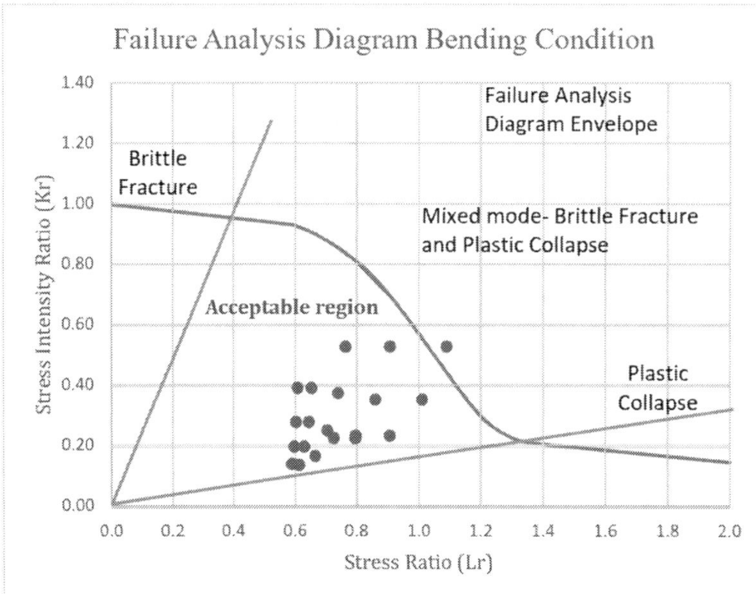

Figure 3. Failure analysis diagram at bending condition.

5. Conclusion

This paper examines Engineering Critical Assessment (ECA) for evaluating subsea pipeline integrity, determining maximum allowable flaw sizes for weld defects under hydrotest and installation conditions according to API 579-1/ASME Fitness for Service standard to categorize defect using failure analysis diagram. Based on the calculation analysis, the defect depth (a) has more influence on crack growth. In bending conditions, weld joint number 20 with a defect depth of 4 mm, defect width of 200 mm is unacceptable. While in hydrotest conditions, weld joints number 13 to 20, with a defect depth (a) ranging from 3 mm to 4 mm, are said to be unacceptable.

References

[1] Det Norske Veritas, RP-F108: Fracture Control for Pipeline Installation Methods Introducing Cyclic Plastic Strain, Det Norske Veritas, 2006
[2] Feng Q, Failure Assessment for Girth Weld Defects of Pipeline, 2010
[3] The American Society of Mechanical Engineers, API 579-1: Fitness for Service, 2007
[4] Feng Q, Failure Assessment for Girth Weld Defects of Pipeline, 2010
[5] Anderson T L, Fracture Mechanics: Fundamental and Application, CRC Press, 1995.
[6] Hval M, Lamvik T, Hoff R, Engineering Critical Assessment of Clad Pipeline Installed by S-Lay for The Operation Phase, dalam Proceedings of the 20th European Conference on Fracture, 2014
[7] Gouveia J, Chaves P, Engineering Critical Assessment of Production Pipelines in Presence of Sour Environment, dalam Proceedings of Congresso Nacional de Transporte Aquaviario, Rio de Janeiro, 2010
[8] Permana I, A Study on Engineering Critical Assessment (ECA) of Subsea Pipeline Girth Welds for Reeling Installation, University of Stavanger, 2013
[9] Bruschi R, Leonzio V, Torselletti E, Recent Advances on Engineering Criticality Assessment for Offshore Pipeline, dalam Proceedings of the Rio Pipeline Conference and Exposition, Rio de Janeiro, 2013
[10] Shafiri S M H, Kaveh M, Googarchin H S, Engineering Critical Assessment of Offshore Pipeline under Operational Loading Phase According to BS7910 Guideline, International Journal of Coastal and Offshore Engineering, 2016
[11] API 579-1/ASME Fitness for purpose 2007 Figure 9.20 Failure Analysis Diagrams

IOP Conf. Series: Earth and Environmental Science **1464** (2025) 012033 doi:10.1088/1755-1315/1464/1/012033

Prediction of operating offshore wind farms underwater noise from reanalysis of single point underwater noise measurement

Mahardika Firjatullah[1], Irsan Soemantri Brodjonegoro[2*] and Farid Putra Bakti[2]

[1] Ocean Engineering Program, Institut Teknologi Bandung, Bandung, Indonesia
[2] Offshore Engineering Research Group, Faculty of Civil and Environmental Engineering, Institut Teknologi Bandung, Bandung, Indonesia

*E-mail: irsanb@itb.ac.id

Abstract. In renewable energy, offshore wind turbines are increasingly recognized as significant sources of underwater noise, with growing concerns about their impact on marine life. A key noise contributor is the structural vibration from the turbine's gearbox, transmitted through the tower to the water. This study re-analyzes the underwater noise of a single turbine using field measurements from the Utgrunden wind farm. Turbines are modeled as idealized Euler-Bernoulli beams with a lumped mass at the top, and the analytical response serves as a boundary condition for the employed Combined Helmholtz Integral Equation Formula (CHIEF) underwater noise simulation implementation. By matching CHIEF's output to field data using gradient descent, the approach effectively approximates the noise measurements, despite some limitations like proximity, free surface, and bottom reflection. The findings suggest that cumulative noise from the Utgrunden wind farm is detectable several kilometers away, with in- and out-of-phase noise interactions causing fluctuating noise levels near the farm. This method can optimize wind farm layouts to avoid noise peaks in sensitive areas.

1. Introduction

Predicting the underwater noise generated by the operation of a wind farm is a challenging task. Hydrophones are typically placed in several locations to obtain an accurate readout of the underwater noise. However, the amount of data points that one can obtain is highly restricted by the cost of hydrophones, research vessels, and other operational considerations.

Offshore wind farms pose environmental challenges, including increased underwater noise, collision risks, and pollution from vessel traffic. While noise from a single turbine is comparable to typical marine operations and 20–30 dB lower than passing ships [1], [2]. the trend of larger, densely clustered farms will need reassessment. Simulations show noise levels amplify with more turbines [3], [4], highlighting the need to understand wind farms' overall underwater noise impact.

This study proposes a methodology to estimate noise distribution in the vicinity of an operational wind farm using only a single field point measurement. The proposed framework is illustrated in Figure 1. The field noise measurement is conducted based on a sequential shutdown of the wind farm and analyzed to determine the appropriate mode contribution constant at a given structural vibration frequency. Subsequently, the surface velocity distribution is approximated from this information and then utilized as input for the CHIEF boundary element method. The boundary element method is frequently employed in the modeling of underwater acoustics, particularly in military defense research areas, where it is critical for submarines to remain undetected [5]. The same principle is also used to predict noise from propellers, which ultimately contributes to understanding the noise generated by submarines [6].

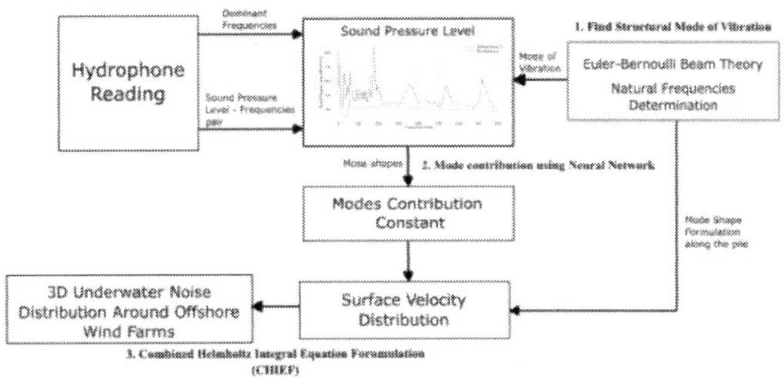

Figure 1. Flow chart describing the framework for predicting 3D underwater noise field.

2. Methodology

2.1 *Estimating the monopile modes of vibration [7]*

The monopile structure is approximated with an undamped Euler-Bernoulli beam with clamped-free boundary condition and a rigid tip mass affixed to the free-top end with mass of M_t and rotary inertia of I_t. The governing equation of motion for the monopile is presented in Equation 1.

$$YI\frac{\partial^4 w(z,t)}{\partial z^4} + m\frac{\partial^2 w(z,t)}{\partial t^2} = 0 \quad (1)$$

With $w(z,t)$ indicates the transverse displacement at point z along the pile with respect to time t. YI is the bending stiffness, and m is the mass per unit length of the monopile. The clamped-free boundary condition with a tip mass can be expressed by Equation 2 and 3.

$$w(0,t) = 0, \quad \left.\frac{\partial w(z,t)}{\partial x}\right|_{x=0} = 0 \quad (2)$$

$$\left[YI\frac{\partial^2 w(z,t)}{\partial z^2} + I_t\frac{\partial^3 w(z,t)}{\partial t^2 \partial z}\right]_{x=L} = 0, \quad \left[YI\frac{\partial^3 w(z,t)}{\partial z^3} + M_t\frac{\partial^2 w(z,t)}{\partial t^2}\right]_{x=L} = 0 \quad (3)$$

The rotary inertia of the tip mass (I_t) can be approximated using a one-mass model of a wind turbine can be calculated by Equation 4. The general response of the beam is the linear combination of all vibration modes as shown in E Equation 5.

$$I_t = I_r + n_g^2 I_g \quad (4)$$

$$w(z,t) = \sum_{k=1}^{\infty} \left[\cos\frac{\lambda_k}{L}z - \cosh\frac{\lambda_k}{L}z + \zeta_k(\sin\frac{\lambda_k}{L}z - \sinh\frac{\lambda_k}{L}z)\right](A_k \cos\omega_k t + B_r \sin\omega_k t) \quad (5)$$

Where A_r and B_r are the unknown constants that can be determined by finding the response to initial conditions $w(z,0)$ and $\partial w(z,t)/\partial t|_{t=0}$, and ω_k is the eigenfrequency of the k-th vibration mode.

2.2 Combined helmholtz integral equation formulation (CHIEF) boundary element method

2.2.1 CHIEF formulation

An arbitrarily shaped geometry is submerged in an infinite ideal homogeneous fluid that fills the exterior region of the solid surfaces. The fluid has a density of ρ, and the speed of sound on the fluid is c. The fluid fills region exterior to the surface region. Previous studies have indicated that the spatial pressure radiated from vibrating bodies and the normal velocity comply with the Helmholtz integral formulas, as shown in Equation 6, where $p(X)$ is the field pressure, $G(\sigma,X)$ is the green function, $\partial p(\sigma)/\partial n$ is the pressure gradient normal to the body's surface S [10], [11].

$$\int_S \left(G(\sigma,X)\frac{\partial p(\sigma)}{\partial n} - p(\sigma)\frac{\partial G(\sigma,X)}{\partial n}\right) dS = \begin{cases} -p(X) & X \text{ exterior to } S \\ -\frac{1}{2}p(X) & X \text{ on } S \\ 0 & X \text{ interior to } S \end{cases} \quad (6)$$

To evaluate the exterior acoustic pressure using Equation 6, one must provide the entire pressure and normal velocity distribution on the surface S. The pressure gradient may be calculated from the surface velocity distribution as can be found in the surface boundary condition shown in Equation 7, and the field pressure must satisfy the radiation condition [12].

$$\frac{\partial p(X)}{\partial n} = \hat{n}_s \cdot \nabla p(X) = -i\omega\rho v(X), \quad X \in S \quad (7)$$

Upon substituting Equation 7 into Equation 6, the external Helmholtz equation can be transformed into Equation 8. Furthermore, an interior Helmholtz relation is also obtained which can offers a supplementary means to compute the surface pressure, this relation is presented in Equation 9 [13]. In this work, the surface velocity distribution is derived by summing the modes while taking into account the forcing to stiffness term factor obtained through the use of gradient descent, which will be elaborated in Section 2.3.

$$\frac{1}{2}p_m - \sum_{n=1}^{N_s}\left(p_n \int_{S_n} \frac{\partial G(\sigma,X_m)}{\partial n} dS(\sigma)\right) = i\omega\rho \sum_{n=1}^{N_s}\left(v_n \int_{S_n} G(\sigma,X_m)\, dS(\sigma)\right) \quad (8)$$

$$0 - \int_{S_n} p(\sigma)\frac{\partial G(\sigma,X)}{\partial n} dS(\sigma) = i\omega\rho \int_{S_n} v(\sigma)G(\sigma,X)\, dS(\sigma) \quad (9)$$

A study have established that the surface Helmholtz integral equation in Equation 8 lacks a unique solution when the wavenumber k coincides or is in proximity to the characteristic wavenumber k' [11]. However, this limitation does not apply to the interior Helmholtz relation in Equation 9. The CHIEF Method resolves this by combining the surface and interior Helmholtz equations, creating an overdetermined system. The interior equation serves as an additional constraint when k is near the characteristic wave number [14]. The overdetermined system of equations mentioned can be written in matrix form equation Equation 10 with matrices A and B defined in Equation 11 and 12 [15]. The overdetermined system of equations is solved by Householder reduction [16].

$$[A]\{P\} = [B]\{V\} \tag{10}$$

$$A_{mn} = \begin{cases} \frac{1}{2}\delta_{mn} - \iint_{S_n} \frac{\partial G(\sigma, X_m)}{\partial n} dS(\sigma); \quad m = 1; N_s \\ -\iint_{S_n} \frac{\partial G(\sigma, X_m)}{\partial n} dS(\sigma); \quad m = (N_s + 1); (N_s + N_I) \end{cases} \tag{11}$$

$$B_{mn} = \begin{cases} i\omega\rho \iint_{S_n} G(\sigma, X_m)\, dS(\sigma); \;\; SHE \;\; m = 1; N_s; 1 \\ i\omega\rho \iint_{S_n} G(\sigma, X_m)\, dS(\sigma); \;\; IHR \;\; m = (N_s + 1); (N_s + N_I); 1 \end{cases} \tag{12}$$

2.2.2 *CHIEF numerical implementation*

The CHIEF algorithm is implemented in FORTRAN language [15]. The inputs of CHIEF are the geometry of the modeled structure, and the boundary condition in the form of surface velocity distribution. The output produced is the near-field pressure.The validity of CHIEF implementation needs to be tested against an analytic formula, in this validation a uniformly vibrating sphere with radius a and radial velocity of V generates field acoustic pressure that can be calculated using Equation 13 [17].

$$p(r) = \frac{i\omega\rho V a^2 e^{ika}}{1 + ika} \frac{e^{-ikr}}{r}, r \geq a \tag{13}$$

In this validation, the radius set to be 1 m, and the radial breathing velocity (V) is $1 m/s$ with wave number (k) of $\pi/3$. The results shown in Figure 2 reveals an average error of 2.22% for various surface discretization size. This validation confirms that the CHIEF numerical tools are suitable for addressing similar pressure radiation issues.

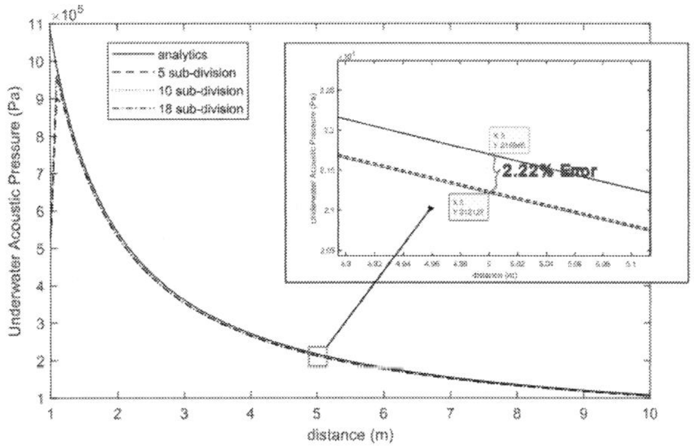

Figure 2. pressure validation for CHIEF implementation to analytical method. The analytical results are compared across various model meshing schemes, demonstrating a relatively consistent error across all schemes.

2.3 Mode contribution determination using gradient descent

In this paper, the problem of determining a hyperparameter of an environmental factor based can be framed as solving an overdetermined system discussed in Section 3.2. To address this issue, the gradient descent algorithm, a first-order optimization method, is proposed to iteratively update hyperparameters in the direction of the objective function's steepest descent.. Minimizing the error between sensor readings and predictions based on environmental factors allows for accurate estimation of the factors governing wind turbine vibrations.

The underlying reason behind this method selection is gradient descent method applies to a wide range of penalty functions especially the non-smooth one [19]. Gradient descent can effectively navigate this challenging landscape by following the gradient direction towards the global minimum of the objective function.

In the process of finding the hyperparameter, the weights W^* that minimized the loss function from Equation 14 needs to be determined. In this work, the weights serve as the forcing-to-stiffness ratio elaborated in Section 3.2.

$$W^* = \min_{W} \sum_{i=1}^{n} L(f(x_i; W), y_i) \tag{14}$$

With $W = w_0, w_1, w_2, \ldots, w_n$ being the collection of weights/hyperparameters to be solved from the system of equation, $f(x_i; W)$ is the predicted output of the model with respect to the initial weights, and y_i is the actual output desired in this case taken from a real measurement campaign.

In solving a system of equations in the form of *Ax=b* overfitting is a highly desired characteristic in order to obtain the parameter that best describes the environmental factors, hence no training datasets used. The formula is shown in Equation 15 [19].

$$Ax = b + r = b_{true} \tag{15}$$

With $A = [a_{ij}] \in R^{mxn}$ is the constant matrix model, in this work the matrix A defined in Equation 11. $b \in R^m$ is a vector containing model output or measurements, $b_{true} \in R^m$ is a vector of true values, and $r \in R^m$ is vector of model error. An energy function that represents model error is defined in Equation 16 that serves as the objective equation that minimizes error with x:

$$E(x) = \sum_{i=1}^{m} \epsilon_i(r_i(x)) \quad (16)$$

To find x that minimizes the energy function, the gradient $\nabla E(x)$ must be calculated as defined in Equation 17. In this study, a modified version of gradient descent called Nesterov Accelerated Gradient Descent (NAG) is applied. NAG is known for accelerating the optimization process and reducing overshoot by incorporating a "lookahead" step in each iteration [20].

$$\nabla E(x) = \left[\frac{\delta E(x)}{\delta x_1}, \frac{\delta E(x)}{\delta x_2}, \ldots, \frac{\delta E(x)}{\delta x_n}\right] = A^T(Ax - b) \quad (17)$$

3. Results and discussion

3.1 Case study

3.1.1 Wind turbine structure's specification overview

This work involves the modeling of a monopile structure with the main source of vibration is originated from gearbox on wind turbine *GE Wind Energy 1.5s Offshore* with its specification shown in Table 1 [21]. The monopile structure surface discretization sketch is shown in Figure 7.

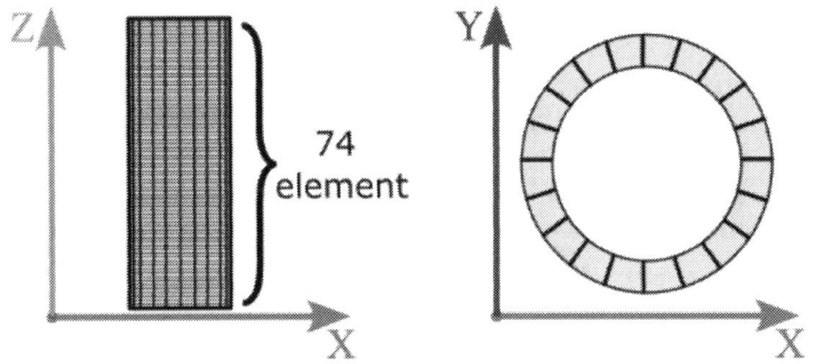

Figure 3. Monopile surface discretization: the structure is divided into 20 equal segments in the circumferential direction and 74 segments in the vertical direction.

3.1.2 Scope of the model

In modeling underwater acoustics using CHIEF, this study imposes several simplifying assumptions to address numerical limitations. The focus is on the structure's interaction with water, specifically the sound pressure transmitted directly to it, while other factors like compressional and shear waves in the sediment [22] are excluded. The model assumes a constant sound speed with depth and simplifies the seafloor profile to a flat 12.9 meters, matching the depth of Hydrophone 3 despite slight actual variation as summarized in Table 2. In this study, only the structure's surface boundary condition is considered, any other boundary conditions, such as bottom and surface reflection, and multiple turbine conditions are disregarded. Additionally, sea surface reverberation has been ignored, as sound waves interacting multiple times with the ocean surface and seabed is beyond the scope of this study.

IOP Conf. Series: Earth and Environmental Science **1464** (2025) 012033 doi:10.1088/1755-1315/1464/1/012033

Table 1.GE wind energy 1.5s offshore wind turbine and monopile structure specification.

Parameter	Value
Rated Power	1.425 MW
Number of blades	3
Rotor Diameter	70.5 m
Hub height	65 m
Cut in speed	3 m/sec
Cut out speed	25 m/sec
Rotor operation speed	11 - 20 rpm
Foundation type	Monopile
Nacelle weight	49,000 kg
Rotor weight	28,000 kg
Tower weight	76,000 kg
Transition piece weight	22,000 kg
Blade weight	5,216 kg
Gearbox ratio	90:1
Monopile dimension	$3.65\ m \times 55\ mm \times 33.7\ m$
Monopile weight	165,000 kg

3.2 *Mode contribution determination*

Three hydrophones are used during the measurement campaign in Utgrunden wind farms [24], the location of each hydrophone relative to turbine 4, and its water depth are shown in Table 2. Hydrophone 3, the closest to the wind farm, was selected for analysis as it captured the highest signal resolution. The pressure level reading from turbine 4 while operating at wind speed 14 m/s is shown in Figure 4.

Table 2. Location of hydrophones relative to turbine 4 and their depth.

Hydrophone	Distance to turbine #4 (m)	Water depth (m)
Hydrophone 1	463	18.0
Hydrophone 2	160	15.2
Hydrophone 3	83	12.9

To calculate the excited vibration of the wind turbine, the dynamic contribution of the mode shape formula is utilized as shown in Equation 18. The term ξ_i/k_i is often named the forcing-to-stiffness ratio. No information regarding the forcing of the environment is available in this study; therefore ξ_i/k_i is assgined as R_iHence, the equation for the vibration velocity of the monopile surface can be expressed by Equation 19. The result of natural frequencies for the monopile structure are shown in Table 3.

The force-to-stiffness ratio ξ_i/k_i obtained from the gradient descent results are summarized in Table 4 and applied to Equation (19) to compare deviations from the reference sound levels shown in Figure 4, the comparison presented in Figure 5 shows a similar sound levels, with the largest differences at low frequencies due to environmental noise captured alongside structural vibrations. To simulate the primary noise contributor at 177 Hz, the results show a slight frequency shift, with 188.7 Hz being used to capture the dominant term instead of the original frequency.

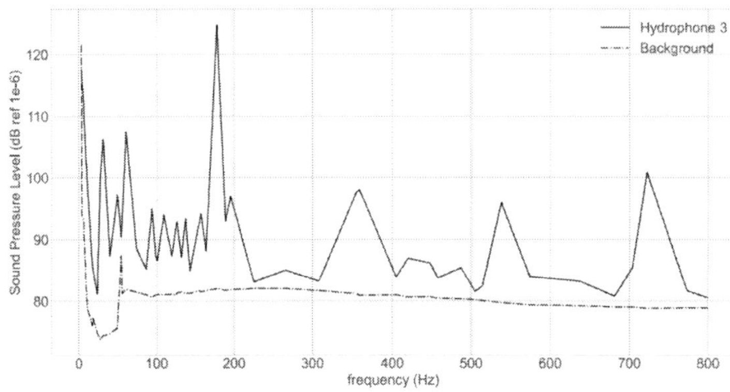

Figure 4. Hydrophone 3 recorded sound pressure levels from turbine 4 at 14 m/s wind speed, compared to background noise with no turbines operating.

$$\{x\} = \sum_{i=1}^{\infty}\left(\{\Phi\}_i \frac{\xi_i}{k_i} DAF_i \cos[\omega_i t]\right) \quad (18)$$

$$\{v\} = \sum_{i=1}^{\infty}(\omega_i \{\Phi\}_i\, R_i DAF_i \cos[\omega_i t]) \quad (19)$$

Table 3. Wind turbine mode frequencies using Euler-Bernoulli beam theory. Frequencies ranging from 1 up to 800 Hz following the basis of this study.

Mode Number	Frequency (Hz)	Mode Number	Frequency (Hz)
1	3.4806	12	225.31801
2	9.53077	13	265.23956
3	17.91222	14	308.47101
4	28.02358	15	355.00885
5	40.39122	16	404.85245
6	56.22984	17	457.99902
7	75.79421	18	514.45022
8	98.92243	19	574.20463
9	125.4884	20	637.26413
10	155.4292	21	703.62561
11	188.7112	22	773.29378

Table 4. The forcing-to-stiffness ratios acquired from the gradient descent algorithm described.

Mode Number	$\frac{\xi_i}{k_i}$	Mode Number	$\frac{\xi_i}{k_i}$
1	0.11350581	12	0.01095502
2	0.04186815	13	0.00213889
3	0.02525101	14	0.00183713
4	0.02715436	15	0.00548724
5	0.0020917	16	0.001278
6	0.00293712	17	0.00044378
7	0.00297696	18	-0.00118743
8	0.00380076	19	-0.00039723
9	0.00290444	20	-0.00031718
10	0.02623104	21	0.00210283
11	0.04467592	22	-0.00097174

3.3 Near-field pressure

3.3.1 Underwater acoustic attenuation per doubled distance comparison

Due to limitations previously described in Section 3.1.2, the outcomes of this study specifically concern to the depth wherein the effect of these two factors is least signifcant, which is in the mid-water depth. To approximate the attenuation rate in the same fashion as the actual measurement [24]. Figure 6 presents the distribution of modeled pressure level with distance. This spatial attenuation rate is applicable in all directions because of the symmetrical noise distribution. As shown in Table 5, the attenuation per doubled distance can be estimated to be around 3 dB to 6 dB which encompasses attenuation obtained from the actual measurement (4 dB) and attenuation in the cylindrical propagation model (3 dB) [23].

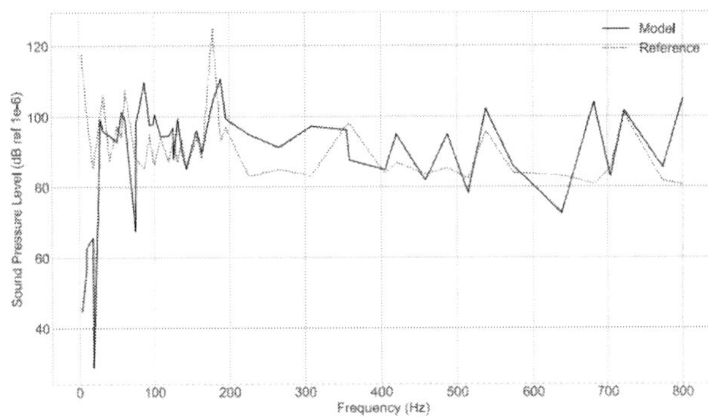

Figure 5. Pressure magnitude and sound pressure level comparison of natural and dominant frequencies compared to reference value from field measurement.

Table 5. Pressure level attenuation rate. Hyd3-Hyd2 denotes the attenuation that occurred between hydrophone 3 and hydrophone 2, and similarly for other pairs.

Frequency (Hz)	Attenuation per doubled distance (dB)		
	Hyd3-Hyd2	Hyd3-Hyd1	Hyd2-Hyd1
31.7	3.469	5.856	5.92
188.7	3.349	6.6	6.341
722.0	4.044	5.9	6
Average (dB)	3.621	6.11	6.087

4. Conclusions

The proposed method has shown a good approximation for the prediction of underwater noise using limited amount of data. Proven by its ability to predict acoustic noise attenuation. However, the achieved result is still far from perfect, and achieving a detailed noise 3D distribution can be challenging due to multiple factors such as sound speed profile, co-existing multiple frequency components, water surface and seabed reflection, and multiple turbine interaction. Therefore, a more complex boundary element model may be needed to be developed to tackle the problem statement. The proposed model improvement aims to account for sound wave interactions among multiple bodies within wind farm clusters and complex seafloor morphology, enhancing the understanding of wave patterns in the vicinity of wind farms.

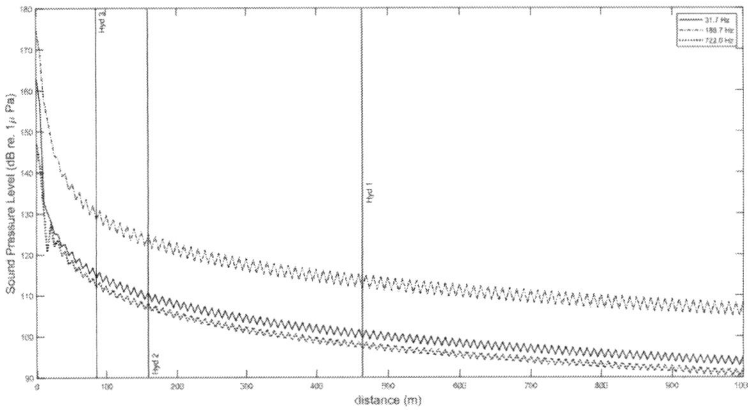

Figure 6. The sound pressure level distribution caused by one turbine in excitation frequency of 31.7, 188,7, and 722.0 Hz. The highlighted points correspond to the location of each hydrophone.

4.1 3D underwater acoustic distribution

The resulting offshore wind farms noise contour from the operation of the whole turbine is shown in Figure 7 [18] and its three dimensional plot is presented in Figure 8.

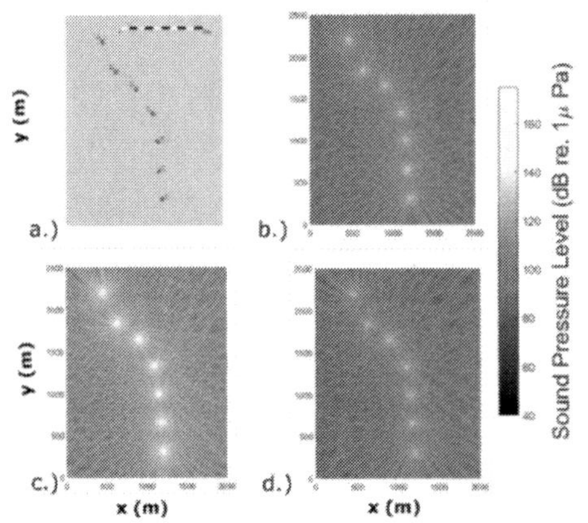

Figure 7. Mid-water depth sound pressure level distribution. (a) Actual wind farm layout, (b) Noise distribution at 31.7 Hz, (c) Noise distribution at 188.7 Hz, (d) Noise distribution at 722.0 Hz.

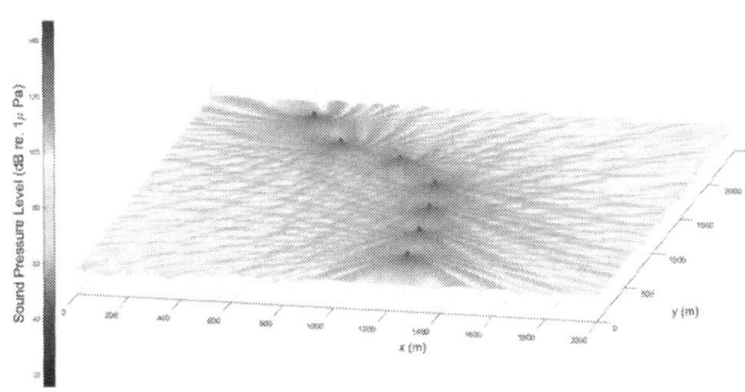

Figure 8. Mid-water depth sound pressure level distribution in isometric view. a) Sound pressure level distribution for dominant frequency 722.0 Hz.

Acknowledgements

Acknowledgement to Faculty of Civil and Environmental Engineering through P2MI (Research and Innovation Program), ITB Research Grant 2022.

References

[1] J. Tougaard, O. D. Henriksen, and L. A. Miller, "Underwater noise from three types of offshore wind turbines: Estimation of impact zones for harbor porpoises and harbor seals," J Acoust Soc Am, vol. 125, no. 6, pp. 3766–3773, Jun. 2009, doi: 10.1121/1.3117444.
[2] B. Marmo, I. Roberts, M. P. Buckingham, S. King, and C. Booth, "Modelling of Noise Effects of Operational Offshore Wind Turbines including noise transmission through various foundation types," 2013. Accessed: Mar. 16, 2023. [Online]. Available: https://tethys.pnnl.gov/sites/default/files/publications/Marmo_et_al_2013.pdf
[3] J. Tougaard, L. Hermannsen, and P. T. Madsen, "How loud is the underwater noise from operating offshore wind turbines?," J Acoust Soc Am, vol. 148, no. 5, pp. 2885–2893, Nov. 2020, doi: 10.1121/10.0002453.
[4] T. Pangerc, P. D. Theobald, L. S. Wang, S. P. Robinson, and P. A. Lepper, "Measurement and characterisation of radiated underwater sound from a 3.6 MW monopile wind turbine," J Acoust Soc Am, vol. 140, no. 4, pp. 2913–2922, Oct. 2016, doi: 10.1121/1.4964824.
[5] K. Sathish, R. Anbazhagan, R. C. Venkata, F. Arena, and G. Pau, "Investigation and Numerical Simulation of the Acoustic Target Strength of the Underwater Submarine Vehicle," Inventions, vol. 7, no. 4, p. 111, Dec. 2022, doi: 10.3390/inventions7040111.
[6] A. Purwana, I. M. Ariana, D. W. Handani, and W. Wardhana, "Performance and Noise Prediction of Marine Propeller Using Numerical Simulation," IPTEK Journal of Proceedings Series, vol. 4, no. 1, p. 20, Dec. 2018, doi: 10.12962/j23546026.y2018i1.3501.
[7] A. Erturk and D. J. Inman, Piezoelectric Energy Harvesting. Wiley, 2011. doi: 10.1002/9781119991151.
[8] B. Boukhezzar, L. Lupu, H. Siguerdidjane, and M. Hand, "Multivariable control strategy for variable speed, variable pitch wind turbines," Renew Energy, vol. 32, no. 8, pp. 1273–1287, Jul. 2007, doi: 10.1016/j.renene.2006.06.010.
[9] C. Jauch, "A flywheel in a wind turbine rotor for inertia control," Wind Energy, vol. 18, no. 9, pp. 1645–1656, Sep. 2015, doi: 10.1002/we.1784.
[10] D. T. Wilton, "Acoustic radiation and scattering from elastic structures," Int J Numer Methods Eng, vol. 13, no. 1, pp. 123–138, 1978, doi: 10.1002/nme.1620130109.
[11] H. A. Schenck, "Improved Integral Formulation for Acoustic Radiation Problems," J Acoust Soc Am, vol. 44, no. 1, pp. 41–58, Jul. 1968, doi: 10.1121/1.1911085.
[12] C. H. Wilcox, "A generalization of theorems of Rellich and Atkinson," Proceedings of the American Mathematical Society, vol. 7, no. 2, pp. 271–276, 1956, doi: 10.1090/S0002-9939-1956-0078912-4.
[13] O. D. Kellogg, "The Potential," in Foundations of Potential Theory, Berlin, Heidelberg: Springer Berlin Heidelberg, 1967, pp. 160–164. doi: 10.1007/978-3-642-86748-4_3.
[14] A. Lavie and A. Leblanc, "Integral equation methods with unique solution for all wavenumbers applied to acoustic radiation," European Journal of Computational Mechanics, pp. 619–636, Aug. 2010, doi: 10.13052/EJCM.19.619-636.
[15] G. W. Benthien, D. Barach, and D. Gilette, "CHIEF User's Manual," Naval Ocean System Center, 1988.
[16] P. Businger and G. H. Golub, "Linear least squares solutions by householder transformations," Numer Math (Heidelb), vol. 7, no. 3, pp. 269–276, Jun. 1965, doi: 10.1007/BF01436084.
[17] I. B. Crandall, "Theory of Vibrating Systems and Sound," 2015.
[18] J. Nocedal and S. J. Wright, Numerical Optimization, vol. 2. Springer New York, 2006. doi: 10.1007/978-0-387-40065-5.
[19] I. Kidil, "Neural Networks Solving Linear Systems - Chair of Scientific Computing," 2021. https://www.cs.cit.tum.de/en/sccs/news/sccs-colloquium/article/iremnur-kidil-neural-networks-solving-linear-systems/ (accessed Mar. 16, 2023).
[20] Y. Nesterov, Introductory Lectures on Convex Optimization, vol. 87. Boston, MA: Springer US, 2004. doi: 10.1007/978-1-4419-8853-9.
[21] "Enron 1.5 s offshore - 1,50 MW - Wind turbine." https://en.wind-turbine-models.com/turbines/91-enron-1.5-s-offshore (accessed Mar. 16, 2023).
[22] J. Amaral, K. Vigness-Raposa, J. H. Miller, G. R. Potty, A. Newhall, and Y.-T. Lin, "The Underwater Sound from Offshore Wind Farms", doi: 10.1121/AT.2020.16.2.13.
[23] W. A. Kuperman and J. F. Lynch, "Shallow-Water Acoustics," Phys Today, vol. 57, no. 10, pp. 55–61, Oct. 2004, doi: 10.1063/1.1825269.
[24] H. Lindell, "Utgrunden off-shore wind farm-Measurements of underwater noise," 2003. [Online]. Available: https://www.researchgate.net/publication/275037654

IOP Conf. Series: Earth and Environmental Science **1464** (2025) 012034 doi:10.1088/1755-1315/1464/1/012034

Dynamic response of a jacket lower module in the decommissioning lifting operation

Lurohman Mamin M[1*], Rildova[1], and Gabril Kautsar Maldini[1]

[1] Ocean Engineering Program, Faculty of Civil and Environmental Engineering, Institut Teknologi Bandung, Bandung, Indonesia

*E-mail: mamin@itb.ac.id

Abstract. Given the economic conditions and rapid population growth in Indonesia, there is an urgent need for a prompt provision of energy resources to meet the energy demands. Since 1972, in Indonesia, oil and natural gas have been major energy sources, driving the development of offshore oil and gas industry. As a result of the offshore developments, numerous oil and gas offshore facilities scatter throughout the Indonesian seas, some of which are nearing the end of their operational time. For this reason, decommissioning related topics have been intensively discussed recently. Offshore decommissioning refers to the systematic removal of offshore oil and gas facilities to ensure the facilities that are no longer operating do not pose a potential threat to the environment or shipping activities. The most common method in the offshore decommissioning stage is the lifting operation. This operation, in the context of offshore decommissioning, is the process of removing or relocating an installed structural module to the final disposal area. In this study, the jacket modules are split in two and lifted out using a heavy lifting vessel (HLV). The dynamic response of the lifted module is influenced by the dynamic response of the HLV caused by hydrodynamic interaction during lifting operation. In engineering practice, the dynamic lifting effect is compensated by incorporating the dynamic amplification factor (DAF) to the static lifting analysis. In this research, the dynamic lifting simulation will be conducted by modelling the HLV carried the lifted module in varying wave parameters to evaluate the dynamic response of the module. The results shown that the recommended practice provides a conservative DAF compared to the calculated DAF resulted from dynamic lifting analysis. In the sensitivity study, the interaction between the HLV motion and the jacket rigging configuration may cause a resonance, as indicated by a significant increase in the amplitude of the jacket module displacement. Furthermore, two typical jacket module responses were identified while conducting lifting operations in different wave parameters and lifting stages, namely, swinging and heaving. These responses occurred in accordance with the frequency lock-in between the HLV motion and the two dominant frequency modes of the rigging arrangement. The remediation action by adding rigging lines tied-back to the HLV is recommended to disrupt the frequency lock-in.

1. Introduction

Offshore platform decommissioning refers to the dismantling process undertaken when an offshore platform has ended its operation life. Decommissioning is carried out by considering various reasons, including environmental, regulatory, and economic constrains. In this study, decommissioning operation involves relocating platform either to the dismantling yard or transforming into an artificial reef. One of the common operations during decommissioning operation is lifting or hoisting. Lifting operations conducted offshore present unique challenges, with one of the key challenges being vessel motion and the hydrodynamic forces acting on the lifted structure. In the relatively small-scale structure, lowering of a subsea structure onto the seabed also considers similar challenges during installation [1]-[3]. The lifting method is commonly performed in this operation, in which the industry standard is established for these operations. The recommended practice suggests the use of an amplification factor in the design process, known as the Dynamic Amplification Factor (DAF). This factor is assumed to account for the additional load due to dynamic effects caused by vessel movement and hydrodynamic forces. Generally, the amplification factors recommended in standards proposed conventional approach, and it is often suggested to perform dynamic simulations to obtain more accurate values. Based on the literature study, there is limited number of studies discussed the dynamic effects on the structure during lifting structure module for decommissioning purposes. By this means, this paper will discuss the effects of vessel motion and hydrodynamic forces during offshore decommissioning operations by using lifting method.

2. Methodology

This section presents analysis methods and assumptions that are used in this study. Both static and dynamic analysis of the decommissioning lifting were performed using the OrcaFlex software [4]. To account the dynamic lifting effect, the vessel responses are determined in the form of RAOs (Response Amplitude Operators) with the assumption that vessel motion is caused solely by wave forces. During the dynamic analysis, the environmental forces considered are limited to waves, wind, and currents under installation conditions. The crane stiffness and damping are neglected, which means that the movement of the hook point is directly influenced by vessel motion induced by the environmental loads. To understand the dynamic behaviour of the model, the study focuses on the effect of regular waves to the lifted module responds.

2.1 Heavy Lifting Vessel Model

The theory of six degrees of freedom refers to the six types of responses a floating structure exhibits when interacting with external forces, such as wind, currents, and waves. These responses can be broadly classified into three translational (surge, sway, heave) and three rotational motions (roll, pitch, yaw). As a floating structure, the lifting vessel moves in response to external forces acting upon it. In each degree of freedom, the movement of such a structure can be modelled using mass-damped-spring system with the external forces represents the time-varying excitation force, which, in the context of this study, is primarily due to harmonic regular waves.

In the modelling input, the Response Amplitude Operator (RAO) is required as a transfer function that quantifies the influence of environmental loads on a structure. The RAO describes the structure's response across various degrees of freedom, wave directions, and wave periods. In a regular wave, the heave displacement RAO can be ration of the 1st order heave motion amplitude to the wave amplitude.

IOP Conf. Series: Earth and Environmental Science **1464** (2025) 012034 doi:10.1088/1755-1315/1464/1/012034

2.2 Dynamic Amplification Factor (DAF)

In lifting operations, all components are affected by dynamic effects from the crane, vessel motion, hoisting speeds, and environmental factors such as wind, currents, and waves. According to the DNV-ST-N001 – Marine operation and marine warranty (2023) [5], dynamic effects during lifting must be considered. These dynamic effects are calculated using the Dynamic Amplification Factor (DAF). Table 1 shows example DAF values applied to the Static Hook Load of a lifted load.

Table 1. Recommended Dynamic Amplification Factor (DAF) based on DNV ST N001.

Static Hook Load	Dynamic Amplification Factor		
	Onshore	Inshore	Offshore
3t* < SHL < 100t	1.10	$1.07 + 0.05\sqrt{100/SHL}$	$1 + 0.25\sqrt{100/SHL}$
100t < SHL < 300t	1.05	1.12	1.25
300t < SHL < 1000t	1.05	1.10	1.20
1000t < SHL < 2500t	1.03	1.08	1.15
SHL > 2500t	1.03	1.05	1.10

* For lifted item less than 3 tonnes, it can be assumed that the item has weight of 3 tonnes.

The DAF values listed above are the recommended practice for the static lifting analysis in the case when no specific dynamic analysis is conducted. There are also conditions that must be met before using these values. The DAF values provided by DNV ST N001 specifically to account for dynamic lifting effect which do not include the consequence factor and skew load factor. In the case where dynamic lifting analysis is performed, the DAF is directly calculated based on the ratio between dynamic hook load and static hook load. In this study, the recommended DAF is compared and evaluate to the calculated DAF whether the recommended DAF is visible to be used for decommissioning lifting analysis.

API RP 2A WSD [6] also addresses the dynamic lifting factor using different approach. API RP 2A proposed single lifting factor for combination of DAF and consequence factor. In DNV, there are separate factors determined for the dynamic effect and consequence. In addition, the DNV provides different DAF according to classification of the module's weight, while API proposed single value of DAF. API specifies combination of the dynamic effect and the consequence factor, such as for members directly connected to the lifting point, the factor is 2.0, while for other members that transmit the load, the factor is 1.35.

3. Modelling and Simulation

3.1 Static Simulation

According to the decommissioning procedure, the jacket structure is cut into two modules at elevation (-) 31.0 m from MSL. The jacket lower module is studied in this paper to simulate the dynamic effect during decommissioning lifting analysis. Prior to simulate the dynamic lifting analysis, the static analysis is performed to determine the lifting configuration as well as to investigate the static lifting weight. In this static simulation, the calculated load is the load experienced by the sling when supporting the bottom jacket. The static simulation is performed under three different conditions: when the structure is fully submerged, in the splash zone, and when the structure is in the air. These conditions are modelled by adjusting the length of the Main Sling (Winches). The model illustration for three lifting conditions is presented in Figure 1.

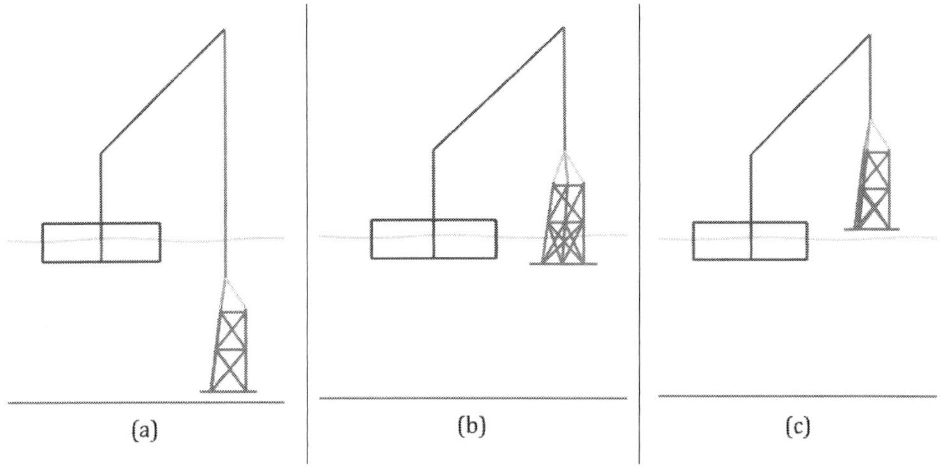

Figure 1. Illustration of the static simulation model for lifting the bottom jacket (a) in the fully submerged condition (b) in the splash zone (c) in air.

From this analysis, the static weight of the jacket lower module in air is around 262.4 MT. Single point lift is employed to lift-up the lower module of the jacket form its original location. The model is assumed that the pile above seabed is removed prior to the jacket removal. Therefore, there is no interaction between the above seabed structure with the pile below mudline. Figure 1a indicates all the structural member is submerged. In this condition, the tension in the main sling is lower because the structure still benefits from the full buoyancy effect as all parts of the structure are underwater. The splash zone lifting is presented in Figure 1b, in which half of the structure remains underwater. The tension in the sling will be higher than in the fully submerged condition since part of the buoyancy contributes to the lifting weight. Once the entire structure is in the air, the tension in the main sling increases because there is no buoyancy that reduces the sing load on the structure.

According to Table 1, based on the static weight, dynamic factor of the lower module is around 1.25. The results of the static simulation for three lifting conditions are shown in Table 2. In addition, the equivalent dynamic load is calculated using recommended DAF proposed by DNV.

Table 2. Recommended Dynamic Amplification Factor (DAF) based on DNV ST N001.

Lifting Case	Crane Tip to Structure CoG (m)	Static Hook Load (kN)	Equivalent Dynamic Hook Load (kN)*
Fully Submerged	121.0	2316.5	2895.6
Splash Zone	71.0	2488.7	3110.9
In Air	56.0	2574.5	3218.1

* Recommended DAF for ~230 MT offshore lifting according to DNV ST N001 is 1.25

IOP Conf. Series: Earth and Environmental Science **1464** (2025) 012034 doi:10.1088/1755-1315/1464/1/012034

3.2 Dynamic Simulation

Dynamic simulations for the bottom jacket lifting operation are also conducted under the three conditions: fully submerged, splash zone, and in air, as illustrated in Figure 2. The properties used for the dynamic simulation are presented in Table 3.

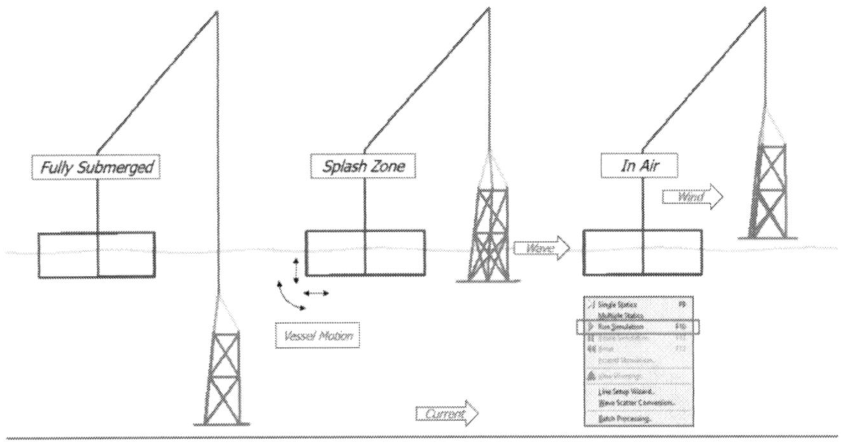

Figure 2. Illustration of the dynamic simulation model for lifting the bottom jacket.

Table 3. Properties used in the dynamic simulation model for the bottom jacket structure. (Table details not provided, but assumed to include parameters such as mass, buoyancy, wave height, and environmental forces).

Parameters	Value
Seabed	
• Water Depth:	66 m
Regular Waves	
• Height:	0.5-1.25m
• Period:	2-60s
• Wave Type:	Single Airy
Current	
• Speed (At Surface):	0,55 m/s
Wind	
• Speed:	16,9 m/s

Note: Simulation durations are 600 sec and 1200 sec for fully submerged and splash zone/in air, respectively.

Dynamic simulations for the bottom jacket lifting operation are also conducted under the three conditions: fully submerged, splash zone, and in air, as illustrated in Figure 2 above. The properties used for the dynamic simulation are presented in Table 3.

4. Results and Discussions

4.1 Decommissioning Lifting Results

The dynamic analysis of the lifting decommissioning simulates using the regular wave with the height of 1.25 m and the period of 6 seconds. Like the static lifting cases, the dynamic simulation focusses on three lifting stages, which are the fully submerged, the splash zone, and the lifting in air conditions. In the fully submerged condition, the tension in the main sling oscillates before reaching a steady state. The simulation duration is 600 seconds. The maximum tension in the main sling is taken from the range of 500 to 600 seconds when the simulation is in the steady state. Based on the dynamic simulation of the fully submerged module, the sling tension reaches a maximum value of 2372.9 kN. In the splash zone, the tension in the main sling reaches the steady state after 1000 seconds simulation. In this case, the maximum sling tension is 2542.4 kN. When the structure is in the air, the tension in the main sling continues to increase. The simulation requires 1000 seconds to reach the steady state, which recorded the maximum sling tension of 2629.03 kN.

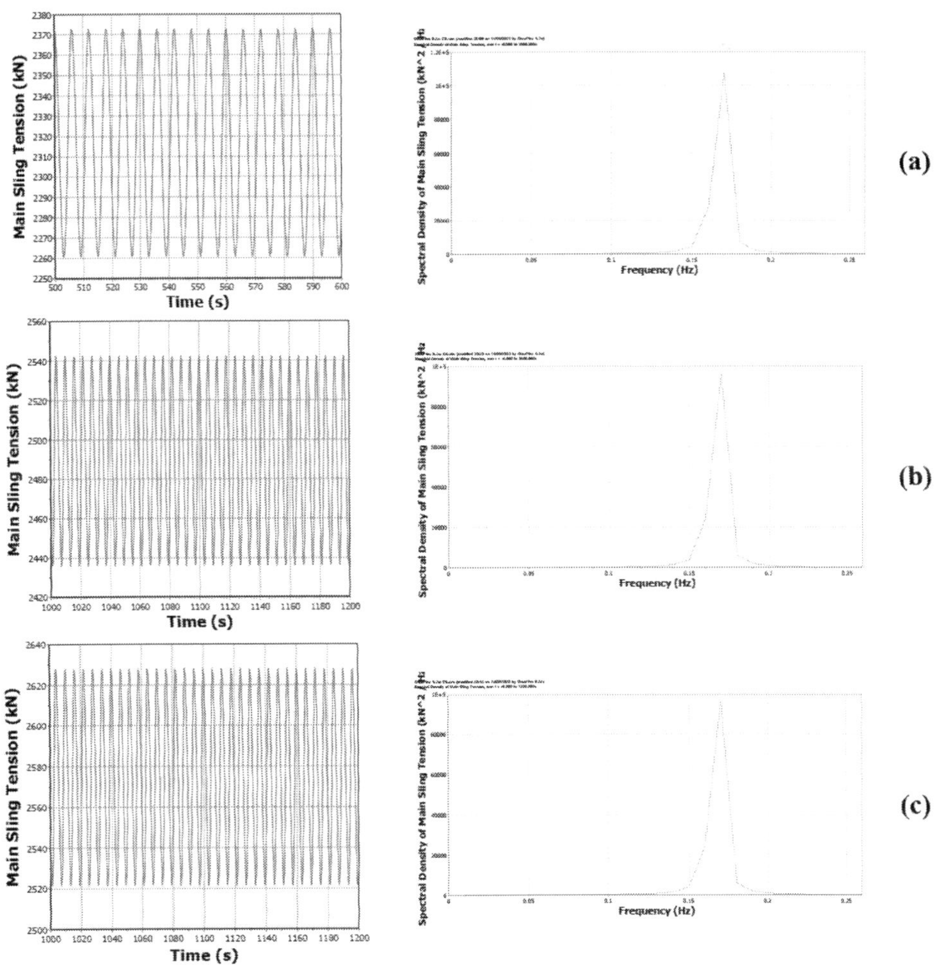

Figure 3. Dynamic main sling load (hook load) for lifting the bottom jacket (a) in the fully submerged (b) in the splash zone (c) in air.

IOP Conf. Series: Earth and Environmental Science **1464** (2025) 012034 doi:10.1088/1755-1315/1464/1/012034

The same pattern is observed between the static and dynamic simulation. The buoyancy affects the simulated hook load in the dynamic analysis. The highest hook load of 2629.03 kN is recorded during lifting in air. The same condition is also observed in the static simulation. The dominant period of the sling also meets with the wave period, which indicates the lifting crane motion dominated by the wave period. Based on the hook load results on the static and the dynamic simulations, the Dynamic Amplification Factor (DAF) can be calculated. Across all the stages, the simulated DAF is around 1.02, which is significantly lower than the recommended DAF by DNV ST N001 of 1.25. The dynamic simulation results of the main sling for all stages are presented in Figure 3.

4.2 Sensitivity Study

The sensitivity analysis for the bottom jacket structure follows the same general approach as the previous structures. The same lifting configuration and stages are identical with the decommissioning lifting analysis for fully submerged, splash zone, and in air. Each of these lifting stages was simulated for 42 different wave periods. Three wave heights are selected, which are 0.5 m, 1.0 m, and 1.25 m. The results of these simulations are summarized in the Dynamic Amplification Factor (DAF) vs. Wave Period graph shown in Figure 4.

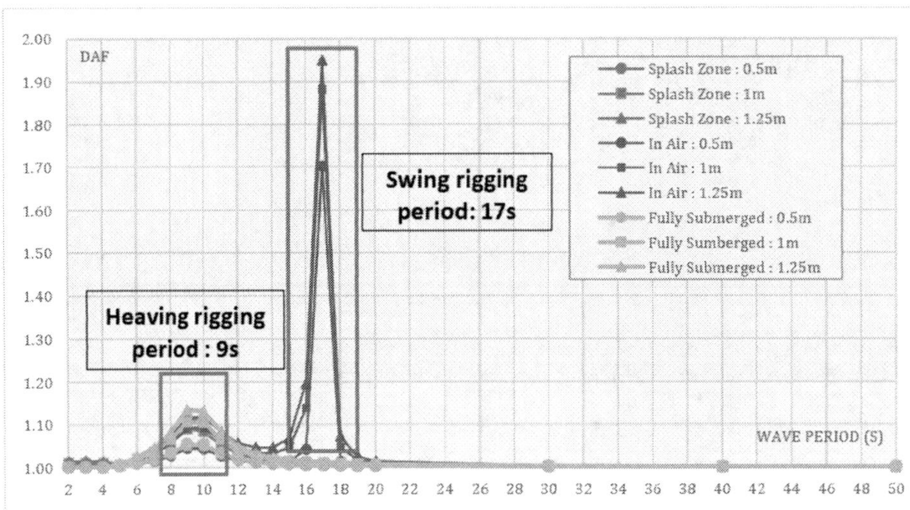

Figure 4. Sensitivity analysis graph for the bottom jacket structure.

In the Figure 4, there are two resonance modes appeared in the sensitivity study of the lifting jacket bottom structure. The first resonance of the lifting module has occurred at period around 17 second. The period is coincided with the natural period of the rigging, as such, the lifting module behaves pendulum-like. This resonance has swing modes due to the natural period of rigging meets the wave period. As a pendulum, the swing natural period of the lifting module has a function of the ringing length only. Therefore, the first resonance of the lifting bottom jacket module is only observed at lifting in air for all wave height cases, where the rigging length is similar. In addition, this resonance might be triggered in the case where there is no damping involved in the system. In the fully submerged and in the splash-zone lifting stage, the resonance is suppressed by the influence of the hydrodynamic damping acting on the lifting module.

The second resonance has heaving responses due to the rigging natural period meets with the wave frequency. The heaving natural period of the system is a function of the mass and stiffness of overall lifting system. In all lifting stages, the structural mass is similar, with a constant stiffness provided by the rigging sling. However, the added mass and hydrodynamic stiffness are varied in the fully submerged and in the splash-zone lifting stages. For this study, the ratio of the added mass and the hydrostatic stiffness for the previously mentioned stages results in a similar resonance frequency with the lifting in air. This is indicated in Figure 4 that the heaving resonance have occurred at the similar period for all lifting stages. The highest magnitude of DAF is observed in the fully submerged where the highest total mass is computed form the added mass and the jacket bottom structural mass.

Figure 4 indicates an excessive DAF for the pendulum-like resonance during lifting in air. For this study case, the maximum DAF recorded is almost 2.0, which is greater than the recommended DAF of 1.25. The DAF is calculated based on the pendulum-like responds in resonance. To mitigate the excessive swing respond, an additional rigging lines is required to minimize the swing of the module. By using the additional rigging line that is tied into the crane vessel (refer to Figure 5b), the responds are damped within an acceptable respond. The additional rigging lines are activated when the structure lifted in air. The detailed DAF analysis graph showing wave height and period using additional slings can be seen in Figure 5a.

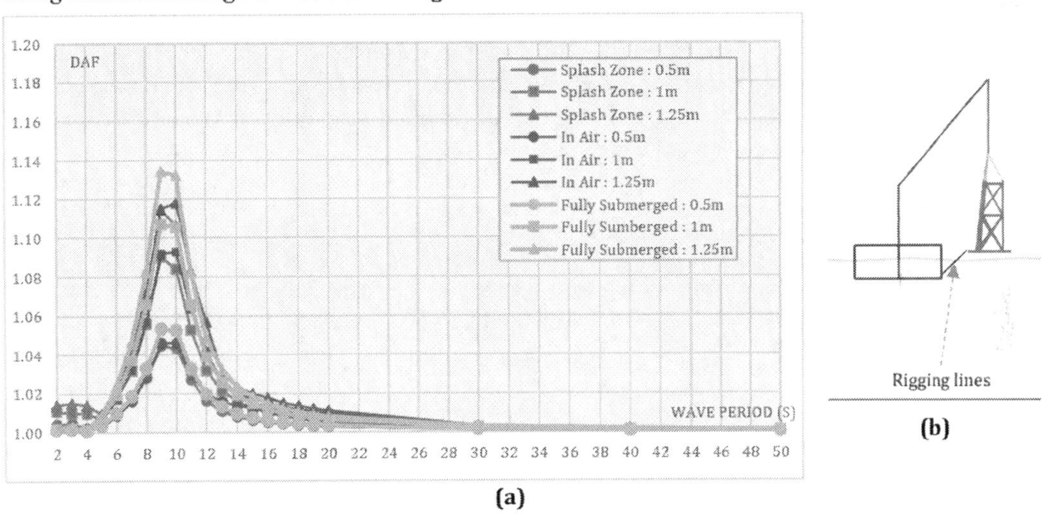

Figure 5. (a) DAF graph for the bottom jacket structure with additional slings. (b) illustration of liting in air with additional rigging line.

According to the optimization results, the highest DAF value across all cases occurs in Case C9 (Wave Height 1.25 m and Period 9 seconds) under the fully submerged condition, with a DAF of 1.13. The main sling reaches maximum sling force at the heaving resonance period around 9.5s to10 s. As previously mentioned, in the fully submerged condition, the lifting structure system has maximum mass magnitude originated from structural mass and hydrodynamic added mass. Since the DAF is lower than the recommended DAF, no mitigation action is required in this case. Based on the dynamic simulation, the maximum main sling force in the fully submerged condition at period 9s and wave height of 1.25m is recorded to be 2627.4 kN.

IOP Conf. Series: Earth and Environmental Science **1464** (2025) 012034 doi:10.1088/1755-1315/1464/1/012034

4. Conclusion

This paper demonstrated that the Dynamic Amplification Factor (DAF) offshore lifting value using vessel crane barge has a function of wave height and wave period. The recommended DAF by DNV ST N001 for offshore lifting of 1.25 is conservatively set for engineering practice. It can be shown by performing the dynamic lifting simulation that the maximum DAF offshore of the lifting jacket is around 1.02 during lifting in air. Based on the sensitivity study, it was observed that two resonance modes might appear during offshore lifting (i.e., heaving modes and pendulum-like mode resonances), which results in higher DAF values compared to the rest of sea-state conditions. In the offshore lifting, the structure is prone to failed in the swing mode resonance (pendulum-like behaviour) which produces higher DAF (i.e., 1.96) compared to the heaving resonance mode (i.e., 1.13). The swing behaviour is only observed when the lifting module is completely lifted in air. This can be mainly due to the hydrodynamic damping in the form of drag force able to suppress the structure lifting system in the swing direction. For lifting in air, this phenomenon can be mitigated by adding the damping or stiffness in the direction of the response in a form of additional rigging line. For the heaving resonance behaviour, it can be observed that when the jacket is fully submerged (DAF = 1.13), the DAF tends to be slightly higher compared to when the structure is in the splash zone (DAF = 1.11) or in air (DAF = 1.12). This is primarily due to the hydrodynamic effects acting on the structure provided the hydrodynamic mass to overall lifting system. In a fully submerged state, the drag force exerted by water on the structure is relatively smaller, while the inertia force becomes more dominant. As a result, the hydrodynamic effects on the lifting structure results in a higher DAF value in the fully submerged lifting stage compared to the lifting in the splash-zone and in the air.

Acknowledgment

The research for this paper is funded by a grant from Faculty of Civil and Environmental Engineering, Institut Teknologi Bandung, through 2024 Research and Community Service Program (P2MI) for Offshore Engineering Research Group. The authors gratefully acknowledge this support.

References

[1] G. Sun, F. Wang, Y. Zhang, K. Tang, and J. Wang (2015). "A Generic Study on Lowering Analysis of Massive Jumper Structures," *Proceedings of ASME 2015 34th International Conference on Offshore Mechanics and Arctic Engineering, Volume 1: Offshore Technology; Offshore Geotechnics*. doi: https://doi.org/10.1115/OMAE2015-41055.

[2] A. Sarkar and O. T. Gudmestad (2010). "Splash Zone Lifting Analysis of Subsea Structures," *Proceedings of ASME 2010 29th International Conference on Ocean, Offshore and Arctic Engineering: Volume 1*. doi: https://doi.org/10.1115/OMAE2010-20489.

[3] T. E. Søfteland, O. V. Skrunes, and D. Karunakaran (2014). "Lifting Analysis of Subsea Framework Structures," *Proceedings of ASME 2014 33rd International Conference on Ocean, Offshore, and Arctic Engineering, Volume 1B: Offshore Technology*. doi: https://doi.org/10.1115/OMAE2014-23816.

[4] *OrcaFlex* ©, Version 11.0; Orcina: Ulverston, UK.

[5] DNV (2023). DNV-ST-N001 Marine operation and marine warranty.

[6] American Petroleum Institute (2000). Recommended Practice for Planning, Designing and Constructing Fixed Offshore Platforms - Working Stress Design (21st ed.).

[7] Chakrabakti, S. K. (2005). Handbook of Offshore Engineering Vol.1. Oxford: Elsevier.

[8] GL Noble Denton 0027/ND Guidelines for Marine Lifting and Lowering Operations

[9] Offshore Lifting Operations and Engineering Analysis. (2022). Retrieved from Marine Insight: https://www.marineinsight.com/offshore/offshore-lifting-operations-and-engineering-analysis/

IOP Conf. Series: Earth and Environmental Science **1464** (2025) 012035 doi:10.1088/1755-1315/1464/1/012035

Wave parameter selection criteria in spectral fatigue analysis of four legs fixed platform

Lurohman Mamin M[1*], Rildova[1], and Annes Rikci Parmonangan Siahaan[1]

[1] Ocean Engineering Program, Faculty of Civil and Environmental Engineering, Institut Teknologi Bandung, Bandung, Indonesia

*E-mail: mamin@itb.ac.id

Abstract. Offshore platforms are crucial facilities for extracting energy resources in the oil and gas industry. In Indonesia, oil and gas exploitation activities are predominantly in shallow water, which requires an offshore fixed structure to provide a working floor for the oil and gas extraction process. One of the design requirements of offshore structure is to fulfil the minimum service/operating life of its platform, which is determined by the fatigue criteria caused by cyclic loads induced by waves. An offshore fixed-steel structure is selected as the subject study for fatigue evaluation. Based on the dynamic mode analysis, the structure has a first-mode natural period of 4.8 seconds, suggesting that the jacket structure is sensitive to dynamic loads. By this means to account for the overall spectrum period range of the waves, spectral fatigue analysis is preferred over deterministic fatigue analysis. In the linear wave theory, to accurately represent the irregular wave loads, the selection of regular wave parameters is crucial. This paper examines two selection criteria for determining these regular wave parameters: employing wave dispersion relationships and discretizing the theoretical wave spectrum. The primary approach, which relies on wave dispersion relationships, is the common wave selection method in practical applications. This method establishes a direct linear correlation between the height of a wave and its period. However, the theoretical wave spectrum does not reflect the linear relationship. Therefore, the discretization of a theoretical wave spectrum is proposed as an alternative to accurately represent irregular wave loads. The wave load effects are then evaluated to determine the fatigue damage in the selected joints caused by the irregular wave loads generated by the two methods. The results indicate substantial disparities between the initial approach and the subsequent approach. The first method yields a higher fatigue damage value than the second method. This indicates that the irregular wave loads generated by the wave dispersion relationship result in conservative cyclic loads on the jacket structure.

1. Introduction

Indonesia, as a maritime country, possesses abundant reserves of oil and natural gas, distributed both onshore and offshore. With two-thirds of Indonesia's territory being water, exploration and exploitation of oil and gas are frequently conducted in the offshore regions of the country. To support these offshore exploration and exploitation activities, offshore platforms are required as infrastructure to facilitate these operations.

An offshore platform (or offshore rig) is a structure built offshore designed to support the exploration and extraction of mineral resources and natural deposits. The primary function of the platform is to provide a working deck for the exploitation and production of oil and natural gas offshore. Typically, offshore platforms operate in areas around oil wells or mining zones with limited spatial extent. The construction process for the structure and components of an offshore platform is not carried out directly offshore; rather, it is assembled at a fabrication yard and then transported to the exploration site [1]. The offshore platform model used in this study can be seen in Figure 1.

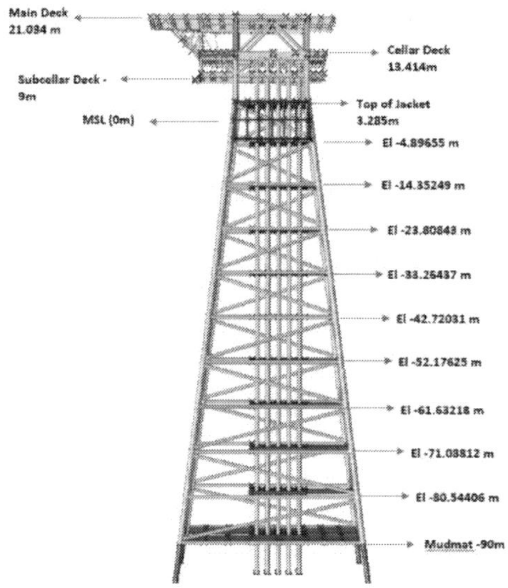

Figure 1. Offshore platforms model [1].

This paper discusses a fixed platform structure with four legs that commonly used in the relatively shallow waters in Indonesia. The structure includes a main deck, a cellar deck, a subcellar deck, and 11 horizontal jacket elevations. It is in the Natuna Sea, positioned at a depth of 90 meters Mean Sea Level (MSL). The platform has met the Inplace and Seismic criteria, and the design safety factor of the pile [2]. Therefore, this paper focuses on evaluating fatigue life of the platform.

According to API RP-2A [2], the Fatigue criterion must be evaluated to assess the resilience of the jacket joints against cyclic loads resulting from wave interactions with the jacket structure. Fatigue is a structural failure phenomenon typically occurring at structural joints due to repetitive loading over a certain number of cycles. For offshore platforms, the cyclical load in question is primarily due to waves. Thus, in the fatigue criterion review, only wave loads will be considered in determining the service life of the jacket structure.

The determination of wave loads for fatigue analysis is crucial as it affects the service life of the structure. Irregular waves are used to represent actual wave events in nature. In linear wave theory, irregular waves are a superposition of several regular waves [3]. Each regular wave must have different wave heights, periods, wavelengths, and phases. Typically, around 20-30 regular waves are required to form an irregular wave [4]. Thus, the selection of individual waves can influence the wave loads applied to the structure.

From the wave parameter determination procedure, the wave dispersion relation (Equation 1) can be used to determine the wavelength (λ), assuming the installation site of the platform is in deep-water conditions. By using a uniform wave steepness (S), the wave height (H) can be determined. The value of S will remain constant for each individual wave.

$$\lambda = \frac{gT^2}{2\pi} \quad (1)$$

$$S = \frac{H}{\lambda} \quad (2)$$

This method is commonly used for determining individual wave parameters, as outlined in the SACS manual [4]. From the relationships provided in Equations 1 and 2, it is evident that longer wave periods correspond to higher wave amplitudes. However, in the case of irregular waves, it is possible for a wave to have a small amplitude with a long period. This phenomenon cannot be accurately determined using only the wave dispersion relationship. Therefore, this study will explore the determination of wave parameters through the discretization of the wave spectrum. It is anticipated that with the latest computational methods, the determination of individual wave parameters will more accurately represent the irregular wave phenomena observed in nature.

2. Methodology

There are two methods for calculating the service life of structural connections using fatigue criteria: deterministic fatigue and spectral fatigue. The choice of method is based on the sensitivity of the structure to dynamic loads [6]. The more flexible a structure, the more sensitive it is to dynamic loads. To understand the dynamic characteristics of a structure, a structural modal analysis is performed.

The natural period of the structure is the result of this analysis. A higher natural period indicates greater sensitivity to dynamic loads. Structures with a natural period greater than 3.0 seconds are categorized as flexible structures [5]. The modal analysis yielded a natural period of 4.8 seconds for the structure, thus categorizing it as sensitive to dynamic loads. Therefore, the structure with the configuration shown in Figure 2 is recommended to use the spectral fatigue method for determining the service life of the structure.

2.1 Wave Data

The cyclic loads applied in fatigue analysis is originated from waves acting on the structure. One year wave data is required to wave distribution for all seasons. The characteristics of irregular waves at a given location are summarized in a table or graph known as a scatter diagram. The scatter diagram displays the distribution of waves for each significant wave height (H_s) and peak period (T_p). Additionally, to determine the wave distribution in each direction, the distribution of waves by direction is required. This table also illustrates the variation in wave height distribution across different directions. The scatter diagram is presented in Table 1, while the wave distribution by direction is shown in Table 2.

To model irregular waves, JONSWAP theoretical wave spectrum determines the irregular wave characteristics in Natuna [7]. The main spectrum parameters is represented by the significant wave height and the peak period. The significant wave height is the average of the one-third highest wave height data. The peak period of the wave spectrum provides information about the dominant period of the water body. The theoretical expression for the JONSWAP wave spectrum is given in Equation 3 [8].

Table 1. Wave scatter diagram.

Peak Periode (s)	Significant Wave All Month & Years (m)													
	0	0.25	0.75	1.25	1.75	2.25	2.75	3.25	3.75	4.25	4.75	5.25	5.75	Total
0.00 - 0.99	0	0	0	0	0	0	0	0	0	0	0	0	0	0
1.00 -1.99	0.0008	0	0	0	0	0	0	0	0	0	0	0	0	0.0008
2.00-2.99	0.0097	0.0115	0	0	0	0	0	0	0	0	0	0	0	0.0212
3.00-3.99	0.0023	0.1859	0.0092	0.0031	0.0007	0.0001	0	0	0	0	0	0	0	0.2013
4.00-4.99	0.0031	0.0829	0.0993	0.0001	0	0	0	0	0	0	0	0	0	0.1854
5.00-5.99	0.001	0.0175	0.0714	0.0349	0	0	0	0	0	0	0	0	0	0.1248
6.00-6.99	0.0002	0.0083	0.0373	0.0354	0.0083	0	0	0	0	0	0	0	0	0.0895
7.00-7.99	0.0003	0.0052	0.0154	0.0377	0.0262	0.0022	0	0	0	0	0	0	0	0.087
8.00-8.99	0.0009	0.0040	0.0067	0.0185	0.03	0.0249	0.0038	0.0001	0	0	0	0	0	0.0889
9.00-9.99	0.0004	0.0017	0.0052	0.0112	0.0155	0.024	0.0208	0.0051	0.0003	0	0	0	0	0.0842
10.00-10.99	0	0.0020	0.0043	0.0052	0.0084	0.0103	0.0115	0.0114	0.0039	0.0005	0	0	0	0.0575
11.00-11.99	0.0001	0.0008	0.0038	0.0033	0.0039	0.005	0.0056	0.0048	0.0031	0.0029	0.001	0.0001	0	0.0344
12.00-12.99	0	0.0008	0.0024	0.002	0.0011	0.0018	0.0015	0.0021	0.0012	0.001	0.0006	0.0007	0	0.0152
13.00-13.99	0	0.0002	0.0013	0.0004	0.0004	0.0006	0.0002	0.0009	0.0005	0.0004	0.0001	0.0001	0.0001	0.0052
14.00-14.99	0	0	0.001	0.0004	0.0004	0.0004	0.0002	0	0	0	0	0	0	0.0024
15.00-15.99	0.0001	0	0.0002	0.0003	0.0005	0.0007	0.0003	0	0	0	0	0	0	0.0021
16.00-16.99	0	0	0	0	0.0001	0	0	0	0	0	0	0	0	0.0001
Total	0.0189	0.3208	0.2575	0.1525	0.0955	0.0700	0.0439	0.0244	0.0090	0.0048	0.0017	0.0009	0.0001	1.0000

Table 2. Wave directional distribution.

Significant Wave (m)	Vector Mean Direction								
	N	NE	E	SE	S	SW	W	NW	Total
0.00 - 0.2499	0.0032	0.0038	0.0021	0.0016	0.0012	0.0027	0.0016	0.0028	0.019
0.25 - 0.7499	0.113	0.047	0.0221	0.0077	0.0135	0.04	0.0217	0.0557	0.3207
0.75 - 1.2499	0.0988	0.0332	0.0076	0.0024	0.013	0.089	0.0027	0.011	0.2577
1.25 - 1.7499	0.0358	0.0141	0.0024	0.0001	0.0127	0.0864	0.0003	0.0005	0.1523
1.75 - 2.2499	0.0078	0.0018	0.0005	0	0.0094	0.0761	0	0	0.0956
2.25 - 2.7499	0.0001	0.0001	0	0	0.0123	0.0575	0	0	0.07
2.75 - 3.2499	0	0	0	0	0.0108	0.0331	0	0	0.0439
3.25- 3.7499	0	0	0	0	0.0077	0.0167	0	0	0.0244
3.75 - 4.2499	0	0	0	0	0.0045	0.0046	0	0	0.0091
4.25 - 4.7499	0	0	0	0	0.0025	0.0023	0	0	0.0048
4.75-5.2499	0	0	0	0	0.0004	0.0012	0	0	0.0016
5.25-5.7499	0	0	0	0	0.0001	0.0007	0	0	0.0008
5.75-6.2499	0	0	0	0	0	0.0001	0	0	0.0001
Total	0.2587	0.1	0.0347	0.0118	0.0881	0.4104	0.0263	0.07	1

$$S(f) = \frac{\alpha g^2}{(2\pi)^4} f^{-5} \exp\left[-1.25\left(T_p f\right)^{-4}\right] \gamma^p \quad (3)$$

The required wave parameters to establish a wave spectrum are the significant wave height (H_s) and the peak period (T_p). Other related parameters, as referenced in Equation 3, can be found in Reference [9]. To determine the values of H_s and T_p for the creation of a theoretical wave spectrum, a representative wave damage calculation was performed using the scatter diagram presented in Table 1. This calculation was conducted to represent the scatter diagram using a single value, H_{sr} and T_{pr}. The wave center of damage was calculated using Equation 4, where D_i represents the damage contribution for each bin in the scatter diagram.

$$H_{sr} = \frac{\sum D_i H_{si}}{\sum D_i} \qquad T_{sr} = \frac{\sum D_i T_{pi}}{\sum D_i} \qquad D_i = \frac{P_i {H_{si}}^m}{T_{pi}} \quad (4)$$

Where P_i is the probability of occurrence for each scatter diagram bin, and the exponent m represents the slope of the S-N curve, which will be explained in Section 2.2. Based on the results of the wave center of damage calculation, the wave parameters used to generate the theoretical wave spectrum are presented in Table 3. Based on the selected wave parameters, the JONSWAP wave spectrum is defined as shown in Figure 2.

Table 3. Selected wave parameters.

No	Parameter	Value	Unit
1	Significant wave height, H_{sr}	2.72	Meter
2	Peak period of wave, T_{pr}	10.70	Second

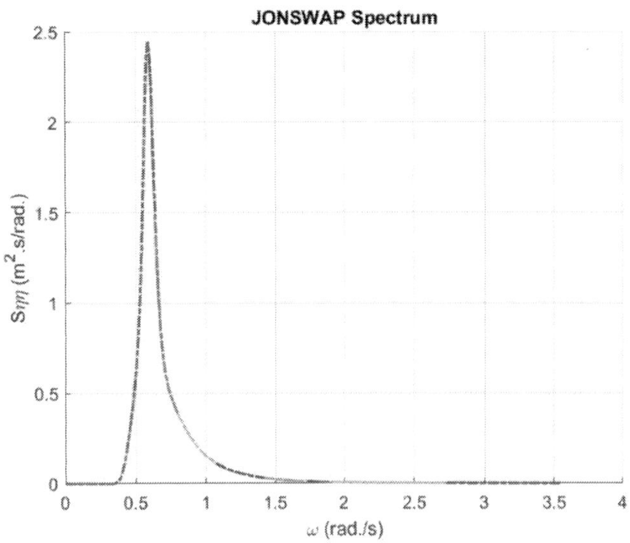

Figure 2. Theoretical JONSWAP wave spectrum for selected wave parameters.

2.2. Fatigue

In this paper, fatigue analysis is conducted using the Wave Spectral method. The nominal stress (σ_{rms}) resulting from wave-structure interaction plays a significant role in determining the Hot Spot Stress (HSS). A structure subjected to wave loads will generate stress in each of its members. Equation 5 presents the general formulation for calculating stress in the member under consideration. The transfer function ($H(f)$) and the wave spectrum ($S(f)$) can be used to determine the stress induced in a member.

$$\sigma_{rms} = \sqrt{\int_0^\infty H_i^2(f).S_i(f)df} \quad (5)$$

The Hot Spot Stress (HSS) used to determine the allowable number of cycles for a joint is calculated from the Stress Concentration Factor (SCF) and the nominal stress (σ_{rms}) resulting from wave loads. There are several methods available for calculating the SCF for tubular joint profiles. API RP-2A recommends using the SCF values determined by the Kuang method. In general, the SCF is a multiplier applied to the nominal stress value, as shown in Equation 6.

$$HSS = SCF \,.\, \sigma_{rms} \tag{6}$$

The stress range is the magnitude required to calculate the allowable number of cycles before fatigue occurred. This allowable number is determined based on the hotspot stress range experienced at a specific joint. The allowable number of cyclic loads is derived from the S-N curve, which represents the relationship between the number of load cycles (N) and the stress (hotspot stress, S). Figure 3 shows the S-N curve for tubular joints in jacket structures, adapted from API RP-2A [2].

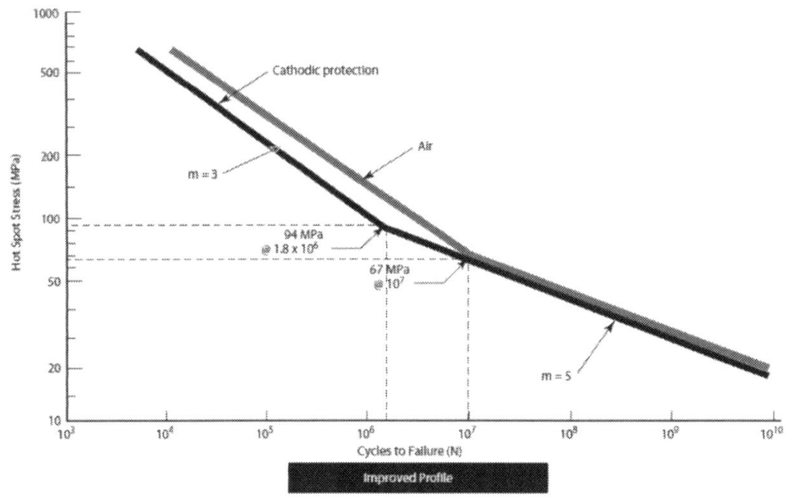

Figure 3. S-N curve [2].

The service life of a structural joint is determined by the cumulative Damage value, *D* (failure), where *D* is calculated as the ratio of the number of wave cycles to the allowable number of cycles at specific joints. The fatigue life (FL) of a structural joint, as shown in Equation 7, is calculated based on the ratio of the Structural Design Life (SL) to cumulative damage. Ideally, the cumulative damage should be less than 1, indicating that the service life of the structural joint exceeds the design life of the structure.

$$FL = \frac{SL}{D.SF} \tag{7}$$

3. Wave Selection Criteria

For linear wave theory, irregular wave loads are determined by summing several regular waves with different wave heights, wave periods, directions and phases. In this study, two approaches are used to determine the properties of regular waves:

1. Dispersion relation
2. Theoretical wave spectrum discretization

3.1 Dispersion Relation

From Equations 1 and 2, the determination of wave height is conducted by specifying a single wave steepness value. A wave steepness of 0.04 was chosen for this study. According to API RP-2A, the minimum wave height in spectral fatigue analysis is 0.33 m (or 1 ft). Thus, for any wave height calculated using the dispersion relation that is less than 0.33 m, the wave height is set to 0.33 m.

The selection of wave periods used in the analysis is not arbitrary. Wave static transfer function analysis is performed to determine the periods that yield maximum or minimum base shear and/or overturning moment. The selection of wave periods is not discussed in this paper. For consistency in comparing the two wave load determination methods, 14 wave periods have been selected for this study: 12, 10, 6.0, 5.6, 5.2, 4.8, 4.4, 4.0, 3.6, 3.2, 2.8, 2.4, 2.0, and 1.6 seconds. A wave steepness S=0.04S = 0.04S=0.04 was used in determining the height of regular waves. According to the dispersion relation, the wave parameters used to form the irregular wave are listed in Table 4. The following is a sample calculation of wave height for the period T=12s.

$$\lambda = \frac{gT^2}{2\pi} = \frac{g12^2}{2\pi} = 224.83\ m\,; \quad H = \lambda\, S = 224.83\,.\,0.04 = \ 8.99\ m$$

3.2 Theoretical Wave Spectrum Discretization

As previously mentioned, the significant wave periods used for the two wave load calculation methods are the same: 12, 10, 6.0, 5.6, 5.2, 4.8, 4.4, 4.0, 3.6, 3.2, 2.8, 2.4, 2.0, and 1.6 seconds. To determine the wave heights of the JONSWAP spectrum shown in Figure 4, the spectrum is discretized into 14 segments, where the midpoint of each segment corresponds to one of the 14 wave periods used in the study. To ensure accurate calculation of the area under the spectrum curve, a representative peak period T_{pr} =10.70 seconds is added to the discretization, making a total of 15 periods used in this method. Figure 4 shows the number of data points used to form the JONSWAP spectrum curve, with different colors indicating the discretization width for each wave period.

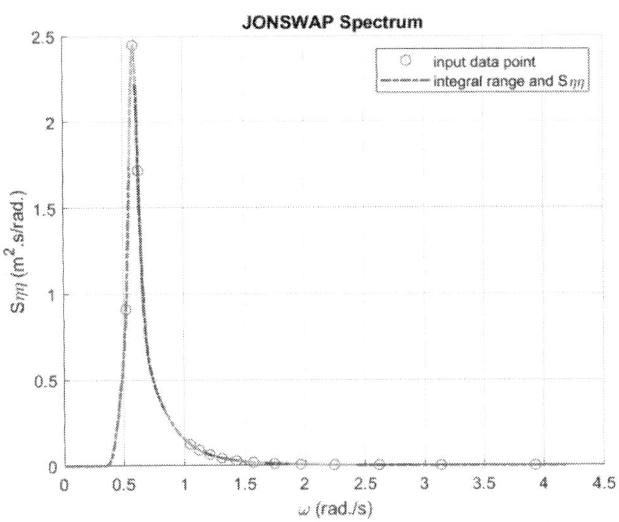

Figure 4. Discretization of the theoretical JONSWAP wave spectrum.

Table 4. Wave parameters using dispersion relation.

Period	Wave Height (m)	Wave Steepness
12.0	8.99	0.04
10.0	6.24	0.04
6.0	2.24	0.04
5.6	1.98	0.04
5.2	1.68	0.04
4.8	1.43	0.04
4.4	1.21	0.04
4.0	0.99	0.04
3.6	0.81	0.04
3.2	0.64	0.04
2.8	0.48	0.04
2.4	0.36	0.04
2.0	0.33	0.04
1.6	0.33	0.04

To determine individual wave heights from the theoretical spectrum, the Inverse Fast Fourier Transform (IFFT) is applied to identify the regular wave components that make up the irregular wave. The irregular wave described by the JONSWAP spectrum in Figure 2 consists of a collection of regular waves with different periods and heights. To determine an individual wave, the difference in the viewing period must be determined in the function$\Delta\omega$. The amplitude of the individual wave is essentially the area under the spectrum for each$\Delta\omega$. Therefore, $S_{\eta\eta}.\Delta\omega$ represents the variance of the wave component. According to Journée & Massie [10], the wave amplitude can be determined using Equation 8. Thus, the wave height for each frequency segment is assumed to be twice the amplitude of the regular wave ζ_a.

$$\zeta_a(\omega) = \sqrt{2S_{\eta\eta}(\omega).\Delta\omega} \tag{8}$$

The minimum wave height criteria set by API RP-2A still apply in this method, where the minimum wave height is 0.33 m. From the wave height calculation results in Table 5, the highest wave height does not correlate with the period; instead, the wave height depends on the shape of the spectrum and the frequency range previously established. The highest wave height occurs at a period of 10 seconds.

Based on the comparison of wave parameters in Tables 4 and 5, the second method generally produces individual wave heights that are much smaller than those calculated using the first method. Notably, the maximum wave height with a one-year return period in the Natuna Sea is 5.6 m. Therefore, the wave height used for fatigue analysis must be limited to the maximum wave height within one year. According to Goda (2008), the maximum wave height can be predicted from the significant wave height using the following Equation 9.

$$H_{max} = (1.6 \sim 2.0)H_s \tag{9}$$

For platform design, it is recommended to use $H_{max} = 2.0\ H_s$ [11]. Thus, for a representative significant wave height $H_{sr} = 2.72$ m, the maximum wave height, which closely matches the wave data in the Natuna Sea, is obtained as $H_{max} = 2.0\ H_{sr} = 5.44$ m. This indicates that the wave calculation approach using the second method more accurately reflects the actual wave conditions.

Table 5. Wave parameters based on discretization of the theoretical JONSWAP wave spectrum.

Period (s)	ω (rad./s)	$S_{\eta\eta}$ (m^2 s/rad.)	Wave Height (m)
12.0	0.52	0.911	1.16
10.7	0.59	2.446	1.01
10.0	0.63	1.717	1.78
6.0	1.05	0.127	0.50
5.6	1.12	0.092	0.33
5.2	1.21	0.065	0.33
4.8	1.31	0.045	0.33
4.4	1.43	0.029	0.33
4.0	1.57	0.018	0.33
3.6	1.75	0.011	0.33
3.2	1.96	0.006	0.33
2.8	2.24	0.003	0.33
2.4	2.62	0.001	0.33
2.0	3.14	0.001	0.33

4. Results and Discussions

The goal of fatigue analysis is to determine the durability of joints under cyclic loads resulting from wave-structure interactions. The Hot Spot Stress (HSS) is defined to determine the allowable number of cycles, which arises from the nominal stress (σ_{rms}), and the Stress Concentration Factor (SCF). SCF is an amplification factor for stress influenced by joint profiles. Joint profiles can cause stress discontinuities and accumulation in structural cross-sections, which is why fatigue failure often occurs at joints. In this study, we focus on K-type joints, and the SCF for K-joints can be found in API RP-2A [2].

4.1 Structural Response

The determination of fatigue life relies on the nominal stress value generated at a particular joint. The nominal stress caused by wave loads on the structure is related to the structural response to the waves. From Modal Analysis results, it was determined that the natural period of the analyzed structure is 4.8 seconds. According to the wave parameter selections, the based shear due to wave loads is reviewed. Figure 5 illustrates the base shear structure produced from two different wave parameter selections.

From these responses, using two different methods of wave parameter determination, it is evident that the maximum structural response to waves is consistent at locations corresponding to the natural period of the structure (4.8 seconds). Based on the wave parameter results shown in Tables 4 and 5, the structural response values using the second method result in smaller responses compared to the first method. This suggests that the magnitude of the structural response in the second method will yield lower stress magnitudes. However, in fatigue analysis, the critical factor is the stress range, which is the difference between the maximum and minimum stress at a joint caused by wave loads.

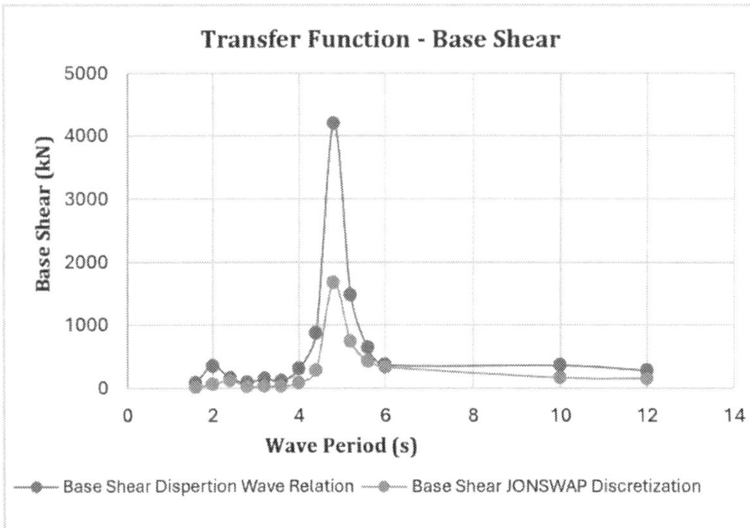

Figure 5. Comparison of the dynamic structure base shear in the 0^0 wave direction.

4.2 Fatigue Damage

To determine Fatigue Life (FL), as shown in Equation 7, the calculation not only depends on fatigue damage (D) but also on the design life of the structure (service life, SL). The design life of the analyzed structure is 15 years. The safety factor (SF) is determined based on the location of the joint, as per the referenced document [2]. For Joint 5004, examined in this paper, the safety factor is SF = 2.0. The location of Joint 5004 is shown in Figure 6a.

From Figure 6, Joint 5004 is located on a jacket leg. The chord of this joint is the jacket leg, and Joint 5004 has four braces, each directly connected to another jacket leg. The joint's details are shown in Figure 6. The joint is a combination of several K-braces, and its SCF adapted from the referenced document [2].

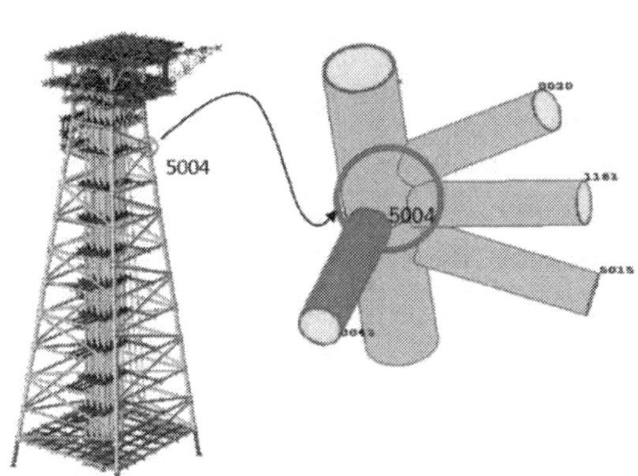

Figure 6. Location of Joint 5004 and the detail joint connection.

Based on the calculated Hot Spot Stress range, the results are plotted on the S-N curve in Figure 3, providing the allowable number of cycles (N). To determine the fatigue damage (D), the number of wave occurrences (n) over one year is calculated based on the scatter diagram in Table 1. The wave occurrences correspond to loads producing a specific nominal stress range. A safety factor of SF = 2.0 is applied as a multiplier to the damage value. Therefore, as damage increases, fatigue life decreases. A comparison of damage and fatigue life using the two wave load determination methods is shown in Table 6.

Table 6. Fatigue damage comparison results.

Chord Joint	Bracing Member	Member Label	Damage	Fatigue Life (years)	Wave Selection Criteria
5004	0043-5004	HB2	0.0954928	157.0799	Wave dispersion relation
5004	5015-5004	DB3	1.503E-06	9980040	
5004	5004-0020	DB3	0.0079004	1898.638	
5004	5004-1161	HB2	0.0118922	1261.331	
5004	0043-5004	HB2	0.0335552	447.0246	JONSWAP discretization
5004	5015-5004	DB3	5.505E-10	2.72E+10	
5004	5004-0020	DB3	0.0015717	9544.049	
5004	5004-1161	HB2	0.0026402	5681.388	

Based on Table 6, it is evident that the fatigue life values produced using the second method are almost three times higher than those obtained using the first method. However, the damage patterns remain similar, with the joint between the Jacket Leg and Horizontal Bracing 0043-5004 exhibiting the highest damage values compared to other joints. The location of this horizontal bracing joint is indicated in red in Figure 6.

5. Conclusion

This paper provides a comprehensive explanation of the spectral fatigue analysis method using the SACS software. The study introduces a novel approach for determining wave parameters to calculate the nominal stress range in structural joints. The proposed method involves discretizing the theoretical JONSWAP spectrum. Accurate calculation of wave loads is a critical factor, as the results directly impact the final fatigue life calculations.

Using the wave load determination method based on the discretization of the wave spectrum, the resulting wave heights are lower compared to calculations using the wave dispersion relation. In fact, the wave dispersion relation tends to overestimate wave heights when compared to the maximum wave height over a one-year return period. Therefore, to represent the realistic irregular wave parameters, the wave spectrum discretization method is recommended. In addition, the wave heights used in calculating the theoretical wave spectrum curve have been verified. This verification showed that the significant wave heights used closely match the significant wave data for a one-year return period.

From these findings, it can be concluded that the discretization of the wave spectrum yields more realistic wave loads compared to the wave dispersion relation method. The fatigue analysis results indicate that the wave dispersion relation method leads to more conservative damage values, resulting in a shorter predicted fatigue life.

Acknowledgment

The research for this paper is funded by a grant from Faculty of Civil and Environmental Engineering, Institut Teknologi Bandung, through 2024 Research and Community Service Program (P2MI) for Offshore Engineering Research Group. The authors gratefully acknowledge this support.

References

[1] https://docplayer.info/, Pengenalan Bangunan Lepas Pantai, accesed 24 Maret 2023
[2] American Petroleum Institute (2000): Recommended Practice for Planning, Designing and Constructing Fixed Offshore Platforms - Working Stress Design 21 st Edition (21st ed.). (2007). Washington DC: American Petroleum Institute.
[3] Dean, R. G., & Dalrymple, R. A. (1984). Water wave mechanics for engineers and scientists. https://doi.org/10.1029/eo066i024p00490-06.
[4] Bentley, SACSTM: Structural Analysis Computer System, Connect Edition.
[5] Siriwardane, S.C.; Adasooriya, N.D.; Pavlou, D. Fatigue Strength Curve for Tubular Joints of Offshore Structures under Dynamic Loading. *Dynamics* 2021, *1*, 125-133. https://doi.org/10.3390/dynamics1010007
[6] Chakrabarti, S. K. (2005). Handbook of Offshore Engineering Vol. I. Oxford: Elsevier.
[7] Saipem, WHP-C Jacket Fatigue Analysis, 2008.
[8] Hasselmann, K.; Barnett, T.P.; Bouws, E.; Carlson, H.; Carwright, D.E.; Enke, K.; Ewing, J.A.; Gienapp, H.; Hasselmann, D.E.; Kruseman, P.; et al. Measurements of windwave growth and swell decay during the Joint North Sea Wave Project (JONSWAP). Dtsch. Hydrogr. Z. 1973, 8, 1-95
[9] Lee, U.-J.; Jeong, W.-M.; Cho, H.-Y. Estimation and Analysis of JONSWAP Spectrum Parameter Using Observed Data around Korean Coast. J. Mar. Sci. Eng. 2022, 10, 578.
[10] Journée, J. M. J., & Massie, W. W. (2008). Offshore Hydromechanics (Second Edition).
[11] Goda, Yoshimi, (2008), Overview on the application of random wave concept in coastal engineering, Proc. Japan Academy, Ser. B, Vol. 84 No. 9, Pages 374-385.

IOP Conf. Series: Earth and Environmental Science **1464** (2025) 012036 doi:10.1088/1755-1315/1464/1/012036

Quantitative risk-based inspection analysis for assessing offshore structure

Arianta[1*], Riska Putri Budiana[1] and Ridha Sena Negara[2]

[1] Civil Engineering, Universitas Pertamina, Jakarta, Indonesia
[2] PT. Kerbau Laut Enjiniring, Tangerang Selatan, Indonesia

*E-mail: arianta@universitaspertamina.ac.id

Abstract. This study examines a quantitative risk analysis based on fatigue life and Reserve Strength Ratio (RSR) to establish inspection intervals for jacket structures within the framework of Risk-Based Inspection (RBI). The methodology involves deterministic fatigue analysis to assess fatigue life and pushover analysis to evaluate the structure's RSR. To estimate the probability of fatigue failure and the structural collapse probability due to existing fatigue damage, Monte Carlo simulations were conducted for 20 identified fatigue points. Offshore platforms play a critical role in oil and gas operations, necessitating robust structural integrity to ensure safety and efficiency in challenging marine environments. Regular inspections are vital for maintaining structural conditions and establishing a comprehensive Inspection, Maintenance, and Repair (IMR) strategy. RBI provides an effective approach to evaluate the structural integrity of these platforms by analyzing the probability of failure (PoF) and the consequence of failure (CoF). Fatigue life and RSR are pivotal parameters in this analysis, with fatigue representing the PoF and RSR indicating the CoF. The results indicate the highest accumulated probability of fatigue failure for the jacket structure is 0.110, occurring in the 503-603 jacket leg member with a fatigue life of 9 years. The highest conditional probability of collapse, at 0.997, is observed when fatigue failure occurs in the jacket leg members, with the structure's damaged RSR approaching zero. The risk matrix shows that seven members are classified as high-risk, and four as medium-risk, distributed across the jacket legs and bracings. For RBI implementation, the inspection interval is set at three years for high-risk members and five years for medium-risk members, with damage surveys recommended.

1. Introduction

Offshore structures play a crucial role in the exploration and production of oil and gas, necessitating robust inspection and maintenance strategies to ensure their integrity and safety. One such strategy is Risk-Based Inspection (RBI), which has gained prominence due to its systematic approach to managing the risks associated with offshore structures. According to Qureshi et al. [1], RBI provides a comprehensive framework for assessing the probability of failure and the associated risks, thereby enabling more informed decision-making in the maintenance of offshore platforms. The safe and optimal operation of offshore activities relies heavily on the integrity of supporting structures and facilities. Fixed jacket platforms are a prevalent type of offshore platform in Indonesia, designed to operate in shallow to medium-depth waters with a specific operational lifespan. Some platforms continue to operate beyond their intended service life, necessitating assessments to ensure structural integrity. Structural integrity can be assessed using probabilistic methods as a framework. Probabilistic analysis can be conducted in the form of structural reliability analysis and/or quantitative structural risk analysis. Through reliability analysis, the probability of failure due to uncertainties can be calculated. Quantitative risk analysis combines this probability of failure with the consequences of potential failures to make optimal decisions [2].

To maintain the integrity of a structure, inspections are essential, particularly at identified fatigue points. Inspections form the basis for assessing structural conditions and are a core part of Inspection, Maintenance, and Repair (IMR) activities. Effective inspection planning requires appropriate methods to identify efficient inspection strategies in terms of time and cost. Risk-Based Inspection (RBI) is a method that can be used to determine inspection intervals based on risk levels [3] . Parameters for determining inspection intervals include the fatigue life of the structure from fatigue analysis and the Reserve Strength Ratio (RSR) from pushover analysis.

Designing offshore platforms involves dealing with uncertainties in the analysis. RBI provides a framework for decision-making under uncertainty. Fatigue failure can be modeled probabilistically, with wave loads as a variable of uncertainty. Failure occurs when a component or system can no longer perform its designated function over a specified period. The probability of failure is calculated based on the relationship between structural resistance and loads within a defined limit state function. The higher the probability of fatigue failure, the greater the likelihood of events influenced by fatigue failure, such as structural collapse [4].

RBI offers distinct advantages over conventional inspection methods. Implementing RBI in inspection planning can reduce IMR costs during operation by prioritizing inspections for structural components with the highest risk levels. Overall, the benefits of applying RBI to offshore platform design include identifying and mitigating risks effectively, ensuring the safety and reliability of the structure [5].

2. Methodology

2.1 Fatigue limit state

When calculating the likelihood of fatigue failure, the limit state function is a key factor. It defines the boundary between the safe and failure states of a structure, helping determine whether a structure or material can withstand applied loads over time or succumb to fatigue. Fatigue failure, characterized by progressive structural degradation due to cyclic loading, necessitates a precise mathematical formulation to predict the timing and nature of failure. The limit state function is expressed by the following Equation 1 [6].

$$g(X) = g\ (\Delta, D) = \Delta - D \tag{1}$$

Where variable Δ describes the variable uncertainty of the fatigue capacity of the structure, and D is the expected total damage over the life of the operation, T, which is calculated in the form of a closed-form fatigue Equation 2 as follows [7].

$$D = T \cdot v_0 \cdot \frac{1}{K_2} A^{m_2} \gamma\left(1 + \frac{m_2}{B}; \left(\frac{S_0}{A}\right)^B\right) + \frac{1}{K} A^{m} \Gamma\left(1 + \frac{m}{B}; \left(\frac{S_0}{A}\right)^B\right) \tag{2}$$

By substituting equation 1 into equation 2, the fatigue limit state function can be rewritten as in equation 3. Table 1 describes the parameters and value taken in Equation 3.

$$g(X) = g\,(\Delta, D) = \Delta - T \cdot v_0 \cdot \frac{1}{K_2} A^{m_2} \gamma\left(1 + \frac{m_2}{B}; \left(\frac{S_0}{A}\right)^B\right) + \frac{1}{K} A^m \Gamma\left(1 + \frac{m}{B}; \left(\frac{S_0}{A}\right)^B\right) \quad (3)$$

Table 1 Fatigue limit state function parameters.

Parameters	Description	Size and Type of Distribution
Δ	Capacity uncertainty variable fatigue	Normal distribution with mean is 1 and CoV is 0.2 [7] [8]
T	Structure operating life	1.45 x 10^9 s (constant)
v_0	Average cycle stress value per second	0.18 [7]
K, K_2	Uncertainty of S-N curve parameters	• $\log_{10}K$ normal distribution with mean mean is 12.98 and StD is 0.25 • $\log_{10}K_2 = \log_{10}N_0 + m_2 . \log_{10}S_0$
m, m_2	Slope line on the S-N curve	$m = 3.0$ *for* $N \leq 10^8$ $m_2 = 5.0$ *for* $N > 10^8$ [7]
A	Weibull distribution scale parameters	Random variables from fatigue analysis result
B	Shape parameter	0.84 [7]
S_0	Stress range level	constant

Table 2 Collapse limit state function parameters.

Parameters	Description	Size and Type of Distribution
ξ	Uncertainty variable in the capacity of structural collapse	normal distribution with mean is 1 and CoV is 0.15 [7]
RSR	Reserve Strength Ratio	constant
L_{100}	100-year period wave load	Pushover analysis result
α	Uncertainty variable in wave load	normal distribution with mean is 1 and CoV is 0.2 [7]
H	Annual maximum wave height uncertainty variable	16-year maximum wave height data with a range of 11.65 ft - 24.51 ft

2.2 Collapse limit state

In the calculation of the probability of structural collapse, the limit state function is expressed by the following Equation 4 [7].

$$g(X) = g\,(CC|\theta\,,\ L|\theta) = CC|\theta - L|\theta \quad (4)$$

where it describes the $CC|\theta$collapse capacity of the structure in the direction θ, and is the annual extreme $L|\theta$base shear load in the direction θ. The collapse capacity of the structure in the loading direction can be calculated using the Equation 5

$$CC|\theta\ = \xi \cdot RSR \cdot L_{100}|\theta \quad (5)$$

The *annual* extreme base shear load in the loading direction θ can be calculated using the Equation 6.

IOP Conf. Series: Earth and Environmental Science **1464** (2025) 012036 doi:10.1088/1755-1315/1464/1/012036

$$L|\theta \; = \alpha \cdot a \cdot H^b|\theta \tag{6}$$

By substituting equations 5 and 6 into equation 4, the collapse limit state function can be rewritten as in equation 7. Table 2 provide a detailed overview of the parameters and corresponding values that are utilized in the calculations within Equation 7.

$$g(X) = g\;(CC|\theta\;, L|\theta) = \;\xi \cdot RSR \cdot L_{100}|\theta - \alpha \cdot a \cdot H^b|\theta \tag{7}$$

2.3 Risk Matrix

A risk matrix is a crucial tool for communicating the outcomes of a risk analysis to decision-makers. It categorizes risks based on accumulated probability of fatigue failure and conditional probability of structural collapse, allowing for informed decisions on inspections, maintenance, and reinforcements. In Figure 1, three distinct colours green, yellow and red are utilized to represent varying levels of risk. The diagonal lines that link the probability of fatigue and collapse failure serve as parameters for delineating these risk levels.

- Low risk (green zone): this category is defined by a diagonal boundary connecting a probability fatigue failure of 0.1 and probability collapse of 0.1. Components in this zone are unlikely to experience failure under typical operational conditions. These members require less frequent inspections, typically visual inspections at longer intervals (e.g., every 5 years [9]).
- Medium risk (yellow zone): this category lies above the low-risk boundary and extends up to the diagonal boundary connecting a fatigue failure probability of 1.1 and a collapse probability of 1.1. Within this risk level, there is a diagonal risk target line that represents the critical risk threshold. Components in this zone may be subject to wear or damage over time, necessitating regular inspection intervals to monitor for potential deterioration. These inspections are more frequent than those for low-risk components and may include both visual inspections and damage surveys.
- High risk (red zone): this category includes structural members with an accumulated probability of fatigue failure and a conditional probability of collapse above 1.1. Components in this zone are critical to the structural integrity of the platform and require immediate attention. These members necessitate more frequent and comprehensive inspections, including Non-Destructive Testing (NDT) and detailed damage surveys, typically at intervals of every 3 years.

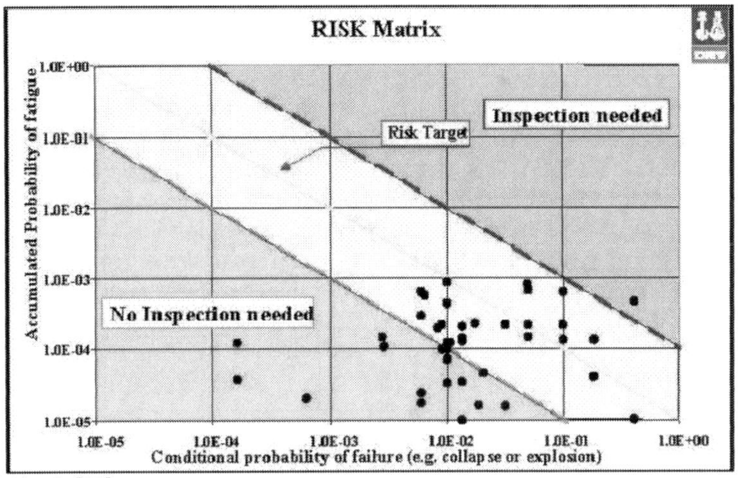

Figure 1. Risk Matrix [10].

3. Result and discussion

The structure studied in this research is a four-legged jacket-type offshore platform categorized as a processing platform (manned evacuated). The platform is located in the Java Sea at a water depth of 145 feet. It has two deck levels: the main deck at an elevation of +36 feet and the cellar deck at an elevation of -20.0 feet. Horizontal bracings are placed at elevations of +10.0 feet, -24.0 feet, -64.0 feet, -106 feet, and -145 feet from Mean Sea Level (MSL). Figure 2 shows the platform.

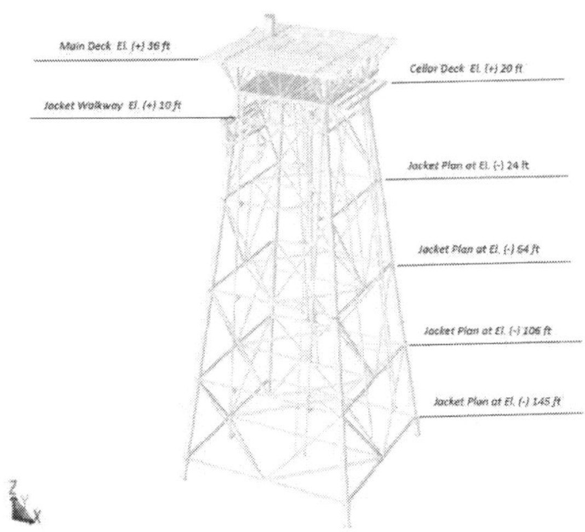

Figure 2. Four legs fixed jacket platform.

A Design Fatigue Factor (DFF) of 5 is recommended for this analysis, as it enables a comprehensive inspection of all structural members [11]. This criterion is consistent with the standards outlined in API RP 2A [12], DNVGL-RP-C203 [13], and ISO 19902 [8], which provide guidelines for fatigue safety in offshore jacket structures. These standards endorse a DFF of 5 as an acceptable threshold for jacket structures with fully accessible members for inspection.

By adopting a DFF of 5, members with a fatigue life that is less than five times their intended operational life are flagged as fatigue-critical points that warrant closer inspection and monitoring. This threshold is based on potential risks identified through fatigue analyses, where members with DFF values below 5 may exhibit higher risk under cyclic loading conditions. In this analysis, 7 joints and 11 members were identified with DFF values below 5, signifying them as critical areas needing inspection. Figure 3 illustrates the locations of these fatigue-critical points (marked by red dots), which are concentrated in key areas of the supporting structure. These identified critical members will undergo a more detailed examination in the quantitative Risk-Based Inspection (RBI) analysis to evaluate their risk levels and establish appropriate inspection intervals

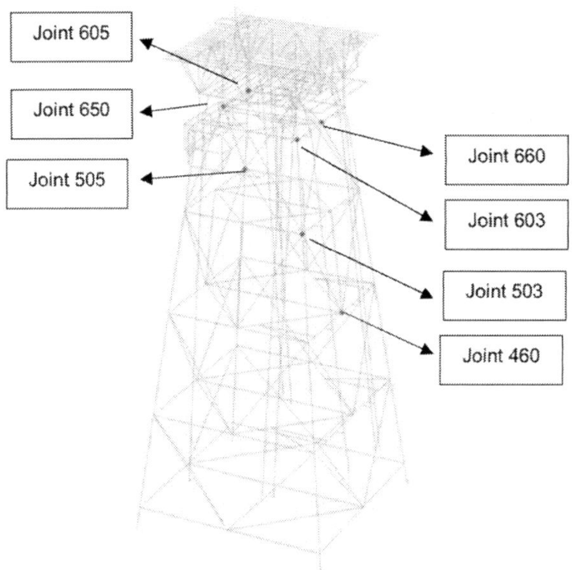

Figure 3. Fatigue critical point.

The geometry or configuration of a structure influences the distribution of stress across its joints. For instance, joints located at the intersection of legs and braces, such as Joint 503 and 603, tend to experience higher stress concentrations. This is often due to changes in load direction or the transfer of load from one element to another. These transitional points can become weak spots in the structure, as elevated stress concentrations cause the material in these areas to undergo more significant stress cycles. This can lead to accelerated fatigue damage over time.

Table 3. Accumulated probability of fatigue failure.

	Fatigue Points			Fatigue Life Calculation		DFF	$P_f^{fatigue}$
No.	Joint	Member	Component Type	Damage	Service Life (years)		
1	460	424- 460	*Main brace*	0.63973	71.9	1.6	2.84E-03
2	503	503- 524	*Main brace*	0.85984	53.5	1.2	7.64E-03
3	503	518- 503	*Main brace*	1.73535	26.5	0.6	5.37E-02
4	503	503- 603	*Leg*	5.12875	9.0	0.2	2.90E-01
5	505	505- 572	*Main brace*	0.43750	105.1	2.3	7.40E-04
6	505	505-2405	*Leg*	1.17866	39.0	0.8	2.07E-02
7	505	568- 505	*Main brace*	0.30183	152.4	3.3	8.00E-05
8	603	603- 703	*Leg*	0.79668	57.7	1.3	6.40E-03
9	605	605- 705	*Leg*	0.23916	192.3	4.2	6.00E-05
10	650	634- 650	*Main brace*	0.35509	129.5	2.8	2.80E-04
11	660	660- 664	*Main brace*	1.25526	36.6	0.8	2.38E-02

To calculate the probability of fatigue failure over the operational lifetime, the random numbers generated for the random variables K, Δ, A, and N_0 are generated and transformed based on their respective distributions. These transformed random variable values are then substituted into the fatigue limit state function (equation 3) for a simulation process. After simulating the limit state function 50,000 times for each point, the cumulative probability of fatigue failure can be determined. The results of the fatigue points are presented in Table 3.

Table 3 shows that DFF is inversely related to the probability of failure, with higher DFF velues generally correlating with lower $P_f^{fatigue}$. This relationship highlights the importance of the maintaining adequate DFF levels in design to mitigate fatigue risks. The positioning of joints, along with exposure to substantial cyclic loads, significantly contributes to the rapid accumulation of fatigue. This phenomenon can lead to a decreased service life of the components involved. When joints are placed in locations that must endure repetitive stress or strain, the intricate balance of forces acting upon these joints can accelerate wear and tear. Consequently, understanding the impact of joint orientation and load conditions is crucial for enhancing durability and extending the lifespan of mechanical systems.

To estimate the magnitude of the consequences in the form of the probability of structural collapse, the random numbers calculated for the random variables ξ, α, and H are transformed according to the type of distribution. The random variable is substituted into the collapse limit state function (equation 7) to then be simulated. After the collapse state function simulation process is carried out 50,000 times for each component type at the fatigue point, the large conditional probability of collapse, $P_f^{collapse}$, can be seen in Table 4.

Table 4. Conditional probability of collapse.

No.	Component Type	$P_f^{collapse}$
1.	Leg	9.97E-01
2.	Main Bracing	6,34E-02
3.	Secondary Bracing	2,03E-02

The data shows that the jacket legs have the highest probability of collapse at 9.97E-01, meaning failure in this critical load-bearing component would almost certainly lead to structural collapse. This highlights the importance of the legs in maintaining overall structural stability, as their failure poses the greatest risk. In contrast, the main bracing has a much lower probability of collapse at 6.34E-02, and the secondary bracing at 2.03E-02, indicating these components have a smaller impact on structural failure. While their failure compromises integrity, the structure's redundancy allows for load redistribution, reducing the likelihood of immediate collapse.

From the results of the accumulated probability of fatigue and the consequences shown in the form of conditional probability of collapse of each point of structural fatigue reviewed, the risk of failure can be calculated. Risk is the result of multiplication between accumulated probability of fatigue and conditional probability of collapse as shown figure 4. The determination of the inspection interval is then carried out referring to API RP 2SIM [9].

The risk matrix analysis of the eleven structural members indicates that four members in the main bracing component and three members in the jacket leg component are classified as high risk. This classification suggests that these members are more susceptible to fatigue failure and structural issues, necessitating closer monitoring and possibly more frequent inspections. These high-risk members are depicted in the upper section of the risk matrix chart, indicating both a high accumulated probability of fatigue and an elevated conditional probability of collapse.

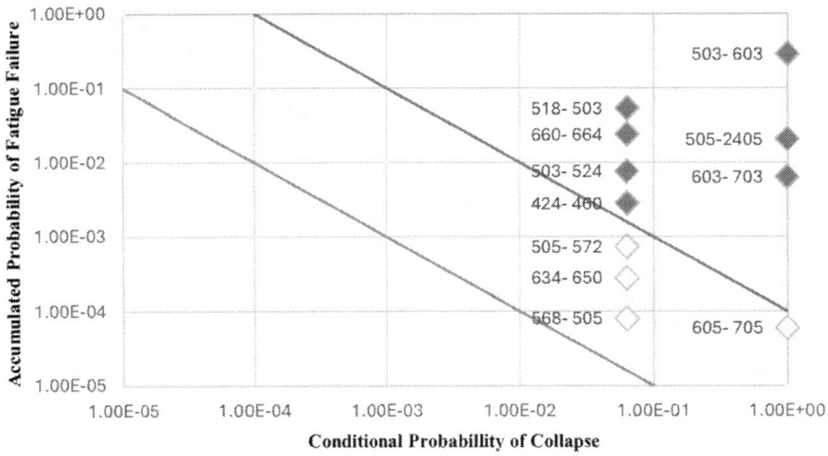

Figure 4. Risk results.

Table 5. Inspection interval.

No.	Joint	Member	Component Type	Risk Category	Exposure Category	Inspection Interval
1	460	424- 460	Main brace	High	L-1	3 Years
2	503	503- 524	Main brace	High	L-1	3 Years
3	503	518- 503	Main brace	High	L-1	3 years
4	503	503- 603	Leg	High	L-1	3 Years

No.	Joint	Member	Component Type	Risk Category	Exposure Category	Inspection Interval
5	505	505- 572	Main brace	Medium	L-2	5 years
6	505	505-2405	Leg	High	L-1	3 Years
7	505	568- 505	Main brace	Medium	L-2	5 years
8	603	603- 703	Leg	High	L-1	3 Years
9	605	605- 705	Leg	Medium	L-2	5 Years
10	650	634- 650	Main brace	Medium	L-2	5 years
11	660	660- 664	Main brace	High	L-1	3 years

Conversely, the remaining structural components fall into the medium-risk category. While they do present some risk, the likelihood of failure and structural collapse is significantly lower. These medium-risk components can be managed with standard inspection intervals, as their structural stability is deemed adequate under normal operational conditions. This classification allows for a more targeted and efficient inspection strategy, directing resources toward high-risk areas while maintaining routine checks on medium-risk components. This approach ultimately supports both safety and cost-effectiveness in maintenance planning. With the application of RBI, the inspection interval time for each connection component that has been previously analysed is obtained in Table 5.

Based on the information above, inspection levels can be determined according to API RP 2SIM. For high-risk members in the L-1 exposure category with a 3-year inspection interval, the required inspection level is Level II. Level II inspections involve general underwater visual assessments conducted by divers or remotely operated vehicles (ROVs) to identify fatigue damage. If significant structural damage is found during a Level II survey, this should prompt a Level III inspection. For medium-risk members in the L-2 exposure category with a 5-year inspection interval, the required inspection level is also Level II. While Level II inspections are the initial requirement, the timing for these inspections varies. This approach prioritizes critical members based on their risk levels, ultimately saving time and reducing costs during the inspection process.

4. Conclusion

This study presents a quantitative risk-based inspection (RBI) analysis aimed at optimizing the inspection strategy for offshore jacket structures. It focuses on two critical parameters: fatigue life and the Reserve Strength Ratio (RSR). The study successfully identifies high-risk components and establishes appropriate inspection intervals. The findings emphasize the importance of probabilistic methods in evaluating both the probability of failure (PoF) and the consequences of failure (CoF) for offshore structures. By utilizing Monte Carlo simulations, the study estimates probabilities of fatigue failure and structural collapse, providing a robust framework for risk assessment. A risk matrix categorizes structural components into high, medium, and low-risk groups, guiding the prioritization of inspection and maintenance activities. This approach aligns with industry standards, such as API RP 2SIM, and supports informed decision-making for the maintenance of offshore platforms. By concentrating on high-risk components and implementing suitable inspection techniques, the study illustrates how RBI can extend the operational life of structures while ensuring safety and cost efficiency. The conclusions drawn from this analysis offer valuable insights for the implementation of RBI strategies in the offshore oil and gas industry.

References

[1] Qureshi, H.A, et all, "Risk-Based Inspection in Offshore Structures:," in *MEST 2022*, 2023.

[2] Mochammad Fathurridho Hermanto, "Kehandalan Struktur Anjungan Lepas Pantai Terpancang Empat Kaki di Perairan Natuna," *REKAYSA*, vol. 15, no. 3, pp. 2083-291, 2022.

[3] F. Guédé, "Risk-based structural integrity management for offshore jacket platform," *Marine Structure*, vol. 63, pp. 444-461, 2019.

[4] Tawekal, R.L, et all, "Development of Risk-Realiability Based Underwater Inspection for Fixed Offshore Platforms in Indonesia," in *MATEC*, 2018.

[5] Tawekal, R.L, et all, "Risk Based Underwater Inspection (RBUI) For Existing Fixed Platforms In Indonesia," *International Journal of Research in Engineering and Science (IJRES)*, vol. 5, no. 7, pp. 25-31, 2017.

[6] R. Heriana, *Analisis Kehandalan Anjungan Lepas Pantai Tipe Jacket Berdasarkan Beban Pushover Statik dan Fatigue Spektral*, Bandung: Thesis, Institut Teknologi Bandung (ITB), 2006.

[7] DNV, "Guideline for Offshore Structural Reliability Analysis: Examples for Jacket Platform," Det Norske Veritas, Norway, 1996.

[8] ISO 19902, "Petroleum and Natural Gas Industries - Fixed Steel Offshore Structure," 2020.

[9] API RP 2 SIM, "Structural Integrity Management of Fixed Offshore Structure," American Petroleum Institute, Washington DC, 2014.

[10] A. G. Ghoneim, *Structural Integrity Management and Risk Based Inspection - Recent Development*, DNV, 2009.

[11] W. Zhao, "Calibration of Design Fatigue Factors for Offshore Structures based on Fatigue Test Database," *International Journal of Fatigue*, 2020.

[12] A. R. P. 2A-WSD, Planning, Designing, and Constructing Fixed Offshore Platforms-Working Stress Desing 22nd Edition, Washington: American Petroleum Institute, 2014.

[13] DNVGL-RP-C203, "Fatigue Design of Offshore Steel Structures," 2020.

[14] D. R. G-101, "Risk Based Inspection of Offshore Topside Static Mechanical Equipment," DNV, Norway, 2010.

IOP Conf. Series: Earth and Environmental Science **1464** (2025) 012037 doi:10.1088/1755-1315/1464/1/012037

Numerical assessment of subsea pipeline global performance under dragged anchor pullover forces

Eko Charnius Ilman[1,2*], Farel Muhammad Elan[1], Jessica Rikanti Tawekal[1,2]

[1]Ocean Engineering Program, Institut Teknologi Bandung, Bandung, Indonesia
[2] Offshore Engineering Research Group, Institut Teknologi Bandung, Bandung, Indonesia

*E-mail: eko.ilman@itb.ac.id

Abstract. Subsea pipelines, located in areas with significant maritime activity, are vulnerable to failure risks arising from potential anchorage hazards, specifically those from dragged anchors. In this paper, a numerical assessment of NPS 6-inch with X52 grade subsea pipelines under dragged anchor pullover forces is performed. This paper presents a pipeline finite element with elastic plastic material model. The soil is represented mathematically as a rigid component, with non-linear pipe-soil interaction. A parametric study is performed assessing dragged anchor for various pullover force angle from 70 to 110 degrees with increment of 10 degrees. Different configuration of expansion spool in each pipeline end is provided. The results obtained provide valuable insights into the unique reactions of subsea pipelines across a range of scenarios. The maximum displacement for the pipeline von Mises stress to pass the ultimate strength can be up to 1.2% of pipeline length. However, the associated variations in displacement are not solely attributable to the angle of the dragged anchor but are also influenced by different configurations of the expansion spools. In conclusion, this study provides a significant contribution through the enhancement of comprehension regarding the global response demonstrated by subsea pipelines under the influence of the forces linked to dragged anchor pullover.

1. Introduction

Subsea pipelines in areas with high marine traffic are at risk of failure due to external interference from anchorages, including dragged anchors [1], [2], [3], [4], [5]. A lot of underwater surveys using side scan sonar show several pipeline segments globally displaced along their route, with dragged anchoring marks on the seabed. The displaced pipeline segments may result in failure, such as local buckling or rupture. In consequence, damage to pipelines will disrupt operations and may lead to financial losses, environmental damage, and affect marine safety. Therefore, there is a need to better understand how subsea pipelines experience global displacement behaviour due to anchoring [2], [3], [4]. This understanding is important for quantitatively assessing the consequences of pipeline failure, followed by assessing the risk and defining appropriate mitigation measures if required [6], [7], [8]. Understanding pipeline global displacement behaviour due to dragged anchors enhances safety and helps improve pipeline integrity [9], [10], [11], [12].

The typical sequence of dragged anchor events includes dragged anchor impact, pullover, and hooking [1], [13]. This study assumes that damage due to impact dragging is not significant, considering that the concrete coating absorbs the impact energy. Hence, the focus is on assessing

IOP Conf. Series: Earth and Environmental Science **1464** (2025) 012037 doi:10.1088/1755-1315/1464/1/012037

pipeline global response due to the anchor dragging pullover force, which is rarely assessed in the open literatures, especially in terms of global behaviour [14], [15], [16].

2. Numerical modelling

This study utilises carbon steel pipeline with material grade API 5L X52, NPS 6-inch with 0.5-inch wall thickness. The material is modelled as elastic-plastic based on the Ramberg-Osgood relationship, as described in Figure 1.

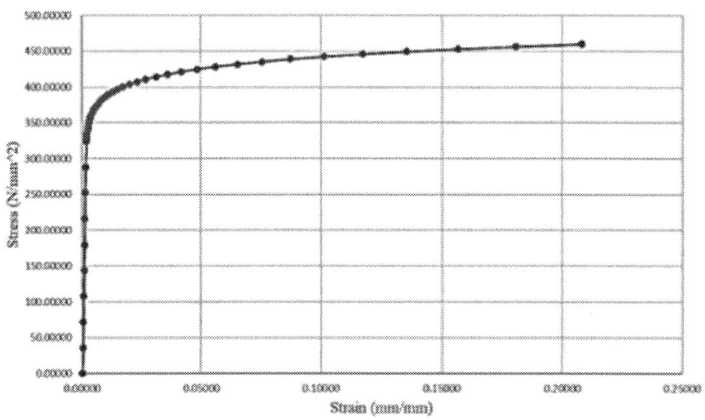

Figure 1. True stress-strain curve of an X52 material.

The pipeline with total length of 16.6 kilometres, is simplified modelled as straight line to focus assessing the behaviour of both end spools. Figure 2(a) shows the pipeline layout, indicating the pipeline midpoint where the pulling force is applied. To provide a more realistic response, an L-shaped expansion spool is connected to both ends of the pipe, as illustrated in Figure 2(a), however, this paper scope does not yet include the parametric effect of expansion spool which will be in further research scope. A parametric study examines the effect of angle of attack of the dragged anchor pullover, ranging from 70 to 110 degrees, with 10-degree increments (shown in Figure 2(b)).

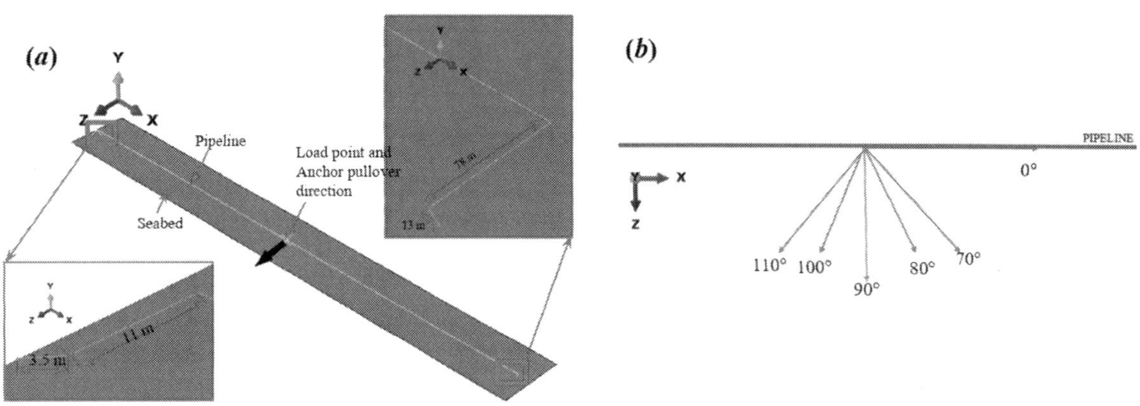

Figure 2. Pipeline-seabed schematic: (a) showing expansion spool configuration and position of load point; (b) dragged anchor pullover angle variation.

Both degrees of freedom in translation and rotation are fixed at the pipeline's ends. The pipeline is modelled as PIPE31H which is a hybrid beam-pipe element in Abaqus. The seabed is modelled as an analytical rigid element. The soil consists of very soft clay, with an undrained shear strength of 9 kPa. The pipeline-seabed normal interaction is modelled using a pressure-overclosure relationships shown in Figure 3 [17], while lateral interaction use friction factor of 0.2.

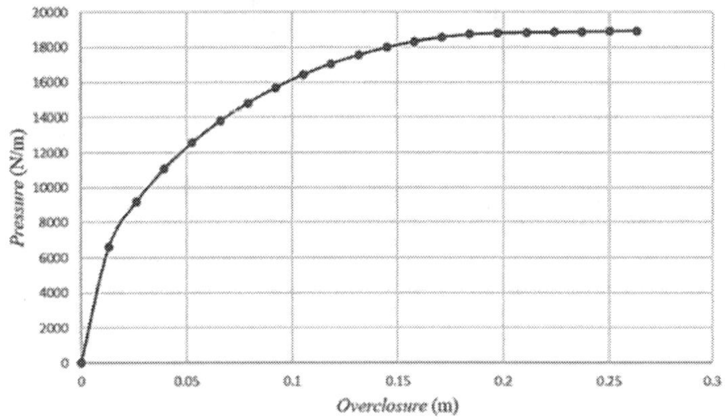

Figure 3. Pressure-overclosure relationship between pipeline and seabed.

To provide a more realistic simulation, a gravity load is applied to establish initial contact between the pipeline and the seabed, followed by dragged anchor pullover at pipeline's centre applied as displacement increments. A mesh convergence study is performed by analyzing a at pullover load point at the pipeline's centre, indicating in Figure 2(a). Figure 4 shows the von Mises stress values at loaded node previously described, while the number of elements are for the mesh refinement of a 500-meter pipe segment at pipeline's centre. Convergence is achieved with 284 elements corresponding to a minimum of mesh size of 2 meters.

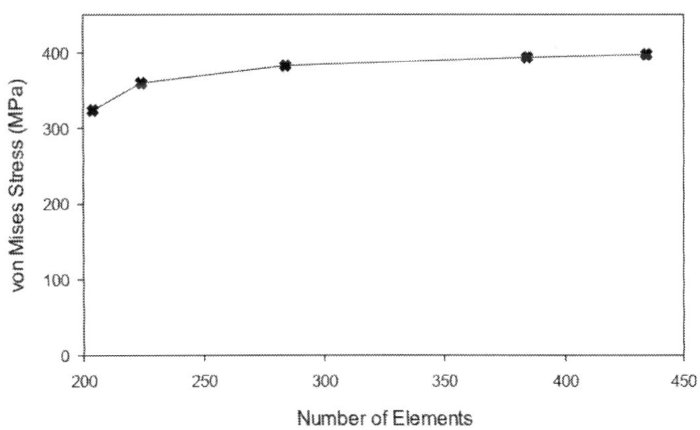

Figure 4. Mesh convergence study.

3. Results and Discussion

The analysis performed based on five cases of dragged angle of attach described in Figure 2(*b*). For example, the dragged anchor pullover displacement of 250-meter at angle of 90 degrees provides pipeline global response as shown in Figure 5.

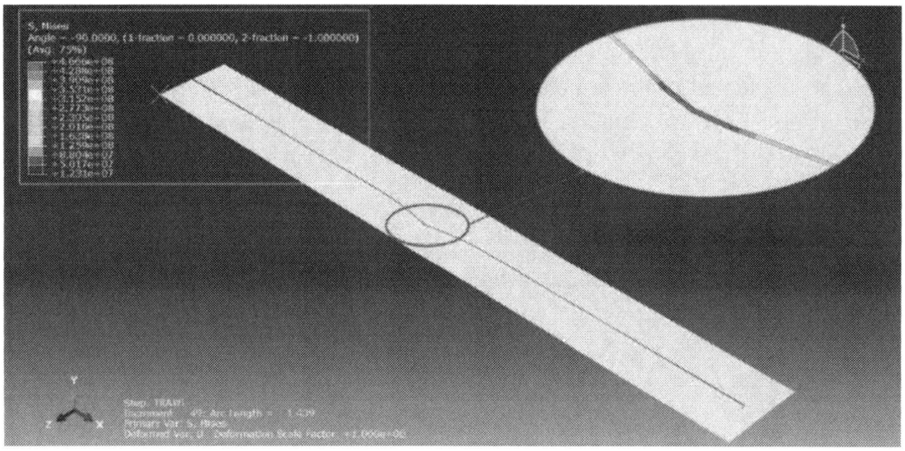

Figure 5. von Mises stress contour along pipeline route after 250 m dragged anchor pull over with 90° angle.

Stress and displacement response are analysed for the five cases, assessing yielding and ultimate strength. The case of dragged anchor pullover with angle of 70 degrees. Figure 6(*a*) shows the analysis when the first yield occurs, the maximum von Mises stress, measuring 363 MPa, located near the load point, with of 3.3 meters displacement. The load point experiences a stress of 330.8 MPa and a displacement of 2.2 meters. After ultimate strength is achieved, the highest von Mises stress value is at one of pipe end of 460.4 MPa, and a displacement of 4.6 meters, as shown in Figure 6(*b*). von Mises stress of 428.8 MPa at the pullover point with 37 meters displacement. Furthermore, because of the rightward force, the stress distribution on the left side of the pipe is greater in magnitude than on the right.

When a dragged anchor pullover with angle of 80 degrees, as shown in Figure 7(*a*), it is apparent that the pipe experiences yielding, resulting in the accumulation of von Mises stress at 365.4 MPa and a displacement of 3.9 meters. Similarly, the von Mises stress value is 360.4 MPa at the pullover point illustrated in the same figure, where the largest displacement is recorded at 4.0 meters. According to the data presented in Figure 7(*b*), the pipe undergoes tensile stress at its left end, which results in the highest von Mises stress of 461.1 MPa and a displacement of 4.7 meters. Simultaneously, the von Mises stress value is 427.5 MPa at the load point depicted in the figure, which is accompanied by a displacement of 67.4 meters. Moreover, as a result of the rightward force, the stress distribution on the left-end of pipeline is greater than on right side, preserving a pattern comparable to the 70-degree angle scenario.

When the pipe undergoes yielding in the scenario of a dragged anchor pullover with a 90-degree angle, as illustrated in Figure 8(a), the von Mises stress value that is most significant is 366 MPa, which is recorded at the pullover point. Concurrently, the maximum displacement magnitude of 6.2 meters is detected at the same location. The figure 8(*a*) illustrates that the left end pipeline experiences maximum von Mises stress of 460 MPa under tensile conditions. This value is accompanied by a displacement of 5.7 meters. Furthermore, von Mises stress magnitude at pullover point is 436 MPa, accompanied by a significant displacement of 192.8 meters. As a result of the anchor pull being perpendicularly oriented at this pullover angle, the stress distribution is relatively symmetrical on both the left and right end of the pipe. The disparity in tension observed at both extremities of the conduit can be ascribed to the impact of the distinct expansion spool configurations.

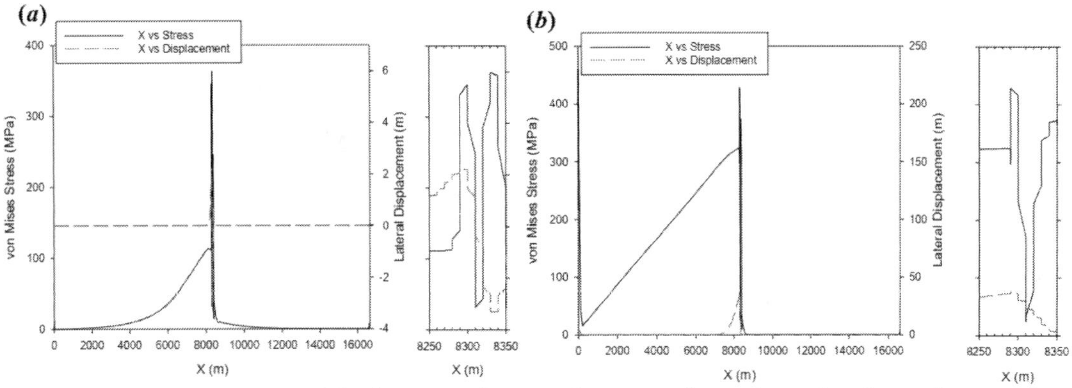

Figure 6. von Mises stress and displacement along pipeline route due to dragged anchor pull over with 70° angle: (*a*) after first yield stress is reached; (*b*) after first tensile stress is reached. X represents pipe length in longitudinal X-coordinate.

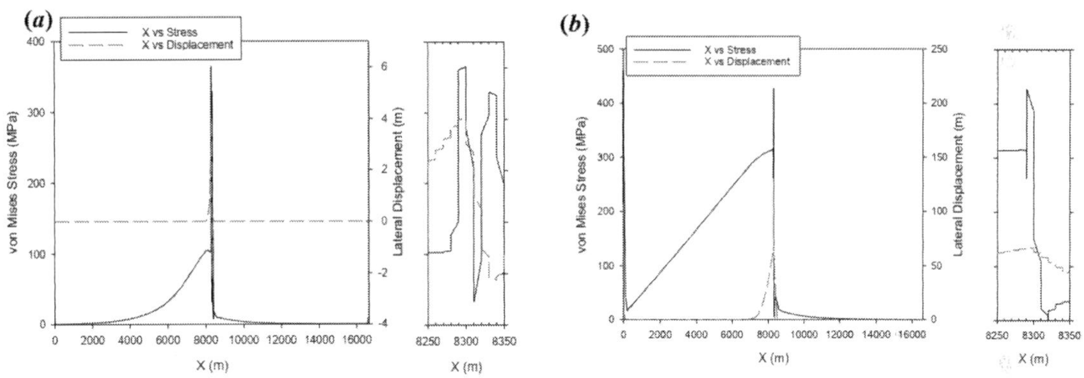

Figure 7. von Mises stress and displacement along pipeline route due to dragged anchor pull over with 80° angle: (*a*) after first yield stress is reached; (*b*) after first tensile stress is reached. X represents pipe length in longitudinal X-coordinate.

When the pipe experiences yielding in the vicinity of the pullover point, as illustrated in Figure 9(*a*) for a dragged anchor pullover at a 100-degree angle, maximum von Mises stress of 365 MPa and a displacement of 3.9 meters are observed. Concurrently, the von Mises stress value at the pullover point is 360 MPa, which corresponds to the maximum displacement of 4.0 meters. As illustrated in Figure 9(*b*), the pipe remains free from tensile stress as a result of the anchor being oriented in the 100-degree direction. The pipe's right end experienced the greatest von Mises stress value, measuring 454.4 MPa, accompanied by a displacement of 16 meters. Furthermore, von Mises stress at the pullover point is 437.3 MPa, which corresponds to the largest displacement of 204 meters, as illustrated in the same illustration. The disparities in stress distributions between this angle pull pattern and the 80-degree corner pull pattern are also due to the effect of different expansion spool configurations.

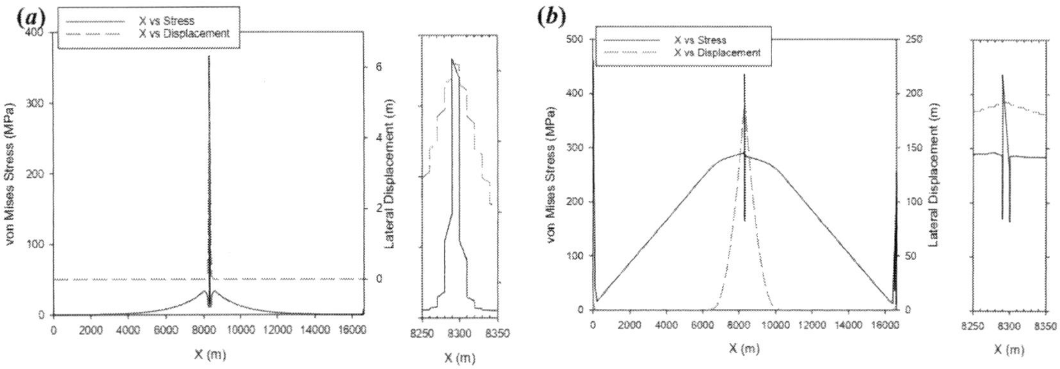

Figure 8. von Mises stress and displacement along pipeline route due to dragged anchor pullover with 90° angle: ***(a)*** after first yield stress is reached; ***(b)*** after first tensile stress is reached. X represents pipe length in longitudinal X-coordinate.

The case study of a dragged anchor pullover at an angle of 110 degrees, as depicted in Figure 10(*a*), reveals that the pipe experiences yielding at a location in close proximity to the pullover point, resulting in von Mises stress of 365 MPa and maximum displacement of 3.3 meters. Simultaneously, von Mises stress magnitude at pullover point is 333 MPa, which is accompanied by a 2.3-meter displacement value. With respect to Figure 10(*b*), the pipe undergoes a localized von Mises stress of 461 MPa at its right end, which coincides with a displacement of 20.1 meters. Likewise, von Mises stress value at pullover point is 440 MPa, which corresponds to the maximum displacement of 133 meters. The disparities in stress distributions between this angle pull pattern and the 70-degree corner pull pattern are also due to the effect of different expansion spool dimensions.

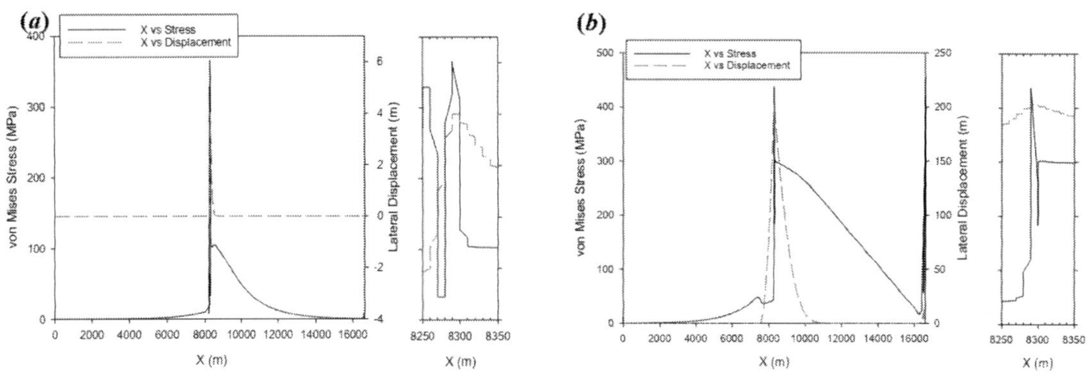

Figure 9. von mises stress and displacement along pipeline route due to dragged anchor pull over with 100° angle: ***(a)*** after first yield stress is reached; ***(b)*** after first tensile stress is reached. X represents pipe length in longitudinal X-coordinate.

As depicted in Figure 11, the summary of modelling variations illustrates the consequences of changes in the angle of the anchor's pullover direction toward the subsea pipeline. A comparative analysis of angle variations of 70° and 80° reveals that under yield conditions, the von Mises stress value increases by 8.93% and the displacement value rises by 79.98%. On the contrary, under tensile conditions, the von Mises stress value experiences a slight decrease of 0.29%, while the displacement value increases by 81.96%. An analysis of the angle variation from

80° to 90° reveals that yield conditions result in a 1.82% increase in von Mises stress and a 54.19% increase in displacement value. On the contrary, tensile conditions result in a significant 186.2% increase in displacement and a 1.97% increase in von Mises stress value. Additionally, upon comparing the angle variation from 90° to 100°, it is observed von Mises stress value decreases by 1.78% and the displacement value decreases by 35.14% under yield conditions. Under tensile conditions, the von Mises stress value exhibits a relatively small increase of 0.29%, whereas the displacement undergoes a substantial surge of 5.66%. Upon analyzing the variation in angle from 100° to 110°, it is observed that von Mises stress decreases by 7.58% and the displacement value decreases by 42.93% under yield conditions. On the contrary, under tensile conditions, there is an observed increase of 0.63% in the von Mises stress value, accompanied by a decrease of 34.65% in the displacement value.

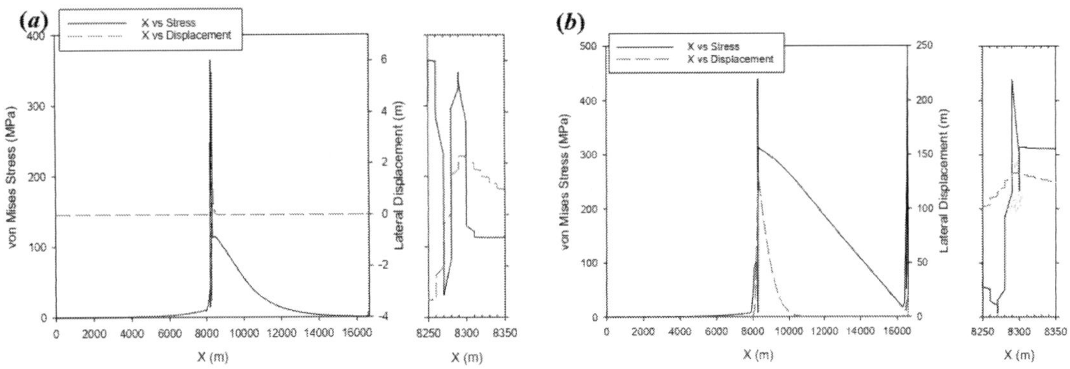

Figure 10. von Mises stress and displacement along pipeline route due to dragged anchor pull over with 110° angle: (*a*) after first yield stress is reached; (*b*) after first tensile stress is reached. X represents pipe length in longitudinal X-coordinate.

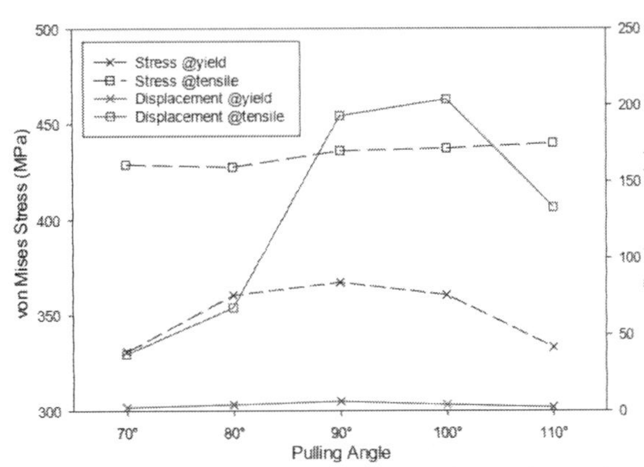

Figure 11. von Mises stress and displacement at pulling point due to dragged anchor for various pulling angle.

4. Conclusions

In conclusion, the numerical investigation into various angles of the dragged anchor reveals a consistent occurrence of first yielding, primarily concentrated at the pullover point. The maximum displacement for the pipeline von Mises stress to pass the ultimate strength can be up to 1.2% of pipeline length. However, the associated variations in displacement are not solely attributable to the angle of the dragged anchor but are also influenced by distinct configurations of the expansion spool. Consequently, in the design of pipes susceptible to the risk of dragged anchor pullover, the incorporation of an expansion spool on the pipe emerges as a measure to improve the pipeline lateral displacement allowance againts yielding and ultimate strength failure. This study proves importance of pipeline global response and expansion spool configurations against yielding and tensile failure.

Acknowledgement

This work acknowledges the support from Research, Community Service and Innovation Program, Faculty of Civil and Environmental Engineering, Institut Teknologi Bandung, Indonesia.

CRediT authorship contribution statement

Eko Charnius Ilman: Conceptualization (lead), Supervision (lead), Methodology (lead), Validation and Investigation (lead), Formal Analysis (equal), Resources (lead), Visualization (lead), Funding Acquisition (co-lead), Writing – Review & Editing (lead). **Farel Muhammad Elan**: Data Curation (lead), Formal Analysis (lead), Validation and Investigation (equal), Visualization (equal), Writing – Original Draft Preparation (lead). **Jessica Rikanti Tawekal**: Supervision (equal), Writing – Review & Editing (equal), Funding Acquisition (lead).

References

[1] F. Davaripour, K. Pike, B. W. T. Quinton, and R. Persaud, 'An Assessment on the Overtrawlability of Small Pipe Sizes Using a Hybrid Shell-Beam Model: an Alternative to the DNV-RP-F111 Model', *Applied Ocean Research*, vol. 105, p. 102413, Dec. 2020, doi: 10.1016/j.apor.2020.102413.

[2] G. Xiang, K. Rao, X. Xiang, and X. Yu, 'Overview and analysis on recent research and challenges of dropped objects in offshore engineering', *Ocean Engineering*, vol. 281, p. 114616, Aug. 2023, doi: 10.1016/j.oceaneng.2023.114616.

[3] V. Monsma, J. Spiekhout, and H. Harthoorn, 'Penetration depth of ships' anchors in regard to pipeline crossing protection: a review.', *Journal of Pipeline Engineering*, vol. 16, no. 2, 2017.

[4] J. Cai, X. Jiang, and G. Lodewijks, 'Residual ultimate strength of offshore metallic pipelines with structural damage – a literature review', *Ships and Offshore Structures*, vol. 12, no. 8, pp. 1037–1055, Nov. 2017, doi: 10.1080/17445302.2017.1308214.

[5] F. Khan, R. Yarveisy, and R. Abbassi, 'Risk-based pipeline integrity management: A road map for the resilient pipelines', *Journal of Pipeline Science and Engineering*, vol. 1, no. 1, pp. 74–87, Mar. 2021, doi: 10.1016/j.jpse.2021.02.001.

[6] J. Yu, Y. Zhao, T. Li, and Y. Yu, 'A three-dimensional numerical method to study pipeline deformations due to transverse impacts from dropped anchors', *Thin-Walled Structures*, vol. 103, pp. 22–32, Jun. 2016, doi: 10.1016/j.tws.2016.02.006.

[7] H.-S. Yoon and W.-B. Na, 'Safety assessment of submarine power cable protectors by anchor dragging field tests', *Ocean Engineering*, vol. 65, pp. 1–9, Jun. 2013, doi: 10.1016/j.oceaneng.2013.03.004.

[8] S. Agbo, K. Roy, S. Adeeb, and Y. Li, 'A Systematic Literature Review on Permanent Ground Deformation Patterns Imposed on Buried Steel Pipelines', *Journal of Pipeline Science and Engineering*, p. 100151, Nov. 2023, doi: 10.1016/j.jpse.2023.100151.

[9] R. Aulia, H. Tan, and S. Sriramula, 'Dynamic reliability model for subsea pipeline risk assessment due to third-party interference', *Journal of Pipeline Science and Engineering*, vol. 1, no. 3, pp. 277–289, Sep. 2021, doi: 10.1016/j.jpse.2021.09.006.

[10] Md. R. U. Kawsar, S. A. Youssef, M. Faisal, A. Kumar, J. K. Seo, and J. K. Paik, 'Assessment of dropped object risk on corroded subsea pipeline', *Ocean Engineering*, vol. 106, pp. 329–340, Sep. 2015, doi: 10.1016/j.oceaneng.2015.06.056.

[11] X. Li, G. Chen, S. Jiang, R. He, C. Xu, and H. Zhu, 'Developing a dynamic model for risk analysis under uncertainty: Case of third-party damage on subsea pipelines', *Journal of Loss Prevention in the Process Industries*, vol. 54, pp. 289–302, Jul. 2018, doi: 10.1016/j.jlp.2018.05.001.

[12] X. Li, G. Chen, Y. Chang, and C. Xu, 'Risk-based operation safety analysis during maintenance activities of subsea pipelines', *Process Safety and Environmental Protection*, vol. 122, pp. 247–262, Feb. 2019, doi: 10.1016/j.psep.2018.12.006.
[13] DNV, 'DNV-RP-F111: Interference between trawl gear and pipelines', *Det Norske Veritas*, 2021.
[14] S. Yan, B. Guo, L. Sun, W. Guo, and Z. Lei, 'Analytical and experimental studies of dragging hall anchors through rock berm', *Ocean Engineering*, vol. 108, pp. 529–538, Nov. 2015, doi: 10.1016/j.oceaneng.2015.08.031.
[15] M.-B. Shin, D.-S. Park, and Y.-K. Seo, 'Response of subsea pipelines to anchor impacts considering pipe–soil–rock interactions', *International Journal of Impact Engineering*, vol. 143, p. 103590, Sep. 2020, doi: 10.1016/j.ijimpeng.2020.103590.
[16] F. Davaripour, B. W. T. Quinton, and K. Pike, 'An Assessment on a Subsea Pipeline Subject to a Diagonal Trawl Impact', *Applied Ocean Research*, vol. 110, p. 102575, May 2021, doi: 10.1016/j.apor.2021.102575.
[17] DNV, 'DNV-RP-F114: Pipe-soil interaction for submarine pipelines', *Det Norske Veritas*, 2021.

AUTHOR INDEX